TECHNIQUES IN
PHOTOMICROGRAPHY

TECHNIQUES IN
PHOTOMICROGRAPHY

E. B. BRAIN
B.Sc., F.I.B.P., F.R.P.S.

Leverhulme Fellow in Histology, Department of Dental Science, Royal College of Surgeons of England.

AND

A. R. TEN CATE
Ph.D., B.Sc., B.D.S.

Leverhulme Fellow in Dental Science, Department of Dental Science, Royal College of Surgeons of England.

D. VAN NOSTRAND Company, Inc.

PRINCETON, NEW JERSEY

TORONTO   LONDON

NEW YORK

Published in Great Britain by
Oliver & Boyd Ltd., Edinburgh and London

First published 1963

Printed in Great Britain by
Oliver and Boyd Ltd., Edinburgh

INTRODUCTION

Since the introduction of automatic recording equipment some of the problems of photomicrography with cameras assembled around standard microscopes have been eliminated. However, the production of high quality prints or transparencies still demands considerable skill and experience. A great deal of information is available on the practice of microscopy and photography but, as far as we are aware, no attempt has been made to collate this in the form of practical step-by-step techniques.

In this book, after introductory chapters on the principles and applications of photomicrography and a discussion of equipment and materials, some seventy techniques are presented which should enable the inexperienced worker to become familiar with the practice of most photomicrographic procedures. It must be pointed out that although practical experience has contributed to their assembly no claim is made for originality.

Details of standard microscopes available have been omitted because more often than not a camera is selected subsequent to the purchase of a microscope. Illustrations and descriptions of photomicrographic equipment represent typical examples of the wide range available.

We wish to acknowledge the help given to us by many friends but we are especially indebted to Mr. John Grant, who conceived the idea for a book of this nature, and to Professor B. Cohen for his encouragement, kindly criticism and for reading and correcting the manuscript. Our thanks are also due to Mr. F. Bradley, Mr. J. G. Dain, Mr. G. D. Powell and Mr. R. Settrington for their help in obtaining many of the blocks used for illustrating Section I; also to Messrs. C. Baker Ltd., R. Beck Ltd., T. F. Blumfield Ltd., R. F. Hunter Ltd., Ilford Ltd., Johnson Ltd., Kodak Ltd., Ernst Leitz and Carl Zeiss for donating or loaning blocks. Finally we wish to express our thanks to Miss A. Ritchie and Miss J. L. Payton for typing the manuscript and to our publishers for their help and patience.

E.B.B.
A.R.T.C.

CONTENTS

CONTENTS

CONTENTS

PREPARATION OF PRINTS

PREPARATION OF MONOCHROME TRANSPARENCIES

PREPARATION OF SOLUTIONS

SECTION III

SECTION I

CHAPTER I

PRINCIPLES AND APPLICATIONS

According to the English edition of a booklet entitled Photomicrography by Dr. Henri van Heurck, published in 1894, the art of reproducing pictures of microscopic images had already at that time been practised for at least fifty years. Van Heurck referred to a publication entitled *Atlas d'anatomie microscopique*, which was compiled from daguerreotypes by Donne and Foucault in 1844. The first paper proofs are thought to have been produced by Bertsch, thirteen years later. For a long time afterwards, photomicrography was practised only by a few who experimented until they had become proficient with difficult and time-consuming techniques. It was only in much more recent times that equipment and methods were evolved to bring photomicrographic techniques within the compass of an ever-increasing body of scientific workers. This can be attributed to advances both in microscopy and in photography; unremitting work is being carried out in both these fields, and as a consequence techniques are constantly subject to modification as improved methods, materials and apparatus become available. Nevertheless it is possible to enunciate certain unchanging principles. Adherence to these makes it possible to attain optimum results within the limits of whatever apparatus is available.

General Principles

A photomicrograph of high quality is one that reproduces the microscopic image accurately and clearly. To achieve this it is necessary to have:

1. Knowledge of the subject matter. This enables the right field to be selected and the important features to be portrayed to advantage.

2. Efficient equipment and suitable materials.

I

3. Knowledge of equipment. The many types of lenses, light sources and condensers used in photomicrography demand some understanding of optics and microscopy.

4. Knowledge of photographic techniques. Experience in general photography is entirely applicable to photomicrography. An understanding of the properties of photographic materials and chemicals is essential.

Finally, care should be devoted to the presentation of a photomicrograph. The method of mounting for display, the use of annotations and similar details are of considerable importance. Although nothing can be done to transform a poor photomicrograph into a good one, the converse is unfortunately true; an accurate record can be marred to the extent that its purpose is defeated by the lack of attention to these details.

Applications

The use of photomicrography as a tool in teaching, research and industry has increased greatly in recent years. Without the use of a projection microscope, which is an elaborate and expensive apparatus, it is difficult to demonstrate microscopical structures to more than one person at a time. In teaching, the reproduction of microscopic images in the form of transparencies suitable for projection, facilitates demonstration to large groups of students.

Many aspects of research requiring microscopy are recorded by means of photography, and there is no better method for communicating results of such research in scientific journals. Moreover, if photomicrographs are freely available they are of immense value for comparative studies; placed side by side they offer a far easier means of detecting differences and similarities than can be achieved by looking at preparations separately under the microscope.

They are also a useful means of recording preparations which cannot be preserved indefinitely. Even where slides can be kept in a mounted state many research workers find that photomicrographs, being easily filed and stored, provide the simplest means for ready reference.

Photomicrographs mounted for display furnish a medium for portraying results of microscopical investigations to large audiences.

2

SELECTION OF SPECIMENS

The preparation of material for microscopy and its subsequent photography are so allied that the closest co-operation is called for between the technician preparing the material, the photographer, and the person requiring the photomicrograph. Each must have an understanding of the others problems if first-class results are to be obtained. For the photographer to have a working knowledge of the principles and methods used to prepare the material encountered in his particular field is an advantage which cannot be stressed too strongly. An experienced technician who has acquired the photographic skill and knowledge necessary to carry out photomicrographic techniques on material which he has himself prepared, is most likely to produce the best results.

Possibly the greatest demand for photomicrography comes from biological workers, especially histologists and pathologists. After visual examinations of a section the histologist or pathologist must select the exact field he requires to be photographed. This calls for considerable thought and care. The best photographic technique will never compensate for, or add to, an injudiciously chosen field. It is essential to have a clear idea of the exact features to be recorded and to select only that field which demonstrates them. Even when such a field is selected it is often possible to enhance the illustrative value of the photomicrograph by careful consideration of composition and magnification.

More often than not histological sections are prepared without any thought to future photography and, unless the section is of the highest quality, of uniform thickness, flatness and staining intensity, it will be of little use for this purpose. There is nothing more frustrating than to find a valuable field marred by a crease or tear in the section. It is of course incumbent on the photographer to ensure that the slide is clean and that all the lenses in the optical system are free of dirt and dust. Equally so, it is the responsibility of the technician to ensure that no foreign matter is present on the slide, section, or cover-glass. Figure 1 illustrates several forms of artefact which could restrict the choice of field.

The final photomicrograph should portray nothing less and nothing more than the image projected on the camera focusing screen. Attempts to remove artefacts or to enhance particular areas by retouching should be discouraged in the interest of scientific accuracy.

FIGURE 1 (*opposite*)

a. Contamination of a microscope slide with yeast cells after prolonged storage in water.
b. The result of mounting a section on one of the slides shown in a.
c. Severe creasing and scoring of the section. Caused by minute serrations along the edge of the microtome knife during section cutting.
d. Droplets of water occurring as a result of imperfect dehydration of the section before mounting. Contamination of the mounting medium with water will produce a similar result.
e. The unstained area in the right half of the field is due to incomplete removal of the paraffin wax from the section before staining.
f. Parts of the field out of focus. Due to the section lifting from the microscope slide.
g. Deposits in the section. In this case due to incorrect treatment of the tissues after mercuric fixation.

4

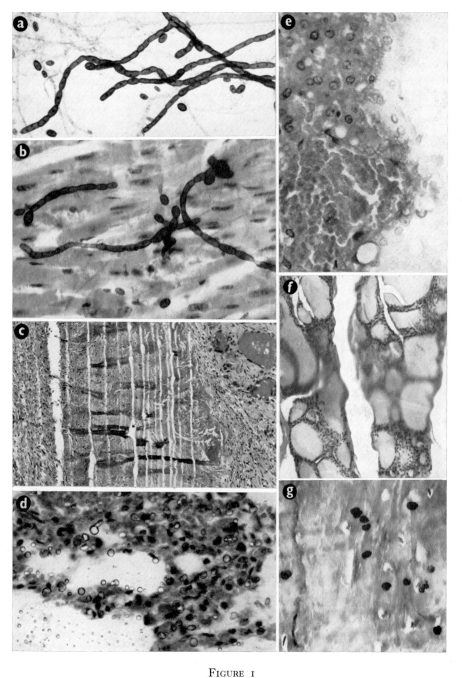

FIGURE 1

Several forms of artefact which could restrict the choice of field for
photomicrography

CHAPTER III

PRINCIPLES OF MICROSCOPY

There are many different methods of illuminating the preparation to be photographed through the microscope. These may be listed as follows

1. *Transmitted illumination*
   Bright field
   Dark field
   Phase contrast
   Polarized light
   Ultra-violet
   Fluorescent
   Infra-red

2. *Incident illumination*
   Vertical
   Oblique

1. *Transmitted illumination*

Bright field. The commonest method used is that in which a semi-translucent object is viewed by transmitted light. In this instance the photograph records a bright background and the subject matter appears as a shadow of varying intensities.

Dark field illumination. For the observation of objects which have a refractive index very similar to that of the medium in which they are mounted, dark field illumination is necessary. This involves the elimination of all direct rays of light passing through the microscope condenser and allows only light scattered or reflected by the preparation to reach the objective lens. There are two methods of obtaining this type of illumination, namely: (1) by inserting an opaque circular stop called a wheel stop in the carrier frame of the substage condenser, or (2) by using specially designed condensers, such as the Cardioid or Paraboloid type of condenser. By these methods a hollow cone of light is formed. The presence of the object at the apex of the cone scatters light into the objective thereby forming a bright image against a dark background.

5

Dark field microscopy is of value for revealing extremely minute particles and its use is therefore largely confined to high magnifications. For maximum efficiency dark field microscopy depends upon the accurate location and adjustment of a suitable condenser, a high intensity light source and a perfectly clean preparation. Minute air bubbles and particles of foreign matter are made obvious in dark field and will obscure details of the subject matter.

This form of photomicrography poses unique problems when used for the study of moving objects. High magnification amplifies apparent speed, and therefore fractional-second exposures are necessary.

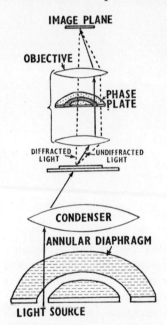

FIG. 2. Principles of phase contrast microscopy

Phase contrast. The differing refractive indices of any specimen produce variations in the optical path of light passing through it. Such differences are indistinguishable with the normal microscope due to the high intensity of illumination, but by relatively simple modifications of the standard compound microscope it is possible to visualize the differences and thus to observe the specimen by phase contrast. Two accessories are required for phase contrast microscopy:

(i) *An annular diaphragm.* The annular diaphragm is fitted under the condenser and produces a thin hollow cone of light (Fig. 2). The diameter of the central portion of the annulus must be less than the diameter of the cone of light accepted by the objective (numerical aperture); therefore different sizes of annular diaphragms are required for objectives of differing numerical apertures. A set of annular diaphragms, or alternatively, a condenser incorporating a rotating mount holding the differing diaphragms, may be used.

(ii) *A phase diffraction plate situated at the rear focus of the objective.* The cone of light, after passing through the specimen, constitutes diffracted and undiffracted light and is picked up by the objective lens. The undiffracted light falls on to an annulus in the diffraction plate and its intensity is diminished. The diffracted light falls on either side of this annulus where its phase is further retarded with a resultant increase in contrast (Fig. 2). By varying the combination of the phase-retarding and light-absorbing material on both the annulus and the surrounding areas of the phase plate, almost any combination of illumination of the preparation can be produced. This system is termed dark contrast and is used in microscopes manufactured by Messrs. Zeiss, Bausch and Lomb and Baker. The Leitz phase microscope employs a radically different type of illumination whereby a Heine mirror condenser functions as an annular diaphragm. The phase plate is based on the same principle as that already described. Altering the position of the condenser varies the annular illumination and obviates the need for a set of annular diaphragms.

Polarized light. Many subjects examined under the microscope are birefringent so that if examined with transmitted polarized light, with the polarizer and analyser in the crossed position, a brightly coloured image on a dark background is produced. Clarity is often enhanced as the resulting coloured image has more contrast than a black and white image. Variations of colour can be produced by rotating the plane of polarization with respect to the preparation and by rotating the analyser with respect to the polarizer. Polarizing microscopy is of particular value in examining crystalline substances. For photomicrography however, the same principles and techniques apply as for other forms of transmitted light. Photography of polarized light effects should be in natural colour wherever possible as correct reproduction of interference colours in black and white is difficult.

The use of reflected polarized light has found considerable application in metallurgy not only for the study of crystal grain orientation but also in decreasing unwanted reflections from any metal surface.

Specially designed polarizing microscopes are manufactured; however, polarizing accessories to fit standard microscopes are available.

Ultra-violet. As the limit of resolution depends upon the wavelength of light, higher resolution of fine detail is possible by working with short wavelength ultra-violet radiation instead of visible light. Ultra-violet microscopy is of value for the examination of preparations which, although transparent in visible light, show differential absorption in the ultra-violet region.

To obtain fine detail a very short wavelength of light (275 mμ) is necessary and this involves the use of special microscopes equipped with quartz optics. For other types of ultra-violet work using light of wavelength 365 mμ, this specialized equipment is not required. The high intensity mercury lamp used in conjunction with special achromatised objectives allows examination of preparations with ultra-violet light. The technique is ingenious and simple. The use of a Wratten (Kodak) filter No. 77A permits examination and focussing of the preparation by visible green light in the 546 mμ line. This is necessary as the insensitivity of the eye to the blue violet region of the spectrum makes focusing virtually impossible. Photography is then accomplished by substituting a Wratten filter No. 18A which absorbs the visible light but freely transmits the mμ 365 line.

If special achromatized objectives are not available, standard objectives may be used. These, however, are not so well corrected for light between the 300-400 mμ region and therefore it can be expected that the focal plane will not be the same as for the shortest blue-violet rays visible to the eye. The correct focus in the shorter wave-length region of any particular objective lens may, however, be determined by one of two methods. The preferable method and by far the quicker, is the use of a strongly fluorescing glass plate of uranium glass which is substituted for the ground glass screen. The image fluoresces on the plate and can then be focused. The other method necessitates a series of test exposures at differing focal levels.

Fluorescent. This type of microscopy is closely allied to ultra-violet illumination. Many substances when excited by the invisible short waves of the ultra-violet region emit light of a longer wavelength. If these longer wavelengths fall within the range of the visible spectrum, they are visible to the eye and the preparation appears luminous. This is known as fluorescence. As the preparation is itself the source of illumination it is possible to obtain detailed images of high resolution. Preparations may also be treated with known fluorescing chemicals which have a selective affinity for differing structures. When viewed with ultra-violet light the result

8

is analogous to differential staining in ordinary microscopical preparations.

For fluorescence microscopy a light source rich in ultra-violet light is essential and an ultra-violet transmitting filter must be used. Most optical glass transmits considerable amounts of ultra-violet light but should maximum illumination be desired all the glass in the illuminating system must be replaced with quartz elements.

For fluorescence photomicrography the ultra-violet light must be eliminated after passing through the preparation because, if allowed to reach the photographic emulsion, it will form an image. This is accomplished by placing an ultra-violet absorbing filter between the preparation and the emulsion.

High speed photographic emulsions must be used as the total light available is small. Even so exposure time is protracted and, because of this, magnifications must be kept low.

Infra-red. The examination of specimens transparent to infra-red radiation but opaque to light in the visible spectrum is an important facet of microscopy. Special stains are available which absorb infra-red radiations and thereby increase the contrast of the resultant photomicrograph. The commonly used light sources, in conjunction with selective filters to confine the radiation to the desired spectral region, may be used. The actual photomicrographic technique differs only slightly from that for other forms of transmitted light. The most important difference lies in focusing the preparation since infra-red radiations are invisible to the eye. To focus it is necessary to make a series of test exposures with different settings of the fine adjustment of the microscope. Special materials sensitised for use in the infra-red region must be used.

2. *Incident illumination*

The examination of small opaque objects with the microscope is possible if incident or reflected light is used. The same principles of illumination however, still apply in that the light reaching the photographic emulsion must be uniform over the entire field and of sufficient intensity.

Two types of incident illumination exist:

(1) *Vertical* or specular illumination whereby the light is projected from the side on to a reflector sited at the back of the objective. This deflects the light into the back lens of the

objective which is thus focused on to the preparation (Fig. 3a). This type of illumination is used mainly in metallurgical work for which specialized equipment is available. Vertical illuminators may be purchased separately and fitted to the standard compound microscope if only occasional examination of opaque objects is required. For work at high magnification with incident light, specialised equipment must be used. With this equipment the illumination passes through the objective but not the optical system (Fig. 3c).

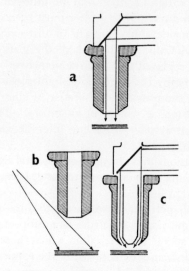

FIG. 3. Types of incident illumination

(2) *Oblique* top illumination where the light is projected at an angle past the objective on to the surface of the preparation (Fig. 3b). As the beam of light has to pass the objective, it is essential that there is a considerable working distance between the preparation and the objective and therefore this method of illumination is only applicable to work at low or medium magnification with the standard compound microscope.

CHAPTER IV

RECORDING EQUIPMENT

The selection of equipment for photomicrography is governed by the nature of the material under examination and by the final purpose of the photomicrograph. Transparent specimens are examined by transmitted light, and a good biological microscope will suffice. The surfaces of hard tissues, such as bone and teeth, metals and geological specimens, demand examination by reflected light for which a more specialized apparatus, the metallurgical microscope, is required. For examination of fresh or wet material, the selection of a vertical unit is obligatory. The influence exerted by the final purpose of the photomicrograph is best illustrated by example. Photomicrographs for publication demand a high degree of definition which is lost if prints are enlarged from 35 mm. film. In this instance it is advisable to use a camera body designed to take $3\frac{1}{4} \times 4\frac{1}{4}$ in. (or larger) plates or cut film, and to prepare prints by contact. Final selection may be governed by cost but a wide range of equipment is obtainable by varying in price from a few pounds for simple attachment cameras to several hundreds of pounds for specialized equipment. From these it should be possible to select, within any given budget, the most suitable photomicrographic set-up.

A warning should be added at this point against improvization with home-made equipment. The amateur enthusiast may delight in improvization but the constant realignment of illumination and optical systems is time-consuming, and the wear and tear caused to accessories, which are usually already worn, precludes the use of such an arrangement for scientific purposes. For first class routine photomicrography only well designed equipment of stable construction will suffice.

A source of illumination, a microscope and a camera constitute the essential components of any photomicrographic apparatus. These may be purchased as:

(1) Separate units;
(2) Microscope with integral light source and camera; or
(3) Light source, microscope and camera assembled in a single unit.

The purchase of separate components allows the use of the light source and microscope for routine microscopy whilst the camera body, if supplied with a detachable lens, may be used for other forms of photography. It is recommended, however, that separate components should be aligned and rigidly mounted on a solid base board so that the set-up can be employed at short notice with the minimum of adjustment. Ideally, two camera bodies should be available, one taking 35 mm. film and the other plates or cut film.

Illumination satisfactory for visual microscopy will not suffice for photomicrography. The latter requires a light source of high intensity to enable (1) observation and focusing on a ground glass screen, (2) the use of colour contrast filters and (3) the use of short exposure times. Many modern microscopes incorporate an integral illumination system adequate for photomicrography.

The following is a list of separate light source units suitable for use with a standard microscope.

1. Carbon arc with mechanical feed
2. Pointalite
3. Tungsten ribbon filament lamps
4. Photoflood lamps coupled to a resistance for intensity control
5. Mercury or sodium vapour lamps for ultra-violet and fluorescence photography.

Illumination for colour photography is more critical. A light source of 3200°K rating is essential if perfect colour balance is to be obtained. Variation from this rating compels the use of compensating filters coupled with an expert knowledge and understanding of a topic fraught with complications.

Different types of photographic apparatus can, for descriptive purposes, be grouped under the general headings of (1) attachment cameras; (2) horizontal units; (3) vertical units, and (4) universal and specialized units.

Attachment cameras

Attachment cameras are designed to fit over the microscope eyepiece thereby allowing the image obtained by the eyepiece lens to project directly on to the film or plate carried within the body of the camera. In some instances cameras designed for general photography may be adapted to fit the microscope by means of a micro-adapter incorporating a focusing and viewing eyepiece (Fig. 4).

FIG. 4. A micro-adaptor supporting a Rolliflex camera

FIG. 5. Projection-attachment camera (Zeiss)

FIG. 6. Attachment cameras (Beck)

FIG. 7. Horizontal assembly (Beck)

As the camera lens is retained the resulting photomicrograph is circular in shape. The additional weight of attachment cameras may cause difficulty when used on microscopes which are focused by altering the position of the objective lens. They should, therefore, be preferably used on microscopes which have a fixed lens turret-tube-eyepiece unit. Such microscopes are focused by movement of the stage in the vertical plane.

With the simplest type of attachment camera the image is focused on to a ground glass screen which is then replaced by a plate holder. The Baker Reflex camera incorporates a reflex mirror. The image is deflected by the mirror on to a ground glass focusing screen so arranged that the focal distance from the object lens to the screen coincides with the focal distance from the object lens to the emulsion surface. The shutter and the reflex mirror are connected so that on release of the shutter, the mirror swings out of the light path allowing the image to fall directly on the emulsion. The shutter is designed for time exposures only.

Messrs. Zeiss have recently produced a projection attachment (Fig. 5) which can be readily converted into a camera by replacing the projection screen with a combined focusing screen and plate holder unit taking 9×12 cm. cut film or plate. A simple shutter is installed in the assembly.

Most attachment cameras, however, have a beam-splitting device which enables the preparation to be focused with the camera already loaded.

The camera illustrated in Figure 6 is equipped with interchangeable bodies permitting the use of 35 mm. film, cut film and $2\frac{1}{2} \times 3\frac{1}{2}$ plates. The camera is attached to the draw tube of the microscope and has an eight-speed shutter. Separate viewing tubes are provided.

The attachment camera manufactured by Messrs. Zeiss consists of three basic components: (1) a connecting ring, (2) a body with a beam-splitting device and an observation system and (3) a shutter and camera. Two types of body are available. Basic body I has a fixed reflecting prism which separates the light passing from the object lens into two unequal components. The major component falls on the emulsion whilst the minor component is deflected into the viewing tube. Basic body II is designed for use in conjunction with the Zeiss exposure meter. Incorporated into the beam-splitting device is a mechanism which controls the light path so that it can be directed either into the focusing eyepiece, into the exposure

13

meter or on to the photographic emulsion. The focusing eyepieces and shutter are identical for both bodies. The eyepiece incorporates a reticule, the setting of which adapts it to the differing film or plate sizes carried within the camera. The shutter is automatic and has seven speeds and provision for time exposures. The cameras are designed for 35 mm. film, $6 \cdot 5 \times 9$ cm. plates, roll film and film pack holders.

A fully automatic 35 mm. attachment camera, the Leitz Orthomat, is now available and is designed for use in conjunction with any modern Leitz microscope. Determination of exposure and operation of the camera are controlled by pressing a single button. Exposure range is between 1/100 sec. to half an hour and partly exposed film magazines can be readily interchanged. The only pre-setting required is an adjustment for film speed.

The following attachment cameras are also available:

Messrs. Beck. Eyepiece $2\frac{1}{2} \times 3\frac{1}{2}$ in. and $3\frac{1}{4} \times 4\frac{1}{4}$ in.
Messrs. Bausch and Lomb. Model N. 35 mm. and $3\frac{1}{4} \times 2\frac{1}{4}$ in.
Messrs. Leitz. 35 mm. (Mikar) and 9×12 cm. (Makam).
Messrs. Watson. Eyepiece 35 mm. and $3\frac{1}{2} \times 2\frac{1}{2}$ in.

Attachment cameras taking 35 mm. film have been described in the above text and it is convenient, at this point, to consider some of the merits and demerits of this unit. The small working and storage space, the ease of assembly and manipulation, and the apparent economy, have led to considerable increase in its popularity for photomicrography. However, it must be appreciated that this is not a universal tool. For the reproduction of similar or consecutive sections and the preparation of monochrome or colour transparencies for projection, the 35 mm. camera is ideal, but the following, and at present insurmountable, limitations restrict the use of this equipment.

1. It is impossible to prepare a single negative without wastage of film.
2. Different preparations recorded on any given length of film must each suffer the restriction imposed by the particular emulsion as each must receive identical treatment during development.
3. The preparation of prints involves enlargement and a consequent loss of definition.
4. Magnification can be altered only by changing the objective or ocular or both.

14

This last limitation has been partially overcome with the introduction of a microscope (Gillett and Sibert, London) with an adjustable draw tube which can be firmly clamped in any position. Full extension increases magnification by approximately one third for any given combination of ocular and objective.

Horizontal equipment

The development of vertical and specialized equipment, the excessive working and storage space required, and the inability to photograph wet specimens have led to a fall in demand and manufacture of horizontal units. However, because of their many advantages they are still favoured by many photomicrographers. Advantages which recommend horizontal units are:

1. The straight line of the optical axis.
2. Auxiliary equipment can be readily accommodated.
3. The light source can be changed.
4. Direct viewing and focusing is possible.
5. The ability to employ large size plate or cut film.
6. If equipped with long bellows, magnification can be adjusted with the same object/eyepiece combination.
7. It allows maintenance of a standard technique.

If self-assembly of a horizontal unit is contemplated, purchase of the basic components is still possible. It is desirable that the lamp housing, condenser and filter holder are all mounted on an optical bench which allows separate adjustment of each unit in the same long axis. The microscope must be securely fixed and the camera so mounted that it can be moved aside or backwards allowing direct viewing of the preparation through the microscope eyepiece. An example of an horizontal assembly is shown in Figure 7.

A 35 mm. adaptor (Kodak) may be substituted for the plate holder unit of the bellows camera allowing the horizontal assembly to be used for 35 mm. photography in monochrome or colour. Alternatively the bellows camera may be replaced by most of the 35 mm. attachment cameras.

Vertical equipment

Vertical units follow a general design the main features of which are a solid base unit supporting a standard microscope and a vertical metal pillar to which is attached a bellows type camera. The assembly must incorporate an intense light source and the camera should be fitted with a reflex mirror to simplify viewing and focusing and a shutter to ensure accurate exposures.

The advantages of vertical equipment are robustness, small storage space and versatility. In contrast to horizontal equipment, the limited bellows length of any vertical camera may necessitate frequent changes of the objective, the eyepiece, or both in order to obtain optimum field size or magnification.

The vertical camera illustrated in Figure 8 is of simple design. The base consists of a heavy iron casting with an adjustable toe for added stability. The vertical pillar can be rotated in its locating boss enabling the camera to be moved to one side for access to the microscope eyepiece. The camera bellows extend over a range of 12·5 cm. to 38 cm. The shutter is actuated by a cable release and has eight speeds from 1 second to 1/250 second plus time and brief exposure settings. Two cameras are available taking either $4\frac{1}{4} \times 3\frac{1}{4}$ in. or 9×12 cm. plates. The former can also take a roll film holder. A ground glass focusing screen and $45°$ reflex mirror are supplied together as a single unit which interchanges with the metal plate holders.

A similar type of vertical camera is manufactured by Messrs. Baker (London) (Fig. 9) and is designed for use with the Projectolux illuminating base. The $3\frac{1}{4} \times 4\frac{1}{4}$ camera is supported by a steel bar and set screw to a chromium plated column bolted to the base. The camera can thus be moved vertically or rotated on the column. The camera bellows extend over a range from 10 in. to approximately 20 in. and the shutter speeds are from 1 second to 1/250 second with time and brief exposure settings. The image is projected on to a ground glass screen at the rear of the camera body.

The Zeiss universal camera is capable of photomicrography over a wide range of magnification and illumination. The heavy base plate is fitted with anti-vibration supports, a stout supporting column and a light source. A connecting sleeve on the lower carrier plate normally unites microscope and camera; a compound shutter can be supplied on request. The bellows extend to 60 cm. and an added refinement is a reflex attachment allowing observation and focusing whilst seated. Exposure is determined with the aid of exposure slides of which two types are available. The Series exposure slide I is attached to the upper carrier plate, and the photographic plate is fitted into a sliding holder which allows it to be moved in a series of steps across a fixed diaphragm. In this manner a series of test

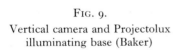

Fig. 8.
Vertical camera (Beck)

Fig. 9.
Vertical camera and Projectolux
illuminating base (Baker)

FIG. 10. Aristophot vertical unit (Leitz)

FIG. 11. Vertical assembly incorporating a metallurgical microscope (Baker)

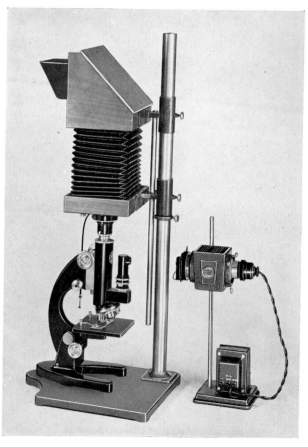

exposures is obtained with each exposure varying from its neighbour by a predetermined amount. The series exposure slide II incorporates a vacuum photo-electric cell which measures directly the amount of light falling on the emulsion.

The Aristophot (Fig. 10) is designed for use with the Leitz Ortholux microscope. The base plate and twin camera holders are of stout construction and easily portable. A centre stop on the base plate ensures correct alignment of microscope and camera. The microscope has a built in high intensity illumination system. Alternative forms of illumination are available. The bellows type camera, extending to 26 or 30 in., depending on whether a reflex attachment is present, has a time and instantaneous shutter operated by a cable release. The outfit includes two 9×12 cm. metal dark slides with $3\frac{1}{4} \times 4\frac{1}{4}$ in. or $6 \cdot 5 \times 9$ cm. adaptors. Focusing is through monocular or binocular observation tubes but final focusing must be on the ground glass screen of the reflex attachment.

The photomicrography of polished or etched surfaces of metals, bone and teeth and other opaque objects may be carried out with a vertical camera unit provided a metallurgical microscope and incident illumination are used. A simple arrangement for such photomicrography is illustrated in Figure 11. The metallurgical microscope which incorporates incident illumination is placed upon the base unit and the vertical camera is positioned over the microscope eyepiece.

The following vertical cameras are also available:

Bausch and Lomb K. Camera. $4\frac{1}{2} \times 3\frac{1}{4}$ in.

„ „ „ Model H. Camera 5×7 in.

„ „ „ Model L. Camera 5×7 in.

Watson Vertical $4\frac{1}{4} \times 3\frac{1}{4}$ in. and $6\frac{1}{2} \times 4\frac{3}{4}$ in.

Universal and specialized equipment

The types of photomicrographic equipment described so far have all consisted of equipment attached to or assembled around standard types of microscopes. Messrs. Leitz first introduced an instrument specifically for photomicrography in which the illumination, microscope and camera were all built into a single compact unit. The Panphot (Fig. 12) incorporates a revolving mirror for transmitted and incident illumination. The extendable mirror reflex camera takes plates of 9×12 cm. or $3\frac{1}{4} \times 4\frac{1}{4}$ in. and is supplied with two dark slides. The design of the microscope allows correct and speedy attachment of accessories. These include monocular and binocular

17

bodies with a combined photographic tube and different object stages and substages. Standard microscope accessories may also be fitted to convert the apparatus for routine microscopy. Supplementary equipment is available for phase contrast, dark field, polarized light, fluorescence, and metallurgical microscopy.

The Ultraphot II, manufactured by Messrs. Zeiss, (Fig. 13) can be used for all forms of microscopy. It is of compact design with all the controls readily accessible and the camera is fully automatic. The camera shutter is opened by a push button, the exposure automatically determined and the shutter then closes itself.

Messrs. Beck have recently introduced the Beck No. 50 Universal microscope and camera which can be used as a bright field, dark field, phase contrast and polarizing microscope with both transmitted and incident illumination (Fig. 14). The microscope and camera are supported on a heavy and rigid casting. The microscope has both binocular and monocular eyepiece bodies. The binocular eyepiece has no magnifying factor and can be interchanged with the monocular without alteration to magnification or focus. The monocular eyepiece has a focusing adjustment so that the focus of the image can be synchronized with that obtained in the binocular or on the focusing screen of the camera; it is used mainly in conjunction with the camera for observing the preparation prior to exposure. Withdrawing the monocular tube to a set stop removes the prism which normally deflects the image into the eyepiece and allows the light to enter the camera. Above the bodies is a rotating turret containing four eyepieces for use either for projection or photography. The camera is of the reflex type and has a bellows extension of 520 mm. and takes $\frac{1}{4}$ plate. The shutter has eight speeds from 1 to 1/250th second and " T " and " B " settings. Provision is made for use of a 35 mm. camera. The bellows may be detached from the eyepiece, compressed and secured to the camera casing. This leaves sufficient space above the eyepiece for a 35 mm. camera to be inserted.

The following universal and specialized cameras are also available:

Watson Holophot $3\frac{1}{4} \times 4\frac{1}{4}$ in.

Vickers Projection Microscope.

Recording accessories: Exposure meters

1. *The Sekonic Microlite Meter.* The microlite meter utilises cadmium sulphide, a semi-conductive material in which the resistance level changes with the intensity of the light falling on it.

18

Fig. 12. Panphot unit
(Leitz)

Fig. 13. Ultraphot unit
(Zeiss)

FIG. 14. Universal unit No. 50 (Beck)

Designed primarily for general photography, its sensitivity is several hundred times greater than a selenium photo-cell. This increased sensitivity allows more accurate measurement and, set for low light conditions, it has been satisfactorily used by taking readings direct from the ground glass focusing screen of horizontal and vertical bellows type cameras.

2. *The Baldwin photometer type M.N.D.* This is a reliable and accurate instrument. The exposure is determined by applying the photo-cell directly to the ground glass focusing screen.

3. *The Leitz Microsix.* This versatile instrument may be used with most types of equipment by virtue of the fact that readings may be taken at any position in the set-up.

4. *The Zeiss exposure meter.* Designed for use with Zeiss miniature attachment cameras and the series exposure slide II. It is a sensitive and reliable instrument.

Filters

These are an essential accessory in photomicrography. A set of contrast colour filters for black and white work, colour compensating filters for colour work and neutral density filters for reducing illumination intensity will be required.

Field markers

1. *The diamond field marker.* After selecting the field the objective is replaced with this accessory which inscribes a circle on the cover-slip around the chosen field. The radius of the circle may be varied to correspond with differing magnifications. The disadvantage of this accessory is that the slide is permanently marked.

2. *The England Finder.* This consists of a 3×1 in. glass slide marked with 1 mm. squares. Each square has a serial letter and number. The square is further sub-divided into five parts, each part identified by a further number. Relocation of a given field is simple if the instructions are followed accurately.

Magnifiers

These are used for accurate focusing of the image on a ground glass screen.

Stage micrometer

This accessory, consisting of a graduated scale etched on a 3×1 in. slide, is used for determining the magnification of the lens-eyepiece-bellows extension set-up.

CHAPTER V

PHOTOGRAPHIC EQUIPMENT AND MATERIALS

1. *Equipment*

Darkroom equipment for photomicrography does not differ essentially from that required for general photography. A wide range of equipment is therefore available.

Safe lights. Figure 15 illustrates the hanging type of safelight and this provides:

> (i) direct illumination via a filter fitted to the under surface of the lamp;
> (ii) indirect illumination, where the filtered safelight is reflected from the ceiling;
> (iii) combined direct and indirect illumination.

The safelight illustrated in Figure 16 can be fitted on a wall, bench or suspended from the ceiling. It is compact in design and the filters are readily interchangeable.

Two timers are necessary in any darkroom. One, of the alarm type, for use when processing negatives, and the other for use when printing. A suitable alarm timer is illustrated in Figure 17. Setting the hand to the number of minutes required winds both the clock and alarm mechanisms. For printing, the large dial second timer (Fig. 18) is ideal. It can be placed on the bench or attached to the wall. The face is graduated in seconds with a smaller dial indicating minutes. Depressing the knob on top of the timer returns both second and minute hands and rewinds the mechanism. The side lever is depressed to start the clock.

Thermometers for use with dish development or for clipping to the side of a processing tank are illustrated in Figures 19, and 20.

The enlarger should be of the vertical type for economy of space and should be designed to take negatives from 35 mm. to $\frac{1}{4}$ plate. The enlarger illustrated (Fig. 21) provides for the interchange of lenses and condensers and for negatives ranging from 35 mm. to $3\frac{1}{4} \times 4\frac{1}{4}$ in. in size.

FIG. 15. Universal safelight Model 2 (Kodak)

FIG. 18. Large dial seconds timer (Kodak)

FIG. 16. Beehive safelight (Kodak)

FIG. 19. Dish thermometer (Kodak)

FIG. 17. Minute timer with alarm (Kodak)

FIG. 20. Tank thermometer (Kodak)

FIG. 21. Enlarger (Blumfield)

FIG. 24. Plate processing rack No. 3 (Kodak)

FIG. 22. Enlarger timer switch (Johnson)

FIG. 25. Sheet film processing rack No. 3 (Kodak)

FIG. 23. Masking frame (Johnson)

FIG. 26. Film hanger bar No. 9 (Kodak)

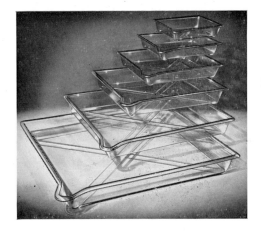

FIG. 27. Transparent plastic processing dishes (Kodak)

FIG. 28. 35 mm. developing tank Model II (Paterson)

FIG. 30. Hard rubber washing tank No. 3 (Kodak)

FIG. 29. Processing tank No. 3 (Kodak)

FIG. 31. Professional washing tank
(Ilford)

FIG. 32. Drying cabinet
Model B (Kodak)

Enlarger timer switch. (Fig. 22). This is a useful accessory and ensures uniform exposures. Exposures between 1 second and 5 minutes may be selected. A separate switch allows manual control of the illumination for focussing.

Enlarging masking frame is required to immobilize the printing paper during exposure. Adjustable masking slides (Fig. 23) allow precise masking of unwanted areas of the negative.

Plate processing rack. This should be adjustable in order to adapt plates of varying sizes (Fig. 24).

Sheet film processing rack is used for tank processing of sheet film in batches. The racks (Fig. 25) are designed to accommodate film hangers.

Film hangers (Fig. 26) are required to avoid handling of sheet film during processing and drying.

Processing dishes. Transparent plastic processing dishes (Fig. 27), in sizes from $6\frac{1}{2} \times 8\frac{1}{2}$ in. to 20×24 in., are both durable and easy to clean. Enamelled steel and stainless steel dishes are obtainable in similar sizes.

35 mm. and roll film tanks. The processing of 35 mm. and roll film should be undertaken in tanks specially designed for this purpose The model illustrated in Figure 28 is designed to take 35 mm. films only. For processing reversal colour films the transparent ends of the spiral allow the second exposure to be made without removing the film. Larger tanks incorporating an adjustable spiral which allows processing of a range of roll films are available.

Processing tanks. The unit shown in Figure 29 is made of vulcanized rubber and consists of a tank with a capacity of three gallons, a cover and a floating lid.

The washing tank shown in Figure 30 is made of hard rubber. A perforated metal tube circulates the water which overflows through a slot near the top. It will accept both plate and sheet film racks. The washing tank shown in Figure 31 is constructed of stainless steel and the water is circulated through a central spray inlet. By removing the plug at the top of the overflow the unit is converted into a constant level stand pipe with intermittent siphoning.

A drying cabinet is essential for rapid drying of negatives and the model illustrated (Fig. 32) is both efficient and compact. The cabinet occupies 21×21 in. bench space and is 3 ft. 6 in. high. A silent running electric fan mounted over a cylindrical heater provides a steady stream of warm air between 110° and 118° F. The cabinet is thermostatically controlled.

Print washers. The automatic dish siphon is clipped to the side of a processing dish. A stream of fresh water is directed into the dish at a pressure sufficient to ensure constant agitation and contaminated water is discharged into the sink. This type of washer works continuously and needs no attention. The stainless steel washer (Fig. 33) has an automatic siphon which empties the tank every three minutes. Removal of the siphon cap isolates the siphon and maintains a constant water level in the tank. This type of washer will take prints up to 16 × 20 in. in size. The enlargement siphon washer (Fig. 34) has a diameter of 24 in. and is 4 in. deep with the siphon fitted on the side. Prints and enlargements up to 12 × 15 in. in size can be satisfactorily treated with this washer.

A print squeegee is used to remove excess moisture from prints. The one illustrated (Fig. 35) has a 15 in. long durable rubber roller.

Print trimmers. The guillotine type of trimmer (Fig. 36) is adequate for trimming prints but a heavier model should be used for cutting and trimming mounting board.

Glazers. The flat bed type of glazer shown in Figure 37 may be used if the output of glazed prints is small. The aluminium glazer bed is electrically heated and the covering cloth is carried on a spring roller. A polished stainless steel plate (14 × 20 in. or 10 × 14 in. in size) is provided to support and glaze the prints. If there is a large output of prints a glazing machine will be required. The model illustrated in Figure 38 is automatically controlled and is fitted on a tubular stand with a shelf to carry the print tray.

A fixing iron is required for tacking dry mounting tissue to print and mount before transfer to the mounting press.

The mounting press illustrated (Fig. 39) is thermostatically controlled and can mount prints of 12 × 15 in. in one operation.

Immersion heaters may be used to raise the temperature of developer and are available for both tank and dish development.

Printing frames. These may be obtained in wood or plastic and are used for printing by the contact method.

A water filter may be fitted either between the tap and washer or

FIG. 33. Professional print washer
(Ilford)

FIG. 34. Enlargement siphon
washer (Kodak)

FIG. 35. Print roller (Kodak)

FIG. 36. " Merrett " guillotine

FIG. 37. Flat-bed glazer Model 2 (Kodak)

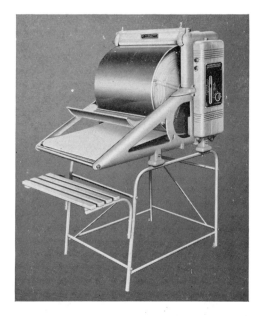

FIG. 38. Glazing machine Model 24TC (Kodak)

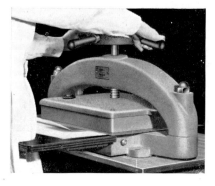

FIG. 39. " Ademco " dry mounting press Model J

between the tap and the mains inlet pipe, and ensures the removal of foreign matter from the water supply.

Wall brackets for storing film hangers. Ideally these brackets should be attached to the wall over the loading bench in the processing room.

2. Materials

Black and white negatives. Two types of emulsion are in general use, namely panchromatic and orthochromatic. Emulsions may be supported on glass, on roll film, or on sheet film.

A panchromatic emulsion is sensitive to all colours of the spectrum and may therefore be used for recording preparations in colour, and the reproduction of any colour may be controlled by the use of filters. An orthochromatic emulsion is sensitive only to blue and green and is best employed for recording preparations which do not contain a predominance of red.

With the exception of live preparations the use of high speed emulsions and fractional sections exposure is unnecessary. In some instances long exposures are to be preferred because the correct exposure time becomes less critical. High speed emulsions are generally associated with large grain size and low contrast, whilst those of low speed have a fine grain size and a high contrast, and therefore require longer exposure times. While it may be necessary to use both types of emulsion, one of medium speed provides satisfactory results for most techniques.

The choice between plate or sheet film is purely personal, but for handling and storage sheet film is to be preferred because less storage space is required and no breakage can occur. The manufacturers' processing instructions should be strictly followed if the best results are to be obtained. It is not good practice to mix filters, plates or films and processing formulae from differing sources of supply.

A guide to the choice of materials suitable for black and white photomicrography is listed in Technique 28.

Colour film. The study of biological preparations frequently demands differential staining of the various tissue elements to enhance contrast. This probably explains the enthusiasm for colour photomicrography. Injudicious use of colour is not recommended however, as it is not possible to assess the accuracy of any colour print or transparency because of the inability to compare it with the original preparation under identical conditions. This allows too much latitude in the acceptance of prints or transparencies as satisfactory and places too much reliance upon faithful reproduction with the colour process used. Indifferent colour prints or transparencies are no substitute for their counterparts in black and white; furthermore colour blocks are costly and the results can be most disappointing.

Colour photomicrographs are frequently used for teaching but their use may be criticized on the grounds that students tend to memorise the colour rather than the tissue's essential structural characteristics. The use of colour is fully justified when recording preparations where it is difficult or impossible to separate colours in black and white and its use is necessary in fluorescence and polarizing photomicrography.

A list of reversal and colour negative material is given in Technique 29. Some manufacturers supply processing kits with their film so that processing can be undertaken by the purchaser. Colour prints from separation negatives made from transparencies, or direct from the preparation through three colour filters, can be produced with the dye transfer process (Kodak Ltd.). The transparent dyes used in this process are particularly suitable for the preparation of photomicrographs and furthermore it is possible to prepare a number of prints from the same set of matrices. Special training facilities exist for professional photographers who wish to use this process

Colour prints on Ektacolor paper (Kodak Ltd.) from colour negatives can be prepared by the contact or enlargement methods of printing. Colour compensating filters must be used, and it is recommended that these be placed between the light source and the negative carrier. A number of enlargers are now fitted with a " colour head " incorporating a filter drawer for the purpose, and the unit may be substituted for the normal lamp housing.

Small quantities of prints may be processed in dishes, provided the temperature of the solutions is carefully maintained at 75° F. and that of the washing water at 75° F.$\pm$2°.

Monochrome transparencies are prepared by printing the negative on a plate or sheet film instead of paper. Special plates are manufactured for this purpose in sizes of 2×2 in. or $3\frac{1}{4}\times3\frac{1}{4}$ in. The emulsion is similar to that of bromide paper and transparencies may be prepared by the contact, the projection and by the reduction methods of printing.

Whilst similar results will be obtained with either plate or sheet film the latter has the advantage that if breakage occurs only the cover glasses need replacing whereas with plates complete renewal is necessary. A disadvantage of sheet film is its tendency to buckle in the heat of the projector but this problem is gradually being eliminated by the introduction of improved projection equipment.

Bromide papers. These are manufactured specifically for the reproduction of a positive image from a black and white negative. They are made in different grades to suit negatives of varying density ranges. The contrast of the paper may only be varied within very narrow limits. To increase or reduce contrast a different grade of paper must be selected. Bromide papers are made with a variety of surface textures but for photomicrography, where preservation of maximum detail is essential, glossy paper, subsequently glazed, should be used. When prints are intended for demonstration it is often advantageous to use a paper with a matt surface, which eliminates surface reflections.

SECTION II

NOTES ON THE PREPARATION OF SECTIONS

1. *Slides*

The slide, cover glass and mounting medium must be free of dust, fragments of textile fibre or any other extraneous matter. Standard 3×1 in. glass slides are suitable for most purposes. The sections should be mounted centrally. If mounted close to the ends of the slide it will be difficult to centre the sections on the microscope stage. Extra thin slides are often troublesome not only because of breakages but because they tend to slip under the lateral stops of the mechanical stage. Slides of 1 mm. in thickness used in conjunction with No. 1 coverglasses are satisfactory for high power work.

2. *Section thickness*

The thickness of the section is an important and frequently disregarded factor in photomicrography. A section of 5 microns or less is suitable for high power and oil immersion work, but accounts in no small part for the notorious difficulties encountered in producing good quality low-power photomicrographs. Low-power work calls for thicker sections if satisfactory results are to be obtained in monochrome and colour. Where serial sections of a specimen are being prepared it is as well to cut occasional sections at 12-15 microns to meet this need. These thicker sections take up more stain, and the attendant disadvantage of layered cells is not apparent at low magnification. It is worth reflecting that the intensity of staining of a section fixed to a slide is regulated far more by section thickness than by the duration of immersion in the stain, the latter being by no means critical.

3. *Staining*

The staining technique applied to a section is of importance in black and white photography. Stained sections of one colour are not difficult to reproduce in black and white. The reproduction of material stained with two contrasting stains is more difficult and necessitates the judicious use of contrast filters for satisfactory results. The reproduction of trichrome stained material in monochrome necessitates the sacrifice of one of the colours. (Technique 14).

4. Mounting

Air bubbles under the cover glass must be avoided. A minimum of mounting medium should be used as an excess of medium increases the distance between the objective and the section and thereby decreases the working distance of high power objectives. After mounting, slides should be allowed to dry slowly. Overheating may cause discoloration of the mounting medium and subsequent difficulty in photomicrography for producing clear background, especially in colour.

T E C H N I Q U E   N O.   2

RESOLUTION

By resolution is meant the ability of the microscope to separate the details of an object so that they are reproducible and visible in the image. The smallest linear separation of two object points which can be distinguished as separate is the limit of resolution. The resolving power of any microscope depends upon:

(a) the wavelength of the light;
(b) the numerical aperture (N.A.) of the object lens; and
(c) critical illumination.

If illumination is critical (see Technique 3)

$$\text{resolution} = \frac{\text{wavelength of light}}{2 \text{ N.A.}}.$$

For example, using a green light with a wavelength of 0·55 μ, an objective lens of 16 mm. focal length with a N.A. of 0·25, the limit of resolution would be $\frac{0·55}{2 \times 0·25} = 1·1$ microns.

The limit of resolution determines magnification. Approximately

28

500 times the N.A. is the highest magnification possible within the limits of resolution. Further magnification, which may be obtained by using higher power oculars or by increasing the lens plate film distance, constitutes " empty magnification ". In other words, the image will be larger in size but will not reveal any finer structural detail. For 35 mm. photomicrography the magnification should not exceed a quarter of the limiting resolution. This allows an enlargement of ×4 when printing from the negative. Any further enlargement will result in empty magnification.

TECHNIQUE NO. 3

CRITICAL ILLUMINATION

Critical illumination demands:

a. That the illuminating rays be symmetrically disposed about the optical axis of the microscope and that they should coincide in area with the back lens of the objective,

b. that the diameter of this area should cover the entire field being photographed, and

c. that the illumination should possess uniform intensity throughout.

If these conditions are met, illumination is critical.

1. Switch on the microscope illumination and ensure that the light intensity is safe for visual work.

2. Half close the iris diaphragm (field diaphragm) in front of the light source.

3. Place a slide on the microscope stage and rack the substage condenser up until it nearly touches the slide.

4. Open the substage diaphragm to its full extent.

5. Focus the preparation.

6. Close the substage diaphragm.

7. Move the lamp condenser back and forth until a sharp image of the light source is obtained on the substage diaphragm.

8. Open the substage diaphragm to its full extent.

9. Check that the object is still in focus.

10. Move the substage condenser until the circle of light, limited by the field diaphragm, is in focus.

29

11. By manipulating the centring screws on the substage condenser, centre the illuminated circle.

12. Adjust the field diaphragm so that the circle of light is just larger than the field to be photographed.

13. Remove the eyepiece and view the back lens of the objective. Close the substage diaphragm until its diameter coincides with the diameter of the back lens of the objective.

14. Replace the eyepiece.

15. Illumination is now critical.

Note

If the microscope has an integral illumination, steps 6-9 may be omitted.

TECHNIQUE NO. 4

MICROSCOPY BY TRANSMITTED LIGHT

Low Power × 10 Objective

1. Remove the top lens of the substage condenser.

2. Switch on the illumination and ensure that the light intensity is safe for visual work.

3. Half close the iris diaphragm (field diaphragm) in front of the light source.

4. Place a preparation on the microscope stage and rack the substage condenser up to its limit.

5. Open the substage diaphragm to its full extent.

6. Swing the × 10 objective into position and focus the preparation.

7. Partially close the substage diaphragm.

8. Move the lamp condenser back and forth until a sharp image of the light source is obtained on the substage diaphragm.

9. Open the substage diaphragm to its full extent.

10. Check the focus.

11. Move the substage condenser until the circle of light, limited by the field diaphragm, is in focus.

12. By manipulating the centring screws on the substage condenser, centre the illuminated circle.

13. Adjust the field diaphragm so that the circle of light is just larger than the field.

14. Remove the eyepiece and view the back lens of the objective. Close the substage diaphragm until its diameter coincides with the diameter of the back lens of the objective.

15. Replace the eyepiece.

16. Refocus the preparation, select the field and record. (Techniques 31-35).

Note

If a lower power objective is used the middle lens must be taken out of the substage condenser in addition to the top lens.

TECHNIQUE NO. 5

MICROSCOPY BY TRANSMITTED LIGHT
Medium Power with ×40 Objective

1. Switch on the illumination and ensure the light intensity is safe for visual work.

2. Half close the iris diaphragm (field diaphragm) in front of the light source.

3. Place the preparation on the microscope stage and rack the substage condenser up until it nearly touches the slide.

4. Open the substage diaphragm to its full extent.

5. Swing the × 10 objective into position and focus the preparation. Then swing the × 40 objective into position and refocus.

6. Partially close the substage diaphragm.

7. Move the lamp condenser back and forth until a sharp image of the light source is obtained on the substage diaphragm.

8. Open the substage diaphragm to its full extent.

9. Check the focus.

10. Move the substage condenser until the circle of light, limited by the field diaphragm, is in focus.

11. By manipulating the centring screws on the substage condenser, centre the illuminated circle.

12. Adjust the field diaphragm so that the circle of light is just larger than the field.

13. Remove the eyepiece and view the back lens of the objective. Close the substage diaphragm until its diameter coincides with the diameter of the back lens of the objective.

14. Replace the eyepiece.

15. Refocus the preparation, select the field and record (Techniques 31-35).

TECHNIQUE NO. 6

MICROSCOPY BY TRANSMITTED LIGHT

HIGH POWER (OIL IMMERSION) WITH ×100 OBJECTIVE

1. Switch on the illumination and ensure that the light intensity is safe for visual work.

2. Half close the iris diaphragm (field diaphragm) in front of the light source.

3. Place a preparation on the microscope stage and rack the substage condenser up until it nearly touches the slide.

4. Open the substage diaphragm to its full extent.

5. Swing the ×10 objective into place and focus the preparation. Then swing the ×40 objective into position and refocus.

6. Partially close the substage diaphragm.

7. Move the lamp condenser back and forth until a sharp image of the light source is obtained on the substage diaphragm.

8. Open the substage diaphragm to its full extent.

9. Check the focus.

10. Move the substage condenser until the circle of light, limited by the field diaphragm, is in focus.

11. By manipulating the centring screws on the substage condenser, centre the illuminating circle.

12. Adjust the field diaphragm so that the circle of light is just larger than the field.

13. Remove the eyepiece and view the back lens of the objective. Close the substage diaphragm until its diameter coincides with the diameter of the back lens of the objective.

14. Replace the eyepiece.

15. Check the focus.

16. Swing the ×40 objective partially aside.

17. Remove the slide carefully from the stage.
18. Place a drop of immersion oil on the top lens of the condenser.
19. Carefully replace the slide.
20. Place a drop of immersion oil on the coverglass of the slide.
21. Swing the ×100 objective into place.
22. Focus the preparation using the fine adjustment and select the field.
23. Record (Techniques 31-35).

TECHNIQUE NO. 7

MICROSCOPY—DARK FIELD

1. Remove the eyepiece, objective and condenser from the microscope.
2. Switch on the high intensity light source.
3. Ensure that the light passes centrally up the microscope tube.
4. Replace the eyepiece and insert an objective of medium power.
5. If a special dark field condenser is to be used, insert this. Focus the objective on to the centring mark sited on the upper lens of the condenser and centre the condenser with the centring screws.
6. If an achromatic substage condenser is to be used insert this. Partially close the substage diaphragm and centre the illuminated circle by means of the centring screws. Interpose a central stop underneath the condenser.
7. Open the substage diaphragm to its full extent.
8. Place a drop of immersion oil on the top lens of the condenser.
9. Place the preparation on the microscope stage.
10. Focus the preparation.
11. Rack the substage condenser up and down until the illuminated area on the slide is of minimum diameter.
12. Place a drop of immersion oil on the coverglass.
13. Bring the oil immersion objective into position.
14. Refocus the preparation and select the field.
15. The preparation may now be recorded (Techniques 31-35).

Notes

1. For moving objects use high speed plate or film.

2. As the intensity of illumination is low, exposure is best assessed by the strip method (Technique 17).

3. For preparations with little detail select the strip which shows the greatest contrast.

4. Where detail is required select the strip which registers detail best in the darkest part of the negative.

TECHNIQUE NO. 8

MICROSCOPY—PHASE CONTRAST

Manufacturers provide phase contrast accessories, consisting of a viewing telescope, special objectives and annular diaphragms, to fit their own microscopes. Except for minor differences in design, these accessories work on the same principle. The following technique, therefore, is applicable to most microscopes.

For correct phase contrast technique, it is essential that (1) the illumination is critical and (2) the phase components are perfectly centred. It is most important that the annulus of the condenser and that of the objective be aligned in the optical axis. To do this it is necessary to observe the back lens of the objective and for this purpose a viewing telescope eyepiece (also known as the auxiliary microscope) is provided for insertion in the microscope tube in place of the standard eyepiece. Centring must be checked for every change of objective.

Procedure

1. Attach the phase contrast condenser to the microscope.

2. Fit the objective, conjugate with the annular diaphragm, in the condenser.

3. Place the preparation on the microscope stage.

4. Insert a green filter into the condenser filter holder.

5. Switch on the illumination.

6. Focus the preparation.

7. Remove the eyepiece and insert the viewing telescope.

8. Adjust the eye lens of the telescope until the phase plate in the objective is in focus.

34

9. Centre the image of the condenser annular diaphragm until it coincides with the annulus of the phase plate.

10. Remove the auxiliary microscope and replace with the observation eyepiece.

11. Focus the preparation.

12. Close the lamp diaphragm to its full extent.

13. Adjust the substage condenser until the image of the aperture in the lamp diaphragm is in focus.

14. Open the lamp diaphragm until the field is fully illuminated.

15. Refocus the preparation, select the field and record (Techniques 31-35).

Note

Filters of any colour may be used. Green is recommended as this gives the best contrast.

TECHNIQUE NO. 9

MICROSCOPY—INCIDENT LIGHT

This technique is based on the use of the standard metallurgical microscope.

Procedure

1. Switch on the illumination.

2. Place a preparation on the microscope stage.

3. Focus the preparation.

4. Open the field diaphragm until the aperture is slightly larger than the field.

5. Remove the eyepiece and adjust the aperture diaphragm so that the circle of light fills approximately three quarters of the back lens of the objective. It may be necessary to increase or decrease this adjustment for certain preparations.

6. Refocus the preparation and select the field.

7. The preparation may now be recorded (Techniques 31-35).

A prism attachment, in conjunction with an intense source of illumination, may be used if the illumination is insufficient for photomicrography.

Procedure

1. Switch on the built-in illumination.
2. Place the preparation on the microscope stage and focus.
3. Remove the lighting unit.
4. Replace with the prism attachment.
5. Align the external light source with the prism.

   (i) Focus the lamp filament on a white card placed against the prism aperture.

   (ii) Close the diaphragm of the lamp condenser.

   (iii) Manipulate the lamp unit until the image of the condenser appears in the centre of the microscope field.

   (iv) Open the diaphragm until the aperture is slightly larger than the field.

6. Remove the eyepiece and adjust the aperture diaphragm so that the circle of light fills approximately three-quarters of the back lens of the objective.
7. Refocus the preparation and select the field.
8. The preparation may now be recorded (Techniques 31-35).

TECHNIQUE NO. 10

MICROSCOPY—POLARIZED LIGHT

Although microscopes specifically designed as polarizing microscopes are available, the following technique is based on the use of a standard microscope and polarizing accessories.

1. Switch on the illumination and ensure that the light intensity is safe for visual work.
2. Half close the iris diaphragm (field diaphragm) in front of the light source.
3. Place a preparation on the microscope stage and rack the substage condenser up until it nearly touches the slide.
4. Open the substage diaphragm to its full extent.
5. Swing the objective into position and focus the preparation.
6. Partially close the substage diaphragm.
7. Move the lamp-condenser back and forth until a sharp image of the light source is obtained on the substage diaphragm.

36

8. Open the diaphragm to its full extent.

9. Check the focus.

10. Move the substage condenser until the circle of light, limited by the field diaphragm, is in focus.

11. By manipulating the centring screws on the substage condenser, centre the illuminated circle.

12. Adjust the field diaphragm so that the circle of light is just larger than the field.

13. Remove the eyepiece and view the back lens of the objective. Close the substage diaphragm until its diameter coincides with the diameter of the back lens of the objective.

14. Replace the eyepiece.

15. Refocus the preparation and select the field.

16. Insert the polarizer into the filter holder of the substage condenser

17. Fit the cap analyser over the eyepiece.

18. Observe the preparation and rotate the cap analyser until the required polarized image is obtained.

19. The preparation may now be recorded (Techniques 31-35).

Note

1. Whenever possible record in colour.

2. Exposure is critical and must be accurately determined. The strip method (Technique 17) is suitable.

In some instances it may be more convenient to rotate the polarizer instead of the analyzer (Step 18).

TECHNIQUE NO. II

MICROSCOPY—ULTRA-VIOLET ILLUMINATION AND FLUORESCENT MICROSCOPY

The techniques for ultra-violet and fluorescent microscopy are almost identical.

Microscopy. Objectives corrected for 365 and 546 line radiations.

1. Switch on the illumination which must emit both ultra-violet and visible radiation. For this purpose use a mercury vapour lamp.

2. Insert a Wratten filter number 77A in front of the light source.

3. Place the preparation on the microscope stage and rack the substage condenser up until it nearly touches the slide.

4. Open the substage diaphragm to its full extent.

5. Swing the objective (corrected for 365 and 546 line radiations) into position and focus the preparation.

6. Partially close the substage diaphragm.

7. Move the lamp condenser back and forth until a sharp image of the light source is obtained on the substage diaphragm.

8. Open the substage diaphragm to its full extent.

9. Check the focus.

10. Move the substage condenser until the circle of light, limited by the field diaphragm, is in focus.

11. By manipulating the centring screws on the substage condenser centre the illuminated circle.

12. Adjust the field diaphragm so that the circle of light is just larger than the field.

13. Remove the eyepiece and view the back lens of the objective. Close the substage diaphragm until its diameter coincides with the diameter of the back lens of the objective.

14. Replace the eyepiece.

15. Focus the preparation visually and select the field.

16. Replace the Wratten 77A filter by a Wratten 18A filter.

17. Record the preparation (Techniques 31-35).

Notes

1. If ultra-violet microscopy is being used to improve resolution, a microscope fitted with quartz optics must be used.

2. If achromatized objectives are not available ultra-violet photomicrography may still be undertaken. However, the focus level of the preparation in ultra-violet light will differ from that in visible light. It is necessary, therefore, to determine the focus in ultra-violet light by a series of test exposures at differing focal levels.

3. An alternative method of focusing in ultra-violet light is by substituting a uranium glass plate for the focusing screen. The image fluoresces on the uranium glass and may be focused.

4. For fluorescent microscopy the same technique is used. After checking the focus (Step 15) a Wratten 2B filter is inserted in the microscope eyepiece. The preparation may then be recorded (Techniques 31-35).

MICROSCOPY—INFRA-RED ILLUMINATION

1. Switch on the illumination which must be a tungsten lamp. Ensure that the light is safe for visual work.
2. Half close the iris diaphragm (field diaphragm) in front of the light source.
3. Place the preparation on the microscope stage and rack the substage condenser up until it nearly touches the slide.
4. Open the substage diaphragm to its full extent.
5. Swing the objective into position and focus the preparation.
6. Partially close the substage diaphragm.
7. Move the lamp condenser back and forth until a sharp image of the light source is obtained on the substage diaphragm.
8. Open the substage diaphragm to its full extent.
9. Check the focus.
10. Move the substage condenser until the circle of light, limited by the field diaphragm, is in focus.
11. By manipulating the centring screws on the substage condenser, centre the illuminated circle.
12. Adjust the field diaphragm so that the circle of light is just larger than the field.
13. Remove the eyepiece and view the back lens of the objective. Close the substage diaphragm until its diameter coincides with the diameter of the back lens of the objective.
14. Replace the eyepiece.
15. Focus the preparation visually and select the field.
16. Insert an infra-red filter in the lamp filter holder.
17. Record the preparation (Techniques 31-35).

Note

1. The practice of infra-red photomicrography differs only slightly from the techniques for photomicrography with visual light. The main difference is the focusing of a preparation no longer visible to the eye. Focusing is accomplished by carrying out a series of test exposures at differing focal levels, noting for each exposure the graduation on the fine adjustment of the microscope.
2. Infra-red photography requires plates sensitive to infra-red radiation.

3. Ensure that the plate holders are opaque to infra-red radiation.

4. Exposure must be determined by the strip method (Technique 17).

MARKING AND RELOCATION OF SELECTED FIELDS
THE ENGLAND FINDER

1. Place the preparation on the microscope stage with the slide *label to the left*. Ensure that its bottom edge is firmly in contact with the base stops of the mechanical stage. It is important that the initial location of the preparation or finder is always made on the base stops.

2. If both lateral stops of the mechanical stage are moveable, secure one stop. This becomes the fixed stop.

3. Slide the preparation laterally until its edge is firmly in contact with the lateral fixed stop.

4. Move the remaining lateral stop (moveable lateral stop) to the slide edge and tighten lightly.

5. Select and centre the field for photomicrography. (A crosswire in the eyepiece will facilitate this.)

6. Without altering the fixed lateral stop remove the preparation and substitute the England Finder, ensuring that its label is to the bottom left. Slide the finder firmly home against the base stops and then ease it laterally so that its edge contacts the fixed lateral stop. Secure the preparation with the moveable lateral stop.

7. Focus the reference pattern on the finder and record the number of the main square and the number of the segment which appears in the centre of the field.

8. To relocate the selected field place the England Finder, with the label at the bottom left-hand corner, on the microscope stage locating it accurately against the base and fixed lateral stops.

9. Find the reference square and segment and centre.

10. Without altering the fixed lateral stop remove the finder and substitute the preparation with its label to the left. If the procedure has been carefully carried out the selected area should occupy the centre of the field.

Note

It is advisable, if using a high-power objective, to locate and obtain the references with a low-power objective.

A small sketch pinpointing the area to be recorded is advisable with high magnifications.

TECHNIQUE NO. 14

CHOICE OF FILTERS

In black and white photomicrography filters are used to make colours appear either lighter or darker, or one colour darker and another lighter at the same time.

If a filter of nearly the same colour as the preparation is employed, structures will be reproduced in lighter tones with enhancement of detail within the coloured area. The use of a filter approximately complementary in colour to the preparation will produce a print of darker tones, and thereby increase contrast.

Filters absorb some light and it is therefore necessary to increase the exposure time. The difference between exposure time with a filter and exposure time without a filter is known as the filter factor. This varies with the colour and density of each filter. The factors for the more commonly used filters are included in the instruction sheet accompanying every batch of film and plates. Ideally, however, the filter factor for any given set-up should be determined by a series of test exposures (Technique 15).

Filters are only of value when used with colour-sensitive materials, namely orthochromatic and panchromatic plates or films. Yellow filters must be used with orthochromatic material because only a partial control of colour rendering can be obtained. The fullest control of colour rendering occurs with panchromatic materials, and a wide range of colour filters may be employed.

Filters are placed between the microscope illumination and the substage condenser and if the illumination is sufficient the focusing may be carried out with the filter in position.

Visual examination of the preparation under the microscope is the most satisfactory method of selecting the correct filter or combination of filters. The resulting effect is carefully studied before the final choice is made. The accompanying table serves as a rough guide for selection of filters.

Common Stains	Filters Required	
	For lighter results	*For darker results*
Acid Fuchsin *Basic Fuchsin*	Deep Red; Deep Orange; Magenta	Deep Blue; Green
Orange G.	Orange; Red	Blue; Light Yellow; Green
Eosin	Magenta	Green
Light Green	Light Yellow Green; Green	Deep Blue; Deep Yellow; Red
Methylene Blue *Aniline Blue*	Light Blue; Blue	Yellow; Deep Yellow; Red
Haematoxylin	Deep Violet; Deep Blue	Deep Yellow
Methyl Violet	Deep Violet; Deep Blue	Deep Yellow

Note:

It is impossible with preparations in which red and blue stains appear together to make both appear lighter at the same time. A combination of red and blue filters would result in no light whatever being transmitted.

Contrast Filters

Colour	*Filter*
Red	Green
Orange	Blue
Yellow	Dark Blue
Green	Red
Blue	Yellow
Violet	Yellowish-Green

Neutral density filters

These filters are made in graded values of optical densities which will transmit a known amount of light to within 5 per cent accuracy and are used to reduce light intensity. Their principal use in photomicrography is to allow direct viewing at the microscope eyepiece or to increase exposure times.

DETERMINATION OF A FILTER FACTOR

To ascertain the exposure factor of a filter select a black and white preparation, for example, a micrometer scale, and record at a medium magnification

1. Select the technique for the equipment in use and prepare a negative without a filter.

2. Using the correct exposure for the non-filtered negative as a basis, make a further series of negatives with a filter in position. Increase the exposure for each negative by a predetermined amount until a filtered negative is produced which matches exactly the unfiltered one. The recording and processing must be standardised throughout.

3. The filter factor is the ratio of the exposure difference between the unfiltered negative and the filtered negative.

THE JUDGMENT OF CORRECT EXPOSURE

1. For black and white photomicrography.

Examine a series of test exposures and select that which gives the best reproduction of detail in the darker parts of the preparation. This should be adjudged as having had the correct exposure.

2. For colour photomicrography using reversal film.

Examine a series of test exposures and select that which gives the best reproduction of detail in the brightest parts of the preparation. This should be adjudged as having had the correct exposure.

The second criterion should be applied if, with black and white photomicrography, a fixed or constant background density is required, as when recording dust samples, pigments or bacteria.

DETERMINATION OF EXPOSURE

Test Exposure; Strip Method

1. Load the plate holder with a plate or sheet film.

2. Switch on the microscope illumination and bring the preparation on the microscope stage into focus.

3. Position the camera over the microscope eyepiece and set the shutter to time exposure and open by pressing the cable release.

4. Focus the image on the ground glass focusing screen.

5. Replace the focussing screen with the plate holder. Close the camera shutter.

6. Withdraw the dark shield until the film or plate is completely uncovered. Expose for one unit of time, for example, 1 second.

7. Slide in the shield approximately 1 inch and repeat the exposure for 1 second.

8. Slide in the shield a further inch and expose for two units of time, for example, 2 seconds.

9. Continue this process so that each exposure is twice that of the preceeding one until the shield is completely closed.

The exposed strips of the film or plate will have received exposures of one unit, $1+1 = 2$ units, $1+1+2 = 4$ units, $1+1+2+4 = 8$ units and so on. The exposure times would be 1, 2, 4 and 8 seconds.

10. Process the film or plate (Technique 41).

11. Examine the negative and select the strip which gives the best reproduction of the preparation.

12. This exposure is valid for the equipment, the subject, the photographic material and the processing employed. If any change in these conditions occurs, another Test Exposure must be prepared.

13. Record for reference the conditions used

Light source	Magnification
Eyepiece	Photographic material
Objective	Exposure time
Camera extension	Processing details

Note

This method will be satisfactory provided the preparation is uniform in thickness and structure over the whole field.

DETERMINATION OF EXPOSURE
Meter Devices

Exposure time may be ascertained by meter devices. These measure light. For use in photomicrography several factors must be considered when determining correct exposure.

1. The distance between the photo-electric cell and the photographic emulsion.
2. The type of light source and the effective voltage.
3. The photographic material and the method of processing.

To account for these the equipment should be calibrated and the details recorded on a table. Separate calibrations are advised for each user of the equipment.

Exposure measurements are usually carried out without a light filter. If a filter is used the increased exposure is ascertained by multiplying the indicated exposure by the filter factor.

DETERMINATION OF EXPOSURE
The Sekonic Microlite Exposure Meter, Model L-88

Calibration procedure for horizontal and vertical units.
1. Load 6 plates or sheet films into plate holders.
2. Switch on the microscope illumination.
3. Place a preparation with a low contrast on the microscope stage.
4. Swing the required objective into position.
5. Obtain critical illumination (Technique 3).
6. Focus the preparation.
7. Position the camera over the microscope eyepiece.
8. Set the shutter to time and open.
9. Extend the camera bellows for the required magnification.
10. Focus the preparation on the ground glass focusing screen.
11. Place a contrast filter in position.

12. Increase the illumination to maximum brightness.

13. Set the emulsion speed of the plate or film on the speed indicator of the exposure meter.

14. Set the meter for dim light conditions.

15. Place the light window on the ground glass focusing screen.

16. Turn on the current switch.

17. By rotating the outer-most dial follow the needle deflection with the green guide mark. Set the guide mark so that the exposure needle lies exactly between its forked end.

18. Turn off the current switch and note the indicator reading against the aperture scale.

19. The reading may not match up exactly with the aperture scale. If this occurs, reduce the illumination and repeat stages 15-18 until the exposure and aperture readings coincide.

20. Assuming that the following readings were obtained: f2·8, $\frac{1}{8}$ sec.; f4, $\frac{1}{4}$ sec.; f5·6, $\frac{1}{2}$ sec.; f8, 1 sec.; f11, 2 secs.; f16, 4 secs., expose the six plates or films consecutively at these exposures without altering the recording conditions in any way.

21. Process the negatives (Technique 41).

22. Inspect the negatives when dry and select the one which reproduces the preparation best.

23. If the negative selected was the one which received $\frac{1}{4}$ sec. exposure the calibration aperture will be f4. The correct exposure for all future readings will be that which is aligned exactly with the aperture scale reading of f4. This calibration will hold good for variations in lighting intensity, difference in optical combinations and for differing filters, provided that the same equipment and processing technique are employed.

Recording technique

1. Load the plate holder with plates of sheet film and repeat steps 5-18.

2. If the indicated exposure reading does not correspond exactly with the calibrated aperture setting, select the longer exposure time and set this against the calibrated aperture.

3. Turn on the current switch of the meter and place the light window on the focusing screen.

4. Observe the deflection of the indicator needle and reduce the illumination until the needle is contained between the forked end of the green guide mark.

5. Switch off the meter current.

6. Set the camera shutter for the correct exposure time.

DETERMINATION OF EXPOSURE
LEITZ MICROSIX METER

The Leitz Micro Six exposure measuring instrument is designed to meet differing requirements and may be used with most types of photomicrographic equipment. Its main features include:

1. A measuring eye containing a selenium cell.
2. A measuring unit which embodies:
    (i) a zero setting screw
    (ii) a needle deflection scale
    (iii) a fixed outer scale marked in seconds and within this a rotating scale.

The needle deflection scale is numbered 0, 3, 6, 9, etc. up to 27 and the outer " second " scale shows the seconds in red and fractions of a second in black. The inner rotating scale is numbered from 0-30 and these figures represent both the needle deflection values and film sensitivity in tenths of a degree DIN.

Recordings with the measuring eye may be taken:
    (i) Within the microscope tube.
    (ii) At the microscope eyepiece.
    (iii) On the focusing telescope of an attachment camera.
    (iv) On the ground glass or clear glass screen of a horizontal or vertical bellows type camera provided the illumination is sufficient to give a reading on the measurement value scale.

A calibration value for each measuring point must be obtained as each varies in intensity.

Method of obtaining the calibration value

1. Load the camera or plate holders and switch on the microscope illumination.
2. Place a preparation with a low contrast range on the microscope stage and focus.
3. Connect the measuring eye to the unit, taking care that the red plug is inserted into the socket marked with a red circle.
4. The indicator needle in the rest position should be located over the unmarked gradation beyond the measurement value 27. If necessary turn the zero setting screw until the needle occupies this position.

H   47

5. Open the camera shutter.

6. Place the measuring eye to the point selected and read off the value given by the needle deflection.

7. Without altering the conditions in any way, make a series of exposures each half as long as the preceeding one; for example, 1/5, 1/10, 1/25, 1/50, 1/100, 1/250 sec.

8. Record the exposures, measurement values and emulsion speed.

9. Process the plate or film (Techniques 41 and 43).

10. From the series of negatives select the one which reproduces the preparation best.

11. Assume that the figures for this negative were:

A measurement value of 10
An exposure time of 1/50 sec.
Film sensitivity 12/10° DIN.

12. Turn the inner rotating scale so that 10 is placed opposite 1/50 on the " outer " second scale ring. The calibration value is then found on the " second " scale ring opposite the film sensitivity 12/10° DIN. The calibration value is 1/30.

This calibration value is valid for the particular equipment used. It may be used for emulsions of differing sensitivity or after changing to another objective. The measuring eye must always, however, be applied to the same point at which the calibration value was obtained.

The calibration value for a bellows type camera must be determined with the bellows extended to 25 centimetres. Measurements from the ground or clear glass screen at this extension give correct exposure times which are applicable to variations in the length of the bellows.

Where measurements have been made in the microscope tube because of insufficient illumination on the focusing screen, the calibration value applies only to a bellows length of 25 centimetres, and the particular eyepiece employed. Extension of the bellows beyond 25 centimetres alters the calibration value and readjustment is necessary (Technique 23).

DETERMINATION OF EXPOSURE
Leitz Microsix Meter

Measuring procedure for 35 mm. attachment cameras

1. Connect the measuring eye of the unit, making sure that the red plug is inserted into the socket marked with a red circle.

2. If necessary adjust the zero setting screw until the indicator needle lies over the unmarked graduation beyond the measurement value 27.

3. Set the DIN value of the plate or film on the inner rotating scale and the valid calibration value on the outer " second " scale opposite each other.

4. Attach the measuring eye to the focusing eyepiece of the camera.

5. Record the needle deflection.

6. The correct exposure time is the figure on the outer " second " scale, opposite the value on the inner ring corresponding to the needle deflection.

7. Set the camera shutter for the correct exposure time.

If an exposure time is obtained which cannot be set on the shutter of the camera in use, select the nearest setting possible. If filters are employed the exposure time must be multiplied by the filter factor. If the illumination is adequate the exposure time measurement may be made with the filter in position.

DETERMINATION OF EXPOSURE
Leitz Microsix Meter

Bellows type cameras. Measurement from the focusing screen

1. Connect the measuring eye of the unit. Make sure that the red plug is inserted into the socket marked with a red circle.

2. If necessary, adjust the zero setting screw until the indicator

needle lies over the unmarked graduation next to the measurement value 27.

3. Set the DIN value of the plate or film on the inner rotating scale and the valid calibration value on the outer " second " scale opposite each other.

4. Extend the camera bellows to 25 centimetres.

5. Set the camera shutter to time and open.

6. Place the measuring eye on the focusing screen.

7. Record the needle deflection.

8. Read the correct exposure time from the outer " second " scale opposite the measurement value recorded.

9. Set the camera shutter for the correct exposure time.

TECHNIQUE NO. 23

DETERMINATION OF EXPOSURE
LEITZ MICROSIX METER

Bellows type cameras.   Measurement at the eyepiece

1. Connect the measuring eye of the unit.   Make sure that the red plug is inserted into the socket marked with a red circle.

2. If necessary, adjust the zero setting screw until the indicator needle lies over the unmarked graduation beyond the measurement value 27.

3. Compensate the calibration value determined at a bellows extension of 25 centimetres.   This is carried out by adding the figure shown in the graduation correction table to the right of the inner rotating scale.   For example, the calibration value determined with the bellows extended 25 centimetres is ½.   With the bellows extended to 50 centimetres the table shows a correction figure of 6.   By turning the inner rotating scale six graduations to the right of ½ a value of 2 is obtained.   With the bellows extended to 50 centimetres the calibration figure is 2.

4. Place the measuring eye to the eyepiece.

5. Record the needle deflection.

6. Read the correct exposure time from the outer " second " scale opposite the measurement value.

FIG. 40. Amplifier of the Zeiss exposure meter

7. Set the camera shutter for the correct exposure time.

Graduation correction table

Extension length (cm)	25	28	31	35	40	45	50
Graduation correction	0	1	1	3	4	5	6

Note

For measurement within the microscope tube, remove the eyepiece.

DETERMINATION OF EXPOSURE

ZEISS EXPOSURE METER

The Zeiss exposure meter device is designed for use with the Zeiss miniature attachment camera but may also be used with other miniature cameras if supplied with suitable connecting tubes.

The equipment consists of:

(i) A photo-electric cell mounted in a spherical casting.

(ii) A four-step amplifier and meter device.

(iii) An exposure table, used in conjunction with the meter reading and the photographic emulsion speed to assess exposure.

The photo-electric cell unit is attached to the Basic Body II.

The control rod in the base casting adjusts the beam-splitting system:

(i) With the control rod in the central position approximately 20 per cent. of the light passing through the object lens is deflected into the observation eyepiece.

(ii) With the control rod fully extended approximately 50 per cent. of the light passing through the object lens is deflected into the photo-electric cell. It is with the control rod at this setting that exposure measurements are made.

(iii) With the control rod pushed fully home all the light passing through the object lens is directed into the camera.

The amplifier and meter device (Fig. 40) has a range switch (5) which allows the sensitivity of the instrument to be increased in four steps to the power of ten 1, 10, 100, 1,000. The knob marked

51

(6) aligns the recording needle on zero, and knob (4) compensates the dark current of the photo-electric cell.

Preparation of the Exposure Table

1. Load the camera with film and attach the camera unit to the microscope.

2. Switch on the illumination and place a preparation on the microscope stage.

3. With the set screw located at the base of the microammeter (3) set the pointer to zero.

4. Connect the photo-electric cell lead to the bush at the rear of the meter.

5. Slide the control rod on the base casting to the central position. This brings the observation eyepiece into use.

6. Turn on main switch (1); the pilot lamp (2) lights up. Leave the range switch on 0 and allow the instrument to warm up for 10 minutes.

7. Turn knob marked Kompensation (4) to its limit in a clockwise direction.

8. Adjust to zero:

(i) by rotating the knob marked Nullpunkt (6) until the pointer rests exactly on zero;

(ii) by turning the range switch (5) through positions 1, 10, 100, 1,000. The deflection of the pointer produced in the last position is due to the dark current of the photo-electric cell. Compensate by rotating the Kompensation knob anticlockwise until the pointer rests exactly on zero.

9. Return the range switch to 0. The pointer should remain on zero as the switch passes the ranges, 100, 10, and 1.

10. Focus the preparation through the observation eyepiece.

11. Pull out the control rod and turn the range switch to position 1.

12. Read the microammeter scale. If the reading is less than 10, pass to the next range (10) or even further until a reading over 10 is shown on the scale.

13. Assume a reading of 20 is obtained with the range switch on 10. Make a series of exposures at this setting each half as long as the preceeding one; for example 4, 2, 1, $\frac{1}{2}$, 1/5, 1/10, 1/25 and 1/50 sec.

14. Develop the film (Techniques 41 and 43).

15. If the negative exposed for 1 second is considered to be the best, then 1 second is entered on the Exposure Table (Fig. 41) in the

column marked " Scale reading 12-25 " and along the line of Range 10. Because the meter sensitivity is 1/10 of that of Range 10, 1/10 of a second is entered in the space immediately above. By the same token in the spaces below Range 10, along the lines 100 and 1,000, 10 and 100 seconds are entered. In the columns to the right half the exposures of the preceding column are entered.

	Scale reading:					
Range:↓	1-3	3-6	6-12	12-25	25-50	50-100
1			$\frac{1}{5}$	$\frac{1}{10}$	$\frac{1}{25}$	$\frac{1}{50}$
10			2	1	$\frac{1}{2}$	$\frac{1}{5}$
100			20	10	5	2
1000			200	100	50	20

Fig. 41. Exposure table (Zeiss)

This completed exposure table is valid for any preparation regardless of the magnification and type of illumination. It is only necessary to prepare a new table when the photo-electric cell, photographic material or processing technique is changed.

TECHNIQUE NO. 25

DETERMINATION OF MAGNIFICATION

35 MM. CAMERAS

1. Switch on the microscope illumination.
2. Place a stage micrometer of known calibration value on the microscope stage.
3. Use a × 10 objective and × 10 eyepiece and focus.

53

4. Attach the 35 mm. camera, open the camera back and lay a piece of fine ground glass plate, ground surface downwards, over the film guides.

5. Set the camera shutter for time and open the shutter.

6. Focus the divisions of the micrometer scale on the ground glass plate.

7. Measure a division or divisions with a pair of dividers.

8. Lay the dividers on a millimetre ruler and read off the distance between each division on the micrometer scale. This represents the magnification with a ×10 objective and ×10 eyepiece for a particular microscope and attachment camera.

Repeat the procedure for additional objective-eyepiece combinations and tabulate the results for reference.

TECHNIQUE NO. 26

DETERMINATION OF MAGNIFICATION
HORIZONTAL AND VERTICAL CAMERAS

1. Switch on the microscope illumination.

2. Place a stage micrometer of known calibration value on the microscope stage.

3. Bring a ×10 objective into position.

4. Set the camera shutter to time and open the shutter.

5. Extend the camera bellows to 10 inches, or to the shortest extension possible.

6. Focus the ruled lines or squares of the micrometer scale on the focusing screen.

7. Measure the distance between the ruled lines with a pair of fine dividers.

8. Check the measurement on a ruler graduated in millimetres.

9. Multiply the known calibration of the micrometer scale by the measurement obtained from the focusing screen.

10. This is the magnification for the particular objective, eyepiece and bellows extension.

11. Repeat steps 3-9, extending the bellows in stages of one inch until the maximum extension is reached.

12. The magnification figures are noted and incorporated into a chart, (Fig. 42) for reference.

Note that each objective-eyepiece combination will require a separate calibration.

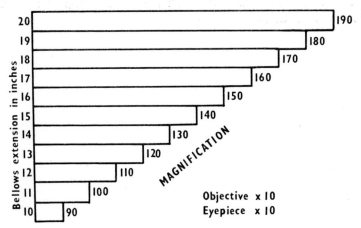

FIG. 42. Magnification chart

DETERMINATION OF MAGNIFICATION
ENLARGING EQUIPMENT

1. Insert a transparent rule graduated in millimetres in the negative carrier of the enlarger.
2. Place a sheet of white card, of the same thickness as the plate or film, in the masking frame.
3. Switch on the enlarger lamp and focus the image of the rule on the card.
4. Measure the distance from the front of the lens to the focusing card.
5. Measure the distance between two graduations with a pair of dividers.
6. Measure in millimetres the distance between the points of the dividers.

7. The number of millimetres between the points of the dividers represents the magnification.

8. The magnifications at varying masking frame to lens distances should be charted for future reference.

9. The use of a lens with a different focal length will require separate calibrations.

SENSITIVE MATERIALS: MONOCHROME

Plates	Makers	Colour Sensitivity	Purpose
G30	Ilford	Orthochromatic	Medium to high contrast metallography, colourless preparations
R40	Ilford	Panchromatic	High contrast coloured preparations
R20	Ilford	Panchromatic	Medium to high contrast coloured preparations
O250	Kodak	Orthochromatic	High contrast metallography
P200	Kodak	Panchromatic	High contrast coloured preparations
P300	Kodak	Panchromatic	Medium to high contrast coloured preparations

Sheet Film

G5.51	Ilford	Orthochromatic	Very high contrast metallography colourless preparations
R5.50	Ilford	Panchromatic	High contrast coloured preparations
FP3	Ilford	Panchromatic	Medium contrast coloured preparations
KS5	Kodak	Blue sensitive	Geological and colourless preparations
KO5	Kodak	Orthochromatic	Metallography
KP5	Kodak	Panchromatic	Coloured preparations
Commercial Ortho	Kodak	Orthochromatic	Medium contrast metallography
Plus X	Kodak	Panchromatic	Medium contrast coloured preparations
Panchro-Royal	Kodak	Panchromatic	For short exposures—living biological preparations

35 mm. Film

FP3	Ilford	Panchromatic	Medium contrast coloured preparations
Panatomic X	Kodak	Panchromatic	Medium contrast coloured preparations
Tri X	Kodak	Panchromatic	For short exposures—living biological preparations

SENSITIVE MATERIALS: COLOUR

1. Reversal Colour Films

Name of Film	*Type Colour Temperature*	*Sizes*	*Speed*	*Processing*
Agfacolor Germany	K. 3200° K	35 mm. 120 rolls Sheet film	ASA 16 BSI 23° DIN 13°	Manufacturer's Laboratories
Anscochrome ANSCO, U.S.A.	Tungsten 3200° K	35 mm. 828, 127, 120, 620 rolls sheet film	ASA 32 BSI 26° DIN 16°	Processing Kits Available
Super Anscochrome ANSCO, U.S.A.	Tungsten 3200° K	35 mm. 120, 620 rolls	ASA 100 BSI 31° DIN 21°	Processing Kits Available
Ektachrome High Speed, Kodak, England and U.S.A.	B. 3200° K	35 mm. only	ASA 125 BSI 32° DIN 22°	Processing Kits Available Process E2
Ferraniacolor Ferrania, Italy	Tungsten 3200° K	35 mm. only	ASA 20 BSI 24° DIN 13°	Processing Kits Available
Kodachrome II Kodak, England and U.S.A.	A. 3200° K	35 mm. 828 rolls	ASA 32 BSI 26° DIN 16°	Manufacturer's Laboratories
Ektachrome Kodak, England and U.S.A.	B. 3200° K	Sheet film	ASA 32 BSI 26° DIN 16°	Processing Kits Available Process E3

2. NEGATIVE COLOUR FILMS

Name of Film	Type Colour Temperature	Sizes	Speed	Processing
Ektacolor Kodak, England and U.S.A.	L. 3200° K	Sheet film only	ASA 16 BSI 23° DIN 13°	Professional Kits Available Process C22
Ferraniacolor Ferrania, Italy	Artificial light —	35 mm. 120, 620 rolls. Sheet film	ASA 32 BSI 22° DIN 16°	Manufacturer's Laboratories
Kodacolor Kodak, England and U.S.A.	3200° K	35 mm. 828 120, 620 rolls	ASA 16 BSI 23° DIN 13°	Processing Kits Available. Process C22
Pakolor Pakolor, England	Artificial light —	35 mm. 120, 620 rolls. Sheet film	ASA 40 BSI 27° DIN 17°	Processing Kits Available

COMPARATIVE RATINGS OF EMULSION SPEEDS

BSI and ASA			European	Relative
Arith.	Log.	DIN	Scheiner	Speed
8	20°	10/10	21°	1
10	21°	11/10	22°	1¼
12	22°	12/10	23°	1½
16	23°	13/10	24°	2
20	24°	14/10	25°	2½
25	25°	15/10	26°	3
32	26°	16/10	27°	4
40	27°	17/10	28°	5
50	28°	18/10	29°	6
64	29°	19/10	30°	8
80	30°	20/10	31°	10
100	31°	21/10	32°	12
125	32°	22/10	33°	16
160	33°	23/10	34°	20
200	34°	24/10	35°	24
250	35°	25/10	36°	32
320	36°	26/10	37°	40
400	37°	27/10	38°	48
500	38°	28/10	39°	64

TECHNIQUE NO. 31

RECORDING PROCEDURE

35 MM. ATTACHMENT CAMERAS

1. Load the camera with film.
2. Remove the binocular body, if fitted, and replace with a monocular tube.
3. Transfer the eyepiece to the connecting ring of the attachment camera.
4. Slide the attachment ring over the monocular tube and secure.
5. Switch on the microscope illumination.
6. Focus the preparation through the observation eyepiece.
7. Ascertain the correct exposure time (Techniques 17-24).
8. If a filter is to be used, readjust the exposure time.

9. Set the camera shutter for the correct exposure.
10. Check the focus.
11. Expose the film by gently pressing the cable release.
12. Wind on the film to bring the next frame into position.
13. Record the magnification and any other relevant data.

TECHNIQUE NO. 32

RECORDING PROCEDURE

THE ZEISS 35 MM. ATTACHMENT CAMERA, WITH BASIC BODY TYPE II.

1. Load the camera with film.
2. Remove the binocular body (if fitted) from the microscope.
3. Replace with the monocular tube.
4. Slide the attachment ring (the larger aperture uppermost) over the tube.
5. Raise the tube attachment ring until the perimeter of the larger aperture lies in the same plane as the upper surface of the eyepiece. Secure the attachment ring with the screw.
6. Place the camera on the attachment ring and tighten the clamping screw.
7. Ensure that the index on the side of the focusing eyepiece is adjusted to the mark CL.
8. Connect the photo-electric cell lead to the bush at rear of meter.
9. Bring the observation eyepiece into use by sliding the control rod to the central position.
10. Turn on mains switch—the pilot lamp lights up. Leave range switch on o for 10 minutes to allow the instrument to warm up.
11. Turn knob marked Kompensation to its limit in a clockwise direction.
12. Adjust to zero:

    (i) by rotating knob marked Nullpunkt until the pointer rests exactly on zero;

    (ii) by turning the range switch through positions 1, 10, 100, 1,000. Compensate for the dark current in the last position by turning Kompensation knob anticlockwise until the needle returns to zero.

13. Return the range switch to 0.

14. Switch on the microscope illumination.

15. Turn the focusing eyepiece outwards to its limit and then turn slowly inwards until the twin crosses appear sharply focused.

16. Focus the preparation.

17. Turn the range switch to position 1 and pull out the control rod to its full extent.

18. Observe the reading on the microammeter scale. Return control rod to central position.

19. Match the reading obtained with the scale readings on the exposure table. Select the exposure.

20. If a filter is employed adjust the exposure time.

21. Set the camera shutter for the correct exposure time.

22. Refocus the preparation.

23. Push the control rod right home.

24. Expose the film by pressing the cable release.

25. Wind on the film to the next frame.

26. Record the magnification and other relevant data.

TECHNIQUE NO. 33

RECORDING PROCEDURE

ATTACHMENT CAMERAS TAKING PLATES OR SHEET FILM

1. Load the plate holder with plates or sheet film and place in the camera.

2. Remove the binocular body, if fitted, and replace with a monocular tube.

3. Transfer the eyepiece to the connecting ring of the attachment camera.

4. Slide the attachment ring over the monocular tube and secure.

5. Switch on the microscope illumination.

6. Focus the preparation through the observation eyepiece.

7. Ascertain the correct exposure time (Techniques 17-24).

8. If a filter is to be employed readjust the exposure time.

9. Set the camera shutter for the correct exposure.

10. Remove the dark shield from the plate holder.

11. Refocus the preparation if necessary through the observation eyepiece.

12. Expose the plate or film by gently pressing the cable release.

13. Replace the dark shield and remove the plate holder from the camera.

14. Record the magnification and any other relevant data.

RECORDING PROCEDURE

HORIZONTAL CAMERAS

1. Load the plate holder with plates or sheet film.
2. Bring the camera into position over the microscope eyepiece.
3. Set the camera shutter to time exposure and open.
4. Switch on the microscope illumination.
5. Focus the preparation on the focusing screen.
6. Extend the bellows to the required length and secure.
7. Refocus the preparation.
8. Ascertain the correct exposure time (Techniques 17-24).
9. Select and place the filter in position.
10. Readjust the exposure time (Filter factor).
11. Refocus the preparation.
12. Close the camera shutter and reset it for the correct exposure time.
13. Remove or swing out the focusing screen.
14. Insert the plate holder carefully avoiding movement of the assembly.
15. Withdraw the dark shield of the plate holder.
16. Expose.
17. Replace the dark shield.
18. Remove the plate holder and replace the focusing screen.
19. Note the objective-eyepiece combination and the bellows extension.
20. Refer to the prepared magnification chart to obtain the magnification, which should be recorded with any other relevant data.

RECORDING PROCEDURE
VERTICAL CAMERAS

1. Load the plate holder with plates or sheet film.
2. Swing the camera over the microscope eyepiece.
3. Set the camera shutter for time exposure and open.
4. Switch on the microscope illumination.
5. Focus the preparation on the focusing screen.
6. Extend the bellows to the required length and secure.
7. Refocus the preparation.
8. Ascertain the correct exposure time (Techniques 17-24).
9. Select and place the filter in position.
10. Readjust the exposure time (Filter factor).
11. Check the focus.
12. Close the camera shutter and reset for the correct exposure time.
13. Remove or swing out the focusing screen.
14. Insert the plate holder.
15. Withdraw the dark shield of the plate holder.
16. Expose.
17. Replace the dark shield and remove the plate holder.
18. Replace the focusing screen.
19. Note the objective-eyepiece combination and the bellows extension.
20. Refer to the prepared magnification chart to obtain the magnification, which should be recorded with any other relevant data.

RECORDING PROCEDURE
ENLARGING EQUIPMENT

It is often necessary to record the entire preparation, or a large portion of it, to show the surrounding structures or location of

which a photomicrograph at higher magnification has previously been prepared.

This can be achieved by use of an enlarger if appropriate lenses are available. An enlarging lens of 50 mm. and one of 75 mm. focal length produce enlargement from 2 to approximately 15 diameters, depending on the working distance between the lens and photographic emulsion. At these magnifications reproduction of cellular detail is not possible but shape and gross structures can be readily identified. To obtain the best results preparations should be 12-14 microns in thickness and stained with Weigert's iron haematoxylin and van Gieson's stain. Such a preparation is enlarged on to a process film. Preparations stained by other methods can be recorded on orthochromatic and panchromatic materials and filters may be used to increase or reduce contrast.

Enlarging procedure

1. Place the preparation in the 35 mm. negative carrier of the enlarger with the coverglass of the slide facing the photographic emulsion.

2. Place a sheet of white card the same thickness as the film beneath the masking frame.

3. Switch off the main lighting and turn on the correct safelighting for the emulsion selected.

4. Arrange the masking frame to the size required.

5. Switch on the enlarger illumination and adjust the height until the desired enlargement is obtained. Focus the preparation.

6. Readjust the focus with the aid of a focus finder.

7. Close the lens diaphragm by one or two steps. This increases the depth of focus and improves definition.

8. Raise the masking frame and remove the white card used for focusing. Substitute a sheet film, emulsion side uppermost, slightly larger than the dimensions of the masked image.

9. Make a test exposure (Techniques 52 and 53).

10. Process the test strip (Technique 41, steps 1-12).

11. Switch on the mains lighting and wash in water for 5 minutes.

12. Inspect the test exposure for the correct exposure time.

13. Switch off main lighting and turn on safelight.

14. Switch on the enlarger illumination and check the focus with the focus finder.

15. Set the exposure timer (if fitted) for the correct exposure time.

16. Insert a sheet film in the masking frame with the emulsion side uppermost.

17. Expose.

18. Process the film (Technique 41).

DEFECTS IN RECORDING

Figure 43

(a) *Correctly* exposed negative.

(b) *Insufficient* exposure.

(c) *Excessive* exposure.

(d) *Halation.* The result of over-closure of the diaphragm of the substage condenser of the microscope.

(e) *Blurred image.* In this case due to the preparation being out of focus. A similar result may be produced by vibration of the equipment or bench during the exposure of a plate or film.

(f) *Uneven illumination.* Caused by maladjustment or malalignment of the light source, optical system, or camera.

(g) *Fogging.* In this instance due to light reaching the emulsion through a fine crack in the dark shield of the wooden plateholder. May also occur as a result of small holes in the camera bellows.

(h) *Double image.* The result of exposing the same film twice.

(i) *Flare.* Due to reflected light from inside the microscope tube. May also be produced by light reflected from the edge of the camera shutter.

Note

These illustrations have been reproduced in print form in order that they may be compared with those showing processing defects.

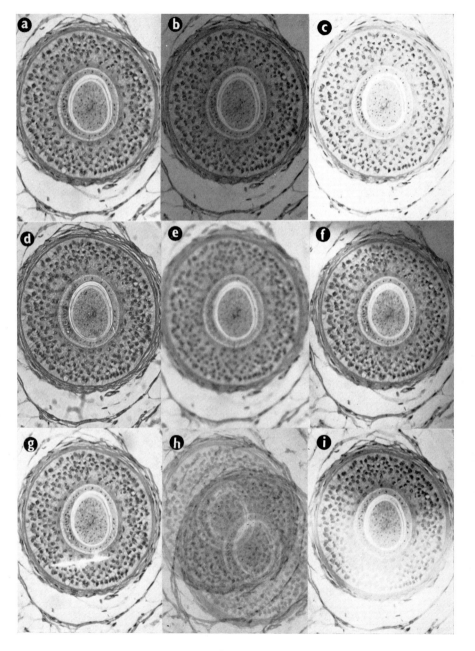

FIG. 43.

PRINCIPLES OF PROCESSING

Processing consists of development, rinsing, fixing, washing and drying.

Development

The greater proportion of negative material used in photomicrography will consist of panchromatic emulsions which should be developed in complete darkness and therefore tank development is advisable. Tank development is also to be preferred as a number of negatives can be developed simultaneously and the risk of damage to the emulsion is reduced.

The correct development of plates or film may be carried out by one of two methods:

    1. The inspection method
    2. The time and temperature method.

With the former, development is observed by the light of a suitable safelight and the process stopped when the negative is considered to have the correct densities. This procedure requires experience and is mainly used for processing blue sensitive materials.

Development by the time and temperature method relies on (i) the use of a standard formula for the developer and (ii) a set time for development which produces a known degree of contrast. The exact time and temperature for development may be the one recommended by the manufacturer for the particular emulsion used, or it may be based on experience. This method will not, however, compensate for variations in the tone range of the microscopical preparation or for errors in exposure.

Factors affecting the development time are:

    1. Development rate of the plate or film.
    2. Variations in the composition of the developer or the time needed to obtain a particular contrast range.
    3. Agitation of the developer.
    4. Temperature of the developing solution. The developing time is reduced as the temperature is increased.

The optimum development temperature is between 65° and 70° F. but development may be carried out between 55° and 75° F. Below

55° F. development is extremely slow and above 75° F. precautions must be taken to avoid softening of the emulsion.

Rinsing

It is advisable to employ an acid stop bath to remove and neutralise the developer. Solutions of potassium metabisulphite or acetic acid are suitable.

Fixing

The action of the fixing bath is to remove the unexposed silver halide from the emulsion. These baths must always contain a silver halide solvent and must not harm the gelatin emulsion or affect the silver image to any great extent. The most commonly used solvent is sodium thiosulphate. It is possible that prolonged immersion in a bath containing this substance may reduce the density of the image, especially if the solution is acid.

The addition of potassium metabisulphite or acetic acid to the fixing bath is preferable to plain solutions.

The time required for fixation depends on :

1. The thickness of the emulsion. Thicker emulsions require a longer time in the bath to remove the unexposed silver halide.

2. The concentration of sodium thiosulphate and the degree of exhaustion of the solution.

3. Temperature. High temperatures accelerate the rate of fixation.

4. The amount of agitation. The more agitation, the more rapid is fixation.

5. Exposure. Heavy exposures decrease the amount of unexposed silver and fixation is therefore more rapid.

Washing

It is necessary to remove all the soluble salts contained in the emulsion after fixing. A negative should be washed in running water for about 30 minutes.

Drying

The negative should be treated in water containing a few drops of wetting agent before drying in order to reduce the formation of water marks. The negative may be allowed to dry in a dust-free atmosphere at room temperature or, preferably, in a drying cabinet.

PROCESSING DATA: DEVELOPMENT

The developing procedure for Kodak material is based on time, temperature and the type of agitation employed, irrespective of the use of a dish or tank.

    C = Continuous agitation

    I = Intermittent agitation—Commencing with thorough agitation and then at one minute intervals for duration of the development.

    Developer dilutions = Developer+water in each instance.

Plate	Safelight	Developer	Dilutions for Use at 68° F.	Time in Minutes
G30	ISO No. 906	ID2	Dish 1+2 Tank 1+5	2 4
		ID36	Dish 1+3 Tank 1+7	$1\frac{1}{2}$ 3
R40	GB No. 908	ID2	Dish 1+2 Tank 1+5	$2\frac{1}{2}$ 5
		ID36	Dish 1+3 Tank 1+7	2 4
R20	GB No. 908	ID2	Dish 1+2 Tank 1+5	$2\frac{1}{2}$ 5
		ID36	Dish 1+3 Tank 1+7	2 4
		Microphen	Dish } Tank } Undiluted	$4\frac{3}{4}$ 6
O250	Wratten Series 2	D61a	1+1 C Normal contrast 1+3 I	$4\frac{1}{2}$ 6
		D76	Undiluted C Fine grain I	$5\frac{1}{2}$ 7
P200	Wratten Series 3	D8	2+1 C Maximum contrast	2
		D158	1+1 C Very high contrast	3
P300	Wratten Series 3	D76	Undiluted C Normal I	5 $6\frac{1}{2}$
		Microdol	Undiluted C Extra fine grain I	6 8

Sheet Film	Safelight	Developer	Dilutions for Use at 68° F.	Time in Minutes
G5.51	ISO No. 906	ID2	Dish 1+2 Tank 1+5	1¾ 3½
		ID36	Dish 1+3 Tank 1+7	1½ 3
R5.50	G No. 907	ID2	Dish 1+2 Tank 1+5	2½ 5
		ID36	Dish 1+3 Tank 1+7	2 4
FP3	GB No. 908	ID48	Dish } Tank } Undiluted	9½ 12
		Microphen	Dish } Tank } Undiluted	8 10
KS5 KO5 KP5	Wratten OB 2 3	D8 D11	2+1 C Maximum contrast Undiluted C Very high contrast	2 5
Commercial Ortho	Wratten Series 2	D61a	1+1 C 1+3 I } Normal	5 9½
		D76	Undiluted $\frac{C}{I}$ Fine grain	10 13
Plus X	Wratten Series 3	D61a	1+1 C 1+3 I } Normal	6 11
		D76	Undiluted $\frac{C}{I}$ Fine grain	13 17
		Microdol	Undiluted $\frac{C}{I}$ Extra fine grain	17 21
Panchro-Royal	Wratten Series 3	D61a	1+1 C 1+3 I } normal	5 10
		D76	Undiluted $\frac{C}{I}$ Fine grain	11 14
		Microdol	Undiluted $\frac{C}{I}$ Extra fine grain	14 17

35 mm. Roll Film	Safelight	Developer	Dilutions for Use at 68° F.	Time in Minutes
FP3 Series 2	GB No. 908	ID48	Undiluted Dish / Tank	$9\frac{1}{2}$ / 12
		Microphen	Undiluted Dish / Tank	8 / 10
Pano-tomic X	Wratten Series 3	D76	Undiluted $\frac{C}{I}$ Fine grain	7 / 9
		Microdol	Undiluted $\frac{C}{I}$ Extra fine grain	7 / 9
Tri X	Wratten Series 3	D76	Undiluted $\frac{C}{I}$ Fine grain	7 / 9
		Microdol	Undiluted $\frac{C}{I}$ Extra fine grain	7 / 11

STOP BATHS, HARDENING BATHS AND FIXING BATHS

Ilford Formulae	Purpose and Materials		Dilution	Period of Immersion
IS1	Stop bath	Plates, films, papers	Undiluted	5 seconds
IH4	Hardener	Plates, films, papers	Undiluted	3 minutes
IH6	Hardener	Plates and films	Undiluted	3 minutes
IF2	Fixer	Plates and films	Undiluted	10-20 minutes
IF2	Fixer	Papers	1+1	5-10 minutes
IF13	Fixer	Papers	Undiluted	5-10 minutes
IF15	Fixer	Plates and films	Undiluted	5-10 minutes
Kodak Formulae				
SH1	Hardener	Plates and films	Undiluted	3 minutes
SB1	Stop bath	Plates, films papers	Undiluted	5 seconds
SB3	Stop bath	Plates and films	Undiluted	3-5 minutes
F5	Fixer	Plates, films, papers	Undiluted	10 minutes
F7	Fixer (Rapid)	Plates, films, papers	Undiluted	Do not prolong treatment
F24	Fixer	Plates, films, papers	Undiluted	10 minutes

PROCESSING OF NEGATIVES

TANK METHOD

Development

1. Remove the lid of the tank and check the temperature of the developer.

2. Assemble the plate rack. (Fig. 24) or film hanger (Fig. 26) nearby.

3. Set the timer for the period of development.

4. Switch off the main lighting.

5. Remove the plate or film from the plate holder in the correct safelighting or in complete darkness.

6. Mark or number the top left hand corner of the plate or film with a diamond pencil.

7. Place the plate in the developing rack, or clip the film on to a film hanger.

8. Place the negative in the developing tank and start the timer. Agitate for 30 seconds..

9. Lift the negative from the tank and allow to drain. Return to the tank and replace the lid.

10. Repeat stage 9 every two minutes until development is complete.

11. Remove the negative from the developing tank and transfer it to the acid stop bath for 2 minutes.

12. Remove the negative from the acid stop bath and place in the fixing bath. Agitate for 30 seconds then leave to fix for 10 minutes.

13. Switch on the main lighting and take the negative from the fixing bath and place in the washing tank. Leave washing for 30 minutes.

14. Transfer the negative to a dish of water containing a few drops of wetting agent for 1 minute.

15. Place the plate in the rack or replace the film on the hanger. Transfer to the drying cabinet or place in a dust-free atmosphere to dry at room temperature.

TECHNIQUE NO. 42

PROCESSING OF NEGATIVES
Dish Method

If dishes are employed it is advisable to process a single plate or film at a time. This applies particularly to panchromatic material which should be processed in complete darkness.

1. Dilute the developer if necessary and mix thoroughly. Ensure it is at the correct working temperature. Leave it in its container until required.

2. Set out three dishes for the developing solution, the acid stop bath, and the acid fixing bath.

3. Pour the stop bath and the fixing bath into their respective dishes.

4. Set the dark room timer for the correct period of development.

5. If processing is to be carried out in complete darkness memorise the positions of the developing dish, the developing solution and the timer. If not, turn on the correct safe light and switch off the main lighting.

6. Remove the plate or sheet film from the plate holder.

7. Place it, emulsion side uppermost, in the empty dish.

8. Start the dark room timer.

9. Pour on the developing solution and agitate the dish to avoid uneven development. The direction of agitation should be varied.

10. Remove the negative from the developer and transfer to the acid stop bath for 2 minutes.

11. Transfer to the acid fixing bath for 10 minutes. Agitate the dish in all directions for the first two or three minutes. The main lighting may be switched on after 5 minutes.

12. Transfer to the washing tank. Leave for 30 minutes.

13. Place in a dish of water containing a few drops of wetting agent, and leave for 2 minutes.

14. Dry in drying cabinet or in a dust-free atmosphere at room temperature.

TECHNIQUE NO. 43

PROCESSING 35 MM. FILM
THE PATERSON MODEL II TANK

Loading the film into the spiral

1. Remove the lid from the tank and take out the spiral.

2. Turn the two halves of the spiral until the openings are opposite each other.

3. Remove the film from the cassette in total darkness or in the recommended safe lighting.

4. Cut between the perforations of the film to remove the tapered end.

FIG. 44. Loading a 35mm. film into the spiral

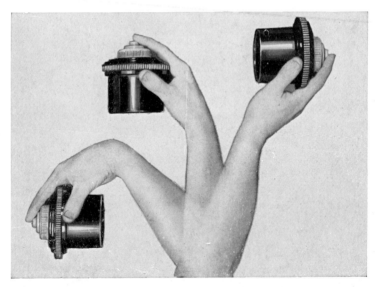

FIG. 45. Inverse agitation

5. Insert and gently feed the cut end of the film into the spiral opening until it engages in the ratchet.

6. Hold the spiral so that the coiled film rests in the left hand, Fig. 44.

7. With the left hand resting on the back of the film, steadily rotate the right hand spiral about a quarter of a turn, backwards and forwards, until the spiral is loaded. When the spiral is approximately half loaded the right hand spiral may stick. At this stage the inner ratchet has come into operation, and by turning the right-hand spiral backwards loading will continue.

Note

A few preliminary loading trials with a length of exposed film in daylight will provide valuable practice and experience for working under dark-room conditions. Never attempt to force the film into the spiral as this may result in tearing of the perforations.

8. Return the loaded spiral to the tank, ensuring that the top spiral (which carries the spring clip) is uppermost. Replace the lid of the tank by rotating it anticlockwise and then turn clockwise to lock. Cover the central hole with the polythene cap if development is to be delayed.

Developing procedure

1. 245 ml. are required to fill the tank. This quantity must on no account be reduced otherwise air bubbles may form on the upper edge of the film.

2. Remove the polythene cap from the lid of the tank. Pour in the developer, holding the tank at an angle to prevent the formation of an air lock in the lid.

3. Place the agitator rod into the hole in the lid and press firmly home. Rotate the spiral by twisting the rod two or three times. This dislodges any air bubbles adhering to the emulsion surface of the film.

4. Turn the agitator rod at regular intervals and occasionally lift up and down to eliminate air bubbles and to ensure even development.

Inversion agitation procedure

1. Fill the tank with developer.

2. Insert the agitator rod and turn two or three times.

3. Leave the rod in position and replace the polythene cap securely on the lid. Leave for one minute.

4. Hold the cap with the forefinger and invert the tank. Return immediately to the upright position (Fig. 45).

5. Repeat stage 4 every half minute until development is complete.

After development the solution may be returned to its container by removing the top of the polythene cap, inverting it, and using the cap as a funnel.

Washing the film

Fill the tank with water and then empty it. Repeat two or three times.

Fixing

1. Fill the tank with fixing solution.
2. Insert the agitator rod and agitate periodically during fixation.
3. Pour off fixing solution.

Final wash

1. Remove the lid of the tank.
2. Insert the end of a rubber tube (attached to the cold water tap) into the central column of the spiral and wash for 30 minutes.

Removing the film from the spiral

Arch the free end of the spiral by slightly bending the edges together. Pull gently on the free end, allowing the spiral to rotate in the other hand. The film will run easily from the spiral.

Drying

(1) Place the film in a dish of water containing a few drops of wetting agent for one minute.

(2) Remove and clip on a hanger and transfer to a drying cabinet or place in a dust-free atmosphere at room temperature.

Thoroughly wash the tank and spiral in cold water. Place upside down in a drying cabinet and dry. When dry assemble for use. If left damp the film emulsion may adhere to the spiral during subsequent loading.

TECHNIQUE NO. 44

DEFECTS IN PROCESSING

These illustrations have been reproduced in print form because some experience is necessary to assess correctly the quality of a negative and to detect some of the defects which may be present.

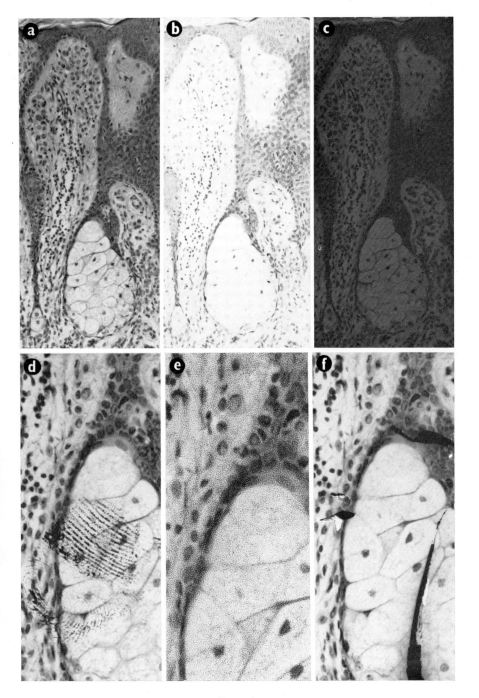

Fig. 46.

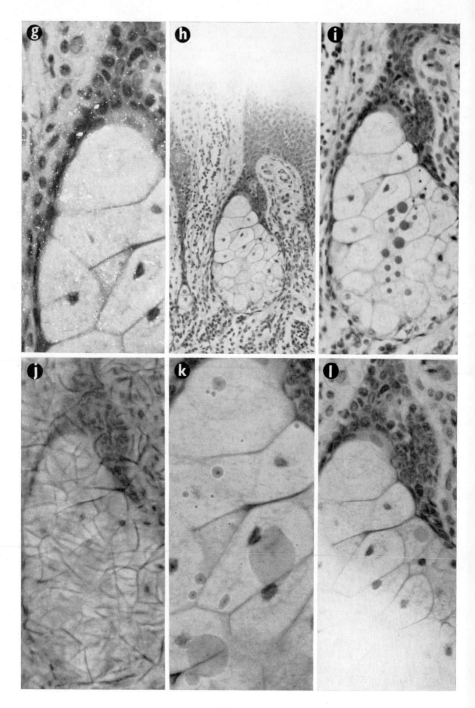

Fɪɢ. 47.

There is also a tendency in the initial stages of photomicrography to prepare prints from all negatives produced.

Figure 46

(a) *Correctly* developed negative.

(b) *Excessive* development.

(c) *Insufficient* development.

(d) *Fingerprint* caused by handling the emulsion prior to development.

(e) *Reticulation*, the result of swelling and a subsequent contraction of the gelatin, and caused by transferring a plate or film from a warm to a cold solution.

(f) *Abrasions* caused by finger nails while handling the plate or film during processing.

Figure 47.

(g) *Dust particles*, deposited on the emulsion while drying at room temperature.

(h) *Fogging*, due to unsafe light falling on the emulsion after exposure.

(i) *Air bubbles* trapped beneath the film during development. Due to placing and leaving the film emulsion side downwards in the dish and a failure to agitate the developing solution in the early stages.

(j) *Irregular density* caused by immersing the film, emulsion face downwards, in the dish and failing to agitate. Also produced by unmixed developer.

(k) *Water droplets* due to splashing the negative after drying.

(l) *Insufficient fixing*. This produces a slightly opaque brownish tint on the portion of the negative retaining partly undissolved silver halide.

TECHNIQUE NO. 45

PROCESSING 35MM. FERRANIACOLOR FILM
JOHNSON'S PROCESSING KIT

Processing

This can be divided into three stages.

    1. First development, hardening and washing.

2. Second exposure, in which the film is exposed to artificial light.

3. Colour development, bleach-hardening and fixing, with intermediate washing, carried out in artificial light or very subdued daylight. The processing must be carried out in a 35 mm. processing tank and the film must receive adequate agitation.

The recommended temperature for processing is 65° F. but a slight increase up to 70° F. can be tolerated. For the standardisation of results the temperature control is important during first development and colour development. The temperature of the other solutions is less critical but should be similar to that of the developing solutions. Washing is very important. Extremes of temperatures should be avoided and the water must be completely replaced during the stages of washing.

Processing Solutions (Technique 73)

1. Load the film in the 35 mm. developing tank (Technique 43).
2. *First development: 65° F. 14 minutes.*
Fill the tank with developer and rotate the spiral to remove air bubbles. Agitate for about 10 seconds in every minute throughout development.
3. Return the developer to its container.
4. *Wash: 65° F. 5 minutes*
Fill the tank with water without removing the lid. Agitate frequently.
5. *Harden: 65° F. 5 minutes*
Fill the tank with hardening solution and agitate frequently. Increase to 8 minutes if solution has been used previously.
6. Return the hardening solution to its container.
7. *Wash: 65° F. 5 minutes*
Fill the tank with water without removing the lid. Agitate frequently and empty and refill the tank with fresh water every minute.
8. *Re-expose*
It is important to re-expose the film evenly and sufficiently. The film cannot be overexposed.
One of two methods may be adopted:

(i) Remove the film from the spiral and place a clip on each end. Extend the film to its limit and pass to and fro in front of, and about 12 inches from, a 100 watt lamp, or 2½ feet from a No. 1 Photoflood lamp. Expose each side for a minimum period of 1½ minutes. Replace the film in the spiral by immersing both the spiral and film beneath water to avoid damaging the emulsion.

(ii) If the spiral possesses transparent ends do not remove the film but place the spiral in a white bowl filled with cold water (55-65° F.). Place a No. 1 Photoflood lamp approximately 15 inches away and expose for at least 5 minutes, presenting all sides of the spiral to the light to obtain even treatment.

9. *Colour development:* 65° F. 8 *minutes*
Return the spiral to the tank and fill with colour developer. Agitate frequently. Return colour developer to its container.

10. *Magnesium Sulphate bath:* 65° F. 3 *minutes*
This is only necessary in areas with a soft water supply. Immediately following colour development, immerse the film in a 2 per cent. solution of magnesium sulphate.

11. *Wash:* 65° F. 15 *minutes*
Wash thoroughly in water by emptying and refilling the tank every 2 minutes. The colour developer must be completely removed at this stage.

12. *Bleach Hardening:* 65° F. 10 *minutes*
Fill the tank with the bleaching solution and agitate frequently.

13. Return the bleach hardener to its container.

14. *Wash:* 65° F. 5 *minutes*
Wash thoroughly in water by emptying and refilling the tank every minute.

15. *Fix:* 65° F. 8-10 *minutes*
Fill the tank with fixing solution and agitate frequently. Return fixing solution to its container.

16. *Wash:* 65° F. 20 *minutes*
Thoroughly wash the film to remove the fixing solution.

17. *Dry*
Remove the film from the spiral and place it in a dish of water containing a few drops of wetting agent. Remove and dry at room temperature in a dust-free atmosphere. Do not employ heat to hasten drying. This results in curling of the film.

PROCESSING 35 MM. FERRANIACOLOR FILM
MAY AND BAKER FORMULA

Processing Solutions (Technique 74)

1. Load the film in the 35 mm. developing tank (Technique 43).
2. *First development:* 65° F. 14 *minutes*

Fill the tank with developer and rotate the spiral to remove air bubbles. Agitate eight times in the first 2 minutes and then at 1 minute intervals.

3. Return the developer to its container.
4. *Wash:* 55-65° F. 15 *minutes*

Fill the tank with water without removing the lid. Agitate the spiral frequently. Empty and refill the tank with water every two minutes. Agitate continuously. Switch on the main lighting.

5. *Re-expose*

It is very important to re-expose the film sufficiently and evenly. One of two methods may be adopted.

> (i) Remove the film from the spiral and submerge in cold water (55-65° F.). Pass the film to and fro approximately 30 inches from a No. 1 Photoflood lamp. Expose each side of the film for 1 minute.

> (ii) If the spiral possesses transparent ends, submerge in cold water (55-65° F.) in a white bowl. Place a No. 1 Photoflood lamp approximately 15 inches away and expose the film for at least 5 minutes, presenting all sides of the spiral to the light.

6. *Colour development:* 65° F. 8 *minutes*

Return the spiral to the tank and fill with the colour developer. Agitate eight times during the first 2 minutes and thereafter at one-minute intervals.

7. Return the colour developer to its container.
8. *Wash:* 55-65° F. 15 *minutes*

Fill the tank with water and agitate frequently. Empty and refill the tank every 2 minutes; agitate continuously.

9. *Bleach-harden:* 55-65° F. 10 *minutes*

Fill the tank with the bleach-hardener. Agitate 8 times within 2 minutes and then at one-minute intervals.

10. Return the bleaching solution to its container.

11. *Wash:* 55-65° *F.* 5 minutes
Fill the tank with water and agitate frequently. Empty and refill the tank every 2 minutes. Agitate continuously.

12. *Fix:* 55-65° *F.* 8 *minutes*
Fill the tank with fixing solution and agitate 8 times within 2 minutes, then at one-minute intervals.

13. *Wash:* 55-65° *F.* 20 *minutes*
Fill the tank with water and agitate frequently. Empty and refill the tank every 2 minutes. Agitate continuously.

14. *Dry*
Remove the film from the water and place it in a dish of water containing a few drops of wetting agent. Remove and dry at room temperature in a dust-free atmosphere.

TECHNIQUE NO. 47

PROCESSING 35 MM. EKTACHROME FILM TYPE B

Processing Equipment
1. A developing tank, preferably having a plastic spiral with transparent ends (Fig. 28).
2. An accurate and clearly scaled thermometer.
3. A luminous alarm timer.
4. A fixture to hold a No. 1 Photoflood lamp faced with a sheet of glass to prevent water splashes shattering the lamp during reversal exposure.

Temperature
The first developer is used at 75° F. Variations of more than $\frac{1}{2}$° F. must be avoided as this temperature is critical. The remaining solutions and washing water should be between 74-77° F.

Timing
The time recommended for each processing step includes 10 seconds for draining the film. Each step must be commenced on time.

Agitation
Correct agitation is important, particularly with the first developer.

81

Processing Solutions (Technique 76)

1. Adjust the temperature of the first developer to 75° F. and set the timer for 10 minutes.

2. Turn off the main lighting.

3. In total darkness load the film into the spiral of the developing tank (Technique 43)

4. *First development: 75° F. 10 minutes*

(i) Place the spiral in the tank and replace the lid. Turn on the main lighting.

(ii) Check the temperature of the developer and pour into the tank. Start the timer.

(iii) Place the agitator rod into the central channel and press firmly home. Rotate several times during the first 15 seconds to agitate and dislodge air bubbles. Leave the rod in position and replace the polythene cap. Continue agitation by the Inversion Method (Technique 43).

(iv) 12 seconds from the end of development remove the polythene cap, pour off the developer and allow the film to drain.

5. *Rinse in water: 73-77° F. 1 minute*

(i) Fill the tank with water.

(ii) Agitate as in Step 4 (iii).

(iii) 10 seconds from the end of washing pour off the water and allow the film to drain.

6. *Harden: 73-77° F. 3 minutes*

(i) Fill the tank with hardening solution.

(ii) Agitate as in Step 4 (iii).

(iii) Pour off the hardening solution and remove the lid of the tank.

7. *Wash in running water: 73-77° F. 3 minutes*

8. *Reversal exposure*

(i) Remove the spiral from the tank and place in a white dish. Expose one end about 1 foot from a No. 1 photoflood lamp, moving it constantly (1 minute). Turn the spiral over and repeat the exposure (1 minute).

(ii) The film must be drained for at least 1 minute after this step. This is normally accomplished during the reversal exposure. Do not allow the film to dry.

9. *Colour development: 73-77° F. 15 minutes*

(i) Fill the tank with developer. Replace the spiral and the lid.

     (ii) Agitate as in Step 4 (iii).

     (iii) Pour off the colour developer.

10. Wash in running water: 73-77° F. 5 minutes.

11. *Clear 73-77° F. 5 minutes*

     (i) Fill the tank with clearing solution. Replace the lid.

     (ii) Agitate as in Step 4 (iii).

     (iii) Pour off the clearing solution.

12. Rinse in running water: 73-77° F. 1 minute.

13. *Bleach: 73-77° F. 8 minutes*

     (i) Fill the tank with bleaching solution. Replace the lid.

     (ii) Agitate as in Step 4 (iii).

     (iii) Pour off the bleaching solution.

14. Rinse in running water: 73-77° F. 1 minute.

15. *Fix: 73-77° F. 6 minutes*

     (i) Fill the tank with fixing solution. Replace the lid.

     (ii) Agitate as in Step 4 (iii).

     (iii) Pour off the fixing solution.

16. Wash in running water: 73-77° F. 8 minutes.

17. *Stabilize: 73-77° F. 1 minute*

     (i) Fill the tank with stabilizing solution. Replace the lid.

     (ii) Agitate as in Step 4 (iii).

     (iii) Pour off the stabilizing solution. This step is essential to prevent fading of some of the dyes. *Do not* rinse in water after treatment with this solution.

18. Remove the film from the spiral. Dry in a dust-free atmosphere at a temperature not exceeding 110° F.

TECHNIQUE NO. 48

PROCESSING EKTACHROME SHEET FILM TYPE B

Processing Equipment

  1. An accurate thermometer with a clear scale.

  2. An alarm timer with a luminous dial.

  3. A fixture to contain two No. 1 photoflood lamps faced by sheets of glass to prevent accidental shattering of the lamps due to splashing with water during the reversal exposure.

  4. Processing tanks. Although the film may be processed in

dishes, the maintenance of correct temperature, complete immersion and agitation present many problems. For this reason a set of seven porcelain tanks (Kodak 2L) is recommended for use with the 2 litre kit.

5. film hangers (Kodak No. 3) up to size $4\frac{3}{4} \times 6\frac{1}{2}$ in.

Processing

It is essential to avoid contamination of solutions. The tanks should be used for the same solutions on all occasions and should be thoroughly washed before refilling.

Agitation

Correct agitation is important, especially during first development.

Temperature

The first developer should be used at a temperature of 75° F. and this must not be allowed to vary by more than $\frac{1}{2}$° F.

Timing

Each step in the processing must be carefully timed and this must include 10 seconds to drain the film prior to placing in the next solution.

Processing solutions (Technique 75)

1. Adjust the temperature of the first developer to 75° F.

2. Place the plate holder and hanger close to the tank containing the first developer.

3. Set the timer for 10 minutes.

4. Switch off all lights.

5. In total darkness remove the film from the plateholder and attach to the hanger.

6. *First development : 75° F. 10 minutes*

    (i) Start the timer.

    (ii) Immediately lower the film into the developer.

    (iii) Tap the hanger sharply on the side of the tank to remove air bubbles from the film.

    (iv) Remove the film from the developer and replace.

    (v) Repeat four times during the first 15 seconds.

    (vi) Leave for the remainder of the first minute.

    (vii) Lift the film clear of the developer and tilt 60° or more in one direction. Replace in the developer. Remove again and tilt 60° or more in the opposite direction. Replace in the developer. No pause for draining should be made during the tilting and the whole process should be performed smoothly in about 7 seconds.

(viii) Repeat this cycle each minute.

(xi) Ten seconds before the alarm is due to ring lift the film from the developer and allow to drain.

7. *Rinse in running water:* 73-77° *F.* 1 *minute*

(i) When the alarm rings place the film into running water and restart the timer.

(ii) Remove and replace the film in the water 4 times in the first 15 seconds.

(iii) Ten seconds before the alarm is due to ring remove the film and allow to drain.

8. *Harden:* 73-77° *F.* 3 *minutes*

(i) When the alarm rings place the film in the hardening solution and restart the timer.

(ii) Remove and replace the film in the hardening solution four times in the first 15 seconds.

(iii) Remove the film and allow to drain for 5 seconds in each minute.

9. Wash in running water: 73-77° F. 3 minutes

10. Switch on the main lighting.

11. *Reversal exposure*

(i) Place the photoflood lamps 2 feet apart.

(ii) Pass the film between the lamps and expose for 5 seconds.

(iii) It is essential that a draining time of at least 1 minute should be allowed before commencing the next step. Do not let the film dry out.

12. *Colour developer:* 73-77° *F.* 15 *minutes.*

(i) Remove and replace four times during the first 15 seconds.

(ii) Remove and allow to drain for 5 seconds in each minute.

13. Transfer to running water: 73-77° F. 5 minutes.

14. *Clearing solution:* 73-77° *F.* 5 *minutes.*

(i) Remove and replace four times during the first 15 seconds.

(ii) Remove and allow to drain for 5 seconds in each minute.

15. Rinse in running water: 73-77° F. 1 minute.

16. *Bleaching solution:* 73-77° *F.* 8 *minutes.*

(i) Remove and replace four times during the first 15 seconds.

(ii) Remove and allow to drain for 5 seconds in each minute.

17. Rinse in running water: 73-77° F. 1 minute.

18. *Fixing solution:* 73-77° *F.* 4 *minutes.*

(i) Remove and replace four times during the first 15 seconds.

(ii) Remove and allow to drain for 5 seconds in each minute.

19. Wash in running water: 73-77° F. 8 minutes.

20. *Stabilizing solutions:* 73-77° F. 1 *minute.*
    (i) Remove and replace four times during the first 15 seconds.
    (ii) Do not rinse in water.
The omission of this step will result in fading of some of the dyes.

21. Dry the film in a dust-free atmosphere at a temperature which does not exceed 110° F.

Note

All processing solutions should be returned to the storage bottles after use.

PROCESSING EKTACOLOR SHEET FILM TYPE L

Temperature control is critical. The temperature of the developer must not be allowed to vary by more than $\frac{1}{2}$° F. from 75° F. The remaining solutions, including the water used for washing, may be used at a temperature of 73-77° F. If the washing water is used at a temperature of 50-73° F. the washing time must be increased by 50 per cent.

Processing solution (Technique 77)

1. Adjust the temperature of the developer to 75° F.
2. Set the luminous alarm timer for 14 minutes.
3. Place the plateholder and film hanger near the developing tank.
4. Switch off the main lighting.
5. In total darkness remove the film from the plate holder and attach to the hanger.
6. *Develop:* 75° F. 14 *minutes*
    (i) Place the film in the developer and start the timer.
    (ii) Agitate by lifting the film from the developer several times during the first 15 seconds.
    (iii) Continue agitation at intervals of 20 seconds by lifting from the tank and tilting the film and allowing it to drain from one corner for 5 seconds. Alternate the directions of the tilt and draining.
    (iv) 10 seconds before the end of development remove the film and allow to drain.

7. *Stop bath:* 73-77° F. *4 minutes*
    (i) Agitate as in step 6 (ii) and (iii).
    (ii) Allow the film to drain for 10 seconds before the next step.
8. *Harden:* 73-77° F. *4 minutes.*
    (i) Agitate as in Step 6 (ii) and (iii).
    (ii) Switch on the main lighting.
9. *Wash in running water:* 73-77° F. *4 minutes.*
    (i) The flow of water should be adjusted so that it is completely replaced every 2 minutes.
10. *Bleach:* 73-77° F. *6 minutes.*
    (i) Agitate as in step 6 (ii) and (iii).
11. *Wash in running water:* 73-77° F. *4 minutes.*
    (i) Step 9 (i).
12. *Fix:* 73-77° F. *8 minutes.*
    (i) Agitate as in Step 6 (ii) and (iii).
13. *Wash in running water:* 73-77° F. *8 minutes.*
    (i) Step 9 (i).
14. Place the film in Photo-Flo solution, diluted to one quarter of the normal recommended strength, for 1 minute at 73-77° F.
15. *Dry* the film in a dust-free atmosphere at a temperature not exceeding 110° F.

TECHNIQUE NO. 50

PROCESSING REVERSAL MATERIALS

By this process a black and white negative is transformed into a positive transparency. A negative image is produced in the usual manner and then treated in a bleaching solution. The silver halide remaining is exposed to a white light and then developed to produce a positive image. Reversal processing may be applied to many materials, but great care must be exercised to ensure that the photomicrograph is correctly exposed in the first instance in order to produce a transparency of high quality.

Reversal procedure using Ilford FP3 roll films. Processing solutions (Technique 71).
    1. Fill a spiral type developing tank with reversal developer and

maintain at a temperature of 68° F. Set the dark room timer for 12 minutes.

2. In the correct safelighting load the film into the spiral.

3. Lower the spiral into the developer; start the timer; lift to remove air bubbles from the film and then replace the tank cover.

4. Agitate for 5 seconds every 15 seconds throughout development.

5. On completion of development quickly remove the spiral and wash in running water for 3 minutes.

6. Return the developer to its container and retain for the second development.

7. Fill the tank with the bleaching solution. Replace the spiral and bleach the film for 5 minutes in the tank. Agitate continuously.

8. Switch on the main lighting.

9. Wash in running water for 2 minutes.

10. Clear for 2 minutes.

11. Wash in running water for 2 minutes.

12. Remove the film from the spiral and expose at 18 inches from a 100 watt tungsten lamp for 30-120 seconds.

13. Return the film to the spiral and redevelop for 6 minutes at 68° F. in the developer retained from the first development until maximum density is obtained.

14. Place in an acid hardening fixing bath for 10 minutes.

15. Wash in running water for 30 minutes.

16. Rinse in water containing a few drops of wetting agent.

17. Remove the film from the spiral. Dry in a dust-free atmosphere or in a drying cabinet.

Procedure for Kodak Panatomic X 35 mm. film. Processing solutions (Technique 71)

1. Fill the tank with D19b developer (undiluted) and maintain at a temperature of 68° F. Set the dark room timer for 6 minutes.

2. Switch off main lighting. In complete darkness load the film into the spiral (Technique 43).

3. Place the spiral in the developer, start the timer and replace the lid of the tank.

4. Place the agitator rod into the central channel in the lid, press firmly home, and rotate several times to agitate and remove air bubbles.

5. Continue agitation by rotating the rod two or three times at 30 second intervals during development.

6. Pour the developer from the tank and wash in running water for 5 minutes.

7. Fill the tank with bleaching solution (R21A). Bleach 5 minutes. Agitate steps 4 and 5.

8. Pour the bleaching solution from the tank and wash in running water for 5 minutes.

9. Fill the tank with clearing solution (R21B). Clear 2 minutes. Agitate steps 4 and 5.

10. Pour the clearing solution from the tank and wash in running water for $\frac{1}{2}$ minute.

11. Expose both ends of the spiral at about 18 inches from a 100 watt tungsten lamp for $2\frac{1}{2}$ minutes.

12. Refill the tank with D19b developer (undiluted) 68° F. Redevelop 4 minutes. Agitate steps 4 and 5.

13. Pour the developer from the tank and wash in running water for 5 minutes.

14. Fill the tank with fixing solution (F.5). Fix 5 minutes. Agitate steps 4 and 5.

15. Pour the fixing solution from the tank and wash in running water for 30 minutes.

16. Fill the tank with water containing a few drops of wetting agent and leave for 1 minute. Remove the film from the spiral and dry at room temperature in a dust-free atmosphere or in a drying cabinet.

Note

A series OB safelight may be used after step 7.

Procedure for Kodak KS5, KO5, and KP5 sheet film. Processing solutions (Technique 71).

1. Fill a dish or tank with D8 diluted 2 volumes of developer and 1 volume of water and maintain at 68° F. Set the dark room timer for 5 minutes.

2. In the correct safelighting (Technique 28) remove the film from the plateholder.

3. Start the timer and place the film in the developer. For dish development (Technique 42): for tank development (Technique 41). Apply continuous agitation.

4. On the completion of development rinse in running water for 1 minute.

5. Bleach in R21A for 2-3 minutes. Apply continuous agitation.

6. Turn on the Wratten series OB safelight.

7. Wash in running water for 5 minutes.

8. Clear in R21B for 2 minutes.

9. Rinse in running water for 1 minute.

10. Expose both sides of the film to a 25 watt lamp at 12 inches for 1 minute.

11. Redevelop in D158* diluted with an equal volume of water for 2 minutes at 68° F.

12. Rinse in running water.

13. Fix in F54a for 5 minutes.

14. Wash in running water for 30 minutes.

15. Rinse in water containing a few drops of wetting agent. Dry in a dust-free atmosphere or a drying cabinet.

* With KO5 and KP5 films 0.25gm. of potassium iodide should be added to each 1000 ml. of working strength developer.

TECHNIQUE NO. 51

BROMIDE PAPER

KODAK

Surface	Grades	Safelight	Developer, Dilution and Development at 68° F. in Minutes	
White Smooth Glossy	Extra soft Soft Normal Hard Extra hard	OB	D163, 1 part + 3 parts water	1½-2
White Fine Lustre	Soft Normal Hard	OB	D163, 1 part +3 parts water	1½-2
ILFORD				
White Glossy	Soft Normal Hard Extra hard Ultra hard	No. S 902	ID20, 1 part +3 parts water	1½-2
Matt	Normal Hard	No. S 902	ID20, 1 part +3 parts water	1½-2

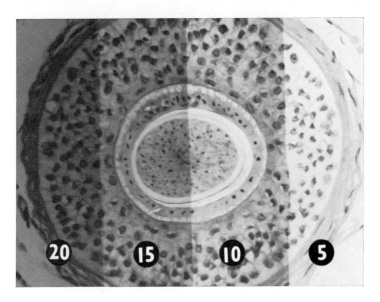

FIG. 48. Test exposure. Strip method

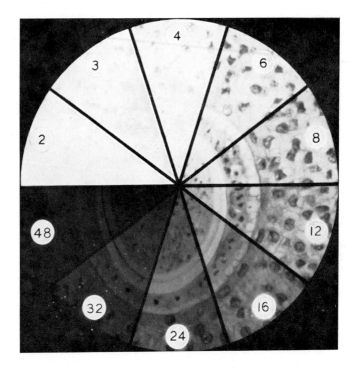

FIG. 49. Enlarging exposure scale (Kodak)

PRINTING: DETERMINATION OF EXPOSURE

Test exposure. Strip method. For printing by contact (Technique 61)
or by projection (Technique 62).

1. Cover three quarters of the bromide paper with an opaque card and expose for 5 seconds.

2. Without turning off the light, quickly move the card approximately half way across the paper and expose for a further 5 seconds.

3. Move the card quickly to uncover three-quarters of the paper and expose for another 5 seconds.

4. Finally remove the card and expose the whole paper for a further 5 seconds.

The test paper records four strips of 20, 15, 10 and 5 seconds exposure (Fig. 48).

5. Process the test paper (Technique 57).

6. Wash in water for 2 minutes. Switch on the main lighting and inspect the test paper.

7. Select the strip which produces the best reproduction of the preparation and in which the background is white or very slightly grey.

Note

The times given are purely arbitrary. The actual range selected will depend upon the density of the negative to be printed.

PRINTING: DETERMINATION OF EXPOSURE
KODAK ENLARGING EXPOSURE SCALE

Contact method

1. Insert the exposure scale (Fig. 49) in the printing frame, glossy surface to the glass.

2. Place the negative, emulsion surface uppermost, on top of the scale.

3. Set the enlarger timer switch for 30 seconds.
4. Turn on the safelight, switch off main lighting.
5. Place a sheet of bromide paper in contact with the negative.
6. Replace the back of the printing frame and secure.
7. Place the frame, glass surface uppermost, on the enlarging easel.
8. Expose.
9. Process the test paper (Technique 57).
10. Inspect the print and select the wedge which produces the best reproduction of the preparation and in which the background is white or very slightly grey.
11. The correct exposure time in seconds is ascertained by halving the number which appears on the wedge selected.

Projection method
1. Set the enlarger timer switch for 1 minute.
2. Turn on the safelight and switch off the main lighting.
3. Place the exposure scale, with the glossy side uppermost, on the emulsion surface of the bromide paper.
4. Expose.
5. Process the test paper (Technique 57).
6. Inspect the print and select the wedge which produces the best reproduction of the preparation and in which the background is white or very slightly grey.
7. The correct exposure time in seconds is ascertained from the number which appears on the wedge selected.

TECHNIQUE NO. 54

PRINTING BY CONTACT

Printing by contact is achieved by placing a negative and a sheet of sensitized paper together, emulsion to emulsion, and exposing to light. This produces a positive image the same size as the negative.

Contact prints are made with a printing frame or printing box. The box printer has the advantage of a speedier technique and a light source placed at a constant distance from the negative. A printing frame is a simpler and cheaper piece of equipment and can be used in conjunction with an enlarger as a light source.

The choice of paper will depend a great deal on experience and to some extent on the result required. As a rough guide, a slightly under-exposed negative will print best on a high contrast grade of paper, a slightly over-exposed negative on a soft grade of paper, and a correctly exposed negative on a normal grade. A single weight paper with a glossy surface is generally preferred for photomicrographs.

Procedure

1. Remove the spring back of the printing frame and clean both sides of the glass.

2. Hold the negative by one corner to avoid finger marks and place it in the frame with the emulsion surface uppermost.

3. Switch on the safelight and turn off main lighting.

4. Place a sheet of bromide paper, emulsion side downwards, in contact with the emulsion of the negative.

5. Replace the back of the printing frame and secure firmly.

6. Place the frame, glass surface uppermost, on the enlarging easel.

7. Raise the enlarger so that the illumination will completely cover the frame.

8. Prepare a test exposure (Techniques 52 and 53).

9. Set the enlarger timer switch to the correct exposure time.

10. Turn on the safelight, switch off the main lighting and place a sheet of paper of the same grade used for the test exposure in contact with the negative.

11. Replace the back of the printing frame and secure.

12. Place the printing frame, glass surface uppermost, on the enlarging easel.

13. Expose.

14. Remove the paper from the printing frame.

15. Process the paper (Technique 57). Develop for the same time and at the same temperature used for the test paper.

TECHNIQUE NO. 55

PRINTING BY PROJECTION

Enlarged photomicrographs are primarily employed for display. Unfortunately, enlargement is always accompanied by some loss in definition, because of the increase in grain size of the negative, and this tends to restrict the use of enlarged prints for publication. If a condenser type of enlarger is used grain size may be controlled within limits by:

1. Using a paper with a rough surface.
2. Using a diffuser between the lens and the paper.
3. Adjusting the enlarger so that it produces a print very slightly out of focus.

Enlarged prints can be made on bromide paper of any grade or surface texture, but for display a matt surface is preferable because it does not reflect light.

Procedure

1. Place the negative, with the emulsion side facing the enlarging easel, in the negative carrier.
2. Switch on the safelight and turn off the main lighting.
3. Place a sheet of paper the same thickness as the bromide paper beneath the masking frame. Adjust the masking frame to the size of the print required.
4. Switch on the enlarger and adjust the height until the desired enlargement is obtained.
5. With the lens diaphragm fully open focus the image with a focus finder (Fig. 50).
6. Close the lens diaphragm by at least one stop. This improves definition by increasing the depth of focus. Switch off the illumination.
7. Select a sheet of bromide paper at least half an inch larger than the masked image.
8. Remove the paper used for focusing from the masking frame and substitute the bromide paper with the emulsion side facing the enlarger lens.
9. Prepare a test exposure (Techniques 52 and 53).
10. Set the enlarger timer switch for the correct exposure.
11. Turn on the safelight, switch off the main lighting and check the image for focus.
12. Remove the focusing paper and substitute a sheet of bromide paper of the same grade used for the test exposure.
13. Expose.
14. Remove the paper from the masking frame.
15. Process the paper (Technique 57). Develop for the same time and at the same temperature used for the test exposure.

Focus finder   *Magnifier*

FIG. 50. Paterson focus finder and magnifier

MAKING MONOCHROME PRINTS FROM
35 MM. COLOUR TRANSPARENCIES

Whilst it is possible to produce a satisfactory monochrome print from a positive colour transparency the quality will not match up to a print prepared from a black and white negative made from the same preparation. The technique involves the preparation of an intermediate monochrome negative from the transparency by either the contact or the projection printing method. If the print required is to be larger than the colour original, use the projection method to produce a negative and print by contact. Orthochromatic materials will in most instances produce good results. However, if the predominant colours are red and orange, a panchromatic emulsion in conjunction with a light yellow green filter should be used. This combination will provide the necessary colour correction for a tungsten light source.

It is essential to ensure that all light leaks from the enlarger lamp house and the negative carrier are obliterated.

Procedure

1. Place the colour transparency, emulsion surface facing the lamp, in the negative carrier.

2. Place a sheet of white card of the same thickness as the plate or film selected in the masking frame.

3. Turn on the safelight and switch off main lighting.

4. Switch on the enlarger lamp and adjust the height of the enlarger to produce the desired image size.

5. Focus the image with the lens diaphragm fully open.

6. Insert the filter selected in the holder and refocus.

7. Reduce the lens diaphragm by one stop and switch off the illumination.

9. Make a test exposure (Techniques 52 and 53). Process (Technique 41).

10. Select the strip which produces the best reproduction of the transparency.

11. Set the enlarger timer switch for the correct exposure time.

12. Turn on the safelight, and switch off main lighting.

13. Check the focus.

14. Place another plate or film in the masking frame.

15. Expose.

16. Process the plate or film (Technique 41). Develop for the same time and at the same temperature used for the test exposure.

17. Print the negative by the contact method (Technique 54).

PROCESSING BROMIDE PAPER

Processing involves developing, rinsing, fixing, washing and drying. It is preferable to develop bromide paper by the time temperature method and the instructions provided by the manufacturers should be followed. If this method of development is carried out correctly the resulting image should possess a range of tones from white, through the greys, to black. Moreover, it should be possible to produce additional prints at a later date which will closely match the original.

Insufficient or uneven development will produce defects in the print (Technique 58).

Procedure

1. Use the developer recommended for the paper selected.

2. Dilute if necessary and ensure it is well mixed (Technique 51).

3. Select three dishes larger than the print to be processed and fill with developer, water and acid fixing solution respectively.

4. Set the timer for the recommended period of development.

5. Check that the developer is at the recommended working temperature.

6. Turn on the safelight and switch off the main lighting.

7. Place the paper, emulsion side downwards, into the developer.

8. Start the timer. Agitate by rocking the dish backwards and forwards and from side to side alternately throughout development.

9. Rinse the print in water for 15 seconds. This removes the developer and avoids contamination of the fixing bath.

10. Transfer the print to the acid fixing bath for the recommended period. Agitate to avoid staining. Prolonged fixation will cause loss of detail in the highlight areas. Insufficient fixation will result

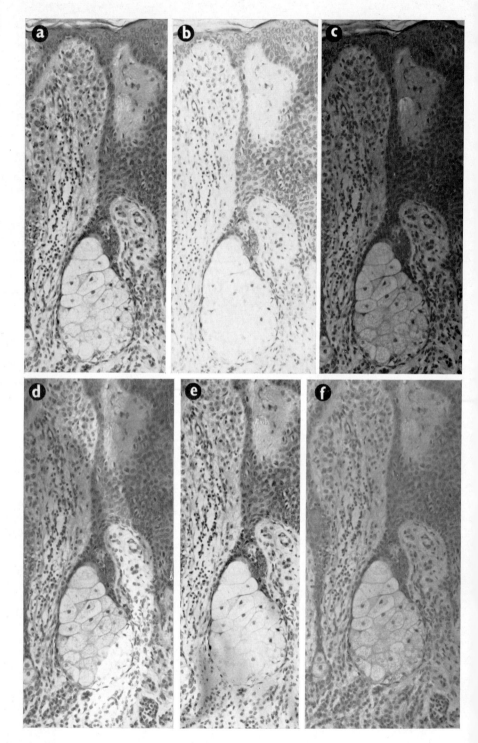

Fig. 51.

in a subsequent fading of the image. Ten minutes immersion in the fixing bath is usually adequate.

11. Wash in running water for 30 minutes to remove the fixing solution and silver salts carried over from the fixing bath.

12. Dry the print by placing it between Fotonic blotting paper. Glossy prints may be glazed (Technique 65).

TECHNIQUE NO. 58

DEFECTS IN PRINTING

The prints (Fig. 51) were made from the same correctly exposed and processed negative.

a. Correct exposure.

b. Insufficient exposure.

c. Excessive exposure.

d. Uneven development. Due to careless immersion of the paper and inadequate agitation.

e. Staining. The result of careless immersion of the print in the fixing bath and inadequate agitation.

f. Lack of contrast due to insufficient development and excessive exposure.

TECHNIQUE NO. 59

PROCESSING EKTACOLOR PAPER
Process P-122

Safelight

A Wratten series 10 H safelight filter with a 25 watt lamp, at a distance of 4 feet from the paper, may be used. The paper must not be exposed to the safelight for more than 4 minutes from the time it is removed from the packet.

Timing

The time for each step is critical and includes an allowance of 20 seconds for draining the paper. As the first steps in processing are carried out in total darkness a luminous timer must be used.

Temperature

The developer must be used at a temperature within $\frac{1}{2}°$ F. of 75° F. The other solutions and washing water may be used at temperatures between 73-77° F. Cold water may be substituted in the washing steps provided the temperature does not fall below 50° F. and the washing times are increased by 50 per cent.

Processing solutions (Technique 78)

1. Adjust the temperature of all the solutions to 75° F.
2. Set the timer for 12 minutes and place it well away from the processing bench.
3. Switch off the main lighting.
4. *Develop the paper at 75° F. for 12 minutes*
    (i) Start the timer.
    (ii) Immerse the paper, emulsion side downwards, in the developer. Make sure it is completely immersed in the solution.
    (iii) Agitate by raising the left hand side of the dish about 2 inches and lowering gently. Repeat with the near side and then the right side of the dish. Return to the near side and complete the cycle by raising the left side. Continue this cycle throughout development.
    (iv) 20 seconds before the end of development lift the paper from the solution to drain.
5. *Transfer to the stop bath at 73-77° F. for 2 minutes*
    (i) Agitate (Step 4 (iii)).
    (ii) Allow the paper to drain for 20 seconds before proceeding to the next step.
6. *Place the paper in the fixing bath at 73-77° F. for 2 minutes*
    (i) Agitate (Step 4 (iii)).
    (ii) Switch on the main lighting.
7. Wash in running water at 73-77° F. for 2 minutes. The flow of water must be sufficient to provide a complete renewal every minute.
8. *Transfer to the bleaching solution at 73-77° F. for 4 minutes.*
    (i) Agitate (Step 4 (iii)).
9. Wash in running water at 73-77° F. for 2 minutes observing the same conditions as Step 7.

98

10. *Transfer to the second hardener fixing bath at 73-77° F. for 2 minutes.*

    (i) Agitate (Step 4 (iii)).

11. Wash in running water at 73-77° F. for 8 minutes observing the same conditions as Step 7.

12. *Immerse the paper in the hardening solution at 73-77° F. for 3 minutes.*

    (i) Agitate (Step 4 (iii)).

13. Wash in running water at 73-77° F. for 2 minutes observing the same conditions as Step 7.

14. *Transfer to the buffer solution at 73-77° F. for 3 minutes.*

    (i) Agitate (Step 4 (iii)).

    (ii) *Do not* replace the print in water.

The print may be cold glazed on a chromium plate as follows:

1. From Step 10, place the paper in

                Buffer solution   900 ml.

                Glycerine   100 ml.

2. Leave in this solution for 3 minutes.

3. Transfer to a chromium plate and remove the excess moisture with a print roller.

4. Discard the solution after use.

The print may be hot glazed by one of the following methods:

1. After drying the print at room temperature place it in water for 1-2 minutes. Transfer to the hot glazer at 180° F.

2. Increase the immersion time in Step 12 to 19 minutes, and transfer to the hot glazer directly after Step 14.

MONOCHROME TRANSPARENCIES

SENSITIVE MATERIALS

Plates	Maker	Sizes	Safelight	Emulsion	Developer, Dilution and Development at 68° F.
L10	Kodak	2 and 3¼ in. square	Series OB	Bromide (Hard)	D163, 1 part +3 parts water 1½-2 minutes
L15	Kodak	2 and 3¼ in. square	Series OB	Bromide (Hard)	D163, 1 part +3 parts water 1½-2 minutes
Contact Lantern	Ilford	2 and 3¼ in. square	Vs2 No. 109	(Slow)	ID36, 1 part +3 parts water 45-60 seconds
Special Lantern	Ilford	2 and 3¼ in. square	F No. 904	Soft Normal Contrast	ID36, 1 part +3 parts water 1¾ 1¾ 2½ minutes

Sheet Film

Process	Kodak	3¼ × 4¼ inch	Series OB	Medium Contrast	D158, 1 part +1 part water 3 minutes
N5.30	Ilford	3¼ × 4¼ inch.	F No. 901	Medium Contrast	ID2, 1 part +2 parts water 1¾ minutes

Roll Film

Positive Safety	Kodak	35 mm.	Series OB	Normal Contrast Softer Contrast	D163, 1 part +3 parts water ½-2 minutes D76 undiluted 5 minutes
Fine grain Safety Positive	Ilford	35mm.	S No. 902	High Contrast	ID2, 1 part +2 parts water 1¾ minutes

PRINTING MONOCHROME TRANSPARENCIES BY CONTACT

Procedure

1. Remove the back of the printing frame and clean the glass on both sides. Dust and fibre particles will be enlarged on projection of the transparency.

2. Place a mask, a little larger than the area required, on the glass.

3. Place the negative on top of the mask with the emulsion side uppermost.

4. Turn on the safelight and switch off the main lighting.

5. Place a plate or film, with the emulsion side downwards, in contact with the emulsion of the negative.

6. Replace the back of the frame and secure.

7. Place the printing frame, glass surface uppermost, on the enlarging easel.

8. Raise the enlarger to ensure the illumination will completely cover the frame.

9. Prepare a test exposure (Technique 52, Steps 1-6) and process Technique 42 (Steps 1-12).

It is possible, after a little experience, to judge with reasonable accuracy the correct exposure from a wet plate or film, but it is better to assess the exposure of the slide after it has been dried and projected. If speed is essential, wash the test plate or film in water for 2 minutes, and then immerse in methylated spirit for 2-3 minutes, dry in a drying cabinet, and project the transparency without binding.

10. Note the correct exposure.

11. Turn on the safelight and switch off the main lighting.

12. Set the enlarger time swtich for the correct exposure time.

13. Place another plate or film in contact with the negative.

14. Replace the back of the printing frame and secure.

15. Place the printing frame, glass surface uppermost, on the enlarging easel.

16. Expose.

17. Remove the plate or film from the printing frame.

18. Process the plate or film (Technique 42, Steps 1-14). Develop for the same time and at the same temperature used for the test exposure.

19. Dry in a dust-free atmosphere or in a drying cabinet.

PRINTING MONOCHROME TRANSPARENCIES BY PROJECTION

This method is used for obtaining an image of larger size than the negative.

Procedure

1. Place the negative in the carrier of the enlarger with the emulsion side facing the lens.

2. Mask the unwanted areas of the negative with a sheet of black paper. This prevents possible fogging by scattered light during exposure.

3. Turn on the safelight and switch off the main lighting.

4. Place a white card of the same thickness as the plate or film in the masking frame. Switch on the enlarger.

5. Focus the image with the diaphragm lens fully open.

6. Close the diaphragm by one or two stops and switch off the illumination.

7. Substitute a plate or film, with the emulsion side facing the lens, for the card used for focusing.

8. Prepare a test exposure (Technique No. 52, Steps 1-6).

9. Assess the correct exposure (Technique No. 61, Step 9).

10. Set the enlarger timer switch for the correct exposure time.

11. Turn on the safelight. Switch off the main lighting and check the image for focus.

12. Remove the focusing card and substitute another plate or film.

13. Expose.

14. Process (Technique 42, Steps 1-14). Develop for the same time and at the same temperature used for the test exposure.

15. Dry in a dust-free atmosphere or in a drying cabinet.

PRINTING MONOCHROME TRANSPARENCIES
BY REDUCTION

It may be necessary to reduce the dimensions of a negative to produce a transparency. If the bellows extension of the enlarger is insufficient for this purpose, or a lens of a longer focal length is not available, reduction may be achieved by one of two methods.

Enlarger extension tubes may be used. After removing the lens these screw into one another to produce different lengths. The tube is attached to the lens panel and the lens is fitted to the lower end of the tube. It is possible by this method to reduce the overall dimension of a negative to very small proportions. The transparency is prepared by the projection method (Technique 55).

Alternatively, the negative can be photographed while it is illuminated by transmitted light. It must be suitably masked to exclude extraneous light and is photographed with a bellows type camera or miniature camera with suitable supplementary close-up lenses. The positive image may be produced on process sheet film or 35 mm. positive safety film.

MASKING AND BINDING OF TRANSPARENCIES

Masking eliminates unwanted portions of the image and also produces straight edges to the framed picture. Binding the transparencies between coverglasses is essential to protect the emulsion surface from damage. The technique of masking and binding, whilst straightforward, requires practice for neat and proficient results. A bench illuminator, consisting of a lamp covered by a sheet of opal glass fitting flush with the bench surface, is particularly useful as it allows not only ready inspection of the transparency but also simplifies the process of masking.

Masking $3\frac{1}{4}$ in. plates or films.

Masking paper, gummed on one side, may be obtained in lengths of $3\frac{1}{4}$ in. and in various widths.

Procedure

1. If a bench illuminator is not available, place the transparency, emulsion surface uppermost, on a white card.

2. Moisten a masking strip and carefully place in position on one side of the transparency. Care must be taken to avoid transferring gum to the unmasked part of the transparency.

3. Repeat Step 2 for the three remaining sides.

4. Orientate the transparency as it will appear when projected.

5. Place a gummed white spot in the lower left hand corner of the mask. This will indicate to the projectionist the correct orientation of the slide for projection.

6. Cover the emulsion surface with a clean dry, coverglass. When mounting a film two coverglasses are necessary.

7. Keep together with a spring clip until ready for binding.

Binding $3\frac{1}{4}$ in. plates and film

It is advisable to adopt a standard sequence for binding, practice making for perfection. Gummed binding strips are available in $13\frac{1}{2}$-14 in. or $3\frac{1}{4}$ in. lengths. In the following technique $3\frac{1}{4}$ in. strips are used as they are more convenient to handle.

Procedure

1. Moisten a binding strip and lay it flat on the bench.

2. Take the transparency between the thumb and forefinger of the right hand, place the edge away from the spring clip on the centre of the binding strip and apply a little pressure.

3. Lift the transparency and transfer to the left hand. With the forefinger of the right hand quickly press the whole length of the gummed strip into contact with the coverglasses.

4. Remove the spring clip.

5. Moisten a second strip and repeat Steps 2 and 3 applying the strip opposite the bound edge.

6. Repeat Steps 2 and 3 for the remaining two sides of the transparency.

7. The corners of the upper and lower strips may be lifted before drying is complete and mitred. This reduces the risk of raised or frayed corners.

8. Clean the glass surfaces to remove gum and finger marks.

Masking 35 mm. transparencies

A variety of preformed masks may be purchased.

Procedure

1. Remove the transparency carefully from its cardboard frame and place within the mask. Secure two sides of the transparency with adhesive tape to prevent movement.
2. Fold the mask over the transparency. Orientate the transparency as it will appear when projected and stick a gummed spot on the lower left hand corner of the mask.
3. Cover both sides of the transparency with clean 2 in.-square coverglasses.
4. Place in a spring clip until ready for binding.

Binding 35 mm. transparencies

Use the same technique as for $3\frac{1}{4}$ in. square transparencies.

Note

Special jigs are available for the binding of transparencies.

TECHNIQUE NO. 65

GLAZING PRINTS

Three methods of glazing prints are available and the choice of method depends on the number of prints for glazing at any given time. The simplest method suitable for glazing a small number of prints, involves the use of chromium-plated metal sheets. For large numbers of prints the flat bed (Fig. 37) or rotary type of glazer (Fig. 38) should be used.

CHROMIUM-PLATED METAL SHEET: COLD GLAZING

Procedure

1. Clean the chromium surface with a soft fluffless cloth moistened with methylated spirit. Follow this with a cloth moistened with glazing solution.
2. Fill a dish with glazing solution diluted according to instructions.
3. Remove the prints from the print washer and place in the glazing solution for 5 minutes. Agitate to remove air bubbles.

4. Take the prints singly from the glazing solution and lay face down on the glazing sheet. The prints should not overlap.

5. When all the prints are in position, place the sheet between Fotonic paper and remove the surplus fluid by passing a print roller over the paper. Excess pressure will cause the prints to slide out of position.

6. Dry the prints by removing the glazing sheet from between the Fotonic paper and stand upright away from draughts at room temperature. The prints will fall from the sheet when dry. Do not attempt to hasten the process by applying heat or using an electric fan.

Flat bed glazer: hot glazing

1. Switch on the machine and allow it to reach the required glazing temperature.

2. Clean the surface of the glazing plate with methylated spirit followed by glazing solution.

3. Take the prints from the print washer and place in glazing solution for 5 minutes. Agitate to remove air bubbles.

4. Remove the prints singly from the glazing solution and place on the chromium plate ensuring the prints do not overlap.

5. Place the plate between Fotonic paper and remove excess fluid with a print roller.

6. Remove the glazing sheet from the Fotonic paper and place face uppermost on the glazer. Secure the apron over the glazing sheet.

7. As the prints dry they will leave the glazing sheet. Lift the apron and remove prints.

Rotary glazer: hot glazing

1. Switch on the machine and allow the drum to reach the required glazing temperature. This is indicated by the red light cutting out.

2. Place the prints in glazing solution and leave for 5 minutes. Agitate to remove air bubbles.

3. Adjust the pressure roller to the glazing position.

4. Set the rotation speed for either single- or double-weight paper.

5. Take the prints singly from the glazing solution and place face upwards on the rotating cloth of the glazer.

6. Remove the glazed prints from the tray as they fall from the drum.

Notes

The chromium surface should be thoroughly cleaned after use.

Lime deposits from hard water, or chemical contamination as a result of insufficient washing of the prints after processing, can produce serious defects in the chromium surface. Abrasive polishers should be avoided as scratches will be transferred to the glazed print. Uneven glazing occurs as a result of dirt, irregular contact of the print, or uneven soaking of the print. Dust particles produce unglazed pits on the print. A dirty surface, insufficient glazing temperature, or damage of the chromium surface may cause sticking of the prints.

TECHNIQUE NO. 66

MOUNTING PRINTS

A number of adhesives may be used to attach photomicrographs to a mount and include preparations consisting of gelatin, starch, rubber solutions, dextrine and dry mounting tissue.

Rubber solutions

1. Mark out on the mount in pencil an area slightly smaller than the dimensions of the trimmed print. Cover this area with a thin even layer of rubber solution, allow to dry.

2. Treat the back of the print with rubber solution, allow to dry.

3. Cover the rubber solution on the mount with a sheet of paper, leaving the upper half inch uncovered.

4. With the forefinger and thumb of each hand, hold the top edges of the print and carefully press into contact with the exposed area on the mount.

5. Holding the lower end of the print in left hand, carefully remove the sheet of paper.

6. With a print roller bring the print into complete contact with the mount.

Starch and Dextrine mixtures

These may be used to attach wet or dry prints to the mount.

1. Mark out in pencil an area on the mount slightly smaller than the dimensions of the trimmed print.

Wet prints

1. Remove the print from water and place face downwards on a clean sheet of glass. Remove the excess water with a print roller.

2. Brush the adhesive on to the back of the print.

3. Remove the print from the glass support and attach it to the mount.

4. Place the mounted print between two sheets of Fotonic paper and apply slight and even pressure with a print roller.

Dry prints

1. Apply the adhesive with a brush to the back of the print.

2. Place the print carefully on the mount.

3. Place the mounted print between Fotonic paper and apply slight pressure with a print roller or, alternatively, place in a cold press.

Gelatin

1. Mark out in pencil an area on the mount slightly less than the dimensions of the trimmed print.

2. Apply the adhesive quickly and evenly to the back of the print with a soft brush.

3. Quickly place the print on to the mount.

4. Place the mounted print between Fotonic paper and then into a cold mounting-press.

Dry mounting

This process is by far the most satisfactory and is now widely used. It consists of placing a sheet of dry mounting tissue between the print and the mount and pressing into contact by means of a heated press (Fig. 39). This melts the tissue to effect a perfect contact between print and mount.

1. Switch on the mounting press.

2. Fix a sheet of tissue a little larger than the print to the back of the print by touching the central area of the tissue with a heated fixing iron.

3. Trim the print and attached tissue.

4. Mark out in pencil an area on the mount slightly smaller than the dimensions of the print.

5. Place the print on the mount and immobilise with two fingers.

6. With the index finger of the same hand raise a corner of the print but not the tissue.

7. Touch the tissue with the heated fixing iron to attach it to the mount.

8. Attach a second corner in the same manner. This localizes the print on the mount.

9. Check that the mounting press has reached a temperature of 140°-170° F.

10. Place the mount and print in the press and cover with a metal plate.

11. Lower the top plate of the press and clamp lightly. Leave for approximately 20 seconds (30 seconds for prints prepared on double-weight paper).

12. Remove the mounted print from the press and leave to cool.

Note

If the tissue adheres to the print and not to the mount the mounting temperature is too low. If the tissue adheres to the mount and not to the print the temperature is too high. The cover plate must be kept free of dust and other matter which may become embedded in the print surface during mounting.

TECHNIQUE NO. 67

LETTERING OF PRINTS

It is frequently necessary in a caption or manuscript to draw attention to a particular structure shown in the photomicrograph. Freehand lettering of inferior quality can completely mar an otherwise excellent print. If this operation cannot be performed satisfactorily it is better left to the printer. However, two methods are available, which, after a little practice, provide satisfactory results for the lettering of photomicrographs. They are:

1. Uno Stencils
2. The Letraset type-lettering system.

Uno Stencils

Uno Stencils (Fig. 52a) comprise alphabets and numerals accurately cut in thin but tough blue transparent plastic material. Two narrow parallel edges raise the stencil from the working surface so that undried letters are not smudged. Standard alphabets, including capitals, lower case letters and numerals are made in sizes from

3/32 in. to 1 in. and in a wide range of styles. Special pens, made in different sizes, fit accurately into the guides. A good quality waterproof drawing ink should be used.

Procedure

1. Place two drops of ink in the pen reservoir. Agitate the wire plunger three or four times to ensure that the ink is flowing freely to the point of the pen.

2. Place the stencil on the surface of the photomicrograph so that the required letter or numeral is in position. Stabilize the stencil with the left hand. Hold the pen upright, insert the point in the guide. Gently lower the pen to the surface of the print and quickly complete the figure.

3. Remove the pen from the guide and lift stencil from the print surface.

Notes

Black or white inks may be used. The pens and stencils must be cleaned after use. Small arrows are sometimes required for drawing attention to structures. These are not included in the stencil but are easily improvised by using a lower case I or the figure one for the stem and a lower case v for the point of the arrow (Fig. 52a).

Letraset type transfers

These are a more recent innovation. Two methods of transfer are available.

1. A dry transfer method (Fig. 52b).
2. A wet transfer method (Fig. 52c).

With both methods the type, consisting of either capitals, lower case letters or numerals, is transferred to the photomicrograph without the film or background support.

1. *Dry method*

The dry method is instantaneous, requires no equipment, and leaves no traces of adhesive.

The type is bonded on a thick plastic sheet coated with a wax-rubber compound adhesive, and for transfer this bond must be broken. This is done by lightly rubbing the back of the supporting plastic material with a smooth rounded instrument.

A protective backing paper, treated so that the letters will not adhere to it, is supplied with the sheet for protection. It should always be used to prevent the transfer of unwanted letters by placing between the working surface and the type sheet. The type is

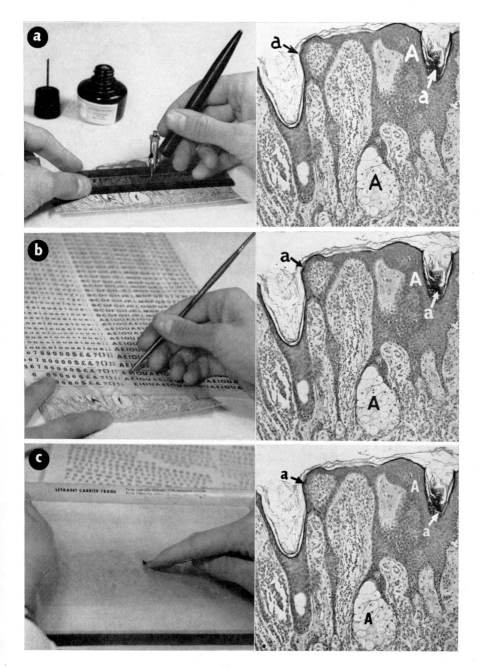

Fig. 52.

produced in many styles and different sizes in black or white and the letters are sharply defined.

Procedure

1. Place the photomicrograph face upwards on a flat surface.
2. Cover the photomicrograph with the protective backing paper, except for the area required for lettering.
3. Place the letter to be transferred in position and lightly rub with a smooth rounded instrument. Excessive pressure may fracture the letters.
4. Carefully lift the type sheet. The letter is transferred.
5. Replace the backing sheet on to the print and rub again to ensure perfect adhesion of the letter.
6. Place the protective backing paper in contact with the adhesive side of the type sheet and store away from dust.

2. *Wet method*

The wet method requires a special frame covered with silk and a sharp knife. The type is printed on a sheet consisting of two layers, a supporting layer and thin upper layer carrying the letters.

Procedure

1. Damp the silk on the back of the frame with cotton wool moistened with water.
2. With a sharp knife score through the gummed tissue around the letter selected.
3. Insert the point of the knife beneath one corner of the tissue and peel from the paper support.
4. Place the letter face downwards on the damp silk. Moisten the tissue on the back of the letter and leave for 1 minute.
5. Place the photomicrograph face upwards on a flat surface.
6. Slide gummed tissue away from letter.
7. Turn the frame over and remove excess water from the upper surface with a soft dry cloth.
8. Hold the frame in the left hand and lower to within an eighth of an inch of the print.
9. Position the required letter over the area selected for transfer and press the letter into contact with the right forefinger.
10. Lift the frame away and gently dry the transferred letter with a soft dry cloth.

FILING AND STORAGE OF NEGATIVES
AND PRINTS

The time and cost involved in the preparation of negatives demands that they should be protected from damage, readily available, and easily identified. It is therefore essential to have an efficient system for filing and storing negatives. It is not possible to be dogmatic about the best system but the following method, employed for a number of years, has proved to be effective and reliable. 5-3 in. cards, printed with the following headings: Photomicrograph No., Preparation No., Date, Staining Method, Magnification, Structure or Lesion recorded, Sensitive Materials, Processing, Printing, Filter No., are kept, numbered in sequence, ready for use in the dark room. The details are inserted at the completion of each stage. Filing the cards in numerical sequence may be suitable but it is usually more convenient to index them under headings such as specific structures, diseases or experiments.

Each negative should be marked with a reference number before development. After development, as soon as the negative is dry, this number should be clearly reinscribed in waterproof white ink in one corner. The negative should then be placed immediately in a transparent envelope to protect it from dust, finger-marks and water splashes.

The top right- or left-hand corner of the envelope should be printed with a rubber stamp containing headings for the photomicrograph number and the date. This will facilitate tracing a particular negative when the occasion arises and will also eliminate the need for its removal from the envelope.

Negatives can be conveniently stored in metal cabinets and they should be filed in numerical sequences for easy reference. Two prints are prepared from each negative. One is filed with the negative with the image side facing outwards for easy recognition and the other may be mounted on the reverse side of the card containing the technical data.

PREPARATION OF DEVELOPERS

To prepare the following developing solutions the chemicals must be (*a*) dissolved in the order listed and (*b*) dissolved in approximately three quarters of the total volume of water, which may be warm but not hot. Each chemical must be completely dissolved before adding the next. The total volume is made up by adding cold water.

KODAK DEVELOPERS

D.8 *Stock solution*

Sodium sulphite (anhyd.)	90·0 gm.
Hydroquinone	45·0 gm.
Sodium Hydroxide	37·5 gm.
Potassium bromide	30·0 gm.
Water to	1000 ml.

D.11 *Stock solution*

Elon	1·0 gm.
Sodium sulphite (anhyd.)	75·0 gm.
Hydroquinone	9·0 gm.
Sodium carbonate (anhyd.)	25·5 gm.
Potassium bromide	5·0 gm.
Water to	1000 ml.

D.61a *Stock solution*

Elon	3·1 gm.
Sodium sulphite (anhyd.)	90·0 gm.
Sodium metabisulphite	2·1 gm.
Hydroquinone	5·9 gm.
Sodium carbonate (anhyd.)	11·5 gm.
Potassium bromide	1·7 gm.
Water to	1000 ml.

D.76 *Stock solution*

Elon	2·0 gm.
Sodium sulphite (anhyd.)	100·0 gm.
Hydroquinone	5·0 gm.
Borax	2·0 gm.
Water to	1000 ml.

D.163 *Stock solution*

Elon	2·2 gm.
Sodium sulphite (anhyd.)	75·0 gm.
Hydroquinone	17·0 gm.
Sodium carbonate (anhyd.)	65·0 gm.
Potassium bromide	2·8 gm.
Water to	1000 ml.

D.165 *Stock solution*

Elon	6·0 gm.
Sodium sulphite (anhyd.)	25·0 gm.
Sodium carbonate (anhyd.)	37·0 gm.
Potassium bromide	1·0 gm.
Water to	1000 ml.

Microdol. Available in packings sufficient for 600 ml., 1 litre, 10 litres.

ILFORD DEVELOPERS

ID2. *Stock solution*

Metol	2·0 gm.
Sodium sulphite (anhyd.)	75·0 gm.
Hydroquinone	8·0 gm.
Sodium carbonate (anhyd.)	37·5 gm.
Potassium bromide	2·0 gm.
Water to	1000 ml.

ID36. *Stock solution*

Metol	2·0 gm.
Sodium sulphite (anhyd.)	50·0 gm.
Hydroquinone	12·5 gm.
Sodium carbonate (anhyd.)	72·0 gm.
Potassium bromide	0·75 gm.
Water to	1000 ml.

ID20. *Stock solution*

Metol	3·0 gm.
Sodium sulphite (anhyd.)	50·0 gm.
Hydroquinone	12·0 gm.
Sodium carbonate (anhyd.)	60·0 gm.
Potassium bromide	4·0 gm.
Water to	1000 ml.

ID11. *Stock solution*

Metol	2·0 gm.
Sodium sulphite (anhyd.)	100·0 gm.
Hydroquinone	5·0 gm.
Borax	2·0 gm.
Water to	1000 ml.

ID48. Available in packings sufficient for 570 ml., 2·25 litres, 4·5 litres, 13·5 litres.

Microphen. Available in packings sufficient for 600 ml., 2·5 litres, 13·5 litres.

TECHNIQUE NO. 70

PREPARATION OF FIXERS, HARDENERS AND STOP BATHS

IF2. *Acid fixing bath*

Sodium Thiosulphate (cryst.)	200 gm.
Potassium metabisulphite	12·5 gm.
Water to	1000 ml.

IF13. *Acid hardener fixing bath*
Stock hardening solution:

(*a*) Warm water	200 ml.
Add sodium sulphite (anhyd.)	50 gm.
Cool, stirring constantly	
Slowly add acetic acid, glacial	75 ml.

115

(*b*) Hot water   600 ml.
    Add potassium alum (cryst.)   100 gm.
    When cool (65° F.) add (*a*)
    Make up to 1000 ml. with cold water.

Working solution:
Warm water   500 ml.
Add sodium thiosulphate (cryst.)   200 gm.
Cool, add stock hardening solution   125 ml.
Cold water to   1000 ml.

IF15. *Acid hardener fixing bath*

(*a*) Warm water   500 ml.
    Add sodium thiosulphate (cryst.)   320 gm.
    Cool
    Add sodium sulphite (anhyd.)   30 gm.

(*b*) Warm water   150 ml.
    Add Boric acid (cryst.)   10 gm.
    Acetic acid, glacial   18 ml.
    Potassium alum (cryst.)   25 gm.
    Cool to 65° F.
    Slowly add the thiosulphate-sulphite
      solution (*a*)
    Cold water to   1000 ml.

F5. *Acid hardener fixing bath*

(*a*) Warm water   500 ml.
    Add Sodium thiosulphate (cryst).   240 gm.
    Cool
    Add sodium sulphite (anhyd.)   15 gm.

(*b*) Warm water   150 ml.
    Add acetic acid (glacial)   17 ml.
    Boric acid (cryst.)   7·5 gm.
    Potassium alum   15 gm.
    Cool to 65° F. Slowly add the thiosulphate-
      sulphite solution (*a*)
    Cold water to   1000 ml.

F7. *Rapid fixing bath*

Warm water (125° F.)	600 ml.
Sodium thiosulphate (cryst.)	360 gm.
Ammonium chloride	50 gm.
Sodium sulphate (anhyd.)	15 gm.
Boric acid (cryst.)	7·5 gm.
Acetic acid (glacial)	17 ml.
Potassium alum	15 gm.
Water to	1000 ml.

F24. *Acid fixing bath*

Warm water (125° F.)	500 ml.
Sodium thiosulphate (cryst.)	240 gm.
Sodium sulphite (anhyd.)	10 gm.
Sodium bisulphite	25 gm.
Water	1000 ml.

IH4. *Formalin hardener*

Formaldehyde (40 per cent solution)	10 ml.
Sodium carbonate (anhyd.)	5 gm.
Water to	1000 ml.

IH6. *Alum hardener*

Chrome alum	30 gm.
Water to	1000 ml.

SH1. *Formalin hardener*

Formalin (40 per cent Formaldehyde solution)	10 ml.
Sodium carbonate (anhyd.)	5 gm.
Water to	1000 ml.

SB1. *stop bath*

Water	1000 ml.
Acetic acid (glacial)	17 ml.

SB3. *Hardener stop bath*

Water	1000 ml.
Potassium chrome alum	30 gm.

IS1. *Stop Bath*

Acetic acid, glacial	17 ml.
Water to	1000 ml.

REVERSAL PROCESSING
PREPARATION OF SOLUTIONS

Developers

D8. (modified)

Water	750 ml.
Sodium sulphite (anhyd.)	90 gm.
Hydroquinone	45 gm.
Sodium hydroxide	37·5 gm.
Potassium bromide	30 gm.
Potassium thiocyanate	4 gm.
Water to	1000 ml.

D158.

Water	750 ml.
Elon	3·2 gm.
Sodium sulphite (anhyd.)	50·0 gm.
Hydroquinone	13·3 gm.
Sodium carbonate (anhyd.)	69·0 gm.
Potassium bromide	0·9 gm.
Water to	1000 ml.

D19b.

Water	750 ml.
Elon	2·2 gm.
Sodium sulphite (anhyd.)	72·0 gm.
Hydroquinone	8·8 gm.
Sodium carbonate (anhyd.)	48·0 gm.
Potassium bromide	4·0 gm.
Potassium thiocyanate	2·0 gm.
Water to	1000 ml.

Note

For second development use this solution, omitting potassium thiocyanate.

Ilford Reversal Developer

ID36. with 8 gm. hypo crystals added to each 1000 ml. of working strength developer.

Bleaching and clearing baths

R21.A

Water	1000 ml.
Potassium bichromate	50 gm.
Sulphuric acid (conc.)	50 ml.

Dissolve the bichromate in cold water. Add the sulphuric acid slowly and stir constantly.

For use—dilute one part of solution to 9 parts water.

R21.B

Water	1000 ml.
Sodium sulphite (anhyd.)	50 gm.
Sodium hydroxide	1 gm.

Use without dilution.

Ilford Bleaching Solution

A.	Potassium permanganate	2 gm.
	Water	500 ml.
B.	Water	500 ml.
	Sulphuric acid (conc.)	10 ml.

For use—equal parts of A and B freshly mixed for each film.

Ilford Clearing Solution

Sodium (or potassium) metabisulphite	25 gm.
Water	1000 ml.

Acid hardening-fixing bath

F54a

Warm water	500 ml.
Sodium thiosulphate	400 gm.
When cold add	
Kodak liquid hardener	75 ml.
Water to	1000 ml.

IF13

Technique 70

PREPARATION OF MOUNTING PASTES

Starch

To approximately 10 ml. cold water add 30 gm. starch. Mix until a very stiff paste results. Add 350 ml. *boiling* water and stir until the mixture becomes partially transparent. Cool and remove the formed skin from the surface. The paste is ready for use.

Dextrine

To 500 ml. water, heated to 160° F. in a water bath, slowly add 345 gm. best white dextrine. Stir until dissolved. Add 0·25 ml. methyl salicylate and 0·25 ml. oil of cloves. Mix well. Cool and transfer to a stoppered wide-mouthed container. Store in a cool place until the mixture has formed into a paste.

Gelatin

To 100 ml. warm water in a beaker, kept warm in a water bath, gradually add 25 gm. of gelatin, stirring constantly. When the gelatin has dissolved add 30 ml. methylated spirit and 6 ml. glycerine. Store in a glass container. For use immerse the container in a dish of hot water to melt the gelatin.

REPARATION OF SOLUTIONS FOR PROCESSING FERRANIACOLOR 35 MM. FILM

JOHNSON'S PROCESSING KIT

The kit consists of two parts.
Part 1. First developer and colour developer.
Part 2. Hardener, bleach-hardener, and fixer.

Preparing solutions

1. Dissolve the chemicals in cold or warm tap water. Do not use hot water.

2. Do not use vessels made of iron, zinc, galvanized iron, or aluminium.

3. The *First Developer* consists of two packets, labelled A and B. Dissolve the contents of B in approximately 500 ml. cold water and then add the contents of packet A. When dissolved make up to a total volume of 600 ml. with cold water.

4. The *Colour Developer* consists of three packets, labelled C, D and E. Prepare two solutions.

Solution (i). Dissolve the contents of packet E in 250 ml. cold water.

Solution (ii). Dissolve the contents of C and D in 250 ml. cold water.

Gradually add Solution (ii) to Solution (i) stirring constantly. Make up the total volume to 600 ml. with cold water.

Hardener. Dissolve the contents of the packet in 500 ml. of water and make up to a total volume of 600 ml. Slowly add the chemicals and stir continuously.

Bleach-hardener. Dissolve A, and then B, in about 500 ml. water. Make up to 600 ml.

Fixer. Dissolve the contents of the packet in 500 ml. water and make up to 600 ml. Slowly add the chemicals and stir continuously.

Allow all solutions to stand for half an hour and filter before use. After filtration, store solutions in tightly stoppered bottles. Label and keep in subdued light.

The first and colour developers do not keep and should not be stored for more than two days after use.

Solutions which have not been used will keep for about a week if stored in full stoppered bottles.

The hardener, bleach-hardener and fixer, even when used, will remain in good condition for about 6 weeks. Filter each time they are used. Two 36 exposure and three 20 exposure films may be processed in 600 cc. of the developers and three 20 exposure 35 mm. films may be taken through the hardener, bleach-hardener and fixer.

PREPARATION OF SOLUTIONS FOR PROCESSING FERRANIACOLOR 35 MM. FILM

MAY AND BAKER

1. *First developer*

Water	750 ml.
Sodium polymetaphosphate	2 gm.
Sodium sulphite (anhyd.)	50 gm.
Amidol	5 gm.
Potassium Bromide	1 gm.
Water to	1000 ml.

2. *Colour developer*

(a)
Water	350 ml.
Genochrome	2·8 gm.
Hydroxylaminehydrochloride	1 gm.
Water to	500 ml.

(b)
Water	350 ml.
Sodium polymetaphosphate	2 gm.
Sodium carbonate (anhyd.)	65 gm.
Sodium sulphite (anhyd.)	2·5 gm.
Potassium Bromide	1·2 gm.
Sodium hydroxide	1 gm.
Water to	500 ml.

For use slowly add (a) to (b) stirring continuously. Filter or decant.

3. *Bleach hardener*

Water	750 ml.
Potassium ferricyanide	50 gm.
Potassium bromide	25 gm.
Sodium acetate (cryst.)	60 gm.
Boric acid	5 gm.
Potassium alum	30 gm.
Water to	1000 ml.

Keep the solution away from light.

4. *Fixer*

Water	750 ml.
Sodium Thiosulphate (cryst.)	200 gm.
Water to	1000 ml.

Preparation of solutions

1. Use water at a temperature not exceeding 30° C. (86° F.).

2. If the formula includes polymetaphosphate dissolve this in three-quarters of the total volume of water required. Add the remaining chemicals to this solution. When the chemicals are completely dissolved, make up to the total volume with water.

3. Prepare all solutions 12 hours before use.

Storage of solutions

The first developer will keep for 3-4 days if stored away from light and in an air-tight container.

The colour developer, if stored under similar conditions, will keep for approximately 12 days.

Capacity of solutions

1000 ml. of either developer will process 5 twenty-exposure 35 mm. films or 3 No. 120 roll films.

1000 ml. of bleach-hardener and fixer will process 20 twenty-exposure 35 mm. films or 12 No. 120 roll films.

TECHNIQUE NO. 75

PREPARATIONS OF SOLUTIONS FOR PROCESSING
EKTACHROME SHEET FILM TYPE B

PROCESS E.3

Packed chemicals are available to produce working solutions of 2 litres and 13·5 litres.

Preparation of solutions.   2-litre kit.

1. Under no circumstances must the quantity of solution be reduced.

2. Do not use containers made of iron, zinc, galvanised iron or aluminium.

3. The first developer consists of three components marked A, B and C.

      (i) To 2000 ml. of water at a temperature of 90° F. add the contents of A. Stir until completely dissolved.

      (ii) To this add the contents of B. Stir until dissolved.

(iii) Add C. Make sure that the entire contents are transferred by rinsing the container with the solution. Stir well to mix.

4. The *colour developer* consists of three components marked A, B and C.

(i) To 2000 ml. of water at a temperature of 70-80° F. add the contents of A. Ensure complete transference by rinsing the container with the solution. Stir until the solution is clear.

(ii) Add the contents of B. Stir until dissolved.

(iii) Add C. Stir until dissolved.

5. *Hardening solution.* To 2000 ml. of water at a temperature not exceeding 90° F. add the hardener. Stir until completely dissolved.

6. *Clearing solution*

(i) Add the reagent slowly to 1500 ml. of water at a temperature not exceeding 80° F. and stir continuously until dissolved.

(ii) Make up to 2000 ml. with cold water and stir to mix.

7. The *bleaching solution* consists of two components marked A and B.

(i) To 1500 ml. of water at a temperature not exceeding 90° F. add the contents of A. Stir until dissolved.

(ii) Add the contents of B. Stir until dissolved.

(iii) Make up to 2000 ml. with cold water. Stir to mix.

8. *Fixing solution*

(i) Slowly add the fixing reagent to 1500 ml. of water at a temperature not exceeding 80° F. and stir constantly until dissolved.

(ii) Make up to 2000 ml. with cold water. Stir to mix.

9. *Stabilizing solution.* To 2000 ml. of water at a temperature of 70-80° F. add the reagent and stir until dissolved.

Storage life of solutions

	Full stoppered bottles	*Tank with floating lid*
First developer		
Unused	14 days	7 days
Partially used	7 days	7 days
Colour developer		
Unused	14 days	7 days
Partially used	7 days	7 days
Other solutions		
Unused or partially used	56 days	56 days

Capacities of processing solutions

This factor need only be considered if the storage periods have not been exceeded.

Volume of solution	First and colour developer	Other solutions
2 litres (2000 ml.)	4 sq. ft. of film	9 sq. ft. of film
13·5 litres (13,500 ml.)	28 sq. ft. of film	60 sq. ft. of film

TECHNIQUE NO. 76

PREPARATION OF SOLUTIONS FOR PROCESSING EKTACHROME 35 MM. FILM

PROCESS E2 AND E2 IMPROVED TYPE

To make 600 ml. of working solution two separate kits are required. One contains the reagents for the preparation of the first and colour developer, and the other the hardener, clearing bath, bleach, fixing bath and stabilizer. Smaller quantities must not be used and the total volume must not be increased. Solutions must not be prepared or stored in tin, copper, aluminium or galvanised vessels.

Preparation of solution

1. The *first developer* consists of three components A, B and C.
    (i) To 600 ml. of water at a temperature of 90° F. add the contents of A. Stir until dissolved.
    (ii) Add B. Stir until dissolved.
    (iii) Add C. Make sure that the entire contents are transferred by rinsing the container with the solution. Stir until dissolved.
2. The *colour developer* consists of three components A, B and C.
    (i) To 600 ml. of water at a temperature of 70-80° F. add the contents of A. Ensure complete transference by rinsing the container with the solution. Stir until the solution is clear.
    (ii) Add B. Stir until dissolved.
    (iii) Add C. Stir until dissolved.

125

3. *Hardening solution.* To 600 ml. of water at a temperature not exceeding 90° F. add the reagents. Stir until dissolved.

4. *Clearing solution*

    (i) To 450 ml. of water at a temperature not exceeding 80° F. slowly add the reagents. Stir until dissolved.

    (ii) Make up to 600 ml. with cold water. Stir to mix.

5. The *bleaching solution* consists of two components A and B.

    (i) To 450 ml. of water at a temperature not exceeding 90° F. add the contents of A. Stir until dissolved.

    (ii) Add the contents of B. Stir until dissolved.

    (iii) Make up to 600 ml. with cold water. Stir to mix.

6. The *fixing bath* consists of two components A and B.

    (i) To 450 ml. of water not exceeding 80° F. slowly add the reagents. Stir until dissolved.

    (ii) Make up to 600 ml. with cold water. Stir to mix.

7. *Stabilizing solution.* To 600 ml. of water at a temperature of 70-80° F. add the reagents. Stir until dissolved.

Storage life of solutions

Do not keep used solutions or unused solutions for longer than the following periods:

First and colour developers   2 weeks in full stoppered bottles

Other solutions   8 weeks in full stoppered bottles

All solutions should be stored in stock bottles when not in use.

Capacity of solutions

600 ml. will satisfactorily process 2·68 sq. ft. of film, approximately equal to 8 rolls of 20 exposure No. 135 film.

Area of 10 sheets of film.

$$2\tfrac{1}{4} \times 3\tfrac{1}{4} \text{ in.} = 0 \cdot 5 \text{ sq. ft.}$$
$$2\tfrac{1}{2} \times 3\tfrac{1}{2} \text{ in.} = 0 \cdot 6 \text{ sq. ft.}$$
$$3\tfrac{1}{4} \times 4\tfrac{1}{4} \text{ in.} = 1 \cdot 0 \text{ sq. ft.}$$
$$4 \times 5 \text{ in.} = 1 \cdot 4 \text{ sq. ft.}$$
$$4\tfrac{3}{4} \times 6\tfrac{1}{2} \text{ in.} = 2 \cdot 1 \text{ sq. ft.}$$
$$6\tfrac{1}{2} \times 8\tfrac{1}{2} \text{ in.} = 3 \cdot 8 \text{ sq. ft.}$$
$$8 \times 10 \text{ in.} = 5 \cdot 5 \text{ sq. ft.}$$

PREPARATION OF SOLUTIONS FOR PROCESSING EKTACOLOR SHEET FILM TYPE L

PROCESS C-22

Processing chemicals to make 13·5 litre or 35 litre solutions are available. Smaller quantities must not be prepared from either of these kits, nor must the total volume be exceeded. Aluminium, tin, copper or galvanised vessels must not be used for mixing or storage of the solutions.

Preparations of solutions. 13·5 litre kit

1. The *developer* consists of three components marked A, B and C.
    (i) To 13,500 ml. of water at a temperature of 70-80° F. add A. Make sure that the entire contents are transferred by rinsing the container with the solution. Stir until dissolved.
    (ii) Add B. Stir until dissolved.
    (iii) Add C. Stir until dissolved.
2. The *stop bath* consists of two components marked A and B.
    (i) To 13,500 ml. of water at a temperature of 70-80° F. add the contents of A. Stir until dissolved.
    (ii) Add the contents of B. Ensure complete transference by rinsing the container with the solution. Stir until mixed.
3. The *hardening solution* consists of two components marked A and B.
    (i) To 13,500 ml. of water at a temperature of 70-80° F. add A. Stir until dissolved.
    (ii) Add the contents of B. Stir until dissolved.
4. The *bleaching solution* consists of two components marked A and B.
    (i) To 13,500 ml. of water at a temperature of 70-80° F. add the contents of A. Stir until dissolved.
    (ii) Add B. Stir until dissolved.

    (The bleach has a corrosive action on metal.)

5. *Fixing solution.* To 13,500 ml. of water at a temperature of 70-80° F. add the reagents. Stir until dissolved.

Capacity of solutions

13·5 litres of the developer and stop bath will process 150 units of Ektacolour film. The hardening, bleach, and fixing solutions will process 300 units.

The table below provides a guide to the determination of the unit value.

Single sheets of film of the following sizes are equal to:

Units	Film Size in inches
$\frac{1}{3}$	$2\frac{1}{4} \times 3\frac{1}{4}$
$\frac{1}{2}$	$3\frac{1}{4} \times 4\frac{1}{4}$
$\frac{3}{4}$	4×5
1	$4\frac{3}{4} \times 6\frac{1}{2}$
3	8×10
$5\frac{1}{2}$	1 sq. ft. of film.

A development-time compensation is necessary if the capacities given above are accepted and replenishing solutions are not employed. The development time must be increased by the following amounts for each successive processing of 50 units.

1st processing	14 minutes
2nd ,,	14 minutes 48 seconds
3rd ,,	15 minutes 30 seconds
4th ,,	16 minutes 15 seconds
5th ,,	17 minutes—use fresh developer and stop bath.

All other solutions have double this capacity.

Storage of solution

Developer	Full stoppered bottles	Tank with floating lid
Unused	6 weeks	3 weeks
Partially used	4 weeks	2 weeks
Other solutions		
Unused or partially used	8 weeks	8 weeks

Contamination of one solution with another must be avoided and the same tanks must be used for the same solutions on each occasion. Tanks should be thoroughly washed before they are refilled with fresh solutions.

PREPARATION OF SOLUTIONS FOR PROCESSING EKTACOLOR PAPER

PROCESS P-122

Ektacolor paper in sheet form can be processed in a dish. The processing chemicals may be obtained in the form of a kit or separately in quantities sufficient to make 4·5 litres of each working solution.

Preparation of solutions. 4·5-litre kit

1. The *developer* consists of four components marked A, B, C1 and C2.

   (i) To 4590 ml. of water at a temperature of 70-80° F., add the contents of A. Ensure complete transference by rinsing the container with the solution; stir until clear.

   (ii) Add the contents of B. Stir until dissolved.

   (iii) Add C1 and C2. Stir until dissolved. The solution may appear cloudy but this is of no consequence.

2. The *stop bath* consists of two components A and B.

   (i) To 4500 ml. of water at a temperature not exceeding 80° F. add the contents of A and B. Stir until dissolved.

3. *Hardener fixer.* Sufficient reagent is supplied in separate containers to provide two 4·5 litre quantities of this solution as it is used twice during processing.

   (i) To 4500 ml. of water at a temperature not exceeding 80° F. slowly add the contents of one bag. Stir until dissolved.

4. The *bleach* consists of two components marked A and B.

   (i) To 4500 ml. of water at a temperature not exceeding 80° F. add the contents of A. Stir until dissolved.

   (ii) Add the contents of B. Stir until dissolved. This is a corrosive solution and will react with metals.

5. The *hardener* consists of two components A and B.

   (i) To 4500 ml. of water at a temperature not exceeding 80° F. add the contents of A.

   (ii) Add B and stir until dissolved.

6. *Buffer*

   (i) To 4500 ml. of water at a temperature not exceeding 80° F. add the reagent. Stir until dissolved.

Capacity of solutions

1. When processing three 8×10 in. sheets of paper use 1 litre of each solution.
2. Use volumes proportional to the area for papers of other sizes.
3. Use the developer once only, then discard.
4. Use other solutions to their exhaustion point (six 8×10 in. sheets per litre).

Storage life of solutions

Developer	Full stoppered bottles
Unused	6 weeks

Other solutions	
Unused and partially used	8 weeks.

Print quality and the life of solutions depend on their purity. The contamination of one solution with another must be avoided and the same dish should be used for the same solution on all occasions. Dishes and storage bottles must be thoroughly washed prior to refilling.

TECHNIQUE NO. 79

CONVERSION TABLES

1. *To convert Fahrenheit into Centigrade.*

   (i) The temperature Fahrenheit $-32 \times 5/9$ = temperature Centigrade.

   (ii) The temperature Fahrenheit $+40 \times 5/9 - 40$ = temperature Centigrade.

2. *To substitute the anhydrous for the crystalline form.*

Sodium sulphite	divide by 2
Sodium carbonate	multiply by 0·37
Hypo	multiply by 0·625

To substitute the crystalline for the anhydrous form.

Sodium sulphite	multiply by 2
Sodium carbonate	multiply by 2·7
Hypo	multiply by 1·6

3. *Plate and sheet sizes*

Inches	Centimetres
$2\frac{1}{2} \times 3\frac{1}{2}$	$6·4 \times 8·9$
$3\frac{1}{4} \times 3\frac{1}{4}$	$8·25 \times 8·25$
$3\frac{1}{4} \times 4\frac{1}{4}$	$8·25 \times 10·8$
4×5	$10·1 \times 12·7$
$4\frac{3}{4} \times 6\frac{1}{2}$	$12·0 \times 16·5$
5×7	$12·7 \times 17·8$
$6\frac{1}{2} \times 8\frac{1}{2}$	$16·5 \times 21·5$
8×10	$20·3 \times 25·4$

4. 1 gram $= 15·4324$ grains
    28·35 grams $= 1$ ounce
    0·4536 kilograms $= 1$ pound
    1 kilogram $= 35·274$ ounces
    or 2·2046 pounds.

5. 28·42 ml. $= 1$ fluid ounce
    1 litre $= 35$ fluid ounces or
    1·7598 pints.

6. 1 micron $= 0·001$ mm.
    2·54 centimetres $= 1$ inch.

7. Formulae given in parts may be prepared by substituting grams for the solids and ml. for the fluids.

8. To dilute stock solution by parts all quantities should be calculated to the same units of volume, for example,

Stock solution 10 parts $= 10$ ml.
Water solution 40 parts $= 40$ ml.

SECTION III

GLOSSARY

Aberration: Defects in the performance of a lens, for example, spherical aberration; chromatic aberration; as a result of which an imperfect image is produced.

Achromatic lens: A lens designed to bring light of all colours to the same focus.

Acid stop bath: Used between development and fixation. It removes not only the developer on the surface of the film, but also neutralizes developer carried over in the emulsion layer. 1% acetic acid is commonly used.

Agitation: By agitation is meant the disturbance of a fluid reagent in contact with an emulsion. Essential for even development. Reduces the time necessary for the reagent to act upon the sensitive material.

Annular diaphragm: A diaphragm which, when inserted into the optical path, confines the light entering the substage condenser to a hollow cone. Such illumination is necessary for phase contrast microscopy.

Analyser: Of the two polarizers used for microscopy by polarized light, that which is rotated is the analyser.

Aperture: The size of the opening of the lens diaphragm. Usually expressed as a ratio of the focal length of the lens. See " f " number.

Aperture scale: A scale indicating the diameter of the aperture.

Artefact: Any apparent structure which is not part of the specimen. Usually due to faulty preparation.

ASA: A measurement of emulsion speed.

Attachment camera: A camera body, incorporating a shutter, designed to fit over the eyepiece of a microscope and attached to the microscope tube.

Axis: A theoretical line passing through the centre of a lens system from front to rear. Indicated in optical diagrams by an elongated arrow.

Back focus: The distance from the back glass of a lens to the point where incoming rays parallel to the lens axis are brought to focus.

Back lens of the objective: The rear component of a compound objective lens.

Beam splitting device: A device for altering the course of the light path.

Bellows camera: A camera incorporating an adjustable bellows which allows the lens-emulsion distance to be varied.

Binocular: Any optical instrument designed for the simultaneous use of both eyes. A microscope with two eyepieces.

Birefringent: The double bending of light by crystalline minerals. The difference between the greatest and the least refractive index for light passing through a mineral is a measure of its birefringence.

133

Bleach hardener: A solution removing the negative image and simultaneously hardening the gelatine.

Bright field: A term used to denote the illumination of an object so that the preparation appears as a dark object on a bright background.

Bromide paper: Photographic paper containing silver bromide.

BSI: A measurement of emulsion speed.

Calibration: The graduation of an instrument with allowance for its irregularities.

Camera: A light proof box with a lens at one end. The lens, when opened by means of a shutter, and properly focused, throws an image of the object at which the lens is directed upon a plate or film sensitive to light placed at the back of the box. In photomicrography the eyepiece lens of the microscope usually serves as the camera lens.

Camera extension: Denotes the bellows length of a bellows type camera.

Carbon arc: The illumination provided by the electrical discharge between two carbon arcs connected to a D.C. supply. Such illumination is of high intensity.

Cassette: Light-proof plate or film holder.

Colour temperature: The temperature to which a black-body radiator (a body that absorbs all radiation falling on it) must be raised so that its visual colour matches that of a particular light source. Usually expressed in degrees Kelvin (°K).

Colour transparency: A transparent, positive image in colour.

Compound lens: A lens consisting of two or more pieces of glass.

Compound microscope: An instrument consisting essentially of two converging lenses called the objective and the eyepiece respectively. The objective, which is nearest the viewed object, forms a real inverted magnified image of the object just inside the focal distance of the eyepiece. This image is then viewed and enlarged through the eyepiece.

Condenser: A lens or lens system principally concerned with the even distribution of light and not with the quality of the image. In microscopy used to converge light on the preparation.

Contact print: A positive print the same size as the negative, the sensitised paper being held in contact with the negative during exposure.

Contrast: Generally taken to mean the density range of a photographic image.

Coverglass: In microscopy: a thin piece of glass mounted over the preparation.

    In photography: a piece of glass used for protecting a monochrome or colour transparency.

Critical illumination: Illumination of uniform intensity, symmetrically disposed about the optical axis of the microscope, of the same area as the back lens of the objective, and covering the entire field.

Cut film: Same as sheet film.

Daguerrotype: Early photographic process, in which the impression was taken upon a silver plate sensitized with iodine, and then developed by mercury vapour.

Dark contrast: Phase contrast microscopy using an annulus and a phase diffraction plate.

Dark current: The current produced by a photo-electric cell in total darkness. This is caused by insulation resistance and must be compensated for.

Dark field: A method of illumination whereby transparent or unstained objects appear as bright particles on a black background.

Dark shield: A component of a plate holder which when withdrawn exposes the plate or film to light.

Definition: In optics: the quality of an image produced by a lens.

In photography: the quality of an image recorded on a photographic emulsion.

Dehydration: The removal of water.

Density: A blackening produced by a deposit of silver. The light-absorbing power of a photographic image. The quantity of silver deposited in a given area.

Depth of field: The distance between near and distant objects which are simultaneously in reasonably sharp focus.

Depth of focus: That region on either side of the exact focus where the quality and sharpness of the image is up to a required standard.

Detail: The minute or subordinate portion of any whole.

Developer: The solution converting exposed silver halide to metallic silver.

Development: Chemical process of converting exposed silver halide to metallic silver.

Diaphragm: A partition in front of or behind a single lens or in between the components of a multiple lens, usually having an adjustable aperture.

Diffracted light: An optical term used to denote the bending of light rays by the edges of an opaque body.

Diffraction: When a beam of light passes through an aperture or past the edge of an opaque obstacle and is allowed to fall upon a screen, patterns of light and dark bands (with monochromatic light) or coloured bands (with white light) are seen near the edges of the beam and extend into a geometrical shadow. This phenomenon, which is a particular case of interference, is due to the wave nature of light, and is known as diffraction.

Diffuser: A piece of ground glass used for scattering transmitted light.

DIN: A measurement of emulsion speed.

Draw tube: A tube, incorporating a graduated scale, which fits into the body of a monocular microscope and which supports the eyepiece.

Dye transfer process: The production of colour prints on paper from three colour separation negatives by the imbibition method of colour printing.

Elon: Trade mark for monomethyl para-aminophenol sulphate. A developing agent.

Empty magnification: The magnification of an image without revealing further detail.

T

Emulsion: A gelatin solution or coating containing, in colloidal suspension, light-sensitive silver halides.

Enlarge: The method of increasing the size of a photographic image.

Enlarger: Apparatus designed to produce a larger image from a negative.

Etched surface: The resulting surface of a hard substance after treatment with acid.

Exhaustion point: A term applied to solutions to indicate the stage at which the active constituents no longer function.

Exposure: Exposure proper is the product of exposure time and light intensity. The term is popularly, but incorrectly, used to mean exposure time.

Exposure meter: An instrument which measures the intensity of light.

Exposure table: A prepared table of exposure times for differing objective-bellows-eyepiece combinations.

Exposure slide: A special plate or film holder for making test exposures.

Exposure time: The length of time the sensitive material is exposed to a light source.

Extension tube: A tube used to increase the lens-emulsion distance (camera) or lens-negative distance (enlarger).

Eyepiece: The lens or system of lenses nearest the observer's eye; generally used to view the image formed by the objective.

" f " number: A number indicating the light-passing power of a lens, and obtained by dividing the focal length by the diameter of the beam of light; the smaller the " f " number the more light is passed by the lens.

Field: The circle of light projected by a lens or system of lenses.

Field diaphragm: The iris diaphragm positioned in front of the light source.

Field marker: A device for marking a selected field.

Field size: The area of the field.

Filter: Transparent material used before, behind or between the components of the camera or microscope lens to alter the composition of light by selective absorption.

Filter factor: That number by which exposure time must be multiplied if a filter is used.

Fixation: Removal of undeveloped silver salts from the emulsion—usually by the action of sodium thiosulphate (hypo).

Fixer: A solution used to remove the unexposed silver halide from a negative.

Flare: Reflection of light from a polished surface in an optical system.

Fluorescence: The property whereby a substance is able to absorb light of one wavelength and to emit, in its place, light of another wavelength.

Focal distance: The distance between the centre of a lens and its focus.

Focal length: See focal distance.

Focal plane: A plane at right angles to the lens axis through the focal point.

Focal point: The position of the focus when the rays of light entering or leaving the lens are parallel to the lens axis.

Focus: The point where the rays of light radiating from a luminous point and bent by a lens meet again, or the position where they come to their most compact form as judged by the size of the illuminated area they produce on a screen placed there. Ideally this light patch is a point.

Focusing screen: A sheet of ground glass placed in the camera in the position of the sensitive plate or film so that by inspection of the image formed upon this focusing of the lens can be carried out. When the image is at its best on the focusing screen, it will also be so on the sensitive plate or film. The image is formed on the ground surface of the glass closest to the lens.

Fog: A visible silver deposit other than that of the desired image.

Fotonic paper: A fluffless blotting paper.

Glazer: An apparatus for glazing prints.

Glazing: The process of imparting a high gloss to the surface of prints.

Glazing solution: Used to treat prints before glazing.

Gloss: Superficial lustre.

Grain: Developed silver particles in the emulsion. The size of the particles and their degree of aggregation determines the granular appearance of the photographic image on enlargement.

Halation: Spreading of the image caused by light which has passed through the emulsion and then been reflected back by the support.

Half tone: Any tone between maximum and minimum density.

Hanger: A metal frame, usually made of stainless steel, to hold film rigid during processing.

Hard: Of high contrast. Harsh blacks against a light background.

High lights: Those areas of an object which reflect most light.

Horizontal unit: A unit where the illumination, microscope and camera are on the same horizontal axis.

Hypo: Sodium thiosulphate. Used for fixation. It reacts with unexposed silver bromide to give a soluble double salt, silver sodium thiosulphate, which is washed away.

Image: In optics: a simulation of an object. In photography: the collection of tones produced by silver deposits of various densities.

Incident illumination: A method of illumination whereby the light falls on to the surface of the object.

Infra-red: Electromagnetic waves possessing wavelengths between those of visible light and those of wireless waves.

Inspection method: A method of development. In this instance, the appearance and growth of the image is watched in a suitable safelight. Development is stopped when the negative is judged to have suitable densities.

Interference colours: Coloured bands of light produced by diffracted light.

Iris diaphragm: An adjustable diaphragm.

K° rating: The measure of colour temperature expressed in degrees Kelvin.

Lens: A term used to mean either a single piece of glass with polished faces, or a number of such pieces of glass mounted together, so that they are capable of bending light rays.

Light path: See optical path.

Limit of resolution: The limit of resolution is expressed as the smallest linear separation of two object points which can be distinguished as separate.

Magnification: The ratio of linear dimensions of the final image to the linear dimensions of the object.

Masking: The elimination of unwanted parts of a negative or print.

Masking frame: An adjustable frame used in projection printing to keep the paper flat and to produce straight edges.

Matt: A term applied to all dull surface prints as distinguished from those with a glossy or glazed surface.

Mechanical stage: A movable microscope stage permitting movement of the slide in the horizontal plane.

Mercury vapour lamp: A lamp emitting a strong bluish light by the passage of an electric current through mercury vapour in a bulb. The light is rich in ultra-violet radiations.

Metallurgical microscope: A microscope designed for observing the surface of metals.

Metol: Monomethyl para-aminophenol sulphate. A developing agent.

Micron: One millionth of a metre. Written as μ.

Monochrome: In microscopy: light of a single spectral colour. In photography: representation of the image in one colour.

Monochrome transparency: A black and white positive image printed on a transparent film or plate.

Monocular: A microscope with a single eyepiece.

Negative: The term applied to an image in which the blacks and the whites are reversed. Also used to denote exposed plates or film.

Negative colour film: Film designed to produce a negative in complementary colours from which a positive print or transparency may be prepared in colour or in black and white.

Neutral density filters: Filters of known optical density which are used to reduce illumination.

Numerical aperture: The numerical aperture (N.A.) equals $n \sin u$, where n is the index of refraction of the medium and u is one-half the angle of the cone of light entering the objective.

Objective lens: Lens or system of lenses nearest the preparation in a compound microscope.

Oblique illumination: A type of incident illumination whereby the preparation is illuminated from the side at an oblique angle.

Ocular: See eyepiece.

Oil immersion lens: Type of objective used for high power work. The lowest lens of the objective lens is immersed in cedarwood oil placed upon the coverglass. This allows more light to enter the lens system.

Optical axis: A line passing through the optical centre of a lens.

Optical bench: A base support designed to carry different optical fittings which, when fitted, lie in the same optical axis.

Optical density: If one medium has a greater refractive index than another for a given wavelength, then it has a greater optical density for that wavelength.

Optical path: The course of light through any optical system.

Orthochromatic: Emulsions sensitive to ultra-violet, blue, green and yellow radiation.

Panchromatic: Emulsions sensitive to ultra-violet, blue, green, yellow and red radiation.

Phase contrast: A method of microscopy which visualizes the differing refractive indices of a preparation.

Phase diffraction plate: A component of the phase contrast microscope situated at the rear focus of the objective.

Photo-electric cell: A cell which, when exposed to light, produces a potential difference proportional to the light intensity.

Photoflood: A tungsten filament lamp designed to give a very high luminous efficiency at the expense of working life. Termed an overrun lamp.

Plate: A term used to describe photographic emulsion supported on glass.

Plate holder: A term used for the holder of the sensitive plate or sheet film which is inserted in the back of the camera in place of the focusing screen. Also known as a dark slide.

Pointalite: A special form of tungsten lamp, consisting of an arc, one pole of which is a tungsten ball. The light given by the Pointalite takes the form of a small spot of intense uniform brightness.

Polarized light: Light waves vibrating in one plane only.

Polarizer: A plate or prism which polarizes light.

Positive: Image in which bright parts represent bright parts of the object. The image reproduced from a negative.

Positive transparency: The image reproduced from a negative on a plate or film.

Print: A positive image on paper.

Printing box: A self-contained unit for printing by contact.

Printing frame: A frame for holding the negative and the paper when printing by contact.

Prism: A transparent triangular body with the refracting surfaces at an acute angle to each other.

Processing: The treatment of sensitive material after exposure.

Quartz optics: Optical components mede of quartz instead of glass.

Rear focus: See back focus.

Reducing: Production of a print or copy smaller than the original negative.

Reflected light: Light reflected from a surface.

Reflex mirror: A movable mirror situated in the light path and used for deflecting the image.

Refractive index: A number denoting the extent to which glass bends light.

Resolving power: As applied to a lens: the limit of ability of a lens to distinguish two small points close together. As applied to emulsion: the ability of an emulsion to record fine detail. Depends on grain size, contrast and turbidity.

Reticulation: A net-like appearance which is generally caused by differences in temperature between the processing baths, the rinsing or washing water. Serious reticulation shows as a roughening of the surface of the emulsion.

Reversal: Initial development of negative image, removal of the metallic silver thus produced, followed by exposure to light and redevelopment of remaining silver halide to produce a positive image.

Reversal colour film: Film which after development and reversal of the negative gives a positive image in colour.

Rheostat: Variable electrical resistance.

Roll film: Lengths of celluloid coated with emulsion rolled on to a spool.

Safe light: A coloured medium, usually dyed gelatine, between glass sheets, which when substituted for, or covering, the window of a lamp, and used with a correct wattage bulb, will transmit light in which specific sensitive material may be handled for a limited period of time.

Selenium photo-cell: A photo-electric cell which contains selenium, a silver grey crystalline solid which varies in electrical resistance on exposure to light.

Sensitivity: Response to action of light; speed.

Serial sections: Consecutive, or consecutively spaced, sections of a tissue block.

Sheet film: Photographic emulsion supported on celluloid cut to various sizes.

Shutter: A device for regulating the duration of exposure time.

Sodium vapour lamp: A lamp emitting an intense yellow line spectrum of wavelength 589 °A by passage of an electrical current through sodium vapour in a bulb.

Soft: Lacking in contrast.

Spectrum: The band of colours into which light is split on passing through a prism. The visible spectrum covers the range from about 400 μ (blue end) to about 700 μ (red end).

Specular illumination: See vertical illumination.

Stage micrometer: A glass slide on which is etched a scale of known value.

Stop: See aperture.

Sub-stage diaphragm: The diaphragm incorporated in the condenser unit of the microscope.

Telescope eyepiece: A phase contrast accessory used for viewing the back lens of the objective.

Test exposure: Plate, film or paper. A series of predetermined exposures.

Time-temperature method: A method of development. The development time at a given temperature is calibrated to produce a particular degree of contrast.

Tone: The differing degrees of subject brightness or the light transmitting properties of a deposit of silver.

Transmitted illumination: In microscopy: that type of illumination in which the light passes through the preparation.

Trichrome: of three colours.

Tungsten ribbon filament lamp: A lamp producing light by passage of an electric current through a tungsten filament.

Ultra-violet light: Electromagnetic waves between visible light waves and X-rays. The longest ultra-violet waves have wavelengths just shorter than those of violet light, the shortest perceptible by the human eye.

Universal apparatus: Equipment in which the illumination, microscope and camera are combined in a single unit.

Uranium glass: Coated glass which fluoresces in ultra-violet radiation.

Vertical illumination: A type of incident illumination whereby the light passes through the back lens of the objective and is focused on the preparation. Also called specular illumination.

Vertical unit: Assembly of microscope and camera in the vertical plane.

Wetting agent: An agent decreasing surface tension.

Wheel stop: An opaque circular stop for insertion in a substage condenser to produce a hollow cone of light.

Working distance: The distance between the coverglass or surface of the preparation and the front lens of the objective.

INDEX

U

PRINTED IN GREAT BRITAIN BY
OLIVER AND BOYD LTD.
EDINBURGH

The Receptors

A COMPREHENSIVE TREATISE

Volume 1

General Principles
and Procedures

The Receptors
A COMPREHENSIVE TREATISE

Series Editor for Volume 1:
R. D. O'Brien, Cornell University

Series Editor for succeeding volumes:
Palmer Taylor, University of California, San Diego

Volume 1: General Principles and Procedures
Edited by R. D. O'Brien

A Continuation Order Plan is available for this series. A continuation order will bring delivery of each new volume immediately upon publication. Volumes are billed only upon actual shipment. For further information please contact the publisher.

The Receptors™

A COMPREHENSIVE TREATISE

Volume 1
General Principles
and Procedures

Edited by

R. D. O'Brien

Cornell University
Ithaca, New York

PLENUM PRESS · NEW YORK AND LONDON

Library of Congress Cataloging in Publication Data

Main entry under title:

The Receptors.

Includes bibliographies and index.
CONTENTS: v. 1. General principles and procedures.
1. Cell receptors. I. O'Brien, Richard D.
QH603.C43R43   574.8'75   78-24366
ISBN 0-306-40100-2

First Printing – June 1979
Second Printing – June 1980

© 1979 Plenum Press, New York
A Division of Plenum Publishing Corporation
227 West 17th Street, New York, N.Y. 10011

Printed in the United States of America

Contributors

E. J. Ariëns, Pharmacological Institute, University of Nijmegen, Nijmegen, The Netherlands

Eric A. Barnard, Department of Biochemistry, Imperial College, London, England

A. J. Beld, Pharmacological Institute, University of Nijmegen, Nijmegen, The Netherlands

Robert Blumenthal, Laboratory of Theoretical Biology, National Cancer Institute, National Institutes of Health, Bethesda, Maryland

D. Colquhoun, Department of Pharmacology, St. George's Hospital Medical School, London, England

Pedro Cuatrecasas, The Wellcome Research Laboratories, Burroughs Wellcome Company, Research Triangle Park, North Carolina

A. De Lean, Medical Research Council of Canada; Endocrinology and Reproduction Research Branch, National Institute of Child Health and Human Development, National Institutes of Health, Bethesda, Maryland

Morley D. Hollenberg, Division of Clinical Pharmacology, Departments of Medicine and of Pharmacology and Experimental Therapeutics, The Johns Hopkins University School of Medicine, Baltimore, Maryland

Christopher Miller, Graduate Department of Biochemistry, Brandeis University, Waltham, Massachusetts

R. D. O'Brien, Section of Neurobiology and Behavior, Cornell University, Ithaca, New York. Present affiliation: Provost, University of Rochester, Rochester, New York

Efraim Racker, Section of Biochemistry, Molecular and Cell Biology, Cornell University, Ithaca, New York

D. Rodbard, Endocrinology and Reproduction Research Branch, National Institute of Child Health and Human Development, National Institutes of Health, Bethesda, Maryland

J. F. Rodrigues de Miranda, Pharmacological Institute, University of Nijmegen, Nijmegen, The Netherlands

Adil E. Shamoo, Department of Radiation Biology and Biophysics, University of Rochester School of Medicine and Dentistry, Rochester, New York

A. M. Simonis, Pharmacological Institute, University of Nijmegen, Nijmegen, The Netherlands

Preface

The following remarks are intended to serve as an introduction to this particular volume as well as to the whole series of volumes of which this is the first. The intent of the series is to provide an authentic and relatively complete statement about the status of our understanding of the receptors. The models we had in mind while developing this series are *The Enzymes*, *The Proteins*, and comparable groups of books. The receptors have received a degree of importance and richness of understanding that makes them deserving of comprehensive and complete coverage. The study of these molecules, which may well include such diverse items as the receptors for hormones, neurohumors, pheromones, taste, and many other chemical signals, have a great deal in common, so that the student of any one of them will wish to know the status of research about the others. This commonality is in part substantive, and in part practical and procedural.

Substantively, the receptors are all macromolecules whose function is to receive some form of chemical signal and transduce it to a form which is usable by the receiving cell. In this way, a chemical signal may lead to a neural response, to the turning-on of a cell's chromosomes, or to the activation of some enzymic apparatus to produce or release a substance. Because most of these processes are noncatalytic, special techniques not previously commonplace in biochemistry have been developed in order to study the receptors. Procedurally, then, there are also a number of common themes. The special technique that is most widely used is that of the binding of the signaling agent itself (hormone, neurohumor, or pheromone) or some analog which may be agonistic or antagonistic, employed as a probe to reveal the existence of the binding macromolecule. Another common characteristic which sets such studies apart from classical biochemistry is the concern to reconstitute some essential features of the intact system which were necessarily lost when the cells were initially disrupted in order to release the receptor.

The function of the first volume in the series is to focus on some of the principles which have applicability to a great variety of receptor systems. In every case the author has some particular interest in a relatively limited number of receptor systems, but in every case he has been asked to discuss the subject matter in a way which is perfectly general. It is hoped that the resultant volume will be of interest to anyone working with any of the receptors and who is anxious to get a broader view of the status of understanding developed elsewhere in the field. In addition, anyone contemplating the exploration of a new receptor should find this an excellent place to begin the search, with the accumulated understanding that has developed as many other investigators have faced the same fundamental problems.

Each subsequent volume in the series will deal with a particular receptor and will examine it from both the biochemical and physiological points of view. The intent is that each volume will provide a definitive and authoritative description of the status of our knowledge with respect to that particular receptor, and thus provide a point of departure for those continuing exploration in this area. For instance, the second volume will deal with the nicotinic receptor, and one would like to think that anyone starting anew in the study of that receptor would find that volume the obvious starting place to read. Taken as a whole, the series of volumes should describe the totality of our understanding of the major receptors.

It is simultaneously a joy and a concern that these volumes are being written at the very time in which new understanding is growing daily. The problem which is posed is the need to prepare volumes which are not only useful today but which will be valuable references for many years to come. Thus we have sought to make each volume not in any way like an "annual review" for the area, but rather (in addition to being the most recent work in the area) a comprehensive account of all that has been done up to the present time. The pleasurable feature of the tremendous activity in the field is of course that one enjoys participating in a search which is moving so fast and with such a high sense of excitement, and in which the rate of publication is so large that a series of the present kind may well be the only one that can save a researcher from being aware only of the research within his own particular system. It is a hallmark of current research that, when the time for exploration of a particular system falls due, the progress of understanding is astonishingly fast. Consequently, we run the risk of laying down, as in the fossil record, a series of outdated statements about each receptor which will serve only as a reminder in the future of how partial our information was. We hope to minimize this risk by selecting the timing of each volume (which we anticipate will appear approximately annually) in such a way as to coincide with the time of maximal elucidation of the particular receptor, knowing that the frenzied activity in any particular system cannot last eternally. The success of this endeavor cannot be measured until at least another decade has gone by. It is our ambition that this series will then constitute a benchmark describing the new understanding of macromolecules whose existence has been accepted for many years. It was the early decades of the twentieth century in which the existence of receptors for these chemical signals was postulated, but it is the closing years of the century in which the actual composition and mechanism of their action will be understood.

After completing this first volume, I accepted the position of Provost at the University of Rochester, simultaneously ending my research career. The function of the editor of this series is, above all, to be aware of areas of timeliness for the preparation of volumes in this series and of the key individuals to be consulted about the contents of each volume. Obviously, this is no task for an ex-researcher. Fortunately, Dr. Palmer Taylor of the Division of Pharmacology of the School of Medicine at the University of California, San Diego, has undertaken the responsibility for continuing this series. His outstanding excellence as a researcher, his experience as an editor, his breadth of knowledge, and his taste for excellence surely guarantee the success of future volumes.

R. D. O'Brien

Ithaca

Contents

Chapter 3

The Link between Drug Binding and Response: Theories and Observations

D. Colquhoun

Chapter 4
Kinetics of Cooperative Binding
A. De Lean and D. Rodbard

Chapter 5

Distinction of Receptor from Nonreceptor Interactions in Binding Studies

Morley D. Hollenberg and Pedro Cuatrecasas

Chapter 6

Incorporation of Transport Molecules into Black Lipid Membranes

Robert Blumenthal and Adil E. Shamoo

Chapter 7
Visualization and Counting of Receptors at the Light and Electron Microscope Levels
Eric A. Barnard

Chapter 8
Problems and Approaches in Noncatalytic Biochemistry
R. D. O'Brien

Reconstitution of Membrane Transport Functions

Christopher Miller and Efraim Racker

1. Introduction

At the very outset of this survey of membrane transport reconstitution, it is fitting to remind the reader that reconstitution is not a technique, nor is it new. Rather it is a basic approach to biological science, and it has been used from the very beginnings of biochemistry. Its prime assumption is that the whole of a system is, if not the sum, the well-constructed integral of its parts. This approach has led to the reconstruction of many integrated systems from their individual parts: of glycolysis and many pathways catalyzed by soluble multienzyme systems, of gene transcription and protein synthesis, and of the mechanisms controlling muscle contraction, to mention a few.

What *is* relatively new is the reconstitution of membranous systems that involve distinct aqueous compartments. During the last two decades, the study of the movement of solutes across biological membranes has progressed from the collection of physiological, phenomenological observations on cellular transport and electrical behavior to the identification of some of the proteins responsible for the phenomena. In a few cases, the proteins have been purified. We are now beginning to witness a further step in the development of the field: a marriage of transport physiology and membrane biochemistry, in which isolated membrane proteins are being inserted into artificial membranes, with the object of reconstructing the very same transport functions seen in the living cell.

There are two basic reasons why it has been difficult to apply the reconstitution approach to membrane transport systems. First, it is only within the past 5 or 10 years that techniques have come into use for isolating and purifying water-insoluble transport systems without so damaging them as to destroy their activities. Without prior isolation, there can be no reconstitution. The second and more profound reason is that the reconstitution of transport function nearly

Christopher Miller ● Graduate Department of Biochemistry, Brandeis University, Waltham, Massachusetts. *Efraim Racker* ● Section of Biochemistry, Molecular and Cell Biology, Cornell University, Ithaca, New York.

always requires the imposition of order, or asymmetry, upon the system. It is not merely a question of adding the right factors into the test tube. An artificial membrane must first be created to define the internal and external aqueous phases, the origin and the destination of the transported species. Then the transport protein must be introduced into this membrane, preferably asymmetrically, so as to mimic the general asymmetry of biological membranes. In some reconstitutions, the establishment of protein asymmetry is absolutely essential for the assay of the reconstituted function. In most of the cases to be discussed below, the asymmetry problem has been solved thanks to nature's fortuitous kindness rather than to our skill, but occasionally the biochemist's hand has intervened to bring about the desired orientation.

Since this is a volume about receptors, it would seem natural for this chapter to deal with the reconstitution of some of the well-known receptor systems. It is impossible to do this, however, since there simply have not been enough successful reconstitutions of this type to warrant a review. Instead, we shall focus on a variety of membrane transport systems to which reconstitution techniques have been applied. We hope to illustrate some of the special problems encountered in membrane protein reconstitution, in the firm belief that these problems will not fail to arise during attempts to reconstitute receptor functions.

As a final introductory point, we should mention that our use of the word *reconstitution* is somewhat more restricted than that usually found in the literature. In our view, it is not a protein or a membrane which is reconstituted but rather a *function*, specifically a transmembrane function for the systems considered here. Isolated membrane proteins, by virtue of their hydrophobic character, associate strongly with artificial membranes. The mere binding of a protein to such a membrane does not fulfill our criteria for reconstitution. Nor would, for instance, the binding of α-bungarotoxin to a liposome-associated acetylcholine receptor, for this activity may not rely upon the protein's transmembrane orientation. Only when a function is reconstructed which depends on the insertion of a protein into the membrane bilayer can we begin to claim that reconstitution has been achieved.

2. Reconstitution of Active and Passive Transport Systems

Solute transport across biological membranes can be divided into two classes, active and passive. In the former class, a species is transported against its transmembrane electrochemical gradient, and the free energy necessary for this movement comes from a chemical reaction directly coupled to the transport (Kedem, 1961). An example of this type of transport is the coupling of Na^+ and K^+ movements to ATP hydrolysis in animal cells (Garrahan and Glynn, 1967a,b). Passive transport can be divided into two subclasses: systems which facilitate the movement of solutes down their electrochemical gradients and systems which transport several solutes in an obligatorily coupled fashion. The glucose transporter of erythrocytes (Wilbrandt and Rosenberg, 1961) and the ion channels of excitable membranes (Armstrong, 1975) are instances of simple facilitated passive transport, while Na^+-linked sugar transport of certain epithelial cells (Crane, 1977) or

K^+-H^+ exchange by the ionophore nigericin (Pressman *et al.*, 1967) are examples of coupled passive transport. In the latter type of transport, one solute may actually be moved uphill against its electrochemical gradient; since the free energy for this movement comes from the movement of the other "coupled" solute down its preexisting potential gradient and not directly from a chemical reaction, this is not classified as active transport, according to the definition of Kedem (1961).

Active transport is conceptually more complicated than passive transport. Not only must the protein responsible allow the solute to move across a membrane, but it must also harvest the free energy released by a chemical reaction and then deliver this harvest to the solute's translocation against thermodynamic gravity. It may therefore seem surprising that the majority of transport systems which have been reconstituted at this early date are primary active transport systems, as shown in Table 1. Why should these apparently more complex systems have been the first to succumb to reconstitution? The primary reason for this situation is that by definition all of these transporters carry with them, in addition to their translocation function, an enzymatic activity of some kind which can be assayed chemically. Thus, in isolating the protein desired, the biochemist can use the chemical assay. Then after a satisfactory isolation, a search for conditions for reconstitution of the system's transport function may begin. In other words, with active transport proteins, the reconstitution problem may be separated into two distinct steps: isolation of the protein and reconstruction of transport.

In the reconstitution of passive transporters, however, such a separation of variables cannot be made. Here, there is no chemical reaction involved, and so the reconstitution itself must serve as the assay for the isolation and purification of the transport protein. This makes the biochemist's task far more difficult, since the parameters entering into both the isolation and the reconstitution must be optimized together. When faced with negative results in such a situation, the experimenter does not have an anchor to hold while trying to locate the problem. However, the reconstitution of passive transporters is possible, as demonstrated with the nucleotide transporter of mitochondria (Shertzer and Racker, 1974, 1976), with the glucose transporter of red blood cells (Kasahara and Hinkle, 1976, 1977), and with the mitochondrial phosphate transporter (Banerjee *et al.*, 1977). In all of these studies a solute flux into liposomes was used as the assay for isolation of the transport protein. Such an assay is not sufficient to demonstrate reconstitution, however, since the insertion of a protein into a liposome may introduce a nonspecific leakiness of the membrane. Therefore, it is necessary to have available either specific inhibitors (e.g., atractyloside for the nucleotide transporter) or a stereospecific substrate, as in the case of D-glucose transport; it is the use of such approaches which gives strongest credence to the above reconstitutions of passive transporters.

The problem of separation of variables will of course apply to the reconstitution of all chemoreceptor ion channels, since these are passive transporters, albeit gated ones. Although in these cases there is a binding activity which may be assayed independently of ion transport, the task of reconstitution is still expected to be more difficult than for active transport proteins, because the enzymatic activities of active transporters generally involve intimate cooperation between

Table I

Transport Systems Reconstituted in Liposomes

Species transported	Protein(s) responsible	Type of transport	Reference
Na^+, K^+, Cl^-	$(Na^+ + K^+)$-ATPase	Active	Goldin and Tong (1974), Hilden et al. (1974), Racker and Fisher (1975)
Ca^{2+}	Ca^{2+}–ATPase	Active	Racker (1972a,b), Meissner and Fleischer (1974), Warren et al. (1974)
$H^+(e^-)$	Cytochrome c oxidase	Active	Hinkle et al. (1972)
	NADH–coenzyme Q reductase	Active	Ragan and Hinkle (1975)
	Coenzyme Q–cytochrome c reductase	Active	Leung and Hinkle (1975)
H^+	NAD–NADPH transhydrogenase	Active	Rydström et al. (1975)
	Rutamycin-sensitive ATPase	Active	Kagawa et al. (1973), Serrano et al. (1976)
	DCCD-sensitive ATPase	Active	Sone et al. (1975)
	Bacteriorhodopsin	Active (photochemical)	Racker and Stoeckenius (1974), Yoshida et al. (1975a,b)
ATP, ADP	Mitochondrial nucleotide transporter	Passive, coupled	Shertzer and Racker (1976)
D-glucose	Erythrocyte glucose transporter	Passive	Kasahara and Hinkle (1976)
	Intestinal sugar transporter	Passive, coupled to Na^+	Crane et al. (1976)
Na^+, K^+, Ca^{2+}	Acetylcholine receptor	Passive, chemically gated	Hazelbauer and Changeux (1974), Michaelson and Raftery (1974)
Alanine	Bacterial alanine transporter	Passive, coupled to protons	Hirata et al. (1976)
Phosphate	Mitochondrial phosphate transporter	Passive, coupled to protons	Banerjee et al. (1977)

different parts of the entire protein molecule (conformational changes, cyclic allosteric binding and dissociation, etc.), while a mere binding function may survive biochemical treatments destructive to other parts of the receptor–channel complex. For this reason, receptors may be even more difficult to reconstitute than simple passive transporters, if the preservation of an effector binding function lulls the biochemist into a false sense of security about the condition of the entire receptor complex.

3. General Techniques of Reconstitution

3.1. Liposomes: Test Tubes with a Difference

Liposomes are to membrane biochemistry what test tubes are to soluble enzyme biochemistry. They are the artificial vessels which are brought into the laboratory to hold the proteins under study. But they are test tubes with a difference; in contrast to their Pyrex cousins, they are anything but inert. Indeed, it is an exceptional reconstituted system which is not extremely sensitive to the liposomes' chemical composition and method of formation.

3.1.1. Multilamellar Liposomes

An early method used to form liposomes for model-membrane studies is to suspend in aqueous solution a pure phospholipid or lipid mixture with vigorous agitation. This procedure yields large ($> 1 \mu$m in diameter) multilamellar liposomes, i.e., concentric layers of bilayer membranes, much like the structure of an onion (Bangham *et al.*, 1965; Papahadjopoulos and Miller, 1967). These large liposomes have not been useful for reconstitution work, but they have been used in the study of the structure and permeability of protein-free artificial membranes (Papahadjopoulos and Miller, 1967; Chapman *et al.*, 1974; Blok *et al.*, 1975; DeKruijff *et al.*, 1975).

3.1.2. Unilamellar Liposomes

It is possible to form single-walled phospholipid membrane vesicles by a variety of methods (see below), and it is these unilamellar liposomes which have been used most often for reconstitution work. One method in particular, the sonication of aqueous suspensions of lipid, has been used to create very small (20–30 nm in diameter) liposomes of uniform size suitable for physical studies (Huang, 1969; Huang and Thompson, 1974). The basic structure of these small liposomes is a spherical shell composed of a phospholipid bilayer enclosing an internal aqueous volume, as has been confirmed by a variety of methods, including electron microscopy (Huang, 1969; Thompson and Henn, 1970), physical methods (Sheetz and Chan, 1972; Jacobson and Papahadjopoulos, 1975), and biochemical approaches (Johnson *et al.*, 1975; Rothman and Dawidowicz, 1975).

Although the overall structure of the phospholipid bilayer of these artificial membranes is similar to that of cell membranes, several properties of the small liposomes used in reconstitution are of particular interest and possibly concern with regard to their use in reconstituting membrane transport functions. First, because of their small size, the membranes are highly curved. Thus, the surface area of the outer lipid monolayer is substantially larger than that of the inner monolayer; accordingly, the membrane should be asymmetric with the respect to the number of lipid molecules in the two halves of the bilayer. This expectation has been confirmed by observations of distinguishable nuclear magnetic resonance signals from the outer and inner phospholipid molecules (Bystrov *et al.*, 1971; Michaelson *et al.*, 1973; DeKruijff *et al.*, 1975; Lange *et al.*, 1975; Sears *et al.*, 1976) and by studies of accessibility of the lipids to phospholipases and phospholipid exchange proteins (Johnson *et al.*, 1975). In liposomes containing mixtures of phospholipids, a second type of asymmetry has been found—asymmetry of lipid composition, with certain phospholipids tending to accumulate on the outer half of the bilayer and others on the inner half (Litman, 1973; Michaelson *et al.*, 1973; Berden *et al.*, 1975).

There is a second property of small liposomes which must give us pause in considering these systems as precise models for biological membranes. Evidence has been steadily accumulating (Sheetz and Chan, 1972; Lichtenberg *et al.*, 1975; Suurkuusk *et al.*, 1976; Lentz *et al.*, 1976—but see DeKruijff *et al.*, 1975; Stockton *et al.*, 1976) that the hydrocarbon interior of small, highly curved liposomes is in a less ordered state than that of membranes with negligible curvature (> 100 nm in diameter). Since reconstitution of transport systems has been largely confined to small liposomes, questions about the effect of membrane curvature and order upon the operation of the reconstituted systems remain open. As discussed below, techniques are now coming into use for creating larger liposomes with negligible curvature for use in reconstitution studies.

3.2. *Methods for Inserting Proteins into Liposomes*

The major problem that must be solved in incorporating a hydrophobic protein into an artificial membrane is to render the preparation sufficiently free of lipid so that it will be thermodynamically inclined to be incorporated into a liposome yet not so free as to damage the protein irreversibly. In this section, we shall discuss several ways in which this problem has been overcome; the examples used are meant to be illustrative of the approach taken, not comprehensive of all the systems to which the method has been applied.

3.2.1. *Cholate Dialysis*

This technique was the first applied to reconstitutions of the mitochondrial proton pump (Kagawa and Racker, 1971; Kagawa *et al.*, 1973). Phospholipids are sonicated in the presence of 20–25 mM cholate and then mixed with the isolated, purified transport protein. The mixture is then dialyzed against cholate-free buffer, so that as the detergent leaves the dialysis bag, liposomes form spontaneously, now with the membrane protein incorporated into the bilayer. Several hours of dialysis are required before transmembrane function begins to appear, with optimal results calling for dialysis times ranging from 10 to 80 hr, depending on the protein (Kagawa and Racker, 1971; Hinkle *et al.*, 1972; Racker, 1972a; Goldin and Tong, 1974) and the sensitivity of the system to residual cholate.

The most likely explanation for the success of this method is that the protein becomes associated with the liposome membrane without prolonged exposure to a damaging hydrophilic environment. Initially, the protein's hydrophobic centers are associated with a large amount of detergent, and, as this becomes depleted by dialysis, it is continually replaced by the excess phospholipids originally added to the system. Then, as the detergent concentration falls below the critical micelle concentration, the protein has become sufficiently coated with lipid that it can participate in the spontaneous formation of liposomes.

A modification of the cholate-dialysis method is to form reconstituted liposomes by quick dilution of the detergent (Racker *et al.,* 1975a). This method is particularly convenient because of its speed and reproducibility. It has become the method of choice for assay of $^{32}P_i$–ATP exchange by the reconstituted rutamycin-sensitive ATPase of mitochondria; this enzyme complex is quite labile and is partly inactivated during the long time required by the dialysis method. The quick dilution method has also been applied to cytochrome oxidase (Racker, 1972a) and the Ca^{2+} pump of sarcoplasmic reticulum (Racker *et al.,* 1975a).

3.2.2. Sonication

This method involves exposing a mixture of protein and excess lipid to sonic oscillation for 5 min to 1 hr (Racker, 1973). It has been applied to a number of mitochondrial transport systems (Racker, 1974b, 1976), Ca^{2+} pumping by the Ca^{2+}–ATPase of sarcoplasmic reticulum (Racker and Eytan, 1973), and Na^+ transport by the $(Na^+ + K^+)$–ATPase of electric eel electric organ (Racker and Fisher, 1975). The results have been less satisfactory for cytochrome oxidase, which is partially inactivated by the exposure to sonication necessary for its insertion into the liposome membrane (Carroll, unpublished).

Two membrane proteins, $(Na^+ + K^+)$–ATPase and bacteriorhodopsin, have responded particularly well to the sonication technique, which is now the method of choice in this laboratory for their reconstitution. $(Na^+ + K^+)$–ATPase, the enzyme responsible for establishing and maintaining the asymmetric distributions of Na^+ and K^+ in animal cells, has been reconstituted in several laboratories. The earlier work applied the cholate-dialysis method to enzyme from dog kidney (Goldin and Tong, 1974) and brain (Sweadner and Goldin, 1975) and from dogfish salt gland (Hilden *et al.,* 1974; Hilden and Hokin, 1975). Whereas the $Na^+/K^+/ATP$ stoichiometries obtained were respectably close to physiological values, the absolute rates of ion pumping were rather low (5–20 nmol/mg-min). With the sonication technique with an enzyme from electric eel, Racker and Fisher (1975) were able to obtain pumping rates about tenfold higher than the previously reported values and than the rates measured on the same enzyme when reconstituted by cholate dialysis (Fisher, unpublished).

The sonication procedure, when applied to the reconstitution of light-driven proton transport by bacteriorhodopsin, has given results superior to those obtained by cholate dialysis (Racker and Hinkle, 1974; Racker and Stoeckenius, 1974; Kayushin and Skulachev, 1974; Racker, 1975; LaBelle and Racker, 1977). Here again the orientation problem arises and is solved fortuitously. Since in this system protons are actively transported upon absorption of light (Oesterhelt and Stoeckenius, 1973; Oesterhelt, 1976), and since light cannot be added on only one side of the liposome membrane, the orientation of the protein is essential for observing net proton transport.

In general, the sonication procedure is desirable because of its convenience and speed. Our experiences with probe-type sonicators have been bad. The best results have been obtained with a bath-type sonicator (Laboratories Supplies Co., Hicksville, New York). It allows sonication of small volumes (0.1–0.2 ml) in test tubes under controlled temperatures. However, to obtain reproducible results, it is necessary to pay careful attention to the control of sonication power, time and temperature, sample volume, and even the type of test tube used.

A feature common to all methods of reconstitution is the small size (20–50 nm in diameter) of the proteoliposomes obtained. Aside from the uncertainty regarding the physical state of the lipids in these membranes (see above), the small size can present problems when the sensitivity of the assay of transport activity is determined by the internal volume of the reconstituted vesicles, as it is, for example, in the demonstration of net active transport by $(Na^+ + K^+)$–ATPase (Hilden and Hokin, 1975). Recently, while developing the reconstitution of the glucose transporter of human red blood cells, Kasahara and Hinkle (1976, 1977) discovered a technique which apparently brings about functional incorporation of protein into larger liposomes. First, small liposomes are formed as usual by sonication, and the protein fraction is added to the liposome suspension. Then the mixture is frozen. Upon thawing, very large ($> 1\ \mu m$ in diameter) aggregates of lipid (and protein?) form, and these are broken up by a short (5–10 sec) sonication. This procedure yields single-walled liposomes of diameters in the range 100–300 nm, with smaller sizes also present. The method allowed reconstitution of the glucose transporter to be used as a reproducible assay of the protein's purification, which has now been accomplished (Kasahara and Hinkle, 1977). The results do not allow us to state with certainty that the reconstitutively functional transporter molecules actually reside in the large liposomes, but an experiment done by Kasahara and Hinkle (1977) suggests that this is the case. They found that the transport activity of a preparation of freeze–thaw reconstituted glucose transport vesicles could be severely inhibited by a 10–30 min sonication which reduced the sizes of the liposomes to the 20–50 nm diameter range; however, refreezing and thawing these small liposomes restored the transport activity and concomitantly created large liposomes once again.

If it can be clearly shown that large-sized liposomes are capable of transport, this will prove a particularly valuable method not only for reconstitution of transport activities but also for studies of single-walled model membranes with negligible curvature. The freeze–thaw sonication technique is applicable to other membrane transport systems such as $(Na^+ + K^+)$–ATPase and cytochrome oxidase (Cohn and White, unpublished).

3.2.3. Incorporation

The most obvious way to attempt to place a membrane protein in a liposome bilayer would be simply to add the purified protein preparation to a preformed liposome suspension, hoping that correct insertion might take place spontaneously.

When tried in the early days of reconstitution work, this procedure invariably failed. However, recently Eytan and co-workers (1976, 1977) have reinvestigated this approach and have obtained successful results with several different transport systems. The procedure is to incubate the purified membrane protein for 10–30 min with preformed liposomes of precisely controlled lipid composition. In general, the prime lipid requirement is the inclusion of an acidic lipid, such as phosphatidylserine, at concentrations not less than 15% and not greater than 30% of the lipid mixture used to form the artificial membranes. With this method, called the *incorporation technique*, a variety of active transport systems has been reconstituted, including cytochrome oxidase, proton-translocating rutamycin-sensitive ATPase, coenzyme Q–cytochrome c reductase, and the Ca^{2+} pump of sarcoplasmic reticulum. The incorporation of all of these systems resulted in a high degree of asymmetry of orientation; this may be related to the distribution of hydrophilic and hydrophobic domains in transmembrane proteins. For instance, cytochrome oxidase has a hydrophobic region apparently sandwiched between two hydrophilic regions (Eytan *et al.*, 1975), as it must have if it is to interact with substrates on both sides of a membrane and at the same time have thermodynamic stability within the bilayer. The correct incorporation into a preformed liposome requires that one of the hydrophilic ends of the protein traverse the membrane. Since such a process should involve considerable kinetic barriers, the least hydrophilic end of the protein may be favored to make its journey through the bilayer to the thermodynamically favored incorporated state. This kinetic selection could give rise to a highly asymmetric orientation of the protein in the membrane. In addition, the high surface curvature of the small liposomes may introduce a favored orientation within the membrane at *equilibrium*. We do not yet know if the liposomes' surface curvature does affect the incorporation process, since we do not have available a satisfactorily uniform system of large single-walled liposomes upon which the incorporation procedure can be tested. The development of such a system would be an important advance for reconstitution studies, particularly if the incorporation procedure proves to be generally useful. Recently, there has appeared a general interest in creating large single-walled liposomes (Lawaczeck *et al.*, 1976; Deamer and Bangham, 1976; Kasahara and Hinkle, 1977), and some of these techniques may be applied to reconstituted systems.

One particularly exciting development to emerge from the incorporation technique concerns the molecular psychology of the insertion process (Eytan and Racker, 1977). When faced with a choice between protein-free liposomes and liposomes into which rutamycin-sensitive ATPase had previously been incorporated, cytochrome oxidase will always incorporate into the former, avoiding the latter. However, in a converse experiment, the ATPase prefers to incorporate into liposomes containing cytochrome oxidase over protein-free liposomes. This type of work was extended to other proteins, including bacteriorhodopsin, Ca^{2+}–ATPase, and coenzyme Q–cytochrome c reductase, and the general conclusion emerged that the incorporation of a given protein will depend on the type of protein residing in the target liposome membrane. At this time the data are too sparse to permit even empirical rules for incorporation selectivity to be constructed, but the phenomenon suggests a possible mechanism for membrane differentiation during membrane biogenesis.

3.2.4. The Use of Superstable Membrane Proteins

In this section we shall discuss an approach to reconstitution which is not so much a technique as a strategy. The notorious lability of membrane proteins is a major difficulty to be overcome in efforts to reconstitute transport function. This problem has been largely circumvented by Kagawa and collaborators, who have used the membranes of a thermophilic bacterium as the source of proteins under study. The organism possesses transport systems similar to those well characterized in *E. coli* and mitochondria, but these transport systems normally operate in mineral hot springs at temperatures near 70°C. Thus, it was reasoned, these proteins must have extreme stability and might be easier to manipulate biochemically. Such has turned out to be the case. With this system, the DCCD-sensitive proton-translocating ATPase complex has been reconstituted and purified (Yoshida *et al.*, 1975a; Sone *et al.*, 1975; Kagawa *et al.*, 1976). In addition, the alanine transporter of this membrane has also been reconstituted, purified, and shown to be driven by the transport of an ion (possibly H^+ or Na^+) down its gradient (Hirata *et al.*, 1976). As expected, the proteins involved have been found to be extraordinarily stable and convenient to work with. With its capability of combining genetic manipulation with reconstitution, the approach with thermophilic bacteria appears potentially very fruitful.

4. What We Can Learn from Reconstitution

4.1. Oxidative Phosphorylation

In this controversial subject (for reviews, see Greville, 1969; Chance and Montal, 1971; Slater, 1971; Racker, 1974a, 1976), the fundamental issue in contention is the form in which free energy released by the reactions of electron transport (in mitochondria, chloroplasts, and bacteria) is stored for delivery to the ATP-synthesizing machinery. On one side of the argument is the chemical hypothesis, originally put forth by Slater (1953), asserting that electron transport reactions store free energy in a chemical intermediate with a "high-energy" bond. The free energy released from the breaking of this bond can be donated to the ATP–synthetase complex. On the other side is the chemiosmotic hypothesis of Mitchell (1966), which proposes that free energy is stored in the form of a trans-membrane electrochemical potential gradient for protons, generated by the electron transport chain operating as a proton active transport system. Furthermore, the hypothesis asserts, the ATP–synthetase complex utilizes the proton gradient's stored energy by functioning as an ATP-driven proton pump in reverse.

Many experiments have been performed on mitochondria and chloroplasts that have been claimed to be decisive to one or the other of the theories of oxidative phosphorylation. These experiments have not failed to elicit counterarguments, which reduce to this: that reactions other than those of immediate interest and pertinence may be occurring, obscuring the clarity of the conclusions to be drawn, i.e., that use of whole organelles, or even of whole membranes derived from those organelles, leads to equivocal results because of their complexity. It may therefore be salubrious to minds confounded by the thrust and parry of such argumentative experimentation to consider the reconstitution approach as another way to attack the problem. The basic idea is to find *minimum conditions* for ATP synthesis by reconstituting purified portions of the complex machinery involved in oxidative phosphorylation. Let us then review what partial reactions can be carried out in reconstituted systems.

First, we now know that an electrochemical gradient for protons can be created by at least three of the four "complexes" (Hatefi *et al.*, 1962) of the mitochondrial electron transport chain in reconstituted systems: NADH–coenzyme Q reductase (Ragan and Hinkle, 1975), coenzyme Q–cytochrome c reductase (Leung and Hinkle, 1975), and cytochrome c oxidase (Hinkle *et al.*, 1972). The absolute size of the proton gradients developed by these complexes has not been measured in the reconstituted liposomes.

Second, we know that the mitochondrial ATPase complex acts as an ATP-driven proton pump in reconstituted vesicles (Kagawa *et al.*, 1973; Serrano *et al.*, 1976). Part of this complex is a water-soluble ATPase (Penefsky *et al.*, 1960), and part is a membrane-bound proton ionophoric pathway (Racker, 1972a; Shchipakin *et al.*, 1976). Again, we do not know the absolute size of the proton gradient that can be sustained by this proton pump in reconstituted vesicles.

Third, we know that *individual sites* of oxidative phosphorylation can be reconstructed by reconstituting a given respiratory chain complex together with the proton-pumping ATPase (Racker and Kandrach, 1973; Ragan and Racker, 1973; Racker *et al.*, 1975a). Assembly of the entire respiratory chain is not necessary for oxidative phosphorylation. Efficiencies of ATP synthesized to electrons transported have been found to be up to 50% of those observed with mitochondria.

Finally, we know that the proton-pumping ATPase can be made to synthesize ATP when supplied with a proton gradient generated by reconstituting, along with the ATPase, the light-driven proton pump bacteriorhodopsin (Racker and Stoeckenius, 1974). This small protein from a halophilic bacterium has no apparent similarity to the electron transport enzymes of the mitochondrial respiratory chain; it catalyzes no redox reactions. And yet when it is incorporated into liposomes together with the ATPase of bovine heart mitochondria, ATP is synthesized, driven by light. This result was originally obtained using an impure preparation of the ATPase complex (Racker and Stoeckenius, 1974), but identical observations have been made (Eytan, unpublished) with the purified preparation of Serrano and co-workers (1976) and with highly purified ATPase from thermophilic bacteria (Yoshida *et al.*, 1975b).

The results on reconstituted systems have led us to the conclusion that the chemiosmotic mechanism of energy coupling is basically correct. As we begin to ask questions about the actual mechanisms of the proton and electron translocating machinery of the mitochondrial membrane, however, we move into another controversy, which has its "chemical" and "chemiosmotic" sides (Boyer *et al.*, 1977). We should point out, however, that this is a question of enzyme mechanism, an entirely separate controversy, and it should not be confused with the problem of the form of free-energy storage.

4.2. Ca^{2+}-ATPase

The Ca^{2+}-ATPase of sarcoplasmic reticulum (SR) has been well characterized as the biochemical entity responsible for sequestering Ca^{2+} in muscle by pumping the ion into the SR at the expense of ATP hydrolysis (Makinose and Hasselbach, 1971; Panet and Selinger, 1972). The resulting concentration gradient of Ca^{2+} between the SR space and the myoplasm is what poises the muscle to be able to release Ca^{2+} into the myoplasm upon electrical stimulation (for reviews, see Ebashi, 1976; Endo, 1977). The enzyme has been purified (MacLennan, 1970) and has been found to consist of a single protein of 100,000 molecular weight. Variable amounts of a small proteolipid were observed to be present in most preparations (MacLennan et al., 1972). The ATP-dependent transport of Ca^{2+} into liposomes containing this enzyme was the first nonmitochondrial transport function to be reconstituted (Racker, 1972b), and the enzyme is now being studied in the reconstituted state in a number of other laboratories (Warren et al., 1974; Meissner and Fleischer, 1974).

We have already learned about several basic properties of the transport system from reconstitution work—properties which were not known from studies of the system in its native membrane. First, it has been possible to rule out definitively all models for the mechanisms of the enzyme which propose that the terminal phosphate released from ATP is cotransported with Ca^{2+} (Martonosi, 1972). This was done (Knowles and Racker, 1975) by showing that γ-[^{32}P]-ATP releases its label to the outside of Ca^{2+}-transporting liposomes. This experiment is not possible in native SR membrane vesicles since they are quite leaky to phosphate (Duggan and Martonosi, 1970).

As in the case of oxidative phosphorylation, reconstitution of a functional Ca^{2+} pump is the only way to attack the question of how many components are needed for active transport of Ca^{2+}. Recent reconstitution work with this system has led to the conclusion that the Ca^{2+}-ATPase complex contains a heat-releasable factor which determines the efficiency of Ca^{2+} transported to ATP hydrolyzed (Racker and Eytan, 1975; Racker et al., 1975b). This "coupling factor" was shown to have Ca^{2+} ionophoric properties. When reconstituted, preparations of Ca^{2+}-ATPase depleted in this factor show low Ca^{2+}/ATP ratios (0.1–0.5); when the factor is added during reconstitution to these preparations, the system displays ratios in the range 1.4–1.7, approaching the physiological value of 2 (Makinose and Hasselbach, 1971; Panet and Selinger, 1972). There are two explanations for the effect of this factor. It may truly be a coupling factor needed to maintain an obligatory tight coupling between Ca^{2+} transport and ATP hydrolysis within the enzyme itself. Or the factor may be acting as a *pilot protein* in the reconstitution system, allowing efficient insertion of the ATPase complex into the liposomes. Evidence has been presented (Racker and Eytan, 1975) that this coupling factor may be identical with the proteolipid known to be associated with the Ca^{2+}-ATPase complex (MacLennan et al., 1972).

The role of the phospholipids has been examined by either replacement (Warren et al., 1974, Warren and Metcalfe, 1977) or removal of the phospholipids (Knowles et al., 1976) from the enzyme. It was shown that the lipid tightly bound to the enzyme ("annular lipid") serves as a gasket to seal the protein into the membrane with minimal leak and does not participate in phase transitions of the bulk membrane lipids in artificial membranes (Hesketh *et al.,* 1976). Removal of phospholipids with retention of transport activity allowed study of the individual steps of catalysis and their specific requirements for phospholipids (Knowles *et al.,* 1976).

4.3. $(Na^+ + K^+)$–ATPase

$(Na^+ + K^+)$–ATPase, first isolated by Skou (1957) from crab nerve, is the system responsible for the Na^+ and K^+ ion gradients maintained between animal cells and their aqueous environments. Much work on the cellular physiology of ion transport has shown that the enzyme couples the free energy of ATP hydrolysis to the active transport of Na^+ out of and K^+ into the cell; a stoichiometry of $3Na^+/2K^+/1ATP$ has been measured in human erythrocytes (Garrahan and Glynn, 1967a,b), but variable coupling schemes have been proposed in other tissues (Harris, 1967; Kernan, 1972).

In recent years, the biochemists have entered into the study of Na^+ and K^+ transport through investigations of the enzyme responsible. ATPase has now been purified from a number of sources (Kyte, 1971; Hokin *et al.,* 1973; Kuriki and Racker, 1976), and biochemical studies have limited the number of possible mechanisms for its ATPase activity (for reviews, see Schwartz *et al.,* 1972; Skou, 1975; Whittam and Chipperfield, 1975). The reconstruction of ion pumping in a reconstituted system has been achieved in the last few years (Goldin and Tong, 1974; Hilden *et al.,* 1974; Sweadner and Goldin, 1975; Hilden and Hokin, 1975; Racker and Fisher, 1975).

One result emerging from reconstitution of this enzyme is the demonstration of ATP-driven *net* movement of Na^+ and K^+ against their concentration gradients in a system containing highly purified $(Na^+ + K^+)$–ATPase and lipids (Hilden and Hokin, 1975). For most, this will seem to be merely a formal demonstration of the enzyme's well-accepted physiological role. However, in view of the repeated claims of a vocal minority of physiologists (Ling, 1969; Ling *et al.,* 1973) that active transport of Na^+ and K^+ across the cell membrane does not exist, it is worthwhile to point out the importance of the reconstitution of active transport in a biochemically minimal system.

4.4. Acetylcholine Receptor

The acetylcholine receptor remains the most extensively studied single system in neurobiology. Its function is to bind acetylcholine at the postsynaptic membrane (or muscle end plate), allowing a rather nonselective ion permeability pathway to open to Na^+ and K^+ ions, which then flow down their electrochemical potential gradients and locally depolarize the membrane. This sequence of events constitutes the initial stimulus for the propagation of a regenerative action potential along the entire nerve or muscle membrane. Work on whole cells over the past two decades has provided us with a wealth of information on the physiological and pharmacological properties of the receptor complex (for reviews, see Katz, 1969; Rang, 1975; Gage, 1976). It is now generally agreed that the fundamental mechanism of the receptor's operation is a ligand-induced conformational change which vastly increases the probability of the complex to exist in its conducting state. Indeed, work of the past few years using the sophisticated technique of fluctuation analysis (Stevens, 1972; Anderson and Stevens, 1973; Katz and Miledi, 1975; Colquhoun et al., 1975) has led to the inference that the conductance pathway of the receptor complex is a channel (rather than a carrier), with a conductance in the open state of about 2×10^{-11} mho. The execution of the measurement of single-channel conductance in this way is truly a *tour de force* of electrophysiology.

There are many questions about the receptor complex, however, which cannot be answered by the electrophysiological approach and which will require biochemical procedures, particularly reconstitution. Possibly the most fundamental question about the system concerns its size and its state of aggregation in the active form. There is general agreement that the purified receptor contains several nonidentical subunits with molecular weights ranging from 35,000 to 55,000 daltons (Gage, 1976). The subunits may be arranged in protomers, which then aggregate to form the active receptor complex. The number of protomers comprising the active complex is unknown, though there have been suggestions that this number is in the range 4–6 (Meunier et al., 1972; Biesecker, 1973; Potter, 1973; Eldefrawi et al., 1975; Reynolds and Karlin, 1978). One approach to this question would be to reconstitute the acetylcholine-induced ion transport function in liposomes and to examine the variation of this function with subunit concentration. Obviously, no such approach would be possible in biological membranes.

A second question concerns the apparent cooperativity of the physiological response of the receptor with respect to variation of ligand concentration (for reviews, see Rang, 1975; Gage, 1976). Various models placing the nonlinearity within different parts of the sequence of events leading to electrical response are difficult to distinguish on the basis of whole-cell studies. The cooperativity in acetylcholine binding which is seen in the solubilized receptor (Eldefrawi and Eldefrawi, 1975; Gibson, 1976) is lost upon purification procedures (O'Brien, personal communication). It would therefore be useful to reinsert the receptor into a reconstituted membrane system to see if the cooperativity seen in the biological membrane and in the solubilized receptor can be restored.

An important question is whether or not the ionophoric portion of the complex exists in only two states, *on* and *off*, as in the case of the model ionophore gramicidin A (Hladky and Haydon, 1970), or whether there are multiple conductance states, as in the case of the model ionophore alamethicin (Eisenberg et al., 1973). This question is crucial because the analysis of the electrophysiological fluctuation data used to determine the single-channel parameters depends on the model assumed for the channel, i.e., whether it fluctuates between only two states or more than two. Because of this, there is a certain circularity to the fluctuation work, although so far self-consistency has prevailed (Gage, 1976). Furthermore, it is now almost certain that the two-state model applies, at least to the receptor of frog muscle, since single channels have been recently observed directly in this tissue (Neher and Sackmann, 1976) and only one conductance value of the on state is observed. In spite of these successes with the physiological approach, it will be important to reconstruct single-channel behavior of the receptor in a reconstituted system in order to probe the ionophore biochemically. This cannot be done in a liposome system, since the assay must be an electrical one, but the planar bilayer system (see below) may provide an artificial membrane in which the problem can be attacked.

One final question which may be amenable to attack by reconstitution (in planar bilayers) concerns an apparently general property of biological ion channels, that of inactivation, or desensitization, as it is called in reference to the acetylcholine receptor. When the stimulus which opens the channel is maintained, the channel closes spontaneously. This phenomenon is seen with maintained application of acetylcholine receptor agonist (Rang and Ritter, 1970), with prolonged depolarization of excitable membranes (Adrian *et al.*, 1970; Armstrong, 1975), and with maintained depolarization in excitation–contraction coupling (Zachar, 1971). The question arises whether this is due to a process separate from the mechanism of opening the channel, as is postulated in the formulation of Hodgkin and Huxley (1952) for Na^+ channel inactivation, or whether the activated state of the channel is a transient state, necessarily preceding the final, equilibrium state in which the channel is closed again; the latter formulation has been developed by Baumann and Mueller (1974) and is discussed at length by Mueller (1975). If the reconstitution of various ion channels can be achieved, it will be most important to address the question of inactivation, especially if the ionophoric proteins are isolated in a reasonably pure state; in such a system, we would have hopes of determining whether inactivation is due to independent systems in the receptor–channel complex, or whether it is an integral property of the channel itself.

It is easy to ask these questions but another matter altogether to answer them. At this writing, there have been two claims of reconstitution of the acetylcholine receptor in liposomes (Hazelbauer and Changeux, 1974; Michaelson and Raftery, 1974) and one report of failure to repeat these reconstitutions (McNamee *et al.*, 1975). The assay for reconstitution was a ligand-dependent increase in the rate of efflux of Na^+ from liposomes containing the receptor protein. It was found that the half-time of Na^+ efflux decreased upon adding carbamylcholine from a background value of about 20 min to values in the range 2–10 min. The dose–response curve for the reconstituted system paralleled that for the same assay performed on the receptor-containing membrane fraction (*microsacs*) from which the protein was purified. Furthermore, the response to carbamylcholine was blocked by α-bungarotoxin and other specific inhibitors of the nicotinic receptor.

Reconstitution of this system is still in its infant stage, and so it is not surprising that there are several severe problems with this work which warrant discussion. Foremost is the extreme irreproducibility of the system, which is generally acknowledged (Michaelson and Raftery, 1974; McNamee *et al.*, 1975). Until this problem has been overcome, the use of reconstituted acetylcholine receptor will have very little value in actually learning about the complex. A second and more fundamental problem is that the rate constants of Na^+ efflux from the liposomes in the presence of carbamylcholine are far too slow. To show this, we note that the half-time of Na^+ leakage from spherical membrane vesicles of volume V is related to the vesicle membrane's specific conductance g for Na^+:

$$t_{1/2} = \frac{F^2 V \ln 2}{RTg} \qquad (1)$$

This expression, which is valid for the zero-voltage condition, follows from the relation of permeability coefficient to zero-voltage conductance according to the Nernst–Planck treatment of electrodiffusion [see MacInnes (1939)]. Now, by calculating the expected half-time for vesicles containing *only one receptor complex each*, we can arrive at the slowest possible ligand-dependent efflux rate to be expected from these liposomes. We can do this since we know an approximate value for the conductance of a single open channel, 10^{-11} mho at 0.1 M Na (Anderson and Stevens, 1973; Neher and Sackmann, 1976). For vesicles with a diameter of 60 nm (an upper limit for cholate-dialysis liposomes), the expected maximum half-time comes out to be 6 msec, four orders of magnitude faster than is actually measured in the Na^+ efflux experiments described above. This discrepancy can be explained in several ways:

1. The receptor is in its open state only 0.01 % of the time.
2. The ionophoric pathway is so damaged by isolation and purification that its open-state conductance is several orders of magnitude lower than in its native condition.
3. The vesicles into which the receptor is inserted are 20-fold larger in diameter than the majority of small vesicles seen in the electron microscope.
4. The reconstituted system does not respond to ligand by forming its open channel, as *in vivo*, but rather by undergoing a conformational change which changes its *fit* in the liposome membrane and induces an additional leakiness through which ions can flow rather nonspecifically. It is possible that the reconstituted system lacks certain components, possibly proteolipids (DeRobertis *et al.*, 1976; Eldefrawi, unpublished), for its correct operation.

Of these four possibilities, the first seems unlikely, given the known properties of the receptor, which *in vivo* appears to stay open for 10–50% of the time that ligand is bound (Adams, 1975). The second possibility is unattractive, not only because it is impossible to test but also because it is very difficult to see how the conductance of an otherwise intact channel could be lowered by a factor of 10^4. The third possibility is ruled out by the observation that the ligand-dependent Na^+ flux comes from at least 50% of the total internal volume of the liposomes, not from a very small fraction of the volume.

The last possibility seems to us the most likely explanation. It is consistent with the observed pharmacology in the reconstituted system as well as with the reasonable hypothesis that the annular lipid associated with a membrane protein determines its fit into the membrane and may be very sensitive to the protein's conformation. It is well known that the insertion of nonfunctional proteins into liposomes makes them nonspecifically leaky (Jilka *et al.*, 1975; Racker, 1972a). Therefore, further studies should be carried out on the reconstituted receptor to test the ion specificity of the ligand-induced permeability. We do, however, have one serious reservation about this explanation: The 10,000-fold discrepancy in efflux half-time is seen also in membrane microsacs, in which the receptor is still in place in its native, albeit isolated, membrane (Kasai and Changeux, 1971). This discrepancy was also noted by Katz and Miledi (1975). Thus, the damage done to the receptor has been done *before* isolation and reconstitution.

Because this is often misunderstood, we wish to emphasize that the calculation of the above discrepancy relies neither on an assumed value for the total surface area of the vesicle preparation nor on the assumption of complete equilibration of Na^+ in the vesicles. It rests solely on the measured size of the vesicles and the value of specific conductance for a single open channel. We have pointed out these problems in detail to illustrate a single point: that artifacts are frequently encountered in membrane reconstitutions and that quantitative as well as qualitative aspects of function need to be considered.

4.5. The Problem of Orientation

Biological membranes are asymmetric, and it would seem that the biochemist, using the membrane reconstitution techniques described above, would have very little chance of mimicking this asymmetry in an artificial system. However, a remarkable degree of asymmetry is displayed in most reconstituted transport proteins. In cytochrome oxidase vesicles made by cholate dialysis, for example, the enzyme is better than 95% oriented in the mitochondrial configuration, with the cytochrome c site exposed on the outside of the vesicles and the O_2 site on the inside (Carrol and Racker, 1977). Unidirectional orientation of protein in reconstituted liposomes has also been seen with the Ca^{2+}–ATPase of sarcoplasmic reticulum (Knowles and Racker, 1975), with proton-translocating rutamycin-sensitive ATPase (Eytan *et al.*, 1976), and with nucleotide transporter (Shertzer and Racker, 1976). We do not know why this asymmetry occurs— whether it is an equilibrium reflection of the membrane lipid asymmetry (see above), or whether it is a manifestation of intermediate states occurring during the formation of the reconstituted liposomes.

The $(Na^+ + K^+)$–ATPase reconstitution supplies an interesting illustration of the orientation problem (Goldin and Tong, 1974; Hilden *et al.*, 1974; Racker and Fisher, 1975). By both sonication and cholate-dialysis techniques, most of the enzyme molecules are bound to the liposomes in a *sideways* orientation, with both the ATP site and the ouabain site exposed to the external aqueous phase. About 10% of the enzyme is oriented with the ATP site external and the ouabain site internal, i.e., in a configuration opposite to that found in the cell plasma membrane. The assay for reconstitution is thus *inward* Na^+ transport driven by externally added ATP. This Na^+ translocation is inhibited by internal but not by external ouabain, as expected for such a configuration. It is fortunate that a fraction of the enzyme is oriented in this inverted configuration, since a measurement of the pump in the *cellular* configuration would require generation of ATP inside the liposomes. While this might be possible, for instance, by incorporation of the mitochondrial nucleotide transporter (Shertzer and Racker, 1976), it would make the biochemist's task more difficult than it now is.

Lest we leave the reader with the impression that all success in achieving net orientation with reconstitution has been due to luck, we shall mention one dramatic example in which the proper protein orientation was essential to success and in which this orientation was produced artificially. This is the case of the reconstruction of site III oxidative phosphorylation (Racker and Kandrach, 1973). In this system, cytochrome oxidase and the mitochondrial ATPase complex are reconstituted together in the same liposomes. The former protein energizes the membrane, while the latter protein complex harvests the free energy thus stored to synthesize ATP from externally added ADP and inorganic phosphate. For the membrane to be energized properly for synthesis of *external* ATP, the proton pumping by cytochrome oxidase must be only *inward*, i.e., in the reverse mitochondrial configuration. This was achieved by removing the external cytochrome c and preventing the binding of remaining traces of the substrate with externally added polylysine. Only when all these orientation-producing conditions were met was site III oxidative phosphorylation observed.

5. Reconstitution in Planar Bilayer Membranes

Liposomes have served well as artificial membranes for reconstituting transport proteins, and as long as the assay for reconstitution is a chemically measurable solute flux, they will continue to be the most useful and easily handled system. However, many transport systems of interest, including a host of receptor systems, are involved in cellular electrical phenomena, and until now liposomes have been inaccessible to electrical measurement because of their small size. While every electrical measurement involves the flux of an electrically charged species, which in principle can be measured, often the fluxes are too small for chemical detection to be practical. Therefore, for some reconstitutions, it is absolutely essential to have an electrical assay, not only because of the superior sensitivity and time resolution but also for the sake of mimicking directly the electrophysiological function of the protein.

Fortunately, an electrically accessible artificial membrane is available, the planar bilayer membrane, known as *black lipid membrane* (Mueller *et al.*, 1962; Pagano and Thompson, 1967; Montal and Mueller, 1972). In this system, a macroscopic bilayer membrane (0.1–3 mm in diameter) is formed on a hole in a well-insulated partition separating two aqueous solutions of milliliter volumes. Electrodes are introduced into the aqueous chambers, and the electrical properties of the artificial membrane are determined. In general, phospholipid membranes unmodified by proteins display very low conductance, around 10^{-8} mho/cm^2, about four orders of magnitude lower than biological membranes (Mueller and Rudin, 1969). This makes for a very low background level of conductance and thus allows the observation of extremely small perturbations of the membrane by transport proteins.

In the past, the planar bilayer has most often been used to study the electrical properties of a class of small peptides which dramatically raise the conductance of the bilayers. These ionophores are produced by microorganisms as lethal agents directed against other organisms (Mueller and Rudin, 1969). Now there are several synthetic ionophores available as well. These compounds have served as models for different types of protein-mediated passive ion transport through membranes. Valinomycin, for instance, is the now-classical example of a highly specific ion carrier, which binds K^+, moves across the membrane, and releases the ion on the other side [for a review, see Eisenman *et al.* (1973)] ; gramicidin A acts as a channel, forming a pore through the membrane (Hladky and Haydon, 1970, 1972; Krasne *et al.*, 1971; Bamberg and Lauger, 1973). Alamethicin also forms a channel, but only above a certain critical voltage (Mueller and Rudin, 1969). The modes of operation of all these ionophores were determined by studies of their electrical properties in the planar bilayer combined with biochemical work on the molecular structures. They are the only transport proteins for which we have a reasonably good picture of the mechanism of operation.

Some of these compounds can be made to mimic in a remarkable way certain aspects of cellular electrical phenomena (Baumann and Mueller, 1974: Mueller, 1975). However, a more direct approach is required, attempting to place the purified transport systems of immediate physiological interest into the planar bilayer membrane. In this section, we shall review the major results in this area and try to point out the likely directions that future work will take. The main obstacle to be overcome is to find methods of placing membrane proteins into these planar bilayers. This is a problem which does not exist for the model ionophores, since many of them are tailor-made to insert into the membranes of target organisms and since they are small molecules not easily denatured. Methods of protein insertion used in liposomes will not generally apply to planar bilayers, since the latter cannot be manipulated by sonication, exposure to detergent, freezing, etc., as can liposomes.

5.1. Sucrase–Isomaltase Complex

It is indeed ironic that this intestinal enzyme complex was the first system to be reconstituted in planar bilayers (Storelli *et al.*, 1972; Semenza, 1976), for neither its function nor its assay is electrical. The isolated enzyme complex was suspended along with lipids in hydrocarbon solution used to form the membrane, and the flux of radioactively labeled sugars through the membrane was measured. It was found that label added to one side of the bilayer as sucrose appeared on the other side as glucose and fructose, while neither free glucose nor free fructose would permeate the membrane. This effect was found to be completely dependent on the addition of the enzyme complex, and so it was concluded that the enzyme couples the hydrolysis of sucrose in the mammalian small intestine to transport of the products of the hydrolysis. While this biochemical work was underway, studies on whole cells and membrane vesicles of intestinal brush border epithelium came to a similar conclusion regarding the coupling of sucrose hydrolysis to transport (Malathi *et al.*, 1973; Ramaswamy *et al.*, 1974, 1976). The reconstitution of the sucrase–isomaltase complex thus identified it as the system responsible for the transport phenomenon seen in cells.

What is not clear from this work is whether the coupling of transport to hydrolysis is obligatory or not. It is possible that sucrose is first transported across the bilayer and then incidentally hydrolyzed by the enzyme complex on the side of the bilayer *trans* to the radioactive sucrose. This question cannot be answered until there is found a way to bring about orientation of the enzyme in the bilayer, either by oriented incorporation or by inhibition of the enzyme from one side only. The experiments do show, however, that hydrolysis of sucrose does not *precede* its transport. This type of experiment is tedious to carry out in the planar bilayer system because of the low surface-to-volume ratio; it would seem useful to continue this work with liposomes on a larger scale (Semenza, 1976).

5.2. Acetylcholine Receptor in Planar Bilayers

As in the case of the reconstitution of this system in liposomes, suggestive results are just beginning to appear. These results hold great promise, however, since a successful reconstitution in planar bilayers would not only provide a way to reproduce in a defined system the electrical characteristics of the postsynaptic membrane but would also allow the experimenter to vary parameters which are inexorably fixed in the native membrane.

The first claimed reconstitution of the receptor's ionophoric function was by Parisi *et al.* (1971), who observed a transient effector-dependent increase in conductance of planar bilayers made from solutions containing a chloroform–methanol soluble fraction from *Electrophorus electricus* electric organ. This group continues to study this proteolipid (Adragna *et al.*, 1975) in planar bilayers, but the physiological significance of the conductance phenomena is unclear, since extremely high concentrations of effectors are used, since antagonists as well as agonists cause the transient conductance increase, and since it is doubtful whether the acetylcholine binding component of the receptor complex is even extracted under the conditions used for the bilayer studies [see DeRobertis *et al.* (1976) and following letters for an exchange of views on the latter point].

Attempts have also been made to reconstitute in planar bilayers the receptor isolated by the more conventional technique of detergent extraction. Acetylcholine receptor partially purified from muscle end plate, when added to the aqueous phase of a planar bilayer system, was reported to cause a steady conductance increase (Kemp *et al.*, 1973), which was stimulated severalfold in the presence of acetylcholine. Later it was reported (Romine *et al.*, 1974) that a nicotinic receptor preparation from brain causes a conductance increase proceeding in discrete steps, as though single units of receptor were being inserted into the membrane. These conductance steps were apparent even in the absence of agonist, but the frequency of their occurrence increased with addition of carbamylcholine, and this increased frequency was counteracted by *d*-tubocurare (Godall *et al.*, 1974). This kind of behavior was also seen with more highly purified preparations of receptor (Bradley *et al.*, 1976).

These promising and preliminary results require and deserve further examination in depth. One problem is that there is a very wide range of values for the conductance jumps, ranging over two orders of magnitude. A second problem with interpreting these data is that the experiments have not distinguished clearly whether the effect of agonists and antagonists is on the state of a channel in the membrane or on the probability of a channel inserting itself into the membrane from the aqueous phase, a different process altogether.

Recently, Shamoo and Eldefrawi (1975) found that a tryptic digest of the purified receptor (retaining, after trypsinization, the ability to bind acetylcholine) raises the conductance of the bilayer in the presence of various ions. In the absence of carbamylcholine, the conductance shows little if any ion specificity, but addition of the agonist induces a further conductance increase. This additional conductance pathway displays an ion selectivity of $Na^+/K^+/Cl^-$ of $1/1/0.2$, agreeing well with the selectivity of the channel measured *in vivo* (Ruiz-Manresa and Grundfest, 1971). It is particularly important that this conductance is seen only if the receptor is added in the absence of carbamylcholine; furthermore, if the antagonist curare is present during addition of protein to the bilayer, subsequent addition of carbamylcholine fails to increase the bilayer conductance. These results indicate that only the nonliganded receptor incorporates properly into the artificial membrane. Once the protein is incorporated, the agonist could exert its effect.

In this system, only the trypsinized receptor gave reproducible effects with agonists; the purified protein before trypsinization produced only nonspecific conductance increases which were not responsive to carbamylcholine. This behavior suggests that functional incorporation requires the presence of groups not normally exposed on the intact receptor surface. It also indicates that even in the absence of functional incorporation the complex *does* enter the bilayer, but with an incorrect orientation.

The above technique of attempting to dissect membrane transport proteins by trypsinization and then to search for ionophores in the digest has also been used with $(Na^+ + K^+)$–ATPase (Shamoo and Albers, 1973; Shamoo and Myers, 1974) and the Ca^{2+}–ATPase of sarcoplasmic reticulum (Shamoo and MacLennan, 1974, 1975; Stewart *et al.*, 1976). It seems likely, as proposed by one of the investigators (MacLennan, personal communication), that the peptide isolated after trypsinization is the hydrophobic fragment of the enzyme which interacts specifically with the cation. The ionophore activity would thus be accidental and unrelated to the mechanism of ion translocation.

5.3. Insertion of Whole Membrane Vesicles

One practical problem with all the reconstitutions mentioned above is that the protein of interest must first be removed from its native membrane before being incorporated into an artificial bilayer, be it liposome or planar bilayer. Recently, we have introduced a gentle method for insertion of electrically assayable proteins into planar bilayers (Miller and Racker, 1976b). The principle of the technique is to cause native membrane vesicles (e.g., sarcoplasmic reticulum) to fuse with the planar bilayer according to a well-documented process of membrane fusion (Papahadjopoulos *et al.*, 1974; Jacobson and Papahadjopoulos, 1975; Miller and Racker, 1976a; Miller *et al.*, 1976). With sarcoplasmic reticulum vesicles, two distinct conductance pathways were observed: one nonspecific in ion selectivity and voltage independent and another cation selective and strongly voltage dependent (Miller, 1978).

5.4. Proton Pumps

Recently, there has been considerable effort expended in attempts to insert various energy-transducing proton pumps into planar bilayers. These include cytochrome oxidase (Drachev *et al.*, 1974a, 1976b; Chien and Mueller, 1976), mitochondrial ATPase (Drachev *et al.*, 1974a, 1976c), chlorophyll complexes (Barsky *et al.*, 1976), and bacteriorhodopsin (Drachev *et al.*, 1974a,b, 1976c; Skulachev, 1976; Shieh and Packer, 1976).

Using various strategies for introducing the proton-transporting proteins into the artificial membranes, the above investigators have all observed the generation of transmembrane voltages up to as high as 200 mV in some cases. While the results are exciting and point in a promising direction to be explored, we consider that there are two serious problems regarding their validity as reconstitutions of transbilayer electrogenic proton pumping.

First, it is necessary to realize that the successful insertion of an operating electrogenic proton pump into a bilayer should produce two basic effects: (1) the generation of a *voltage* at zero current due to the unidirectional transport of the H^+ ion through the membrane and (2) the development of a *conductance* (over and above the background conductance of the bilayer). The additional conductance must appear because the current driven by the electrogenic proton pump must respond to an applied voltage against (or with) which the pumping occurs. [For a demonstration of this effect for the electrogenic Na^+ pump in frog skin, see Vieira *et al.* (1972).] Thus, a typical (and ideally linear) current–voltage $(I-V)$ relation for the membrane should look like Fig. 1. Curve A is the simple ohmic $I-V$ curve for the membrane with the proton pump not operating (i.e., no substrate for cytochrome oxidase or no light for bacteriorhodopsin). Curve B is the $I-V$ curve with the pump operating (with identical salt solutions on the two sides of the membrane). The *I*-axis intercept of curve B is the *short-circuit current*, i.e., the current generated by the pump with zero electrochemical difference for protons across the membrane. The *V*-axis intercept, V_0, is the zero-current voltage generated by the pump in parallel with the leaky membrane; this is the parameter most often measured in the above studies. The point of intersection of the two curves defines the voltage, V_m, at which all the current through the membrane is passive, i.e., the maximum voltage generated by the pump. With no pH difference across the membrane, this will be equivalent to the maximum free-energy storage which the pump can support. As more pump molecules are inserted into the membrane, the short-circuit current, the zero-current voltage, and the slope of curve B will increase, but the point of intersection of the two curves should remain unchanged.

Now, the problem with the experimental results mentioned above is that only one-half of the theoretically expected behavior has been observed. In none of the experiments using planar bilayers was a conductance increase noted. In fact, in those studies which have recorded membrane conductance (Drachev *et al.*, 1974b), the values observed were in the order of 10^{-8} mho/cm^2, very close to if not identical with the conductance of the unmodified planar artificial membrane. The fact that voltages are produced means that charges are being moved by the transport proteins; it is not clear, though, that the charges are being transported *through* rather than *into* the artificial membrane as a result of reactions at the membrane surface. In addition, it is necessary to mention that the studies on planar bilayers did not investigate the $I-V$ curves of the system but rather the voltage at zero current (the *V*-axis intercept of Fig. 1), which has no clear interpretable meaning.

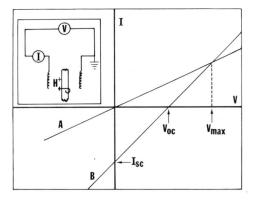

Fig. 1. Expected electrical behavior of a membrane reconstituted with an electrogenic pump. Current–voltage relations are shown for a membrane containing an electrogenic proton pump separating two identical aqueous systems. A: Passive membrane with proton pump not operating. B: Membrane with proton pump operating, with polarity shown in the inset. The following important parameters are indicated on the graph: (1) I_{sc}: Short-circuit current (current at zero voltage). This is the current driven by the pump with no electrochemical potential gradient for protons across the membrane. (2) V_{oc}: Open-circuit voltage (voltage at zero current). This is the parameter most often studied in the experiments discussed in the text. Its value is dependent on both the passive membrane conductance and the number of pumps incorporated in the membrane. (3) V_{max} : Reversal potential of the pump. This is the maximum voltage against which the pump can operate. Its value is independent of both the passive membrane conductance and the number of pumps in the membrane. Inset: Schematic representation of the system under consideration, a membrane containing an electrogenic proton pump, and electrodes for the measurement of the current–voltage relations.

The second problem with these observations is an experimental one. The same phenomenon of voltage generation reported in artificial bilayers is also seen in lipid–hydrocarbon membranes *which have not thinned to bilayers* and are 10–100 nm in thickness (Drachev *et al.*, 1974a, 1976a,b,c; Barsky *et al.*, 1976; Shieh and Packer, 1976). In the recent extensive work of the Russian investigators, the full *I–V* curve was carefully investigated, and a conductance increase upon pumping was observed, as in Fig. 1. However, to our knowledge, none of the systems described above requires that the artificial membrane be a bilayer of 4–5 nm thickness. This complication casts doubt on the results as true reconstitutions of transbilayer proton pumping, and it makes it difficult to interpret in thermodynamic terms the magnitudes of the voltages generated. Furthermore, it has been found (Boguslavsky *et al.*, 1975) that these proton-transporting proteins generate large voltages when placed at the octane–water interface, with substrate present; here, there is no question of transmembrane function. The interpretation of these voltages generated by proton pumps will have to await the clear demonstration of transmembrane electrogenic pumping.

References

Adams, P. R., 1975, Analysis of dose–response curve at voltage-clamped frog endplates, *Pflügers Arch.* **360**:145.

Adragna, N. C., Salas, P. J. I., Parisi, M., and DeRobertis, E., 1975, Curare blocks cationic conductance in artificial membranes containing hydrophobic proteins from cholinergic tissues, *Biochem. Biophys. Res. Commun.* **62**:110.

Adrian, R. H., Chandler, W. K., and Hodgkin, A. L., 1970, Voltage clamp experiments in striated muscle fibers, *J. Physiol. (London)* **208**:607.

Anderson, C. R., and Stevens, C. F., 1973, Voltage clamp analysis of acetylcholine produced end-plate current fluctuations at frog neuromuscular junction, *J. Physiol. (London)* **235**:655.

Armstrong, C. M., 1975, Ionic pores, gates, and gating currents, *Q. Rev. Biophys.* **7**:179.

Bamberg, E., and Lauger, P., 1973, Channel formation kinetics of gramicidin A in lipid bilayer membranes, *J. Membr. Biol.* **11**:177.

Banerjee, R. K., Shertzer, H. G., Kanner, B. I., and Racker, E., 1977, Purification and reconstitution of the phosphate transporter from bovine heart mitochondria, *Biochem. Biophys. Res. Commun.* **75**:772.

Bangham, A. D., Standish, M. M., and Watkins, J. C., 1965, Diffusion of univalent ions across the lamellae of swollen phospholipids, *J. Mol. Biol.* **13**:238.

Barsky, E. L., Dancshazy, Z., Drachev, L. A., Il'ina, M. D., Jasaitis, A. A., Kondrashin, A. A., Samulov, V. D., and Skulachev, V. P., 1976, Reconstitution of biological molecular generators of electric current. Bacteriochlorophyll and plant chlorophyll complexes, *J. Biol. Chem.* **251**:7066.

Baumann, G., and Mueller, P., 1974, A molecular model of membrane excitability, *J. Supramol. Struct.* **2**:538.

Berden, J. A., Barker, R. W., and Radda, G. K., 1975, NMR studies on phospholipid bilayers. Some factors affecting lipid distribution, *Biochim. Biophys. Acta* **375**:186.

Biesecker, G., 1973, Molecular properties of the cholinergic receptor purified from *Electrophorus electricus, Biochemistry* **12**:4403.

Block, M. C., van der Neut-kok, E. C. M., Van Deenen, L. L. M., and DeGier, J., 1975, The effect of chain length and lipid phase transitions on the selective permeability properties of liposomes, *Biochim. Biophys. Acta* **406**:187.

Boguslavsky, L. I., Kondrashin, A. A., Kozlov, I. A., Metelsky, S. T., Skulachev, V. P., and Volkov, A. G., 1975, Charge transfer between water and octane phases by soluble mitochondrial ATPase (F_1), bacteriorhodopsin, and respiratory chain enzymes, *FEBS Lett.* **50**:223.

Boyer, P., Chance, B., Ernster, L., Mitchell, P., Racker, E., and Slater, E., 1977, Oxidative phosphorylation and photophosphorylation, *Annu. Rev. Biochem.* **46**:955.

Bradley, R. J., Romine, W. O., Carl, G. F., Howell, J. H., and Kemp, G. L., 1976, Characterization of a nicotinic acetylcholine receptor from rabbit skeletal muscle and reconstitution in planar phospholipid bilayers, *Biochem. Biophys. Res. Commun.* **68**:577.

Bystrov, V. F., Dubrovina, N. I., Barsukov, L. J., and Bergelson, L. D., 1971, NMR differentiation of the internal and external phospholipid membrane surfaces using paramagnetic Mn^{+2} and Eu^{+3} ions, *Chem. Phys. Lipids* **6**:343.

Carrol, R., and Racker, E., 1977, Preparation and characterization of cytochrome c oxidase vesicles with high respiratory control, *J. Biol. Chem.* **252**:6981.

Chance, B., and Montal, M., 1971, Ion-translocation in energy-conserving membrane systems, in: *Current Topics in Membranes and Transport,* Vol. 2 (F. Bronner and A. Kleinzeller, eds.), pp. 100–156, Academic Press, New York.

Chapman, D., Urbina, J., and Keough, K. M., 1974, Biomembrane phase transitions. Studies of lipid–water systems using differential scanning calorimetry, *J. Biol. Chem.* **249**:2512.

Chien, T. F., and Mueller, P., 1976, The reconstitution of cytochrome oxidase proteolipid into bilayer membrane, *Fed. Proc.* **35**:1599.

Colquhoun, D., Dionne, V. E., Steinbach, J. H., and Stevens, C. F., 1975, Conductance of channels opened by acetylcholine-like drugs in muscle and end-plate, *Nature (London)* **253**:204.

Crane, R. K., 1977, The gradient hypothesis and other models of carrier-mediated active transport, *Rev. Physiol. Biochem. Pharmacol.* **78**:99.

Crane, R. K., Malathi, P., and Preiser, H., 1976, Reconstitution of specific Na^+-dependent D-glucose transport in liposomes by Triton X-100 extracted proteins from brush border membranes of hamster small intestine, *Biochem. Biophys. Res. Commun.* **71**:1010.

Deamer, D., and Bangham, A. D., 1976, Large volume liposomes by an ether vaporization method, *Biochim. Biophys. Acta* **443**:629.

DeKruijff, B., Cullis, P. R., and Radda, G. K., 1975, Differential scanning calorimetry and ^{31}P NMR studies on sonicated and unsonicated phosphatidylcholine liposomes, *Biochim. Biophys. Acta* **406**:6.

DeRobertis, E., DePlazas, S. F., and DeCarlin, M. C. L., 1976, Similarities between "cholinergic proteolipid" and detergent-extracted cholinergic proteins, *Nature (London)* **259**:605.

Drachev, L. A., Jasaitis, A. A., Kaulen, A. D., Kondrashin, A. A., Liberman, E. A., Nemecek, I. B., Ostroumov, S. A., Semenov, A. Yu., and Skulachev, V. P., 1974a, Direct measurement of electric current generation by cytochrome oxidase, H⁺-ATPase, and bacteriorhodopsin, *Nature (London)* **249**:321.

Drachev, L. A., Kaulen, A. D., Ostroumov, S. A., and Skulachev, V. P., 1974b, Electrogenesis by bacteriorhodopsin incorporated in a planar phospholipid membrane, *FEBS Lett.* **39**:43.

Drachev, L. A., Frolov, V. N., Kaulen, A. D., Liberman, E. A., Ostroumov, S. A., Plakunova, V. G., Severina, I. I., and Skulachev, V. P., 1976a, Reconstitution of biological molecular generators of electric current. Bacteriorhodopsin, *J. Biol. Chem.* **251**:7059.

Drachev, L. A., Jasaitis, A. A., Kaulen, A. D., Kondrashin, A. A., Chu, L. V., Semenov, A. Y., Severina, I. I., and Skulachev, V. P., 1976b, Reconstitution of biological molecular generators of electric current. Cytochrome oxidase, *J. Biol. Chem.* **251**:7072.

Drachev, L. A., Jasaitis, A. A., Mikelsaar, H., Nemecek, I. B., Semenov, A. Y., Semenova, E. G., Severina, I. I., and Skulachev, V. P., 1976c, Reconstitution of biological molecular generators of electric current. H⁺-ATPase, *J. Biol. Chem.* **251**:7077.

Duggan, P. F., and Martonosi, A. N., 1970, Sarcoplasmic reticulum IX. The permeability of sarcoplasmic reticulum membranes, *J. Gen. Physiol.* **56**:147.

Ebashi, S., 1976, Excitation–contraction coupling, *Annu. Rev. Physiol.* **38**:293.

Eisenberg, M., Hall, J. E., and Mead, C. A., 1973, The nature of the voltage-dependent conductance induced by alamethicin in black lipid membranes, *J. Membr. Biol.* **14**:143.

Eisenman, G., Szabo, G., McLaughlin, S. G. A., and Ciani, S. M., 1973, Molecular basis for the action of macrocyclic carriers on passive ionic translocation across lipid bilayer membranes, *Bioenergetics* **4**:93.

Eldefrawi, M. E., and Eldefrawi, A. T., 1975, Structure and function of the acetylcholine receptor, *Croat. Chem. Acta* **47**:425.

Eldefrawi, M. E., Eldefrawi, A. T., and Shamoo, A. E., 1975, Molecular and functional properties of the acetylcholine receptor, *Ann. N.Y. Acad. Sci.* **264**:183.

Endo, M., 1977, Calcium release from the sarcoplasmic reticulum, *Physiol. Rev.* **57**:71.

Eytan, G., and Racker, E., 1977, Selective incorporation of membrane proteins into proteoliposomes of different compositions, *J. Biol. Chem.* **252**:3208.

Eytan, G., Carroll, R. C., Schatz, G., and Racker, E., 1975, Arrangement of the subunits in solubilized and membrane-bound cytochrome c oxidase from bovine hearts, *J. Biol. Chem.* **250**: 8598.

Eytan, G., Matheson, M., and Racker, E., 1976, Incorporation of mitochondrial membrane proteins into liposomes containing acidic phospholipids, *J. Biol. Chem.* **251**:6831.

Gage, P. W., 1976, Generation of end-plate potentials, *Physiol. Rev.* **56**:177.

Garrahan, P. J., and Glynn, I. M., 1967a, The stoichiometry of the sodium pump, *J. Physiol. (London)* **192**:217.

Garrahan, P. J., and Glynn, I. M., 1967b, The incorporation of inorganic phosphate into adenosine triphosphate by reversal of the sodium pump, *J. Physiol. (London)* **192**:237.

Gibson, R. E., 1976, Ligand interactions with the acetylcholine receptor from *Torpedo californica*. Extensions of the allosteric model for cooperativity to half-of-site activity, *Biochemistry* **15**:3890.

Goldin, S. M., and Tong, S. W., 1974, Reconstitution of active transport catalyzed by the purified sodium and potassium ion-stimulated adenosine triphosphatase from canine renal medulla, *J. Biol. Chem.* **249**:5907.

Goodall, M. C., Bradley, R. J., Saccomani, G., and Romine, W. O., 1974, Quantum conductance changes in lipid bilayer membranes associated with incorporation of acetylcholine receptors, *Nature (London)* **250**:68.

Greville, G. D., 1969, A scrutiny of Mitchell's chemiosmotic hypothesis of respiratory chain and photosynthetic phosphorylation, in: *Current Topics in Bioenergetics,* Vol. 3 (D. R. Sanadi, ed.), pp. 1–72, Academic Press, New York.

Harris, E. J., 1967, The stoichiometry of sodium ion movement from frog muscle, *J. Physiol. (London)* **193**:455.

Hatefi, Y., Haavik, A. G., Fowler, L. R., and Griffiths, D. E., 1962, Studies on the electron transfer system XLII. Reconstitution of the electron transfer system, *J. Biol. Chem.* **237**:2661.

Hazelbauer, G. L., and Changeux, J.-P., 1974, Reconstitution of a chemically excitable membrane, *Proc. Natl. Acad. Sci. USA* **71**:1479.

Hesketh, T. R., Smith, G. A., Honslay, M. D., McGill, K. A., Birdsall, N. J. M., Metcalfe, J. C., and

Warren, G. B., 1976, Annular lipids determine the ATPase activity of a calcium transport protein complexed with dipalmitoyllecithin, *Biochemistry* **15**:4145.

Hilden, S., and Hokin, L. E., Active potassium transport coupled to active sodium transport in vesicles reconstituted from purified sodium and potassium ion-activated adenosine triphosphatase from the rectal gland of *Squalus acanthias, J. Biol. Chem.* **250**:6296.

Hilden, S., Rhee, H. M., and Hokin, L. E., 1974, Sodium transport by phospholipid vesicles containing purified sodium and potassium ion-activated adenosine triphosphatase, *J. Biol. Chem.* **249**:7432.

Hinkle, P. C., Kim, J. J., and Racker, E., 1972, Ion transport and respiratory control in vesicles formed from cytochrome oxidase and phospholipids, *J. Biol. Chem.* **247**:1338.

Hirata, H., Sone, N., Yoshida, M., and Kagawa, Y., 1976, Solubilization and partial purification of alanine carrier from membranes of a thermophilic bacterium and its reconstitution into functional vesicles, *Biochem. Biophys. Res. Commun.* **69**:665.

Hladky, S. B., and Haydon, D. A., 1970, Discreteness of conductance change in bimolecular lipid membranes in the presence of certain antibiotics, *Nature (London)* **225**:451.

Hladky, S. B., and Haydon, D. A., 1972, Ion transfer across lipid membranes in the presence of gramicidin A. I. Studies of the unit conductance channel, *Biochim. Biophys. Acta* **274**:294.

Hodgkin, A. L., and Huxley, A. F., 1952, A quantitative description of membrane current and its application to conduction and excitation in nerve, *J. Physiol. (London)* **117**:500.

Hokin, L. E., Dahl, J. L., Deupree, J. D., Dixon, J. F., Hackney, J. F., and Perdue, F., 1973, Studies on the characterization of the sodium and potassium adenosine triphosphatase. X. Purification of the enzyme from the rectal gland of *Squalus acanthias, J. Biol. Chem.* **248**:2593.

Huang, C., 1969, Studies on phosphatidylcholine vesicles. Formation and physical characteristics, *Biochemistry* **8**:344.

Huang, C., and Thompson, T. E., 1974, Preparation of homogeneous, single-walled phosphatidylcholine vesicles, *Methods Enzymol.* **32**:485.

Jacobson, K., and Papahadjopoulos, D., 1975, Phase transitions and phase separations in phospholipid membranes induced by changes in temperature, pH, and concentration of bivalent cations, *Biochemistry* **14**:152.

Jilka, R. L., Martonosi, A. N., and Tillack, T. W., 1975, Effect of purified $(Ca^{+2} + Mg^{+2})$-activated ATPase of sarcoplasmic reticulum upon the passive permeability and ultrastructure of phospholipid vesicles, *J. Biol. Chem.* **250**:7511.

Johnson, L. W., Hughes, M. E., and Zilversmit, D. B., 1975, Use of phospholipid exchange protein to measure inside–outside transposition in phosphatidylcholine liposomes, *Biochim. Biophys. Acta* **275**:176.

Kagawa, Y., and Racker, E., 1971, Partial resolution of the enzymes catalyzing oxidative phosphorylation. XXV. Reconstitution of particles catalyzing $^{32}P_i$–adenosine triphosphate exchange, *J. Biol. Chem.* **246**:5477.

Kagawa, Y., Kandrach, A., and Racker, E., 1973, Partial resolution of the enzymes catalyzing oxidative phosphorylation. XXVI. Specificity of phospholipids required for energy transfer reactions, *J. Biol. Chem.* **248**:676.

Kagawa, Y., Sone, N., Yoshida, M., Hirata, H., and Okamoto, H., 1976, Proton translocating ATPase of a thermophilic bacterium. Morphology, subunits, and chemical composition. *J. Biochem. (Tokyo)* **80**:141.

Kasahara, M., and Hinkle, P. C., 1976, Reconstitution of D-glucose transport catalyzed by a protein fraction from human erythrocytes in sonicated liposomes, *Proc. Natl. Acad. Sci. USA* **73**:396.

Kasahara, M., and Hinkle, P. C., 1977, Reconstitution and purification of the glucose transport protein from human erythrocytes, *J. Biol. Chem.* **252**:7384.

Kasai, M., and Changeux, J.-P., 1971, *In vitro* excitation of purified membrane fragments by cholinergic agents. II. The permeability change caused by cholinergic agents, *J. Membr. Biol.* **6**:24.

Katz, B., 1969, *The Release of Neural Transmitter Substances,* Charles C Thomas, Springfield, Ill.

Katz, B., and Miledi, R., 1975, The statistical nature of the acetylcholine potential and its molecular components, *J. Physiol. (London)* **224**:665.

Kayushin, L. P., and Skulachev, V. P., 1974, Bacteriorhodopsin as an electrogenic proton pump: Reconstitution of bacteriorhodopsin proteoliposomes generating $\Delta\Psi$ and ΔpH, *FEBS Lett.* **39**:39.

Kedem, O., 1961, Criteria of active transport, in: *Membrane Transport and Metabolism* (A. Kleinzeller and A. Kotyk, eds.), pp. 87–93, Academic Press, New York.

Kemp, G., Dolly, J. O., Barnard, E. A., and Wenner, C. E., 1973, Reconstitution of a partially purified endplate acetylcholine receptor preparation in lipid bilayer membranes, *Biochem. Biophys. Res. Commun.* **54**:607.

Kernan, R. P., 1972, Active transport and ionic concentration gradients in muscle, in: *Transport and Accumulation in Biological Systems* (E. J. Harris, ed.), pp. 193–270, University Park Press, Baltimore.

Knowles, A. F., and Racker, E., 1975, Properties of a reconstituted calcium pump, *J. Biol. Chem.* **250**:3538.

Knowles, A. F., Eytan, E., and Racker, E., 1976, Phospholipid–protein interactions in the Ca^{++}-adenosine triphosphatase of sarcoplasmic reticulum, *J. Biol. Chem.* **251**:5161.

Krasne, S., Eisenman, G., and Szabo, G., 1971, Freezing and melting of lipid bilayers and the mode of action of nonactin, valinomycin, and gramicidin, *Science* **174**:412.

Kuriki, Y., and Racker, E., 1976, Inhibition of (Na^+, K^+)adenosine triphosphatase and its partial reactions by quercetin, *Biochemistry* **15**:4951.

Kyte, J., 1971, Purification of the sodium and potassium-dependent adenosine triphosphatase from canine renal medulla, *J. Biol. Chem.* **246**:4157.

LaBelle, E. G., and Racker, E., 1977, Cholesterol stimulation of liposome penetration by hydrophobic compounds, *J. Membr. Biol.* **31**:301.

Lange, Y., Ralph, E. K., and Redfield, A. G., 1975, Observation of the phosphatidyl ethanolamine amino proton magnetic resonance in phospholipid vesicles: Inside/outside ratios and proton transport, *Biochem. Biophys. Res. Commun.* **62**:891.

Lawaczeck, R., Kainosho, M., and Chan, S. I., 1976, The formation and annealing of structural defects in lipid bilayer vesicles, *Biochim. Biophys. Acta* **443**:313.

Lentz, B. R., Barenholz, Y., and Thompson, T. E., 1976, Fluorescence depolarization studies of phase transitions and fluidity in phospholipid bilayers. I. Single component phosphatidylcholine liposomes, *Biochemistry* **15**:4521.

Leung, K. H., and Hinkle, P. C., 1975, Reconstitution of ion transport and respiratory control in vesicles formed from reduced coenzyme Q–cytochrome c reductase and phospholipids, *J. Biol. Chem.* **250**:8467.

Lichtenberg, D., Petersen, N. O., Girardet, J.-L., Kainosho, M., Kroon, P. A., Seiter, C. H. A., Feigenson, G. W., and Chan, S. I., 1975, The interpretation of proton magnetic resonance linewidths for lecithin dispersions. Effect of particle size and chain packing, *Biochim. Biophys. Acta* **382**:10.

Ling, G. N., 1969, A new model for the living cell: A summary of the theory and recent experimental evidence in its support, *Int. Rev. Cytol.* **26**:1.

Ling, G. N., Miller, C., and Ochsenfeld, M. M., 1973, The physical state of solutes and water in living cells according to the association–induction hypothesis, *Ann. NY Acad. Sci.* **204**:6.

Litman, P. J., 1973, Lipid model membranes. Characteristics of mixed phospholipid vesicles. *Biochemistry* **12**:2545.

MacInnes, D. A., 1939, *The Principles of Electrochemistry*, Van Nostrand Reinhold, New York.

MacLennan, D. H., 1970, Purification and properties of an adenosine triphosphatase from sarcoplasmic reticulum, *J. Biol. Chem.* **245**:4508.

MacLennan, D. H., Yip, C. C., Iles, G. H., and Seaman, P., 1972, Isolation of sarcoplasmic reticulum proteins, *Cold Spring Harbor Symp. Quant. Biol.* **37**:469.

Makinose, M., and Hasselbach, W., 1971, ATP synthesis by the reverse of the sarcoplasmic calcium pump, *FEBS Lett.* **12**:271.

Malathi, P., Ramaswamy, K., Caspary, W. F., and Crane, R. K., 1973, Studies on the transport of glucose from disaccharides by hamster small intestine *in vitro*. I. Evidence for a disaccharide-related transport system, *Biochim. Biophys. Acta* **307**:613.

Martonosi, A. N., 1972, Biochemical and clinical aspects of sarcoplasmic reticulum function, in: *Current Topics in Membranes and Transport*, Vol. 3 (F. Bronner and A. Kleinzeller, eds.), pp. 83–197, Academic Press, New York.

McNamee, M. G., Weill, C. L., and Karlin, A., 1975, Purification of acetylcholine receptor from *Torpedo californica* and its incorporation into phospholipid vesicles, *Ann. NY Acad. Sci.* **264**:175.

Meissner, G., and Fleischer, S., 1974, Dissociation and reconstitution of functional sarcoplasmic reticulum vesicles, *J. Biol. Chem.* **249**:302.

Meunier, J.-L., Olsen, R. W., Menez, A., Fromageot, P., Boquet, P., and Changeux, J.-P., 1972,

Some physical properties of the cholinergic receptor protein from *Electrophorus electricus* revealed by a tritiated α-toxin from *Naja nigricollis* venom, *Biochemistry* **11**:1200.

Michaelson, D. M., and Raftery, M. A., 1974, Purified acetylcholine receptor: its reconstitution to a chemically excitable membrane, *Proc. Natl. Acad. Sci. USA* **71**:4768.

Michaelson, D. M., Horowitz, A. F., and Klein, M. P., 1973, Transbilayer asymmetry and surface homogeneity of mixed phospholipids in cosonicated vesicles, *Biochemistry* **12**:2637.

Miller, C., 1978, Voltage-gated cation conductance channel from fragmented sarcoplasmic reticulum, *J. Membr. Biol.* **40**:1.

Miller, C., and Racker, E., 1976a, Fusion of phospholipid vesicles reconstituted with cytochrome c oxidase and mitochondrial hydrophobic protein, *J. Membr. Biol.* **26**:319.

Miller, C., and Racker, E., 1976b, Ca^{++}-induced fusion of fragmented sarcoplasmic reticulum with artificial planar bilayers, *J. Membr. Biol.* **30**:283.

Miller, C., Arvan, P., Telford, J. N., and Racker, E., 1976, Ca^{++}-induced fusion of proteoliposomes: Dependence on transmembrane osmotic gradient, *J. Membr. Biol.* **30**:271.

Mitchell, P., 1966, *Chemiosmotic Coupling in Oxidative and Photosynthetic Phosphorylation*, Glynn Research Ltd., Bodmin, Cornwall.

Montal, M., and Mueller, P., 1972, Formation of bimolecular membranes from lipid monolayers and a study of their electrical properties, *Proc. Natl. Acad. Sci. USA* **69**:3561.

Mueller, P., 1975, Electrical excitability in lipid bilayers and cell membranes, in: *Energy-Transducing Mechanisms*, Vol. 3 (E. Racker, ed.), pp. 75–120, Butterworth's, London.

Mueller, P., and Rudin, D. O., 1969, Translocators in bimolecular lipid membranes: Their role in dissipative and conservative bioenergy transductions, in: *Current Topics in Bioenergetics*, Vol. 3 (D. R. Sanadi, ed.), pp. 157–249, Academic Press, New York.

Mueller, P., Rudin, D. O., Tien, H. T., and Wescott, W. C., 1962, Reconstitution of cell membrane structure *in vitro* and its transformation into an excitable system, *Nature (London)* **194**:979.

Neher, E., and Sackmann, B., 1976, Single-channel currents recorded from membrane of denervated frog muscle-fibers, *Nature (London)* **260**:799.

Oesterhelt, D., 1976, Bacteriorhodopsin as an example of a light-driven proton pump, *Angew. Chem. Int. Ed. Engl.* **15**:17.

Oesterhelt, D., and Stoeckenius, W., 1973, Functions of a new photoreceptor membrane, *Proc. Natl. Acad. Sci. USA* **70**:2853.

Pagano, R., and Thompson, T. E., 1967, Spherical lipid bilayer membranes, *Biochim. Biophys. Acta* **144**:666.

Panet, R., and Selinger, Z., 1972, Synthesis of ATP coupled to Ca^{2+} release from sarcoplasmic reticulum vesicles, *Biochim. Biophys. Acta* **255**:34.

Papahadjopoulos, D., and Miller, N., 1967, Phospholipid model membranes. I. Structural characteristics of hydrated liquid crystals, *Biochim. Biophys. Acta* **135**:624.

Papahadjopoulos, D., Poste, G., Schaeffer, B. E., and Vail, W. J., 1974, Membrane fusion and molecular segregation in phospholipid vesicles, *Biochim. Biophys. Acta* **352**:10.

Parisi, M., Rivas, E., and DeRobertis, E., 1971, Conductance changes produced by acetylcholine in lipidic membranes containing a proteolipid from *Electrophorus, Science* **172**:56.

Penefsky, H. S., Pullman, M. P., Datta, A., and Racker, E., 1960, Partial resolution of the enzymes catalyzing oxidative phosphorylation, *J. Biol. Chem.* **235**:3330.

Potter, L. T., 1973, Acetylcholine receptors in vertebrate skeletal muscle and electric tissues, in: *Drug Receptors: A Symposium* (H. P. Rang, ed.), pp. 295–312, University Park Press, Baltimore.

Pressman, B. C., Harris, E. J., Jagger, W. S., and Johnson, J. H., 1967, Antibiotic-mediated transport of alkali ions across lipid barriers, *Proc. Natl. Acad. Sci. USA* **58**:1949.

Racker, E., 1972a, Reconstitution of cytochrome oxidase vesicles and conferral of sensitivity to energy transfer inhibitors, *J. Membr. Biol.* **10**:221.

Racker, E., 1972b, Reconstitution of a calcium pump with phospholipids and a purified Ca^{++} adenosine triphosphatase from sarcoplasmic reticulum, *J. Biol. Chem.* **247**:8198.

Racker, E., 1973, A new procedure for the reconstitution of biologically active phospholipid vesicles, *Biochem. Biophys. Res. Commun.* **55**:224.

Racker, E., 1974a, Oxidative phosphorylation, in: *Molecular Oxygen in Biology: Topics in Molecular Oxygen Research* (O. Hayashi, ed.), pp. 339–361, North-Holland, Amsterdam.

Racker, E., 1974b, Mechanism of ATP formation in mitochondria and ion pumps, in: *Dynamics of*

Energy-Transducing Membranes (L. Ernster, R. D. Estabrook, and E. C. Slater, eds.), pp. 269–281, American Elsevier, New York.

Racker, E., 1975, Reconstitution of membranes, in: *Proceedings of the Tenth FEBS Meeting: Biological Membranes*, Vol. 41 (J. Montreuil and P. Mandel, eds.), pp. 25–34, North-Holland, Amsterdam.

Racker, E., 1976, *A New Look at Mechanisms in Bioenergetics*, Academic Press, New York.

Racker, E., and Eytan, E., 1973, Reconstitution of efficient Ca^{++} pump without detergents, *Biochem. Biophys. Res. Commun.* **55**:174.

Racker, E., and Eytan, E., 1975, A coupling factor from sarcoplasmic reticulum required for the translocation of Ca^{2+} ions in a reconstituted Ca^{2+} ATPase pump, *J. Biol. Chem.* **250**:7533.

Racker, E., and Fisher, L. W., 1975, Reconstitution of an ATP-dependent sodium pump with an ATPase from electric eel and pure phospholipids, *Biochem. Biophys. Res. Commun.* **67**:1144.

Racker, E., and Hinkle, P. C., 1974, Effect of temperature on the function of a proton pump, *J. Membr. Biol.* **17**:181.

Racker, E., and Kandrach, A., 1973, Partial resolution of the enzymes catalyzing oxidative phosphorylation. XXXIX. Reconstitution of the third segment of oxidative phosphorylation, *J. Biol. Chem.* **248**:5841.

Racker, E., and Stoeckenius, W., 1974, Reconstitution of purple membrane vesicles catalyzing light-driven proton uptake and ATP formation, *J. Biol. Chem.* **249**:662.

Racker, E., Chien, T.-F., and Kandrach, A., 1975a, A cholate dilution procedure for the reconstitution of the Ca^{++} pump, $^{32}P_i$–ATP exchange, and oxidative phosphorylation, *FEBS Lett.* **57**:14.

Racker, E., Knowles, A. F., and Eytan, E., 1975b, Resolution and reconstitution of ion-transport systems, *Ann. NY Acad. Sci.* **264**:17.

Ragan, C. I., and Hinkle, P. C., 1975, Ion transport and respiratory control in vesicles formed from reduced nicotinamide adenine dinucleotide coenzyme Q reductase and phospholipids, *J. Biol. Chem.* **250**:8472.

Ragan, C. I., and Racker, E., 1973, Partial resolution of the enzymes catalyzing oxidative phosphorylation. XXVIII. The reconstitution of the first site of energy conservation, *J. Biol. Chem.* **248**:563.

Ramaswamy, K., Malathi, P., Caspary, W. F., and Crane, R. K., 1974, Studies on the transport of glucose from disaccharides by hamster small intestine *in vitro*. II. Characteristics of the disaccharide-related transport system, *Biochim. Biophys. Acta* **345**:39.

Ramaswamy, K., Malathi, P., and Crane, R. K., 1976, Demonstration of hydrolase-related glucose transport in brush border membrane vesicles prepared from guinea pig small intestine, *Biochem. Biophys. Res. Commun.* **68**:162.

Rang, H. P., 1975, Acetylcholine receptors, *Q. Rev. Biophys.* **7**:283.

Rang, H. P., and Ritter, J. M., 1970, On the mechanism of desensitization at cholinergic receptors, *Mol. Pharmacol.* **6**:357.

Reynolds, J. A., and Karlin, A., 1978, Molecular weight in detergent solution of acetylcholine receptor from *Torpedo californica*, *Biochemistry* **17**:2035.

Romine, W. O., Goodall, M. C., Peterson, J., and Bradley, R. J., 1974, The acetylcholine receptor. Isolation of a brain nicotinic receptor and its preliminary characterization in lipid bilayer membranes, *Biochim. Biophys. Acta* **367**:316.

Rothman, J. E., and Dawidowicz, E. A., 1975, Asymmetric exchange of vesicle phospholipids catalyzed by the phosphatidylcholine exchange protein. Measurement of inside–outside transitions, *Biochemistry* **14**:2809.

Ruiz-Manresa, F., and Grundfest, H., 1971, Synaptic electrogenesis in eel electroplaques, *J. Gen. Physiol.* **57**:71.

Rydström, J., Kanner, N., and Racker, E., 1975, Resolution and reconstitution of mitochondrial nicotinamide nucleotide transhydrogenase, *Biochem. Biophys. Res. Commun.* **67**:831.

Schwartz, A., Lindenmeyer, G. E., and Allen, J. C., 1972, The Na^+,K^+–ATPase membrane transport system: Importance in cellular function, in: *Current Topics in Membranes and Transport*, Vol. 3 (F. Bronner and A. Kleinzeller, eds.), pp. 1–82, Academic Press, New York.

Sears, B., Hutton, W. C., and Thompson, T. E., 1976, Effects of paramagnetic shift reagents on the ^{13}C nuclear magnetic resonance spectra of egg phosphatidylcholine enriched with ^{13}C in the N-methyl carbons, *Biochemistry* **15**:1635.

Semenza, G., 1976, Small intestinal disaccharides: Their properties and role as sugar translocators across natural and artificial membranes, in: *The Enzymes of Biological Membranes,* Vol. 3 (A. Martonosi, ed.), pp. 349–382, Plenum Press, New York.

Serrano, R., Kanner, B. I., and Racker, E., 1976, Purification and properties of the proton-translocating adenosine triphosphatase complex of bovine heart mitochondria, *J. Biol. Chem.* **251**:2453.

Shamoo, A. E., and Albers, R. W., 1973, Na^+-selective ionophoric material derived from electric organ and kidney membranes, *Proc. Natl. Acad. Sci. USA* **70**:1191.

Shamoo, A. E., and Eldefrawi, M. E., 1975, Carbamylcholine and acetylcholine-sensitive cation-selective ionophore as part of the purified acetylcholine receptor, *J. Membr. Biol.* **25**:47.

Shamoo, A. E., and MacLennan, D. H., 1974, A Ca^{++}-dependent and -selective ionophore as part of the $Ca^{++} + Mg^{++}$-dependent adenosine triphophatase of sarcoplasmic reticulum, *Proc. Natl. Acad. Sci. USA* **71**:3522.

Shamoo, A. E., and MacLennan, D. H., 1975, Separate effects of mercurial compounds on the ionophoric and hydrolytic functions of the $(Ca^{++} + Mg^{++})$–ATPase of sarcoplasmic reticulum, *J. Membr. Biol.* **25**:65.

Shamoo, A. E., and Myers, M., 1974, Na^+-dependent ionophore as part of the small polypeptide of the $(Na^+ + K^+)$–ATPase from eel electroplax membrane, *J. Membr. Biol.* **19**:163.

Shchipakin, V., Chuchlova, E., and Evtodienko, Y., 1976, Reconstitution of mitochondrial H^+-transporting system in proteoliposomes, *Biochem. Biophys. Res. Commun.* **69**:123.

Sheetz, M., and Chan, S. I., 1972, Effect of sonication on the structure of lecithin bilayers, *Biochemistry* **11**:4573.

Shertzer, H., and Racker, E., 1974, Adenine nucleotide transport in submitochondrial particles and reconstituted vesicles derived from bovine heart mitochondria, *J. Biol. Chem.* **249**:1320.

Shertzer, H. G., and Racker, E., 1976, Reconstitution and characterization of the adenine nucleotide transporter derived from bovine heart mitochondria, *J. Biol. Chem.* **251**:2446.

Shieh, P., and Packer, L., 1976, Photo-induced potential across a polymer stabilized planar membrane, in the presence of bacteriorhodopsin, *Biochem. Biophys. Res. Commun.* **71**:603.

Skou, J. C., 1957, The influence of some cations on an adenosine triphosphatase from peripheral nerve, *Biochim. Biophys. Acta* **23**:394.

Skou, J. C., 1975, The $(Na^+ + K^+)$ activated enzyme system and its relationship to transport of sodium and potassium, *Q. Rev. Biophys.* **7**:401.

Skulachev, V. P., 1976, Conversion of light energy into electric energy by bacteriorhodopsin, *FEBS Lett.* **64**:23.

Slater, E. C., 1953, Mechanism of phosphorylation in the respiratory chain, *Nature (London)* **172**:975.

Slater, E. C., 1971, The coupling between energy-yielding and energy-utilizing reactions in mitochondria, *Q. Rev. Biophys.* **4**:35.

Sone, N., Yoshida, M., Hirata, H., and Kagawa, Y., 1975, Purification and properties of a dicyclohexylcarbodiimide-sensitive adenosine triphosphatase from a thermophilic bacterium, *J. Biol. Chem.* **250**:7917.

Stevens, C. F., 1972, Inferences about membrane properties from electrical noise measurements, *Biophys. J.* **12**:1028.

Stewart, P. S., MacLennan, D. H., and Shamoo, A. E., 1976, Isolation and characterization of tryptic fragments of the adenosine triphosphatase of sarcoplasmic reticulum, *J. Biol. Chem.* **251**:712.

Stockton, G. W., Polnaszek, C. F., Tullock, A. P., Hasan, F., and Smith, I. C. P., 1976, Molecular motion and order in single-bilayer vesicles and multilamellar dispersions of egg lecithin and lecithin–cholesterol mixture. A deuterium nuclear magnetic resonance study of specificially labelled lipids, *Biochemistry* **15**:954.

Storelli, C., Vögeli, H., and Semenza, G., 1972, Reconstitution of a sucrase-mediated sugar transport system in lipid bilayers, *FEBS Lett.* **24**:287.

Suurkuusk, J., Lentz, B. R., Barenholz, Y., Biltonen, R. L., and Thompson, T. E., 1976, A calorimetric and fluorescent probe study of the gel–liquid crystalline phase transition in small, single-lamellar dipalmitoylphosphatidylcholine vesicles, *Biochemistry* **15**:1393.

Sweadner, K. J., and Goldin, S. M., 1975, Reconstitution of active ion transport by the sodium and potassium ion-stimulated adenosine triphosphate from canine brain, *J. Biol. Chem.* **250**:4022.

Thompson, T. E., and Henn, F. A., 1970, Experimental model membranes, in: *Membranes of Mitochondria and Chloroplasts* (E. Racker, ed.), pp. 1–52, Van Nostrand Reinhold, New York.

Vieira, F. L., Caplan, S. R., and Essig, A., 1972, Energetics of sodium transport in frog skin. II. The effects of electrical potential on oxygen consumption, *J. Gen. Physiol.* **59**:77.

Warren, G. B., and Metcalfe, J. C., 1977, What is the phospholipid specificity of a reconstituted calcium pump? *Biochem. Soc. Trans.* **5**:517.

Warren, G. B., Toon, P. A., Birdsall, N. J. M., Lee, A. G., and Metcalfe, J. C., 1974, Reconstitution of a calcium pump using defined membrane components, *Proc. Natl. Acad. Sci. USA* **71**:622.

Whittam, R., and Chipperfield, A. R., 1975, The reaction mechanism of the sodium pump, *Biochim. Biophys. Acta* **415**:149.

Wilbrandt, W., and Rosenberg, T., 1961, The concept of carrier transport and its corollaries in pharmacology, *Pharmacol. Rev.* **13**:109.

Yoshida, M., Sone, N., Hirata, H., and Kagawa, Y., 1975a, A highly stable ATPase from a thermophilic bacterium. Purification, properties, and reconstitution, *J. Biol. Chem.* **250**:7910.

Yoshida, M., Sone, N., Hirata, H., Kagawa, Y., Takeuchi, Y., and Ohno, K., 1975b, ATP synthesis catalyzed by purified DCCD-sensitive ATPase incorporated into reconstituted purple membrane vesicles, *Biochem. Biophys. Res. Commun.* **67**:1295.

Zachar, J., 1971, *Electrogenesis and Contractility in Skeletal Muscle Cells,* University Park Press, Baltimore.

2

The Pharmacon–Receptor–Effector Concept
A Basis for Understanding the Transmission of Information in Biological Systems

E. J. Ariëns, A. J. Beld, J. F. Rodrigues de Miranda, and A. M. Simonis

1. Introduction

The biological effect of chemical agents can only be the result of an interaction of the molecules of the active agent and particular molecules or molecular complexes present in the biological object. The same holds true for endogenous bioactive agents such as hormones, neurotransmitters, etc. Often these *messengers* have their molecular sites of action, referred to as their specific receptors, located on the outer cell membrane (Lefkowitz *et al.*, 1976). They then transmit information from remote or nearby cells to the surface of the target cells where they trigger these cells, thus inducing certain changes in cell function. The interaction between the bioactive agent and its specific receptors initiates the transduction of information received from outside to the cell interior. An example is the generation of intracellular mediators such as cyclic AMP. Similarly, the hormone–receptor complex, in the case of the interaction of steroid hormones and their specific proteins in the cytoplasm, is involved in the transduction of information to the cell nucleus. The presence of specific receptors for the bioactive agents either on the outer cell membrane or in the cytoplasm of only particular tissue cells accounts for a selectivity in action on these tissues.

Exogenous bioactive agents, be they drugs, pesticides, toxics, etc., generally indicated as pharmaca (pharmacon = active principle), also function as messengers, transferring specific information from outside the organism to particular

E. J. Ariëns, A. J. Beld, J. F. Rodrigues de Miranda, and A. M. Simonis ● Pharmacological Institute, University of Nijmegen, Nijmegen, The Netherlands.

cells or even organelles. External messengers often act on receptor sites also involved in the action of endogenous messengers. They may act as activators or blockers of the receptor–effector systems. For those pharmaca which act as activators, there are various similarities in the characteristics of the dose–response curves with those of the endogenous messengers, as, for example, high selectivity and sensitivity of the receptor system for the agents and "on–off" characteristics in which a very small concentration range covers the full response curve. Some of the factors contributing predominantly to the characteristics will be discussed on the basis of an analysis of the various steps in the sequence of events leading from the pharmacon–receptor interaction to the final response.

2. Biological Action

In the sequence of processes which form the basis of drug action three main phases can be distinguished (Fig. 1) (Ariëns, 1974):

1. *The pharmaceutical phase*, comprising the processes involved in the release of the active agent. This is a function of dosage form and its distribution in the extramural compartment, thus determining the concentration of the active agent available for absorption (the pharmaceutical availability).
2. *The pharmacokinetic phase*, comprising the processes playing a part in the absorption, distribution, metabolic conversion, and excretion of the drug, thus determining the concentration of the active agent in the extracellular fluid and therewith at the sites of action in the target tissue (the biological availability).
3. *The pharmacodynamic phase*, comprising the molecular interaction of the active agent with its specific site of action. This initiates the sequence of processes which finally results in the biological effect measured. Pharmacon–receptor interaction in the strict sense is thus restricted to the pharmacodynamic phase.

The three main steps in the sequence of processes of the pharmacodynamic phase are (Fig. 2) (Ariëns and Simonis, 1976; Goldberg, 1975): (1) the interaction of the pharmacon with the receptor, resulting in the induction of an *initial stimulus*; (2) the *transduction, modulation, and amplification* processes, transmitting the initial stimulus to the molecular effector system, as for instance, those found in the contractile proteins in muscle cells; and (3) the *generation of an effect*.

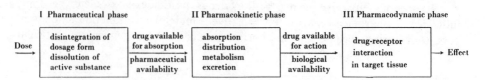

Fig. 1. Main phases in drug action.

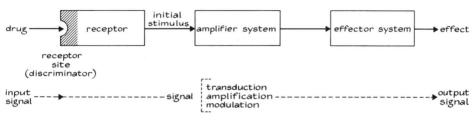

Fig. 2. Main steps in the pharmacodynamic phase of action.

3. Receptors and Receptor Sites

The interaction between the bioactive agent (the pharmacon) and its receptors is much more dynamic than the classical lock-and-key model suggests. It is actually the result of *intermolecular forces mutually molding drug and receptor* (Ariëns, 1964). Conformational changes in the receptor molecule are induced simultaneously, triggering the sequence of biochemical and biophysical events leading to an effect. Although dynamic in nature, the receptors can be regarded as preexisting structural entities. For instance, the optical isomers of an agent often differ largely in their biological activity. It is generally assumed that a certain degree of structural complementarity between the drug and the site of action on the specific receptor molecule is required. Consequently, the chemical properties of an agent determine the type of action as well as its magnitude. Structure–activity relationships find their justification in this basic assumption. However, the structural requirements for action are not always highly specific. For local anesthetic agents, a certain degree of lipophilicity and polarity are the main requirements. These agents probably act by accumulation in the lipid phase of the cell membrane in which proteins functionally involved in ion transport processes are embedded. The local anesthetics probably shift the physicochemical properties of the lipid annulus from the gel to the liquid–crystalline state (Lee, 1977). For the various gaseous anesthetics a certain lipophilicity appears to be the only requirement for action. The action of these general anesthetics is mainly based on a change in the membrane properties due to a diffuse accumulation of these agents in the lipid fraction of the membrane (Haydon *et al.*, 1977). For plasma extenders and osmotic diuretics the term receptor is not applicable anymore since they act mainly by binding water. The receptors for a drug do not necessarily have to be located in the organ in which its response is observed. Strychnine, for instance, causes convulsions in striated muscle, but the receptors on which the effect is induced are located in the central nervous system. In the case of pharmacon–receptor interaction, the conformational changes in the receptor are particularly important in relation to the effects. (In the case of an enzyme–substrate interaction, the conformational changes in the substrate result in an activation with subsequent chemical conversion.) This is analogous to the differentiation between the *active site* on an enzyme and the *enzyme molecule* as a whole. Thus it makes sense to distinguish between the *receptor site* and the *receptor molecule*. Under certain circumstances the receptor site may be constituted from an interface between different molecules. Sometimes the receptor sites for the bioactive agent are the active sites on enzymes involved in the conversion of endogenous substrates. In other cases receptors may be located on sites of the enzyme molecule topographically different from the active site, thus changing enzyme action via allosteric mechanisms. The concept of a pharmacon–receptor interaction as the basis of action is not always applicable in its simplest form. In many cases, the active agent is not the pharmacon applied but its derivative(s) or metabolite(s). This holds true for the various pro-drugs, precursors of the ultimate active agent, which are designed to fulfill particular requirements, as, for example, in maximizing absorption or distribution. In this context, the biotoxification of various agents brought about by reactive alkylating agents with carcinogenic or mutagenic action should be mentioned. In these cases, the pharmacon–receptor concept can be applied to the ultimate pharmacon or ultimate toxic (Heidelberger, 1973; Ariëns and Simonis, 1977). The generation of the ultimate pharmacon can be considered part of the pharmacokinetic phase of action. The pharmacodynamic phase then comprises the interaction of the ultimate pharmacon with its molecular sites of action, its specific receptors in the biological object, and the results it causes. In this regard, indirectly acting agents should be mentioned. They act by virtue of other endogenous agents or mediators. Examples are releasers of neurotransmitters (e.g., norepinephrine), liberators of histamine, and compounds which inhibit the re-uptake of neurotransmitters in the storage sites at presynaptic nerve endings, thus potentiating the action of the neurotransmitter. In cases where the drugs act as inhibitors of enzymes involved in the inactivation of endogenous mediators, such as acetylcholinesterase inhibitors, it is the accumulation of the mediators which lead to the effect. The pharmacon–receptor interaction concept is still applicable here since the active agents have their specific sites of action on the membrane of the storage granules of nerve endings or on the enzymes involved. A borderline case is the situation where the activity of a pharmacon is based on a change in the pharmacokinetics of another foreign-body agent, as, for example, the delay or enhancement of bioactivation or bioinactivation or change in distribution due to displacement from binding sites on plasma albumin. Usually such effects are considered under the heading pharmacokinetics; however, the change in the pharmacokinetics of the first agent in this case is due to the effect induced by the second agent on its sites of action, be it an enzyme or plasma albumin. The various types of binding sites for bioactive agents that are not directly involved in the induction of the effect, such as those on plasma albumin, are called silent receptors or sites of loss. They are important in pharmacokinetics. The same holds true for active sites on enzymes involved in the biochemical conversion of pharmaca.

   Although the receptor concept was mainly of theoretical significance in the past, it has evolved into concrete reality, and receptors can now be studied as tangible molecular entities. The design of new drugs can be based on relationships between structure and receptor affinities measured by direct binding studies (Creese *et al.*, 1976; Raynaud, 1977a,b), receptor density measurements in malignant tissue can tell whether hormonal treatment is indicated or not (McGuire *et al.*, 1975), and alterations in receptor properties and density distributions enhance insight into certain pathological conditions such as myasthenia gravis and diabetes (Grob, 1976; Maugh, 1976).

4. Pharmacon–Receptor Interaction

The efforts to interpret the characteristics of dose–response curves and time–response curves of drugs boil down to some form of application of the mass–action law. A variety of partially overlapping theoretical models has been worked out and can be found in various reviews (Ariëns, 1966b, 1974; Colquhoun, 1973; Monod et al., 1965; Rang and Ritter, 1970a,b; Waud, 1968, 1974, 1976; Belleau, 1964; Paton, 1961; Triggle, 1965; Lüllmann and Ziegler, 1973; Lüllmann et al., 1974; Gosselin and Gosselin, 1973; Burgen, 1970; Rocha e Silva, 1970; Schueler, 1960; Van den Brink, 1973; Thron, 1973; Karlin, 1967; Podleski and Changeux, 1970; Clark, 1937), and an extensive treatise of distinct drug–receptor interaction models can be found in the monograph by Van Rossum (1977). The following sequence of reversible reactions is a condensed form covering essential aspects of these models:

$$D + R \underset{k_{-1}}{\overset{k_1}{\rightleftharpoons}} DR \underset{k_{-2}}{\overset{k_2}{\rightleftharpoons}} DR^* \underset{k_{-3}}{\overset{k_3}{\rightleftharpoons}} D + R^* \underset{k_{-4}}{\overset{k_4}{\rightleftharpoons}} \underline{R} \underset{k_{-5}}{\overset{k_5}{\rightleftharpoons}} R$$

where D = drug, R = receptive receptor, R^* = activated receptor, and $\underline{R}$ = nonreceptive, nonactivated receptor (Ariëns and Simonis, 1976). K_1, K_2, K_3, K_4, and K_5 are the equilibrium constants governing the relative amounts of the different receptor species and are defined in such a way that $K_i = k_{-i}/k_i$. Receptor species DR^* and R^* contribute to the stimulus formation, and species R, DR, and $\underline{R}$ do not. Total concentration of receptors is represented by r. In the *occupation–activation model* it is assumed that the activated state of the receptor exists predominantly as a complex with the pharmacon, i.e., as DR^*. For this model it holds true that at equilibrium

$$\frac{[DR^*]}{[R^*]} = K_3[D] \gg 1$$

In the *hit–activation model* it is assumed that the activated state of the receptor exists predominantly in the noncomplexed form i.e., as R^*. Thus for this model it holds true that at equilibrium

$$\frac{[DR^*]}{[R^*]} = K_3[D] \ll 1$$

For both these models the *stimulus* is, by definition, proportional to the *fraction of receptors present in the activated state*; i.e.,

$$S(:)\frac{(DR^* + R^*)}{r} \approx \frac{(DR^*)}{r}$$

for the *occupation* model, and

$$S(:)\frac{(DR^* + R^*)}{r} \approx \frac{R^*}{r}$$

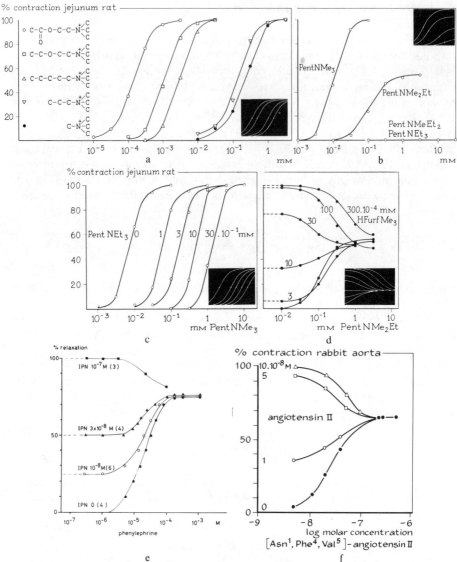

Fig. 3(a). Cumulative log concentration–response curves for a series of quaternary ammonium compounds. Note that the decrease in activity as a result of changes in the chain manifests itself as an increase of the dose necessary to obtain the effect. Compare the experimental curves with the set of theoretical curves (inset). (b) Cumulative log concentration–response curves for a series of pentyl ammonium compounds. Note that the decrease in activity as a result of the gradual ethylation on the ammonium group manifests itself as a decrease in the maximal effect and in the slope of the curves. Compare the experimental curves with the set of theoretical curves (inset). (c) Cumulative log concentration–response curves for the agonistic compound pentyl-NMe₃ and the influence thereon of various concentrations of the "inactive" compound pentyl-NEt₃ [see Fig. 3(b)]. Note the parallel shift in the curves, which indicates a competitive antagonism. Compare the experimental curves with the set of theoretical curves (inset). (d) Cumulative log concentration–response curves for the partial agonist, pentyl-NMe₂Et [see Fig. 3(b)], and the influence thereon of various concentrations of the full agonist furtrethonium (HFurfME₃). Note the dualistic character in the action of the partial agon-

for the *hit* model. The effect as a whole will in general not be proportional to the stimulus, although under comparable circumstances a certain stimulus is postulated to bring about the same effect. In the *rate theory of drug action* originally postulated by Paton (1961), stimulus formation is proportional to the rate of receptor activation. The rate concept can be applied both to the occupation model, i.e., $S(:)k_2(DR)$, and to the hit and run model, i.e., $S(:)k_3(DR^*)$. The presence of the nonreceptive, nonactivated species $\underline{R}$ is not essential in both occupation and hit models; however, it accounts for the degree of fade in the dose–response curves or for the degree of specific desensitization of receptors and thus for tachyphylaxis. The extent of the phenomenon is dependent on whether K_5 is small or large. In rate theory, the regeneration step in which $\underline{R}$ is converted to R is not needed to explain fade or desensitization since here the initial rate of activation is always higher than the rate of activation at equilibrium.

There is growing experimental evidence favoring the occupation model. A classical example is the action of vitamins and vitamin analogues that serve as coenzymes for apoenzymes. The binding of the *steroid hormones* to their receptor proteins in the cytosol followed by the action of the complex as a derepressor in the cell nucleus can also be regarded as a combination of a *coderepressor*, the steroid hormone, with an *apoderepressor*, the receptor protein, to form an *active derepressor* (Ariëns and Simonis, 1976). In fact, the interaction of the pharmacon with the receptor changes the functional state of the receptor molecule.

In pharmacon–receptor interactions one can distinguish between (1) the constants primarily determining the fraction of the receptors involved in the interaction and thus giving the *affinity* of the pharmacon for the receptor, represented by K_1 and (2) the constants for the fraction of the receptors present in the activated state as a result of the drug–receptor interaction and giving the *intrinsic activity* of the pharmacon, represented by K_2. In those cases where the drug–receptor interaction does not result in a receptor activation ($K_2 \ll 1$), the pharmacon acts as a blocker of the receptors, and the intrinsic activity is zero. Compounds with an intermediate intrinsic activity will act as partial agonists; the intrinsic activity is a measure of the fraction of the receptor occupations which give receptor activation (Ariëns, 1974; Ariëns and Simonis, 1964).

The basic characteristics of the dose–response curves obtained with pharmaca with different affinities and intrinsic activities and with combinations of agonists and competitive antagonists as well as agonists and partial agonists are represented in Figs. 3(a)–(f). The characteristics of these curves are in good agreement with the characteristics of theoretical curves calculated for the occupation model with the restriction that $K_5 \ll 1$. This in fact means that tachyphylaxis phenomena are not considered.

In the theoretical models, the operational parameters, affinity and intrinsic

---

ist. Compare the experimental curves with the set of theoretical curves (inset). (e) Cumulative log concentration–response curves for the partial agonist phenylephrine in the presence of various concentrations of the full agonist isoprenaline tested on the guinea pig tracheal chain (Reinhardt and Wagner, 1974). (f) Cumulative log concentration–response curves for the partial agonist [Asn[1], Phe[4], Val[5]]–angiotensin II in the presence of various concentrations of the full agonist angiotensin II tested on the rabbit aorta. Note the dualistic character in the action of the partial agonists (Papadimitriou and Worcel, 1974).

activity, reduce to terms based on rate constants and equilibrium constants. This does not invalidate their practical use, however. On the contrary, the models indicate which constants are predominant for the affinity and which ones for the intrinsic activity. This may be of use in interpreting structure–activity relationships, the basis for understanding drug action on a molecular level and thus drug design.

The receptor in the activated state, as a rule, will have to interact with specific effector molecules. This concerns the first step in the sequence of chemical and physicochemical events of the stimulus–effect relationship. The activation of the receptor in fact implies the formation of an *active site* on the receptor molecule, the site being involved in the interaction with the effector molecule. In the case of an interaction of the receptor with an antagonist molecule, no proper active site is formed on the receptor. In the occupation–activation concept, two cases can be shown:

1. The pharmacon "participates" in the formation of an active site on the receptor molecule (Rudinger *et al.*, 1972). The *participation* of various vitamins and vitamin analogues in the constitution of the active site on redox enzymes is an example. The *receptor site then is an incomplete active site*. Certain smaller polypeptide hormones may serve similarly in completing the active site on enzymes (Rudinger *et al.*, 1972).
2. The pharmacon activates the receptor molecule by a change in its conformation, thus giving *an allosteric generation of an active site* on the receptor as a result of its interaction with the pharmacon.

5. Spare Receptors

When agents with alkylating properties and a capacity to block specific receptor sites by covalently binding to them became available, some remarkable phenomena were noticed. It appeared that in various cases, after irreversible blockade of a high proportion of the receptors available, a maximal effect in the tissue concerned could still be obtained (Ariëns *et al.*, 1960; Furchgott, 1954; Nickerson, 1956). Only with high concentrations of the alkylating compound or only after prolonged incubation with the compound was a depression of the maximal attainable effect observed. Obviously there is no proportionality between the initial stimulus and the effect obtained. Apparently, there is an overcapacity insofar as the induction of the stimulus is concerned, or in other words, only a small fraction of the potentially possible stimulus is required to obtain a maximal response from the effector system.

This phenomenon is now well understood on the basis of *amplifier systems*, which are frequently involved in the pharmacodynamic phase of drug action (Ariëns and Simonis, 1976; Goldberg, 1975; Roach, 1977). Bioactive agents can be extremely potent in an amplifier system. Only a few molecules of the bioactive agent have to interact with its receptors to induce a massive response in which tremendous numbers of molecules are involved. No other mechanism explains this so well. One possible unit for assembling strong amplifier systems is found in the activation of an enzyme by a bioactive agent. The activated enzyme molecule then converts hundreds of substrate molecules to product molecules. Such an

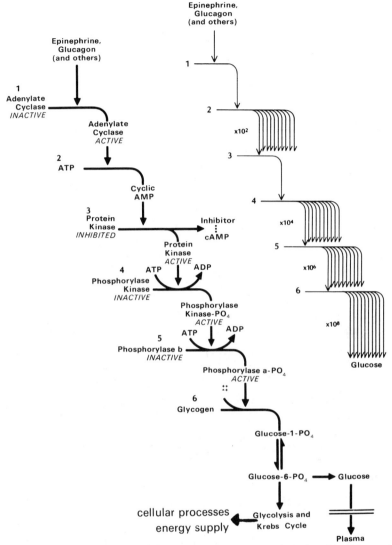

Fig. 4. Schematic representation of the amplifier system involved in the action of various agents acting by activation of adenylate cyclase (Goldberg, 1975).

amplifier unit can be coupled to a second one if the product in its turn activates a second type of enzyme. A sequence of such enzyme activation steps results in tremendous amplification. Such an amplifier system is involved, for instance, in the action of various hormones and drugs that act by activation of adenylate cyclase, an enzyme which converts ATP into cyclic AMP, which in its turn acts as an enzyme activator, etc. (Fig. 4) (Goldberg, 1975). Also illustrative in this respect is the formation of an active derepressor by the binding of steroid hormones to their cytosol receptor. One molecule of the derepressor acting on the DNA in the cell nucleus initiates the synthesis of a large number of messenger RNA molecules. Each messenger molecule in turn can be involved in the generation of a large number of particular protein molecules, giving, for example, an enzyme molecule. This enzyme molecule can then convert large numbers of substrate molecules to product molecules, the generation of which may produce an effect which can be measured.

In the sequence of reactions which form essentially a "cascade amplifier," there will be one step which limits the maximally attainable effect. With a gradual increase in the initial stimulus, one of the enzymes in the sequence will be the first to reach saturation, although the initial stimulus can still be far from its maximal level. A further enhancement of the initial stimulus, although giving an increase in the steps preceding the limiting one, will not result in a further increase in the final effect; the situation will manifest itself as a spare capacity in the receptor system involved. As mentioned previously, irreversible blockade of a certain fraction of the receptors involved is possible without a reduction in the maximal response. The degree of irreversible receptor occupation and therewith functional receptor elimination which is tolerated before the maximal response is reduced is a quantitative measure of the receptor reserve. For the maximal effect of partial agonists a full receptor saturation is required, and there is no spare capacity so that irreversible blockade definitely will reduce the effect (Fig. 5) (Ariëns *et al.*, 1960; Van Rossum, 1968). However, it appears that the existence of a receptor reserve is the general situation and that the absence of a reserve can be regarded as an exception.

The phenomenon of spare capacity is not restricted to the receptor but can be located as well in one of the intermediate steps involved in the generation of an effect. If cyclic AMP is a mediator in a system, there will be in general no proportionality between cyclic AMP generated and the effect measured.

In the analysis of the structure–activity relationship for a number of ACTH

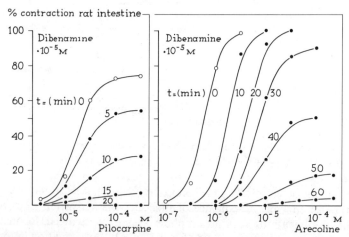

Fig. 5. Cumulative log concentration–effect curves of the cholinergic agents arecoline and pilocarpine before and after exposure of the organ for various times to 10^{-5} M dibenamine, an irreversible blocker of the cholinergic receptors. For the partial agonist, pilocarpine, pretreatment with the irreversible blocking agent results immediately in a depression of the concentration–effect curve, while in the case of the full agonist, arecoline, the concentration–effect curve is first shifted and finally, after longer exposure times, depressed. The interpretation of these results is that there is a receptor reserve for the full agonist arecoline and that there is an absence of a receptor reserve for the partial agonist pilocarpine (Van Rossum, 1968).

analogues, Seelig and Sayers (1973) obtained direct evidence for a spare capacity with regard to cyclic AMP and corticosterone production in adrenal cortex tissue. ACTH analogues differ strongly in their capacity to generate cyclic AMP but not in their capacity to generate corticosterone. Only the ACTH derivative with a minimal capacity for cyclic AMP production behaves as a partial agonist in corticosterone production (Fig. 6) (Seelig and Sayers, 1973).

The operational parameters, affinity (pD_2 and pA_2) and intrinsic activity (indicating the dose at which certain effects are obtained and the capacity of the agents to activate the receptor–effector system—$E_{A\max}$ of the particular drug A), which have been derived from experimental curves, give "apparent" values which may essentially differ from the values of the parameters actually involved in the induction of the initial stimulus. The differences can be a consequence of the involvement of the stimulus–effect relationship. The closer the experimental

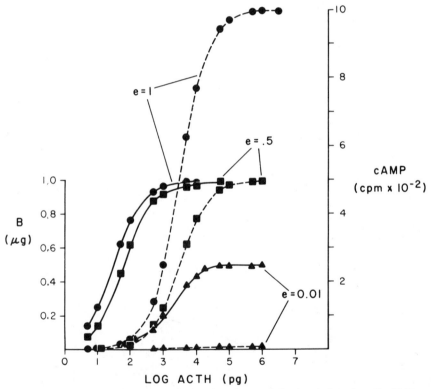

Fig. 6. Log concentration–effect curves for corticosterone production (—) and cyclic AMP production (---) derived from the results obtained from experiments with suspensions of isolated adrenal cortex cells in response to $ACTH_{1-24}$ (●), $ACTH_{5-24}$ (■), and $Trp(Nps)^9ACTH_{1-39}$ (▲). Although the maximal production of cyclic AMP by $ACTH_{1-24}$ and $ACTH_{5-24}$ greatly differs, both compounds show equal maximal corticosterone productions. Although the maximal cyclic AMP production by $Trp(Nps)^9ACTH_{1-39}$ is very small, this compound still gives a relatively high maximal corticosterone production—one that is lower, however, than that obtained with the two other derivatives. $ACTH_{1-24}$ and $ACTH_{5-24}$ behave as full agonists and $Trp(Nps)^9ACTH_{1-39}$ as a partial agonist with regard to the corticosterone production. With regard to the cyclic AMP production, the three compounds markedly differ in their intrinsic activity. For $ACTH_{1-24}$ there definitely is a reserve capacity in the induction of the formation of cyclic AMP (Seelig and Sayers, 1973).

effect parameter is related to the initial stimulus, the smaller the difference between the apparent (experimental) and real values will be. Receptor binding studies will give some perspective in this case, although the receptor activation and thus the intrinsic activity as a rule will not be detectable.

The cascades constituted by a sequence of enzyme activations and serving as amplifier systems play a decisive role in the characteristics of the concentration–effect curve. Figure 7 (Roach, 1977) gives a comparison between a concentration–effect curve based on simple Michaelis–Menten kinetics and one based on a two-step cascade system. In the latter case the concentration range which covers the full response curve is only a fraction of that involved in the former case. The steepness of the curve assumes an "on–off" characteristic, and the concentration at which a 50 % value of the effect is obtained is only a fraction of that needed in the case of the simple system. This implies a high sensitivity.

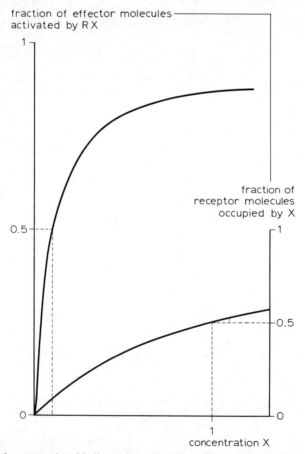

Fig. 7. Calculated concentration–binding and concentration–effect curves. The lower curve represents the binding of pharmacon X ($K_{diss} = 1$) on the basis of simple Michaelis–Menten kinetics. The higher curve represents the activation of an enzyme by binding X ($K_{diss} = 1$), followed by an activation of the effector molecules by the activated enzyme, according to Michaelis–Menten kinetics [in analogy to Fig. 1 in Roach (1977)]. Note the very strong increase in the slope of the curve which results from a first step in the cascade.

As indicated, a large fraction of the receptors may be knocked out chemically with practically full maintenance of the response capacity at only slightly higher concentrations.

One should be well aware of the fact that this holds true only for activators of the receptor–effector systems, the agonists. The action of competitive antagonists follows simple Michaelis–Menten kinetics. For selectivity in action and high affinity, they are dependent solely on the fact that the chemical characteristics of the agents and the receptor sites they act on complement each other. This is known as complementarity.

If the binding of bioactive agents to their specific receptors is studied rather than the effect, the binding curves in the simplest case will be governed by Michaelis–Menten kinetics. As a consequence, the binding constants obtained for competitive antagonists from dose–binding curves and dose–effect curves may be expected to be equal; for agonists where an amplification system is involved in the stimulus–effect relationship the binding constants derived from 50% values in the curves will be substantially lower for dose–binding curves than for dose–effect curves. Indeed this is found to be the case for various anticholinergics and the cholinergic agent acetyl-β-methylcholine (see Sec. 10). The binding constants obtained from dose–effect curves for agonists are apparent binding constants. The values obtained from dose–effect curves for partial agonists are real, since for such agents the pharmacon–receptor interaction is the limiting step in the sequence of events involved. From concentration–effect curves for agonists obtained after irreversible blockade of a proper fraction of the receptors such that the maximal effect obtainable is less than that originally produced (the pharmacon–receptor interaction is the limiting step), real affinity constants can be derived for full agonists. The fraction of receptors that has to be irreversibly blocked in order to reach this situation is a measure for the receptor reserve originally present in the biological system (Ariëns, 1964; Van Rossum and Ariëns, 1962).

An interesting aspect of the on–off characteristics in the concentration–effect curves for agonists acting on receptor–effector systems with a large amplification factor is that only a small fraction of the receptors has to be occupied in order to obtain a full effect. Thus, for the concentration range of the pharmacon involved, there is a proportionality between the concentration of the pharmacon and the concentration of the pharmacon–receptor complex and therewith the initial stimulus. This implies a near proportionality between the concentration of the pharmacon and the effect.

6. Structure and Action

To act as a carrier for selective distribution of specific information the chemical molecule serving as an external *messenger* has to fulfill specific requirements. Many foreign-body bioactive agents are known to act on specific sites of action, the receptor sites, which by nature serve as endogenous bioactive agents (e.g., hormones and neurotransmitters), or on the active sites on enzymes. These receptor sites and active sites discriminate with regard to the molecules in their environment. To warrant specificity, these sites have to contain *recognition points*, i.e., groups capable of forming bonds of different natures based on charge distribution and size with a specific steric configuration. Three different ligands present an optimum number of groups insofar as the discriminating capacity of a small molecule is concerned. One can understand why *stereospecificity* is very common for agents with a selective bioactivity. For the different types of ligands, polar groups (involving point-to-point interaction) are especially suitable recognition points. Hydrophobic ligands are more diffuse in their action, and differentiation in size is a requirement.

For solitary dispersed proteins such as soluble enzymes, the potential for variation in hydrophobic ligands is restricted. For proteins in membranes and for lipoproteins, the size of the hydrophobic ligands as a rule will play no role in discrimination; their steric configuration, however, may do so as long as the lipids are present in a structured, semicrystalline form, as is the case for lipids constituting the annulus around proteins.

The requirements for the ligands in the receptor site or active site and in the bioactive agent are mutual in that they complement each other. This is the basis for their interaction. In the molecules of the agonists and enzyme substrates, besides the chemical requirements essential for the discriminatory capacity for the sites of action, the chemical requirements involved in bringing the receptor molecules in the activated state, and in activating the substrate molecules on the active site of the enzyme so that chemical conversion can take place, must be fulfilled. The antagonists only need a discriminatory capacity with regard to the sites of action. As a rule in the structure of agonists and substrates only a small degree of freedom is allowed, but in the structures of the antagonists usually a large degree of freedom can be tolerated.

The highly active and selective internal and external bioactive messenger molecules contain in their structure the information necessary to fulfill particular functions. They are molecules programmed for the properties required in the various phases of their action. In small agonist molecules, a differentiation is usually difficult to see between the groups in the molecule primarily involved in the receptor binding, the haptophore, and the groups primarily involved in the receptor activation, the actophore. Groups, especially those involved in pharmacon metabolism and pharmacon transport, are more easily recognizable (Ariëns and Simonis, 1974; Ariëns, 1971b). In larger bioactive molecules, especially polypeptides, programming for various aspects of the function may be localized in different amino acid sequences, as shown schematically in Fig. 8 (Hofmann, 1960; Schwyzer, 1970; Hechter and Braun, 1971). Taking into account the large number of recognition points in polypeptides, programming of one molecule for action on a variety of different receptor types is possible. In that case the various components in the spectrum of actions may be separated by suitable molecular manipulation. For example, there are the various actions of arginine–vasopressine (oxypressin). The oxytotic, vasodepressive (chicken), vasopressor (rat), milk ejection, and antidiuretic actions can be separated to a large extent, so that by suitable molecular manipulation, separation and combination of the various different action components are possible (Berde *et al.*, 1961a,b).

An analysis of the relationship between structure and action on a submolec-

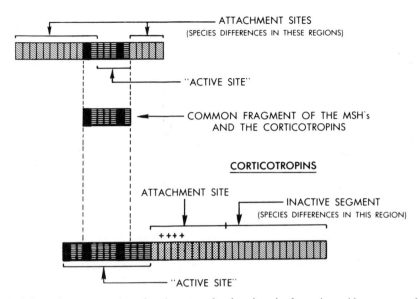

Fig. 8. Schematic representation of various postulated regions in the amino acid sequences of the melanocyte expanding hormones and the corticotropins. Each rectangle represents one amino acid residue.

ular level for agonists and the corresponding competitive antagonists may give insight into the discriminatory factors involved in selectivity of action.

7. Accessory Receptor Sites

Agonists and their competitive antagonists are assumed to act on common receptors and are expected to be chemically complementary to these receptors. They therefore should show a chemical relationship. This indeed holds true for various metabolites and antimetabolites, hormones and antihormones, vitamins and antivitamins, etc., but this is definitely not the case for compounds such as acetylcholine, norepinephrine, histamine, serotonin, and their respective competitive antagonists, anticholinergics, α-adrenergic blocking agents, antihistaminics, and antiserotonins.

A comparison of the various types of agonists and their corresponding competitive antagonists (Fig. 9) (Ariëns, 1974) shows that there is a clear chemical relationship within the various groups of agonists but that there is little or no chemical relationship between the agonists and their competitive antagonists. There is unexpectedly, however, a certain degree of chemical relationship between the various types of competitive antagonists. Furthermore, compounds are known, e.g., chlorpromazine, that have anticholinergic, antihistaminic, and α-adrenergic blocking actions. Such compounds then should be chemically related to different types of agonists such as acetylcholine, histamine, and norepinephrine. This is most unlikely.

An analysis of the structure–activity relationships in a series of compounds

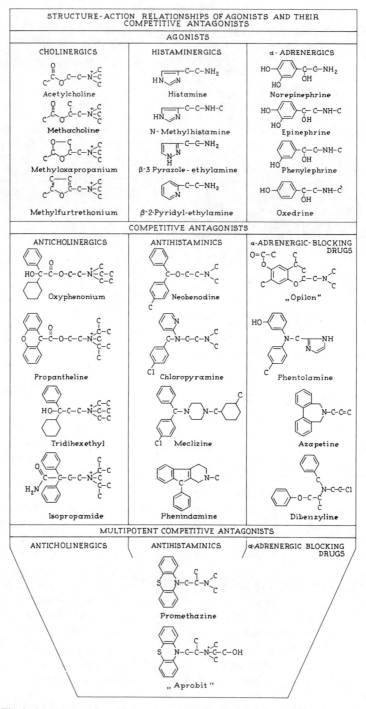

Fig. 9. Structure–action relationships of agonists and their competitive antagonists.

in which a gradual conversion from agonist to competitive antagonist takes place by gradual chemical modification indicates that the hydrophobic double-ring systems common to the antagonists contribute essentially to the affinity of these antagonists for receptors. Their affinities are 1000 to 10,000 times higher than the affinities of the corresponding agonists (Tables I and II) (Ariëns and Simonis, 1960, 1964; Ellenbroek, 1964). The hydrophobic double-ring systems cannot possibly be bound to the same receptor sites to which the highly polar agonists bind (Fig. 10). They must bind to *accessory binding sites* of a predominantly

Table I

Biological Activity of Choline Esters[a]

$$R-\underset{\underset{O}{\|}}{C}-O-C-C-\overset{+}{N}\overset{\diagup C}{\underset{\diagdown C}{}}C$$

R	i.a.	pD$_2$ ± P$_{95}$	pD$_2$ ± P$_{95}$	Est./Org.
H—	1	5.2 ± 0.20		
C—	1	7.0		
C—C—	0.9	5.0 ± 0.30		
(CH$_3$)$_2$CH—	0.4	4.1		
C—C—C—	0.3	3.8		
C—C—C—C—	0		4.7	
C—(C)$_9$—C—	0		5.3 ± 0.20	
phenyl-CH(CH$_2$OH)—	0		8.1 ± 0.06	8/4
phenyl, thienyl-C(OH)—	0		8.2 ± 0.16	78/19
diphenyl-C(OH)—	0		8.6 ± 0.18	26/5
phenyl, cyclohexyl-C(OH)—	0		9.1	4/1

[a] The ± figures give the P$_{95}$ for the mean value; est./org. = number of estimations per number of organs used. The data were obtained by testing the compounds on the isolated gut of the rat.

Table II
Biological Activity of β-2-Pyridylethylamine Derivatives[a]

R	R'	i.a.	$pD_2 \pm P_{95}$	$pA_2 \pm P_{95}$	Est./Org.
—H	—H	1	5.44 ± 0.08		31/24
—C	—H	1	5.46 ± 0.07		36/29
—C—C	—H	0.91 ± 0.03	4.50 ± 0.10		23/27
—C(C)(C) (isopropyl)	—H	0		4.00 ± 0.06	6/2
—C—C—C	—H	0		4.19 ± 0.06	12/5
—C—(C)$_2$—C	—H	0		4.30 ± 0.14	26/13
—C—(C)$_4$—C	—H	0		4.72 ± 0.02	9/4
—C—(C)$_6$—C	—H	0		5.90 ± 0.03	10/3
—C—(C)$_8$—C	—H	0		5.33	3/1
-C-C-C-⬡ (phenyl)	—C	0		5.90 ± 0.13	16/6
C-C-O-C(⬡Cl)(⬡)	—C	0		7.80 ± 0.11	49/10
C-C-N(⬡)(⬡)(C)	—C	0		7.96 ± 0.16	28/11
Histamine		1	6.55 ± 0.13		32/26

[a] The ± figures give the P_{95} for the mean value; est./org. = number of estimations per number of organs used. The data were obtained by testing the compounds on the isolated gut of the guinea pig.

hydrophobic nature located next to the receptor for the corresponding agonist (Ariëns, 1967, 1974; Ariëns and Simonis, 1960, 1964, 1967).

Structure–activity relationship studies in a series of compounds in which acetylcholine, histamine, and norepinephrine are gradually converted by chemical manipulation into their competitive antagonists reveal that with gradual change in structure the affinity of the agonist derivative for its receptor decreases until the hydrophobic double-ring system is introduced. This results in a strong increase in the affinity and indicates that in the competitive antagonists obtained the moiety representing the structure of the original agonist interacts only loosely with the receptor, while the hydrophobic moiety is tightly bound. Elimination of that part of the structure corresponding to the original agonist should be well tolerated without impairing the antagonist action.

This leads to the concept of the accessory receptor areas as major sites of interaction for many of the blocking compounds and is represented schematically in Fig. 11 (Ariëns and Simonis, 1960). It makes feasible the supposition that little if any chemical relationship is required between the agonists and antagonists

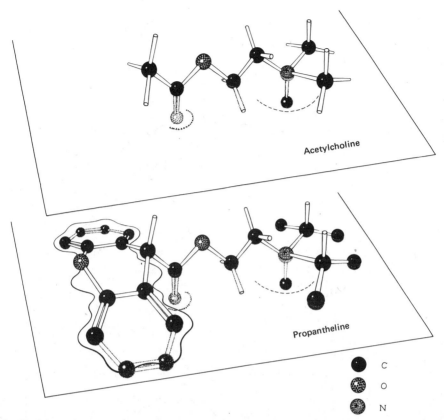

Fig. 10. Schematic representation of acetylcholine and the anticholinergic propantheline on the receptor site of the cholinergic receptor. Note: The hydrophobic double-ring system of propantheline is located on an accessory binding site.

concerned and that one compound can have simultaneously an anticholinergic, an antihistaminic, an α-adrenergic blocking, and an antiserotonin action.

8. *Steric Structure and Action*

Can we obtain further information concerning the interaction of the ring-bearing acyl moiety in the potent anticholinergics with accessory receptor areas differing from but located close to the cholinergic receptors? The question arises, which of the moieties in these anticholinergic drugs is more critical with respect to structural changes, the acyl or the choline moiety? In other words, for which of these moieties is the higher degree of complementarity required? Information on this point can be obtained by introduction of a center of asymmetry in the choline moiety and/or in the acyl moiety. If a *high degree of complementarity* between the moiety concerned and the receptor is required, a *large ratio* for the *activities* of the *stereoisomers* may be expected. As may be seen from Table III, introduction of a center of asymmetry in the choline moiety of acetylcholine, converting this drug to acetyl-β-methylcholine, results in two isomers which

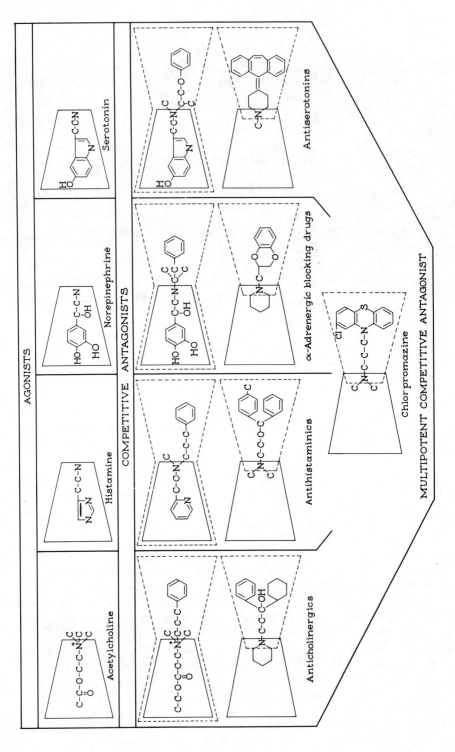

Fig. 11. Structure–action relationships of agonists and their competitive antagonists.

Table III
Stereoisomers and Biological Activity of Choline Esters[a]

Est. Org.	$pD_2 \pm P_{95}$	Config.	Activity ratio	Config.	$pD_2 \pm P_{95}$	Est. Org.
	7.0				7.0	
11/7	6.8 ± 0.14	S_B	320	R_B	4.1 ± 0.23	7/4

	$pA_2 \pm P_{95}$				$pA_2 \pm P_{95}$	
28/9	8.0 ± 0.14	S_B	5/6	R_B	8.1 ± 0.10	31/10
26/5	8.6 ± 0.18				8.6 ± 0.18	26/5
19/18	9.6 ± 0.26	R_A	25	S_A	8.2 ± 0.14	24/23

[a] The $\pm$ figures give the P_{95} for the mean value; est./org. = number of estimations per number of organs used. The data were obtained by testing the compounds on the isolated jejunum of the rat.

show a large l/d ratio, about 200, for the activities. For the stereoisomers obtained by application of the same procedure to the choline moiety of the anticholinergic benzilic acid ester of choline, the ratio for the activities is found to be close to 1. It is apparent that, in contrast to the great influence it has on the activity in acetyl-β-methylcholine, the difference in the steric structure of the choline moiety in the anticholinergic agent has little if any influence on the activity (Ellenbroek, 1964; Ellenbroek *et al.*, 1965).

In the anticholinergic compound benzilyl-β-methylcholine, the choline part of the molecule does not behave as a critical moiety. This indicates that in these ester-type anticholinergic agents the choline moiety has a relation to the receptor which is entirely different from that in the cholinergic compound acetyl-β-methylcholine. If it is true, as suggested before, that the ring-bearing acyl moiety in the potent anticholinergic agents is primarily responsible for the affinity (implying a high degree of complementarity with respect to the accessory receptor areas), introduction of a center of asymmetry in that moiety should result in a large ratio for the activities of the stereoisomers obtained. As can be seen in Table III, this indeed appears to be the case (Ellenbroek, 1964; Ellenbroek *et al.*, 1965).

A further possibility is the introduction of two centers of asymmetry in the potent anticholinergic esters: one in the choline and one in the acyl moiety. The phenylcyclohexyl glycolic acid ester of β-methylcholine is an example (Fig. 12) (Ariëns and Simonis, 1967; Ellenbroek, 1964; Ellenbroek *et al.*, 1965). As a consequence four different isomers are obtained. These can be grouped into two

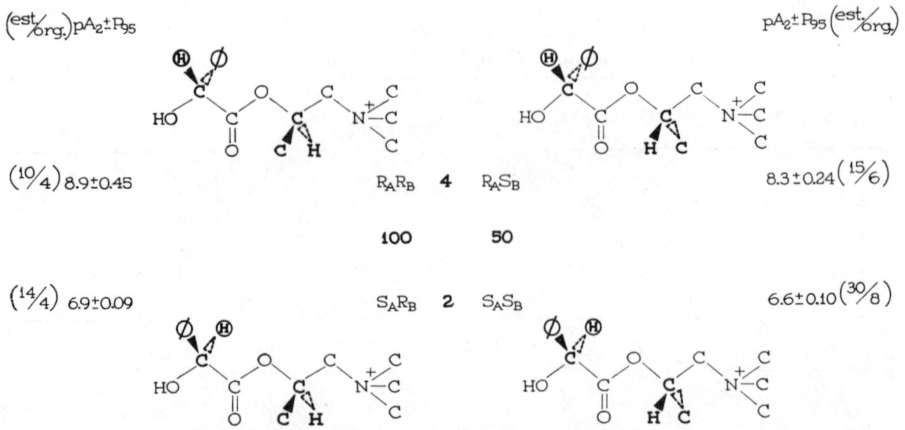

Fig. 12. Stereoisomers and biological activity of choline esters of phenylcyclohexyl glycolic acid activity ratios. The test organ is the rat jejunum. The $\pm$ figures give the P_{95} for the mean value; est./org. = number of estimations per number of organs used (Ellenbroek, 1964; Ellenbroek *et al.*, 1965).

pairs of isomers differing in the steric configuration of the choline moiety and into two pairs of isomers differing on the configuration of the acyl moieties. Experimental results obtained with the phenylcyclohexyl glycolic acid ester of β-methylcholine given in Fig. 12 show that for the pairs differing in the configuration of the choline moiety a low activity ratio is found; for the pairs differing in the configuration of the acyl moiety, large activity ratios are observed (Ariëns and Simonis, 1967; Ellenbroek, 1964; Ellenbroek *et al.*, 1965). This again indicates that in these anticholinergic agents the interaction of the acyl moiety with the accessory receptor area is of greater importance than the interaction of the choline moiety with the cholinergic receptor.

The finding that the activity ratio for optical isomers is large if the center of asymmetry is present in those moieties of the molecule which are tightly involved in drug–receptor interaction and small if present in moieties which are only in loose contact with the receptor is in accordance with a more general phenomenon. *Activity ratios* for *optical isomers* of bioactive agents in general will be *large for highly potent* agents and *small for poorly active* agents, with the restriction that the center of asymmetry must be located in one of the moieties of the molecule involved in the pharmacon–receptor interaction. Pfeiffer (1956) found a highly significant correlation when comparing activities with activity ratios for optical isomers originating from diversified classes of drugs (Fig. 13).

If the activity ratios of pairs of stereoisomers of anticholinergic amino alcohol esters are plotted against the activities of the more active isomers, it is found that, for those compounds in which the center of asymmetry of the acyl moiety differs, the activity ratios depend on the activity of the more active isomer. This pattern is shown in Fig. 14 (Ariëns and Simonis, 1967). This again indicates that the acyl moiety is a critical moiety. The pairs of compounds that differ in the center of asymmetry located in the choline moiety do not follow such a pattern. The ratio of the activities of these isomers is not dependent on the activity of the more active isomer. This indicates that the choline part is a noncritical moiety.

It is interesting that the choline ester of phenyl-2-thienyl glycolic acid does

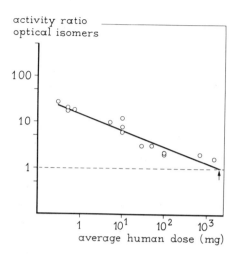

Fig. 13. Decrease in activity ratio of optical isomers with decrease in drug potency. The geometric ratio between the potencies of the optical isomers is plotted in logarithmic units on the ordinate. The points for the 14 drugs plotted may be read from left to right as follows: norepinephrine, atropine, epinephrine, scopolamine, levorphanol, methadone, amphetamine and methamphetamine, isomethadone, ephedrine, paired points 5,5-phenylethylhydantoin and methoin, quinidine, and atrolactamide (Pfeiffer, 1956). Note that large ratios are obtained for highly active drugs and low ratios for weakly active drugs.

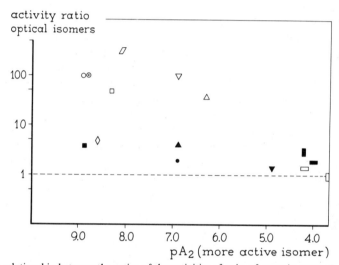

Fig. 14. The relationship between the ratios of the activities of pairs of stereoisomers of various anticholinergic esters of amino alcohols in relation to the activity of one of the isomers, usually the more active one. The pairs of isomers indicated by open symbols differ in the center of asymmetry in the acidic moiety*, while the pairs of isomers indicated by closed symbols differ in the center of asymmetry in the amino alcohol*. The various symbols represent the following: ○, phenylcyclohexyl glycolic acid* ester of β-methylcholine (R); □ , phenylcyclohexyl glycolic acid* ester of β-methylcholine (S); ●, phenylcyclohexyl glycolic acid (R) ester of β-methylcholine*; ■, phenylcyclohexyl glycolic acid (S) ester of β-methylcholine*; ▽, α-methyltropic acid* ester of β-methylcholine (S); △, α-methyltropic acid* ester of β-methylcholine (R); ▼, α-methyltropic acid (R) ester of β-methylcholine*; ▲, α-methyltropic acid (S) ester of β-methylcholine*; ◻, mandelic acid* ester of β-methylcholine (S); ▭, mandelic acid* ester of β-methylcholine (R); ▬, mandelic acid (R) ester of β-methylcholine*; ▮, mandelic acid (S) ester of β-methylcholine*; ◿, α-methyltropic acid* ester of choline; ⊙, α-methyltropic acid* ester of tropanol; ◇, phenylthienyl glycolic acid* ester of choline. (* Indicates location of the center of asymmetry.) Note that for the pairs of isomers that differ in the center of asymmetry in the acyl moiety, the ratio clearly increases with the activity, indicating that this is a critical moiety. For the pairs of isomers that differ in the center of asymmetry in the amino alcohol moiety, the ratio is practically independent of the activity, indicating that this is a noncritical moiety (Ellenbroek, 1964; Ellenbroek *et al.*, 1965).

not fit into the general pattern, although it has a center of asymmetry in the acyl moiety. One should consider, however, the fact that, although in the chemical sense there is a center of asymmetry in the acyl moiety, the phenyl and the thienyl groups are so related that in the biological sense they can be considered as *isosteric* and *isofunctional.* The two groups can substitute functionally for each other, and therefore no potency difference between the enantiomers is found (Ariëns, 1976).

The role of the quaternary ammonium group in cholinergic and anticholinergic compounds is different. This becomes apparent if larger alkyl groups are substituted for methyl groups in both types of compounds. Introduction of ethyl or propyl groups in cholinergics practically abolishes cholinergic action, while such substituents are well tolerated in anticholinergic compounds, as in propantheline. However, there are limits in the anticholinergic agents as well as to what is tolerated with regard to the bulkiness of alkyl substituents. Increasing bulk-repelling forces eventually come into play.

Quaternization of atropine with alkyl groups other than methyl results in two geometrical isomers with the introduced alkyl group either in the endo or exo position to the three-atom bridge of the tricyclic tropinol skeleton. An ethyl group in the endo position gives a more active compound but in the exo position gives a less active compound compared to methylatropine. Introduction of an *n*-propyl group in either position results in a loss of anticholinergic activity. The isopropyl group is most critical in this respect; with the isopropyl group in the endo position no loss of activity compared to methylatropine is found, while almost no activity is left with this group in the exo position (Fig. 15) (Wick, 1972). Evidently the onium group in the anticholinergic agents is not critically involved in the pharmacon–receptor interaction. With an increase in the size of its substituents, however, receptor contact is restored, although in a negative repelling sense and stereochemical factors manifest themselves again.

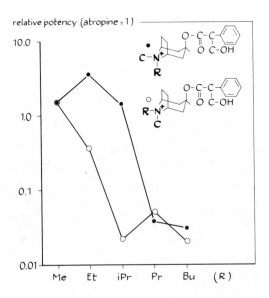

Fig. 15. Structure and action for two series of geometric isomers of atropine. Note: Larger substituents such as propyl and butyl are incompatible in both series, while the ethyl and isopropyl group reduce the activity in only one of the series of isomers (Wick, 1972).

9. Selectivity in Action

With regard to the hydrophobic accessory receptor area involved in the binding of the double-ring systems of the various types of antagonists such as anticholinergics, antihistamines, etc., it can be concluded that the accessory receptor areas belonging to the different receptor types have hydrophobic properties and a high affinity toward double-ring systems in common. They may be expected to differ insofar as their steric characteristics are concerned when the existence of selective anticholinergic and antihistaminic agents is taken into account. *Rigid molecules* with a particular steric configuration and a good fit toward the accessory receptor area of a given receptor will as a rule show a much poorer fit toward the accessory receptor areas of other receptors, because of the demand for a different steric configuration. Such compounds will have a certain degree of *selectivity in their action*. On the other hand, compounds with *flexible structures* and, consequently, a high degree of freedom in conformation may easily fulfill the steric requirements of different types of accessory receptor areas and thus show *multiple action*. Diphenhydramine, a highly flexible molecule, has both an anticholinergic and an antihistaminic action. Introduction of rigidity by suitable ring substitution results in selectivity: a sterically hindering *t*-butyl group in the ortho position gives high anticholinergic and low antihistaminic activity, while a methyl group in the para position gives high antihistaminic and low anticholinergic activity. This indicates that the accessory receptor sites involved are indeed different (Fig. 16) (Harms and Nauta, 1960). One may expect that the ratio of the activities of pairs of optical isomers will be large for highly potent agents, where an optimal fit of the more active isomer with regard to the receptor sites may be assumed. This ratio will be small for less potent agents. This is exemplified by the tertiary butyl derivatives and the *p*-methyl derivatives shown in Fig. 17 (Rekker *et al.*, 1971). Both derivatives have a center of asymmetry. For the tertiary butyl derivative, the ratio for the optical isomers is high for the anticholinergic and low for the antihistaminic activity, while for the *p*-methyl derivative this ratio is high for the antihistaminic activity and low for the anticholinergic activity.

The presence of accessory binding sites in close proximity to the active site is also recognized and made use of in enzymology. In efforts to develop selective inhibitors for the enzyme dihydrofolic acid reductase of mammals and microorganisms, Hitchings (1964) made use of the fact that the active sites on these isodynamic enzymes and on isoenzymes in general are practically identical since they have to be highly complementary to the substrate. An antimetabolite of dihydrofolic acid that is structurally closely related to this metabolite, e.g., methotrexate, cannot distinguish between the active sites on the various dihydrofolic acid reductases. The enzymes differ, however, as is shown by their different immunological properties. These differences must occur, then, outside the active site. This offers possibilities for selectivity in action. The synthesis of compounds only partially related to the substrate structurally in that they can recognize the enzyme and are supplied with hydrophobic groups suitable for binding on possible accessory binding sites resulted in the development of dihydrofolic acid reductase inhibitors. These are highly selective for the enzyme of mammals (rat) on the

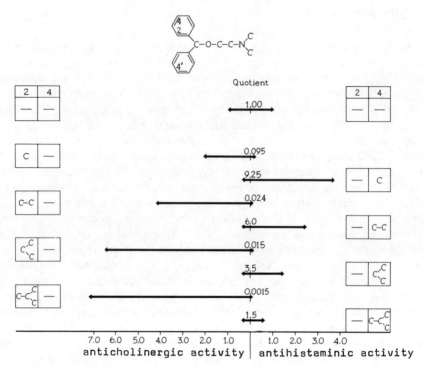

Fig. 16. Increase of specificity in action of diphenhydramine derivatives as a result of a decrease in degree of conformational freedom (Harms and Nauta, 1960).

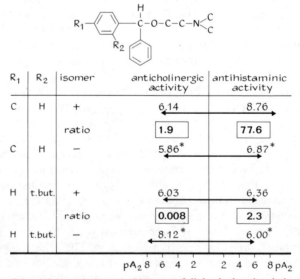

Fig. 17. Increase of the activity ratio in optical isomers of diphenhydramine derivatives as a result of an increase of the potency (Rekker *et al.*, 1971). * Identical configuration (ORD).

one hand and bacteria on the other (Fig. 18) (Hitchings, 1964; Ariëns and Simonis, 1974). In conclusion, the accessory binding sites next to the active site on enzymes of different origin apparently can differ.

10. Differentiation in Closely Related Receptor Types

A spectrum of quite different biological effects can be induced on one particular receptor type, as, for instance, on adrenergic receptors, on cholinergic receptors, or on histaminergic receptors. Activation of cholinergic receptors present in a variety of tissues results in effects such as contraction, secretion, relaxation, generation of action potentials, etc. The question arises whether the receptors or receptor sites involved in the various components of the action spectrum are identical or not. Where they are identical, the differentiation in effects induced is an inherent potential of the cells of the particular tissues involved. In the case of adrenergic actions effected via the generation of cyclic AMP, which in its turn starts the glycolytic and lipolytic processes serving as the basis of the cell's energy supply (Robison *et al.*, 1971), differences in effects obtained in the various tissues are determined by the specific machinery—a result of cell differentiation which constitutes the physiological makeup of the particular cells. The adrenergic agents switch on the energy supply in different types of cells with a variety of responses. Similarly, for agents such as acetylcholine which act via an increase in permeability of the cell membrane for ions, in particular sodium and potassium, thus causing membrane depolarization, the effect depends on the physiological makeup of the cells involved. The *differentiation in effects* obtained with a bioactive agent is, to a major extent, a consequence of a *differentiation of stimulus–effect relations* in the various tissues and is not an indication that different receptor types are involved.

Receptor differentiation is also possible. The receptors for acetylcholine, for instance, can be differentiated into muscarinic and nicotinic receptors. There are α- and β-adrenergic receptors and histamine H_1 and histamine H_2 receptors. As a rule, *with antagonists* —e.g., atropine and curare—a *sharper differentiation between* these *receptor subtypes* is possible than with agonists, which in general act on both subtypes. This is understandable in view of the complementarity principle. The variation tolerated in the receptor sites for relatively small molecules, such as acetylcholine, norepinephrine, and histamine, without abolishing receptor interaction, is much smaller than the variation in the adjacent accessory binding sites involved in the interaction with the antagonist. The composition of membrane constituents constituting the accessory binding sites may very well differ for different types of cells. One should be aware of the fact that receptors are defined by the agents which bind to them. Thus a certain uniformity independent of the tissue in which they are located is expected among the receptors for a particular pharmacon.

The affinity constants of antagonists calculated from binding curves of radiolabeled ligands to subcellular tissue preparations are in general in good agreement with pA_2-values derived from *in vivo* preparations (Table IV) (Beld

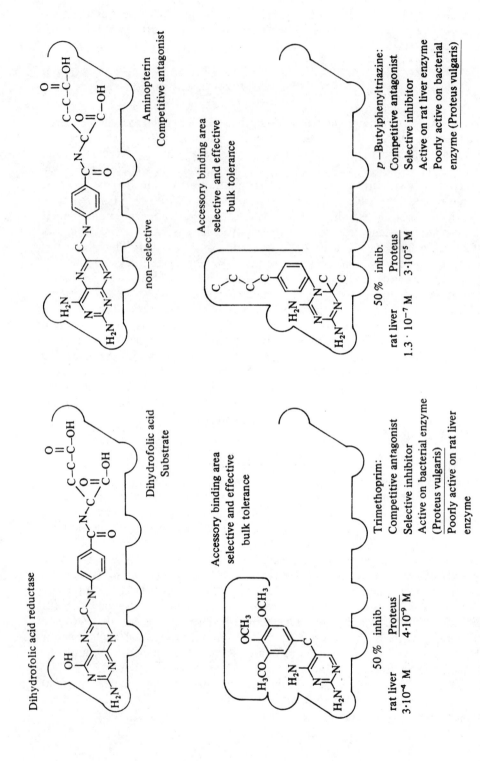

Fig. 18. Design of metabolic inhibitors, binding on accessory binding sites, to obtain selectivity in action (Hitchings, 1964).

Table IV

Comparison of the Affinity Constants Derived from Binding
Experiments on Membrane Preparations of Calf Tracheal
Muscle and from Dose–Response Curves Obtained with the
Isolated Ileum (Guinea Pig[a]; Rat[b])

Compound	Affinity constants (M)	
	Dose–response	Binding
Anticholinergics		
Atropine	4.7×10^{-9} [a]	2.6×10^{-9}
(+)Benzetimide	8.9×10^{-10} [a]	4.2×10^{-10}
Cholinergic		
(+)Metacholine	1.3×10^{-7} [b]	2.3×10^{-6}

[a] For details, see Beld and Ariëns (1974).
[b] For details, see Ellenbroek (1964).

and Ariëns, 1974). The affinities of agonists derived from direct binding studies
or from competition experiments with radiolabeled antagonists sometimes differ
substantially from *in vivo* pD_2-values (Beld and Ariëns, 1974; Beld *et al.*, 1975).
The discrepancy can easily be explained on the basis of the existence of a receptor
reserve in the tissue at hand. A pD_2-value for a particular agonist indicates the
concentration at which half-maximal effect is found. If a larger receptor reserve
is present, this concentration is much lower than the one at which half-maximal
saturation of the receptors, representing the real affinity, occurs (see table IV and
Sec. 5). pD_2-values for agonists are apparent affinities. pA_2-values, due to the
absence of reserve phenomena with regard to antagonists, reflect real affinities.
Moreover, *differences in pD_2-values* for a particular agonist in different tissues
do not necessarily *mean that the receptors are different*. It could well be that
in different tissues the physiological makeup, the amplification factor, and thus
the *receptor reserve* differ. This is a reason why antagonists are more suitable for
detecting a possible differentiation in receptor subtypes than are agonists.

To detect possible differences in muscarinic receptors in various tissues, a
comparison was made for the affinity of atropine to muscarinic receptors in
membrane preparations of bovine tracheal smooth muscle, parotid gland, and
caudate nucleus using as a basis direct binding studies with radiolabeled atropine.
Figure 19 shows specific saturable binding to muscarinic (atropine) receptors in
the three tissues studied (Beld and Ariëns, 1974; Beld *et al.*, 1975). The affinity
constants calculated from the binding curves for the three tissues differ so little
that there is no reason to postulate the existence of more than one muscarinic
receptor type.

The existence of chemically different classes of histamine H_1 antagonists
might well be a consequence of the presence of more than one differentiated
accessory binding area in the direct vicinity of the histamine H_1 receptor. Antihistamines, belonging to one class from a chemical point of view and therefore
binding to an identical accessory receptor area, should compete in binding with

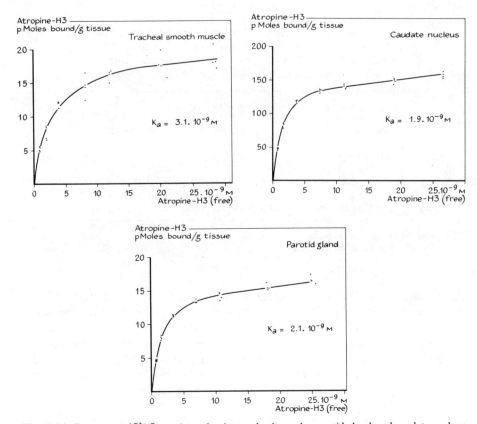

Fig. 19. Binding curves of [³H]atropine to bovine tracheal muscle, parotid gland, and caudate nucleus.

histamine and with each other but possibly not with antihistamines from a different chemical class which bind to another accessory receptor area.

The difference between β_1- and β_2-adrenergic receptors is also intriguing. Although indicated as an adrenergic system—a term originating from the hormone adrenaline released from the adrenal medulla—in actual fact the sympathetic nerve system is *noradrenergic*, which means that norepinephrine is the neurotransmitter. The *adrenergic* receptors for the neurotransmitter norepinephrine and those for the hormone adrenaline (epinephrine) not only differ in their localization and function but were also shaped in evolution to interact with chemically different molecules. Therefore, one may have to consider not only a differentiation of α- and β-adrenergic receptors but also a differentiation between the *receptors for the neurotransmitter*, β-noradrenergic receptors, and *receptors for the hormone*, β-adrenergic receptors. This differentiation of course should be extended to the corresponding types of noradrenergic and adrenergic blocking agents as well.

A variety of organs, such as the heart (frequency and contractile force), the intestinal smooth muscle, the tracheal smooth muscle, the uterine smooth muscle, and the striated muscle fibers, does respond to β-adrenergic agents. Neither the uterine smooth muscle (Driel *et al.*, 1973) nor the striated muscle fibers have a

Table V

*Relative Activity of Epinephrine (Adrenaline) and Norepinephrine
(Noradrenaline) in Different Organs*[a]

Biological object	Animal	Relative activity epin./norepin.
Caecum	Mouse	0.3
Duodenum	Rabbit	0.3
Atrium (freq.)	Guinea pig	0.7
Heart (freq.)	Rabbit	0.6
Atrium (force)	Guinea pig	0.8
Heart (force)	Rabbit	1.4
Trachea	Guinea pig	10
Trachea	Calf	10
Lactic acidemia	Rat	145
Diaphragm	Rat	111
Uterus	Rat	103
Heart	Frog	42

[a] The relative activities may be an expression of the ratio of β-adrenergic hormone and β-adrenergic transmitter receptors in the tissues concerned (Grana *et al.*, 1974).

sympathetic innervation. They are known to respond very well to the hormone adrenaline. As can be seen from Table V (Grana *et al.*, 1974), certain tissues respond primarily to norepinephrine, while other tissues, among them the uterine smooth muscle and the striated muscle fibers, respond primarily to epinephrine. The organs responsive to norepinephrine, such as the heart and the intestine, are well known to be influenced strongly by sympathetic innervation. The uterine smooth muscle and the striated muscle fibers are devoid of sympathetic innervation but respond well to the hormone adrenaline. Thus the data represented in Table V may be easily interpreted on the basis of a differentiation between β-receptors for the neurotransmitter and β-receptors for the hormone. Carlsson *et al.* (1972) showed that the heart is an organ supplied with two types of β-adrenergic receptors, presumably receptors for the transmitter norepinephrine and receptors for the hormone adrenaline.

The role of sympathetic innervation and the hormone adrenaline, i.e., the proportion of transmitter and hormone receptors, differs for different organ systems and tissues. It may even differ for one organ from species to species. Furthermore, the responses obtained with the various types of β-adrenergic blocking agents, especially, will depend on the functional state, namely on the sympathetic tone or the degree of release of epinephrine from the adrenal medulla. An interesting aspect of Table V is that the data from the frog heart do not correspond with those from the mammalian heart. The frog heart apparently is supplied predominantly with hormone receptors since it is sensitive primarily to epinephrine. In contrast to mammalians that have a noradrenergic sympathetic innervation, however, the frog and toad appear to be adrenergic, both with regard to sympathetic innervation and with regard to the hormonal catecholamine (Segura and Biscardi, 1967). The "exceptional" position of the frog heart cannot only be explained but is to be expected.

On the basis of structure–activity relationships for β-adrenergic agents, where alkyl substitution in the carbon atom next to the amino group produces a differentiation in the effects observed, Lands *et al.* (1967) postulated two types of β-adrenergic receptors, namely β_1- and β_2-receptors. The same type of chemical substitution applied to β-adrenergic blocking agents results in an analogous differentiation (Ariëns, 1967). This empirical *differentiation* between β_1- *and* β_2-*receptors* closely *correlates with the differentiation* between the *adrenergic neurotransmitter* and the *adrenergic hormone receptors* and thus may give a natural explanation of this physiological differentiation. The use of the terms *cardioselective* and *bronchoselective* for β_1- and β_2-adrenergic (blocking) agents, respectively, should be reconsidered in the light of the above discussion.

There is also growing evidence for a differentiation in the α-adrenergic receptors (Ariëns and Simonis, 1976; Struyker Boudier *et al.*, 1975; Doxey *et al.*, 1977; Ruffolo *et al.*, 1977; Rosell and Belfrage, 1975).

In contrast to the effects induced on the adrenergic hormone receptors, for effects induced on the transmitter receptors, one would expect an enhanced response (sensitization) after denervation and enhanced responses (potentiation) in the presence of inhibitors of the re-uptake of the neurotransmitter noradrenaline in the nerve endings, e.g., cocaine. Furthermore, indirectly acting adrenergic agents the action of which is based on the release of the transmitter noradrenaline from the nerve endings could be expected to bring about mainly those effects which can be induced on the α- and β-adrenergic transmitter receptors.

11. Receptor Binding and Receptor Isolation

Binding studies of drugs to receptors are advantageous in the sense that they give direct information on receptor occupation and therefore on affinity, but, on the other hand, other aspects of drug action such as receptor activation are largely obscured.

As shown by binding studies, the receptors for various drugs, hormones, and hormonoids are located in plasma membranes (de Robertis, 1976). The receptors for acetylcholine (the cholinergic receptors) appear to be located on the outside of the plasma membrane since application of cholinergic agents intracellularly appears to be ineffective (Del Castillo and Katz, 1955). For other receptor types, viz. those for norepinephrine, histamine, and serotonin, a location at the outside of the cell membrane is highly probable since these receptors are easily accessible to the quaternarized form of the respective competitive blocking agents, compounds known to penetrate intracellularly only with great difficulty or not at all (Ariëns, 1967, 1971a). In addition, for the receptors of various peptide hormones and norepinephrine involved in the activation of adenylate cyclase, a location on the cell membrane has been confirmed by binding studies (Greengard, 1976; Cuatrecasas, 1974, 1975).

The receptor sites for the drugs do not necessarily have to be located on the surface of receptor proteins. The accessory *receptor sites* for the hydrophobic moieties in the various competitive antagonists of membrane-active agents may well be *made up of* an *interface between* a hydrophobic area on the *receptor protein* and *lipid groups in the membrane*. The extremely high affinity constants, 10^9 and 10^{10}, for competitive antagonists (e.g., anticholinergics and antihistaminics) which are based mainly on the contribution of the hydrophobic groups to the affinity can hardly be realized on a mere protein surface. Lipoproteins would be more suitable in this respect. The lipid molecules in a membrane facing the hydrophobic surface of the protein are fixed to a large extent in a quasi-crystalline form, making the high degree of stereospecificity of the antagonists understandable, especially with regard to centers of asymmetry in their hydrophobic moiety. Drug–receptor interaction then results in changes in the interface characteristics. It is even possible that the relatively polar sites for agonists are also located at or made up of an interface, e.g., where the polar groups of certain membrane proteins interact with polar groups of membrane lipids like the phosphate and choline groups in the phosphatidic acids, phosphatidylcholines, and sphingomyelins. The isolation of receptors as molecular entities would become an impossible task if this turns out to be the case. Studies on purified receptors of this nature would be possible only after isolation of the constituent parts and reassembly of these parts into artificial membrane-like structures (deRobertis, 1976).

With regard to receptor isolation and identification there is an essential difference between soluble receptors, e.g., the cytoplasmic steroid receptor proteins and membrane-bound receptors. In the latter case there is a tight interrelation of the receptor molecule and surrounding molecules. Separation of the receptor molecule from its surroundings may well disturb its conformation and thus its specific characteristics. Undoubtedly the isolation techniques, especially those in which detergents (Maddy and Dunn, 1976) are used as solubilizing agents, may influence the conformation of the isolated protein drastically. The variation in binding constants reported for the interaction of drugs with *isolated receptors* therefore does not warrant the existence of various receptor states or conformational states under physiological conditions. A special problem which holds for the soluble as well as for the membrane-bound receptors is that with isolated receptors, no measurable *pharmacological* effect can be induced anymore, making their identification extremely difficult in certain cases.

Since structure–action relationships (SARs) are based on an interaction between the drug molecule and the receptor site, they may give information as to the properties of this site.

Chemical complementarity between the drug and the specific receptor site with which it interacts is a reasonable assumption. Structure-binding relationships therefore can serve as a tool for the identification of isolated receptors. However, only the receptor molecules, not the sites, can be isolated. One may expect that structure-binding relationships for isolated receptors, if they are not denatured to a certain degree in the isolated form, will parallel the SARs observed with tissues. As a rule, however, with the isolated receptors there will be no differentiation between agonists, partial agonists, and competitive antagonists, because only affinity is measured and not intrinsic activity. Opiate receptors are an exception in this sense because agonist binding is dependent on the presence of Na^+ or Li^+ ions while antagonist binding is not, making a differentiation possible

(Pert and Snijder, 1974). In efforts to isolate and identify receptors, use is made of agents which irreversibly bind in a selective way to the receptors, or, better, to the specific receptor sites. However, few such agents are available (Burgen et al., 1974).

12. Dualism in Receptors for Agonists and Their Competitive Antagonists

Anticholinergics, antihistaminics, α-adrenergic blockers, serotonin antagonists, and dopamine antagonists apparently bind through their hydrophobic moieties to accessory binding areas either topically or functionally. They are related to the receptor site in a narrow sense by blocking in some way the activation of the receptor by agonists. They interfere with the binding of the agonist to its receptor site without occupying this receptor site itself. There appears to be a *dualism* in the receptor sites for agonists and corresponding antagonists. On the basis of this concept, the lack of structural similarity between agonists and their corresponding competitive antagonists, the existence of close structural relationships between competitive antagonists blocking different types of receptors, the existence of competitive antagonists blocking receptors for different agonists, and the dependence of selectivity in the action of the competitive antagonists on the steric configuration of the hydrophobic moieties in the molecule can be understood. The receptor sites for agonists and competitive antagonists are not as common as suggested by the original "one-receptor-site concept."

The receptor molecule or macromolecular complex, as discussed in Sec. 4, must be assumed to occur in different states: an activated R^*-form and a nonactivated R-form depending on whether or not an agonist or a competitive antagonist is present. It has been postulated (Monod et al., 1965; Karlin, 1967; Changeux et al., 1967; Wyman, 1967) that even in the absence of drugs there is a dynamic equilibrium between different functional states of the cholinergic receptor. The simplest concept in this respect is the *dual receptor concept*, in which the assumption is made that there is an equilibrium between the receptors in the R^*-form and those in the R-form in the cell membrane. The R^*-form contributes to the effect; the R-form does not. An agonist shifts the equilibrium toward the R^*-conformation and a competitive antagonist toward the R-conformation. The agonist has a relatively high affinity for the receptors in the R^*-conformation, and the competitive antagonist for those in the R-conformation. This in a way fixes the receptors in their respective states (Ariëns and Beld, 1977; Pert, 1976; Raftery, 1975). Partial agonists have an affinity for both states, the ratio of the affinities determining the intrinsic activity of the compound. As long as the rate constants for the equilibrium between R and R^* are large enough, this model will predict phenomena similar to those expected for a competition between drugs based on the one-receptor-site concept. The receptor sites for an agonist and a *competitive* antagonist—although only one receptor protein is involved—are therefore different (Pert, 1976; Raftery, 1975). This is in accordance with the dualism in receptor sites outlined before on the basis of structure–action relationships (Ariëns and Beld, 1977).

More direct evidence for the dualism in the cholinergic receptor, namely receptor sites involved primarily in the binding of the agonist and receptor sites involved primarily in the binding of the antagonist, is available. Reductive alkylation by *N*-ethylmaleimide clearly changes the affinity of muscarinic receptors of neural membranes for agonistic agents without affecting the affinity for the antagonists (Aronstam *et al.*, 1977). Treatment of the cholinergic receptors of the electroplax of *Electrophorus electricus* by dithiothreitol results in a loss of the membrane sensitivity to acetylcholine. Cholinergic agents, agonists, can protect against this chemical desensitization; antagonists, such as *d*-tubocurarine, do not so protect (Bregestovski *et al.*, 1977).

13. The Aggregation–Segregation Concept

For the case where the receptor molecules are embedded in the lipid membrane structure, one may expect that the behavior of such protein molecules, e.g., their tendency to aggregate or segregate, strongly depends on the hydrophobic and hydrophilic properties of the protein surface, especially on that part of the receptor molecule in contact with the lipids in the membrane. The receptor proteins, although hydrophobic in their overall nature (de Robertis, 1975b), have an amphiphilic character. Binding of agonists, usually strongly polar agents, could stabilize a relatively polar conformation and thus shift the equilibrium toward the more hydrophilic R*-form. Binding of the hydrophobic antagonist would then stabilize a hydrophobic R-state. The different character of the R*-form and R-form will imply differences in the quaternary structure of the proteins and the degree and type of aggregation among receptor molecules or among receptor molecules and other proteins (e.g., effector proteins) present in the membrane.

In the hydrophobic state the receptor proteins will tend to remain solitary in the lipid membrane environment. In the hydrophilic state, the tendency to aggregate among themselves or with other hydrophilic macromolecules will be enhanced.

The degree of aggregation influenced by the presence of a hormone or neurotransmitter might determine the effect to a major extent. This concept fits quite well with the fluid mosaic theory of membrane structure (Singer and Nicolson, 1972).

For receptor molecules present in the hydrophilic cytosol, such as the steroid receptors, the tendency to aggregate will be increased in the hydrophobic state. Steroids undoubtedly are hydrophobic in nature. The migration of the steroid-bound receptors via the nuclear membrane to the nucleus may well be related to the degree of hydrophobicity. The proteins in the luminal cell membrane of the toad bladder aggregate in a mosaic pattern under the influence of the antidiuretic hormone (Kachadorian *et al.*, 1975). The protein-walled pores thus formed may well account for the increased water permeability induced by this hormone. Specific receptor proteins for the different hormones activating adenylcyclase are presumed to aggregate with the enzyme molecule under the influence of the hormones, thus activating the enzyme in an allosteric way (Cuatrecasas *et al.*, 1975; Jacobs and Cuatrecasas, 1976; Greaves, 1977).

The transfer of β-adrenergic receptors from turkey erythrocytes in which the enzyme adenylcyclase is inactivated by N-ethylmaleimide to other cells that contain the enzyme but are devoid of β-adrenergic receptors and the possibility of activating the adenylcyclase of these cells after transfer of the β-adrenergic receptors, by means of β-adrenergic agents, provide definite proof of the independence of receptor and effector molecules (the adenylcyclase) (Schramm *et al.*, 1977; Ross and Gilman, 1977).

There are good indications that in the case of the receptors for acetylcholine at the myoneural junctions the cholinergic receptor, after activation by acetylcholine, interacts with an effector protein that is directly involved in ion conductance and is called an *ion conductance modulator* (ICM). ICM is different from the cholinergic receptor protein and possibly is part of the wall of the pores. Competitive antagonists of acetylcholine such as d-tubocurarine bind to the cholinergic receptor protein, as is the case with α-bungarotoxine, a compound that irreversibly inactivates the cholinergic receptor molecule. On the other hand, compounds such as histrionicotoxin and its perhydroderivative, certain local anesthetics, and possibly amantadine bind to the ICM and by doing so interfere with neuromuscular transmission without binding to the cholinergic receptor molecule (Albuquerque *et al.*, 1978; Sobel *et al.*, 1978). They thus act in a noncompetitive way as antagonists of acetylcholine. The relationship between the cholinergic receptor molecule and ICM may well tie in with the aggregation–segregation concept, which implies that only after activation by cholinergic agents does the receptor molecule aggregate in a specific way with ICM, which only then contributes to ionic conductance (Albuquerque *et al.*, 1978; Sobel *et al.*, 1978).

Receptor sites for membrane-active drugs may also be located on proteins or protein complexes constituting the wall of pores (deRobertis, 1975a; Waser, 1975). The passage of ions through such protein-walled pores in a membrane may well change under the influence of drugs as a result of an altered charge distribution in the receptor molecules constituting the pores. The proteins may switch from an open to a closed conformation. Again an equilibrium between both conformations, shifted by agonists to an apparently open conformation and by competitive antagonists to a closed conformation, can be postulated. In the dual-receptor-site model, although only one type of receptor protein or protein complex is postulated, the receptor sites for agonists and those for the corresponding antagonists are functionally distinct, interdependent entities.

14. Dual Receptor Model

The reaction scheme in Sec. 4 starts from the (classic) concept that the pharmacon has to activate the receptor in order to provoke an effect. An agonist is then characterized by its ability to change the conformation of the receptor to that corresponding with the activated state, i.e., $RD \rightarrow R^*D$, while an antagonist will not be able to do this. An alternative way of looking at this process of activation is to assume that the unoccupied receptor already exists in an equilibrium with its activated form (Colquhoun, 1973; Thron, 1973), i.e.,

$$R \rightleftharpoons R^*$$

and that the agonist shifts this equilibrium toward R^*. We are dealing then with a two-state model. The sequence of reaction in Sec. 4 in fact implies three states for the receptor,

$$R \rightleftharpoons R^*$$
$$\underline{R}$$

and the whole sequence can be drawn as a closed figure of equilibria (Fig. 20). For reasons of simplicity we shall confine ourselves to the first two steps in the reaction scheme of Sec. 4, which in fact represents the dual receptor model.

   In this concept it is supposed that the receptor in the R^*-form, i.e., as R^* or R^*D, contributes to the effect; the R-form, i.e., as R or RD, does not. An agonist will show a high affinity for R^* and effectively shifts the equilibrium to the R^*-form; an antagonist will show a high affinity for R and causes the equilibrium to shift to the R-form. The total concentration of receptors is r. In contrast to the classical view in which the agonist has to activate the receptor, it now has to shift the equilibrium in favor of the R^*-form in order to be active:

$$M = \frac{K_1^*}{K_1} \qquad \begin{array}{c} R \xrightarrow{L} R^* \\ +D \qquad +D \end{array} \qquad K_1 = \frac{[R][D]}{[RD]}$$

$$K_2 = LM \qquad \begin{array}{c} k_1 \Big\downarrow \qquad K_1^* \Big\downarrow \end{array} \qquad K_1^* = \frac{[R^*][D]}{[R^*D]} \qquad (1)$$

$$\begin{array}{c} RD \underset{K_2}{\rightleftharpoons} R^*D \end{array} \qquad L = \frac{R}{R^*}$$

$$K_2 = \frac{[RD]}{[R^*D]}$$

Although this alternative route (i.e., via R^*) offered by the dual receptor concept is irrelevant for the concentration of R^*D at equilibrium, it may have much lower activation energies so that equilibrium is reached within experimental time. The affinity in this model, K_{app}, can be defined as the concentration of D at which half-receptor occupation occurs:

$$\text{fraction occupied} = \frac{RD + R^*D}{R + R^* + RD + R^*D}$$

$$= \frac{1}{1 + (K_1/D)[M(1+L)/(1+LM)]} \qquad (2)$$

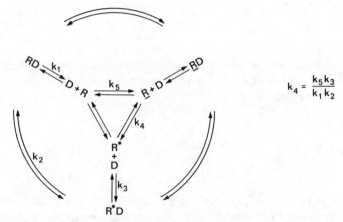

Fig. 20. The three-state kinetics for the "two-state" model.

so that

$$K_{app} = K_1 \frac{M(1 + L)}{1 + LM} = \frac{K_1^* + K_1^*L}{1 + K_2} \tag{3}$$

In the dual receptor model the fraction occupied deviates from the fraction of receptors activated.

As in the case of the simple model, by definition here also the initial stimulus induced is proportional to the fraction of the receptors present in the activated state, $(R^* + R^*D)/r$.

As a rule, due to the involvement of the stimulus–effect relationships [e.g., an amplifier (cascade) system as described previously for the simple model], there will be no proportionality between the initial stimulus and the effect.

If as a first approximation the effect, including the specific resting one, based on the equilibrium $R \rightleftharpoons R^*$ is postulated to be proportional to the initial stimulus, the following equations can be derived:

$$\frac{E}{E_{max}} = \frac{R^* + R^*D}{R + R^* + RD + R^*D}$$

$$= \frac{1 + D/K_1^*}{1 + L + DL/K_1 + D/K_1^*} \tag{4}$$

Even in the absence of a pharmacon there will be some activated receptor resulting in a basic tonus, E_0:

$$\frac{E_0}{E_{max}} = \frac{R^*}{R + R^*} = \frac{1}{1 + L} \tag{5}$$

The intrinsic activity defined as E_{Dmax}/E_{max} no longer is an *ad hoc* parameter but in this model becomes a parameter defined by equilibrium constants; i.e.,

$$\frac{E_{Dmax}}{E_{max}} = \frac{R^*D}{RD + R^*D} = \frac{1}{1 + LM} = \frac{1}{1 + K_2} \tag{6}$$

The effect of D relative to its maximal effect is then given by

$$\frac{E_D}{E_{Dmax}} = \frac{[\text{fract. act.}]_D - [\text{fract. act.}]_{D=0}}{[\text{fract. act.}]_{D=\infty} - [\text{fract. act.}]_{D=0}}$$

$$= \frac{1}{1 + (K_1/D)\,[M(1+L)/(1+LM)]} \tag{7}$$

so that

$$\left(\frac{E_D}{E_{Dmax}}\right)_{50\%} = \frac{K_1 M(1+L)}{1+LM} = \frac{K_1^* + K_1^* L}{1 + K_2} \tag{8}$$

Comparing Eqs. (3) and (8), we see that the 50% points in both the occupation and effect curves are equal, and in both cases we are dealing with normal sigmoid curves with a Hill slope of 1.

It should be noted that, as in the simple model (Sect. 4), the intrinsic activity and apparent affinity are also interdependent parameters here; they can be represented by terms based on the various rate or binding constants. The characteristics of the model are, as demonstrated in Figs. 21–25, similar to those of the classic model (see the black insets in Figs. 3(a), (b), (c), and (d) in Sec. 4). Figure 21 shows occupation (a) and effect (b), curves calculated according to Eqs. (2) and (4), respectively, for a series of full agonists with different affinities. Figure 22 shows in an analogous way curves for a series of compounds ranging from full agonists via partial agonists to antagonists.

In the case of a full agonist, i.e., $K_2 \ll 1$, the apparent affinity approaches $K_1^* L$ (if $L \gg 1$). In the case of a full antagonist. i.e., $K_2 \gg 1$, the apparent affinity approaches K_1 (if $L \gg 1$).

15. Combination of Pharmaca

If pharmaca are combined, the dual receptor model predicts normal sigmoid curves still with a slope of 1; i.e.,

$$\frac{E_D}{E_{Dmax}} =$$

$$= \frac{1}{1 + (K_{11}/D_1)\,[M_1(L+1)/(1+LM_1)]\,\{1 + (D_2/K_{12})\,[(1+LM_2)/M_2(L+1)]\}} \tag{9}$$

or

$$\frac{E_D}{E_{Dmax}} = \frac{1}{[1 + (K_{1app}/D_1)]\,[1 + (D_2/K_{2app})]} \tag{10}$$

with

$$K_{1app} = K_{11}\frac{M_1(1+L)}{1+LM_1}, \qquad K_{2app} = K_{12}\frac{M_2(1+L)}{1+LM_2}$$

D_1 as well as D_2 can be either an agonist or an antagonist.

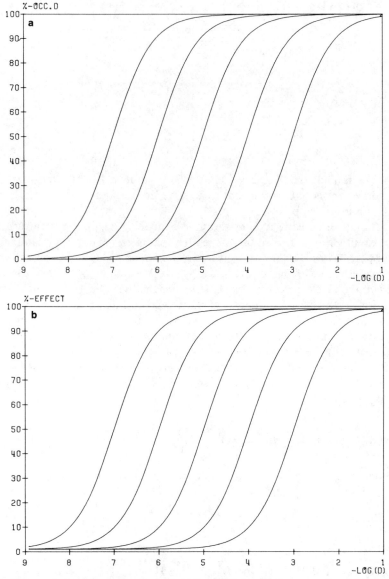

Fig. 21. Curves calculated for occupation (a) and effect (b). Series of full agonists: $L = 100$; $K_2 = 10^{-2}$. K_1-values for the different agonists are 10^{-5}, 10^{-4}, 10^{-3}, 10^{-2}, and 10^{-1} M, respectively. Compare with Fig. 3(a).

Figure 23 shows occupation (a) and effect (b) calculated curves for a partial agonist D_1 in the presence of different doses of a full agonist D_2. Figure 24 shows in a similar way curves for a full agonist D_1 in the presence of different doses of a partial agonist D_2.

The dual receptor model has a certain internal logic, but its simplicity, in spite of being one of its credits, implies that it cannot account for such phenomena

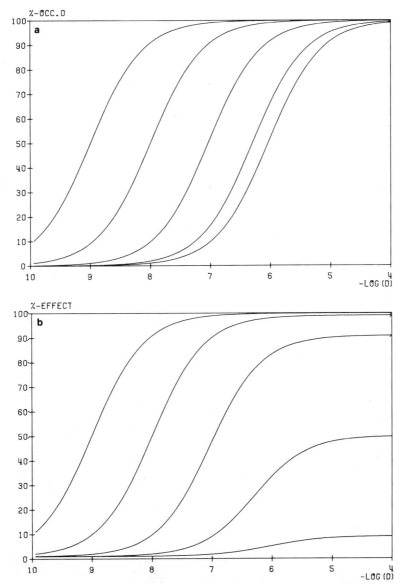

Fig. 22. Curves calculated for occupation (a) and effect (b). Transition from full agonist to antagonist: $L = 100$, $K_1 = 10^{-6}$ M. K_2-values for the different agents are 10^{-3}, 10^{-2}, 10^{-1}, 1, and 10. Note that a change in K_2 affects both the 50% point and the maximum of the curve. Compare with Fig. 3(b).

as receptor reserve and cooperativity. To account for these phenomena the model has to be extended. More complex variants of this model have been worked out in great detail by Changeux *et al.* (1967), Karlin (1967), and Thron (1973).

An interesting aspect of the dual receptor concept is that even in the absence of a pharmacon a certain fraction of the receptors exists in the activated state; this implies the existence of a certain resting stimulus and thus a certain specific

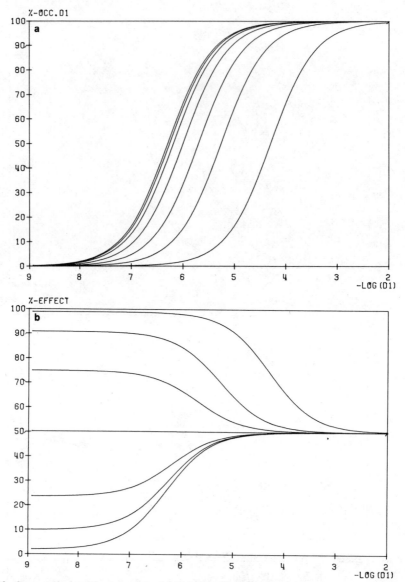

Fig. 23. Curves calculated for occupation (a) and effect (b). D_1 is a partial agonist; i.e., $L = 100$, $K_2 = 1$, $K_1 = 10^{-6}$ M. D_2 is an agonist; i.e., $L = 100$, $K_2 = 10^{-3}$, $K_1 = 10^{-6}$ M. Different doses of D_2: 10^{-11}, 10^{-10}, $3 \cdot 10^{-10}$, 10^{-9}, $3 \cdot 10^{-9}$, 10^{-8}, and 10^{-7} M. Note: At low concentrations of D_2, D_1 increases the effect; i.e., the compound behaves as an agonist. At high concentrations of D_2, D_1 decreases the effect; i.e., the compound behaves as an antagonist. Compare with Fig. 3(d).

resting effect. This could be, for instance, a certain smooth muscle tone related to the cholinergic receptor system. Application of competitive blockers acting on that receptor system (e.g., atropine) should then lead to an annihilation of this resting tone and thus to a relaxation (see Fig. 25). For various smooth muscle preparations, e.g., for the isolated gut, the existence of a resting tone may be due

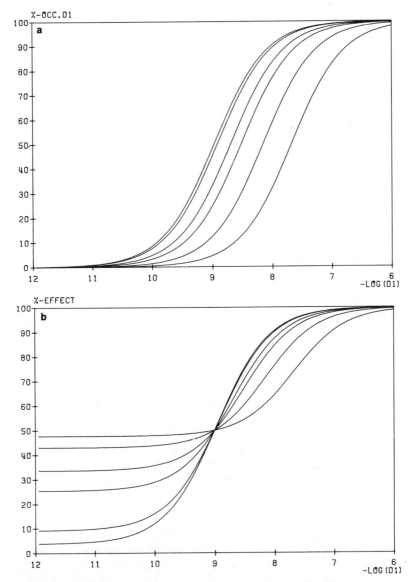

Fig. 24. Curves calculated for occupation (a) and effect (b). D_1 is a full agonist; i.e., $L = 100, K_2 = 10^{-3}$, $K_1 = 10^{-6}$ M. D_2 is a partial agonist; i.e., $L = 100$, $K_2 = 1$, $K_1 = 10^{-6}$ M. Different doses of D_2: $3 \cdot 10^{-8}, 10^{-7}, 5 \cdot 10^{-7}, 10^{-6}, 3 \cdot 10^{-6}$, and 10^{-5} M.

to the presence of acetylcholine originating in the neural tissue of the preparation. In this case atropine should cause a relaxation in the "absence" of cholinergic pharmaca. To check whether or not the resting tone can be ascribed to the presence of an endogenous agonist, e.g., acetylcholine, denervated or aged tissue preparations should be used. (The viability of nerve tissue is lost sooner than that of the smooth muscle tissue.) Under such conditions the rat jejunum does not

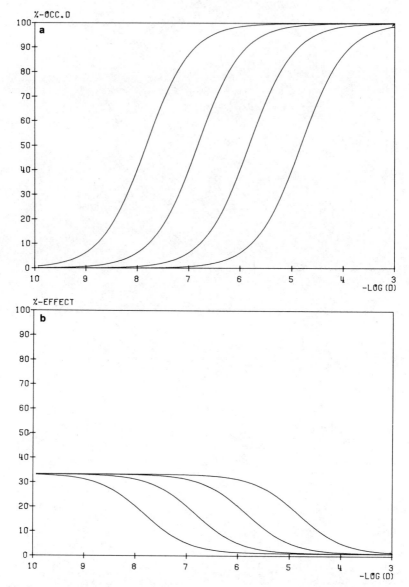

Fig. 25. Curves calculated for occupation (a) and effect (b). Annihilation of a resting tone by an antagonist $D : L = 2$, $K_2 = 100$. K_1-values for the different curves are 10^{-8}, 10^{-7}, 10^{-6}, and 10^{-5} M.

relax under the influence of atropine. This implies that the number or fraction of the cholinergic receptors present in the activated state is negligible, and in fact the original simple receptor model then describes the experimental phenomena adequately.

If the participation model for the pharmacon action holds true (Sec. 4), then it is highly improbable that the dual receptor model is also involved.

16. The Slope of the Concentration–Effect Curves

The slope of the log concentration–effect curves obtained with bioactive agents often deviates from that expected on the basis of simple Michaelis–Menten kinetics or the Langmuir adsorption isotherm. Various mechanisms located at different levels in the receptor–effector system may be involved.

Case 1. On the level of the *receptor system* one may find the following:

1. There may be an allosteric cooperation or autosensitization, which implies that two interdependent receptor sites on one receptor molecule or on a complex of receptor molecules are involved. One receptor site is tied up to the effector system and occupation thereof with an agonist gives an effect. Occupation of the other receptor site by the same drug results in an increase in the affinity of that drug to the receptor site first mentioned. There is an autosensitization. (As a matter of fact, a similar model can also explain an autodesensitization.) In both enzymological and pharmacological literature this type of interaction is extensively discussed. Figure 26 gives an example of this (Ariëns, 1964; Roach, 1977; Ariëns *et al.*, 1956b).
2. There may be a *cooperativity* in binding of the pharmacon based on its interaction with receptor sites on receptors in ternary complexes (aggregates of two or more receptor molecules). A special case of this model has been worked out by Changeux *et al.* (1967), particularly for the pharmacon binding and has been developed for a receptor–effector system by Karlin (1967). Depending on the chosen values for the various constants, log concentration–binding or –effect curves with a varying degree of steepness deviating from that for the simple Michaelis–Menten model are obtained.
3. Between the receptor occupation and activation and the induction of the effect is the interaction of the activated receptor molecules with the effector molecules. This forms the first step in the stimulus–effect relationship. Generally (and certainly if simple Michaelis–Menten kinetics for the receptor occupation hold true) the interaction (binding) of the pharmacon(agonist)-activated receptors with the effector molecules results in a shift of the equilibrium toward further receptor occupation and activation. This does not hold true for competitive antagonists since no receptor activation and thus no interaction of occupied receptors with the effector system takes place:

$$D + R \rightleftharpoons DR \rightleftharpoons DR^* + E \rightleftharpoons DR^*E$$

   DR^* represents the pharmacon-activated receptors, E the effector molecules, and DR^*E the activated effector molecules. The slope of the log concentration–effect curve, relating the fraction of the effector molecules activated to the concentration of the agonist, is determined essentially by the ratio between the total concentration of receptor molecules and the total concentration of effector molecules. A high ratio,

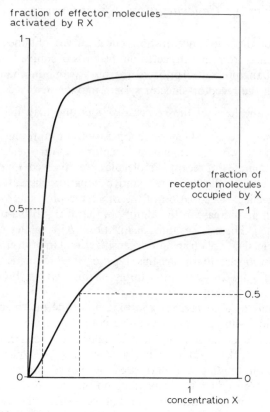

Fig. 26. Calculated concentration–binding and concentration–effect curves. The lower curve represents the binding of pharmacon X on two interdependent binding sites. The binding of X on one site
($K_{diss} = 1$) enhances the binding of X on the second site by reducing K_{diss} for this second to 0.1. The
higher curve represents the activation of an enzyme by binding X on the basis of binding kinetics
represented in the lower curve followed by an activation of the effector molecules by the activated
enzyme according to Michaelis–Menten kinetics [in analogy to Fig. 1 in Roach (1977)]. Note: The
cooperative binding (lower curve) increases the slope of the binding curve (compare with the lower
curve in Fig. 7). Here also one step in the cascade results in a very strong increase in the slope of the
curve.

      i.e., a large receptor surplus, implies steep curves. Even more complex
situations can be postulated (Jacobs and Cuatrecasas, 1976).

4. If the pharmacon(agonist)–receptor interaction involves the activation
of an enzyme molecule, with the conversion by this enzyme of an inactive
molecular species to an activated one as the next step, a plot of this
activation on a log concentration scale for the agonist results in steep
curves, as shown in Fig. 26.

    Case 2. On the level of the *stimulus–effect relationship* one may find the
following:

1. There may be an involvement of an amplifier system in the stimulus–effect
relationship resulting in the phenomenon of a receptor reserve as described

in Sec. 5 (Ariëns *et al.*, 1960; Van Rossum, 1968). Depending on the number of steps in the stimulus–effect sequence and the characteristics of these steps, various degrees of steepness may occur (Van Rossum, 1958).

2. A special case is that of an all-or-none response with a biological variation in the concentration at which the response is triggered. The biological variation then determines the slope of the log concentration–effect curve (Ariëns, 1964; Ariëns and Van Rossum, 1957; Van Rossum, 1966). This model was recognized long ago by Gaddum (1937).

3. If pharmacological effects are generated in physiological systems, especially under *in vivo* conditions, the tendency to maintain homeostasis implies that with an increase in the pharmacological effect, feedback control mechanisms will become more and more active. Such mechanisms, which as a rule will be negative in nature and can be involved in simpler biological systems, will influence the slope of the log concentration–effect curves of pharmaca.

The deviations from a simple Michaelis–Menten curve or Langmuir adsorption isotherm as discussed in case 1 will manifest themselves both in the log concentration–binding and in the log concentration–effect curves. Those discussed in case 2 will not manifest themselves in the log concentration–binding curve. Deviations in the slope of the effect curves will, for both cases 1 and 2, generally imply changes in the shape, e.g., an asymmetry, of the curves.

An interesting question with regard to the mechanisms involved in deviations in the slope of log concentration–binding and log concentration–effect curves is how these curves will be influenced by the presence of competitive antagonists of the agonists concerned. For the cases where the mechanisms involved are at the level of the stimulus–effect relationship (case 2) and where the receptor activation is based on the simple drug–receptor interaction model (Sec. 4) or the dual receptor model (Sec. 14)—taking into account that a certain stimulus always corresponds to a certain effect—there will be a parallel shift in the log concentration curves by the competitive antagonist. The degree of shift will obey Michaelis–Menten kinetics (Ariëns, 1966).

Whether such a parallel shift will occur in 1 and 2 mentioned in case 1 depends on the affinity of the antagonist for the various interdependent receptor sites involved and thus on the degree of displacement of the agonist molecule from these receptor sites. For both instances, under some particular conditions, it is possible for a parallel shift in both the log concentration–binding and –effect curves to occur, obeying Michaelis–Menten kinetics (Ariëns, 1964; Thron, 1973). For 3 and 4 mentioned in case 1, a parallel shift is obviously based on Michaelis–Menten receptor kinetics.

17. The Allosteric Receptor Model

This model, in which agonist A and antagonist B bind on interdependent receptor sites RR′ in such a way that the presence of the agonist on its receptor site R changes the affinity of the antagonist for its receptor site R′ and vice versa in a

strictly mutual sense, has been described in the classic receptor literature (Ariëns
et al., 1956a, 1964). It also, in fact, is a dual receptor model. The characteristics
of this model can be summarized by the following reaction scheme:

$$RR' + A \rightleftharpoons RAR' \qquad K_A = \frac{RR' \cdot A}{RAR'}$$

$$RR' + B \rightleftharpoons RR'B \qquad K_B = \frac{RR' \cdot B}{RR'B}$$

$$RAR' + B \rightleftharpoons RAR'B \qquad K'_B = \frac{RAR' \cdot B}{RAR'B} \qquad \frac{K_A}{K'_A} = \frac{K_B}{K'_B} = (1 - \beta)$$

$$RR'B + A \rightleftharpoons RAR'B \qquad K'_A = \frac{RR'B \cdot A}{RAR'B}$$

This model will be worked out for the situation where there is a mutual inhibitory
relationship. As a matter of fact an enhancement of the affinities can also be
introduced in the model.

The occupation of the binding sites involved can be represented by

$$\text{fraction occupied by A} = \frac{RAR' + RAR'B}{RR' + RAR' + RR'B + RAR'B} \tag{11a}$$

$$= \frac{A}{A + \{K_A(1 + B/K_B)/[1 + B(1 - \beta)/K_B]\}}$$

A crucial term in Eq. (11a) is the intrinsic competition term β, represented by

$$\beta = \frac{1 - K_A}{K'_A}$$

Displacement of agonists by agonists or displacement of antagonists by antagonists is denoted by direct competition. Displacement of agonists by antagonists
and vice versa is denoted by allosteric or cross-competition. For the allosteric competition it can be shown in the case where $\beta = 1$ that Eq. (11a) is reduced to
that for a simple competitive interaction. In the case where $0 < \beta < 1$, the two
agents involved behave only as partial competitors (see Fig. 27). The binding curve
or concentration–response curve for the agonist will be shifted by the antagonist
but only up to a certain limit and vice versa. In the case where $\beta = 0$, B does not
affect the binding of A and vice versa. In the case where $\beta < 0$, the equation represents a mutual allosteric sensitization. These aspects of this model have already
been worked out in detail (Ariëns et al., 1956a, 1964). An interesting characteristic
of this model is that the binding constant of an agonist A_1 as deduced from a direct
competition (i.e., displacement of an agonist A_2) differs from the binding constant
deduced from cross-competition (i.e., displacement of an antagonist B_1). The
same holds true for the binding constant of an antagonist.

The 50% inhibition points for these situations can be deduced from Eq. (11a):

$$(IC_{50})_{A_1} \text{ direct} = K_{A_1}\left(1 + \frac{A_2}{K_{A_2}}\right)$$

$$(IC_{50})_{A_1} \text{ crossed} = \frac{K_{A_1}(1 + B_1/K_{B_1})}{1 + B_1(1 - \beta)/K_{B_1}}$$

$$(IC_{50})_{B_1} \text{ direct} = K_{B_1}\left(1 + \frac{B_2}{K_{B_2}}\right)$$

$$(IC_{50})_{B_1} \text{ crossed} = \frac{K_{B_1}(1 + A_1/K_{A_1})}{1 + A_1(1 - \beta)/K_{A_1}}$$

It should be emphasized that in the case of partial "competition," dependent on the concentration of the agents involved, the receptor sites for agonist and antagonist (located on one receptor molecule) will be simultaneously occupied to a minor or larger extent. With excessively high concentrations of agonist as well as antagonist, there will be a full saturation of both receptor sites.

In this system the following types of compounds must be distinguished:

1. Partial agonist in the classical sense, i.e., compounds which cannot effectuate a maximal effect
2. Partial competitors, i.e., agonists and antagonists with $0 < \beta < 1$ which cannot completely displace each other in a cross-competition experiment
3. Compounds with an autointeraction, i.e., agonists and antagonists which bind to both agonistic and antagonistic binding sites

For the last compounds, as is easily understandable, there will be a positive or negative cooperation, dependent on the type of interdependence. If the affinity of the compound for both sites is equal, the slope of the concentration–response curve will be steeper or more shallow. If the affinity constants of the compound for the two sites differ, the curve will have a biphasic character.

An interesting aspect of this model is that agonists and partial agonists have an intrinsic activity insofar as the receptor activation is concerned, while the

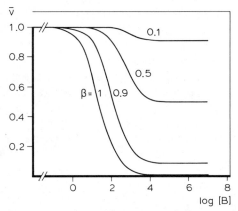

Fig. 27. Theoretical displacement curves for different values of the intrinsic competition constant β. Ordinate: the concentration of the agonistic receptor site occupied by A in the presence of B relative to the concentration of agonistic receptor site occupied by A in the absence of B. Abscissa: the logarithm of the inhibitor concentration (nM). Note: The full competitor ($\beta = 1$) displaces A completely from its binding sites; the partial competitor displaces A to a degree determined by β (e.g., for $\beta = 0.1$, only 10% displacement can be obtained).

partial agonists and full competitors also have an intrinsic activity insofar as their capacity to change (after binding to their receptors) the affinity of the agonist to its receptors is concerned. If this affinity is reduced to zero, the action of the then full "competitor" is indistinguishable from the classical model for competition where simultaneous presence of agonist and competitive antagonist on their receptors is excluded.

In the classical model, this mutual exclusion was based on a topographical displacement, binding to the same site. This is not the case in the allosteric model. On the other hand, the concept of a partial competitor can be interpreted as a case of partial steric hindrance. That is, with the antagonistic agent on its receptor site, due to the mobility of a part of the antagonistic molecule, the agonist still gets a chance, although reduced, to bind. The antagonistic molecule hinders but does not exclude the access of the agonist to its receptor site. The result is a lowering of the binding constant of the agonist in the presence of the antagonist, which then acts as a partial competitor. Such a model was first described by Schaeffer and Johnson (1968) for substrate–enzyme interactions in the presence of particular types of irreversible antimetabolites and has been worked out as the *Charnier model* by Rocha e Silva (1970, 1978) for binding of drugs to pharmacological receptors. Also, in this model the interference in binding of agonist by the antagonist and thus the change in affinity must be mutual, which implies that the agonist changes the affinity of the antagonist in a similar way.

Whereas in the allosteric model both negative and positive interference is feasible, the Charnier model, with its steric hindrance, cannot account for a mutual facilitation of the binding, the sensitization. Also, in the Charnier model a compound with an affinity to the agonistic as well as to the antagonistic receptor site (which does not seem very likely) would show a negative cooperation in its binding curve.

Snyder and his group (Burt *et al.*, 1976; U'Prichard *et al.*, 1977; Bennet and Snyder, 1976) studied the binding and mutual displacement of agonists and corresponding antagonists extensively. Remarkably enough, he found that agonist molecules displace agonists more effectively than antagonists, while antagonist molecules displace antagonists more effectively than agonists. This phenomenon was observed with a number of receptor types such as dopaminergic, α-adrenergic, serotonergic, and morphine receptor preparations (Tables VI and VII). The model described before implies the occurrence of such a phenomenon, which does not mean that it is the only possible way to understand it.

A similar phenomenon is to be expected if two independent receptor populations are involved.

From Sec. 4, which covers both the Michaelis–Menten and the simple receptor model, and from the dual receptor model described in Sec. 14,

$$\text{fraction occupied} = \frac{1}{1 + (K_{1\text{app}}/D_1)[1 + (D_2/K_{2\text{app}})]} \tag{11b}$$

it is obvious that these models cannot account for the phenomenon observed. From Eq. (11b), we see that drug 2 displaces drug 1 according to its $K_{2\text{app}}$. It doesn't depend on whether or not D_1 is an agonist or an antagonist or whether or not D_2 is an agonist or an antagonist, and it is independent of $K_{1\text{app}}$. In general,

Table VI
Inhibition of [³H]Clonidine and [³H]-WB-4101ᵃ Binding by Drugsᵇ

Inhibiting drug, α-adrenergic	K_i for [³H]-WB-4101 binding (nM), A	K_i for [³H]clonidine binding (nM), B	Ratio A/B
Agonists			
(−)Norepinephrine	1000	17	59
(−)Epinephrine	590	5.9	100
Clonidine	430	5.7	75
Oxymetazoline	24	1.9	13
Partial agonists			
Dihydroergotamine	3.5	2.4	1.5
α-Ergocryptine	9	8	1.1
d-LSD	220	220	1.0
Blockers			
Phentolamine	3.6	22	0.16
Phenoxybenzamine	4.0	60	0.07
WB-4101	0.6	200	0.003

ᵃWB-4101 = 2-([2′,6′-dimethoxy]phenoxyethylamino)methylbenzodioxan.
ᵇ After U'Prichard *et al.* (1977).

Table VII
Inhibition of [³H]Haloperidol and [³H]Dopamine Binding by Drugsᵃ

Inhibiting drug, dopaminergic	K_i for [³H]haloperidol binding (nM), A	K_i for [³H]dopamine binding (nM), B	Ratio A/B
Agonists			
Dopamine	670	17.5	38
Apomorphine	51	8.6	6
Partial agonist			
d-LSD	20	30	0.7
Antagonists			
Spiroperidol	0.25	1400	0.0002
(+)Butaclamol	0.54	80	0.007
Fluphenazine	0.88	230	0.004
Haloperidol	1.4	920	0.002

ᵃ After Burt *et al.* (1976).

receptor models in which the different receptor states can interconvert reversibly within the time period of the experiment will not account for the phenomenon.

18. Binding and Displacement on Two or More Independent Classes of Receptor Sites

Drug binding to proteins is most commonly described in terms of independent, noninteracting binding sites (Steinhardt and Reynolds, 1969). Scatchard

(1949) originally formulated this as

$$\bar{V} = \sum_i \frac{n_i}{1 + K_i/D} \tag{12}$$

where V = bound-drug concentration in moles per mole of protein, n_i = number of binding sites in class i, K_i = dissociation constant of class i.

If proteins, e.g., plasma proteins such as albumin, are exposed to pharmaca, the pharmaca will bind to the population of different binding sites on the proteins in accordance with their binding constants for these sites. Different drugs may have certain binding sites in common in the population of sites. The binding constants of the drugs for their common binding sites are fully independent; e.g., sites on which drug A binds strongly may have a weak affinity for B and vice versa, etc. Combination of pharmaca will result in mutual displacement from the common binding sites in accordance with their binding constants.

19. Two-Site Model

For the two-site model, binding can be described as

$$\bar{V} = \frac{1}{1 + K_1/D} + \frac{1}{1 + K_2/D} \tag{13}$$

Combination of two drugs which bind to these sites will generally result in an independent distribution of the drugs over these sites. This can be demonstrated with the binding of fatty acids to bovine albumin. On this protein two sites, a site S which preferentially binds short-chain fatty acids and a site L which preferentially binds long-chain fatty acids, appear to be involved (Rodrigues de Miranda *et al.*, 1976) (see Table VIII). Table VIII shows that butyric acid displaces octanoic acid* (the asterisk indicates the labeled compound) less effectively than butyric acid* and that octanoic acid displaces butyric acid* less effectively than octanoic acid*.

This two-site model offers a simple explanation for the phenomenon re-

Table VIII
Displacement of Fatty Acids Bound to Bovine Albumin (30°C)[a]

Compound	Binding constants (μM)		Displacement constants (μM) for the displacement of	
	Site S	Site L	Butyric* acid[b]	Octanoic* acid
Butyric acid	120	1430	120	770
Octanoic acid	40	2	25	2

[a] For details, see Rodrigues de Miranda *et al.* (1976). Note that at least two binding sites, one preferentially occupied by butyric acid and one preferentially occupied by octanoic acid, are involved.
[b] *: labeled.

ported by Snyder (Tables VI and VII). In the case of the dopaminergic receptor, for example, it can be assumed that two sites are involved there also, namely site 1, for which $K_{1\text{dopamine}} = 17.5$ nM and $K_{1\text{haloperidol}} = 920$ nM, and site 2, for which $K_{2\text{ dopamine}} = 670$ nM and $K_{2\text{ haloperidol}} = 1.4$ nM (see Table VII).

Figures 28(a) and (b) represent curves showing the mutual displacement of dopamine and haloperidol calculated on the basis of these constants. Figure 28(a) is a plot of the sum of the fractional occupation of the two binding sites by dopamine* against the log concentration of this compound. This was done in the absence of a second compound and in the presence of dopamine and haloperidol in concentrations which if given alone would occupy 50% of the sites with the lowest binding constants. Figure 28(b) similarly gives the binding of haloperidol* in the absence of a second compound and in the presence of dopamine and haloperidol. The curves clearly show that at the lower dose ranges dopamine* is more effectively displaced by dopamine [Fig. 28(a)] and haloperidol* more effectively by haloperidol [Fig. 28(b)], as observed by Snyder (Burt *et al.*, 1976). However, at higher dose ranges (not studied by Snyder), the reverse situation obtains.

The receptor sites on which the agonist most effectively displaces agonists, and, in fact, the receptor for which the agonists have the highest affinity, probably are physiologically the most relevant ones. The affinity constants of the agonists on these receptors correlate reasonably well with the activity for these agents tested on biological systems (U'Prichard *et al.*, 1977).

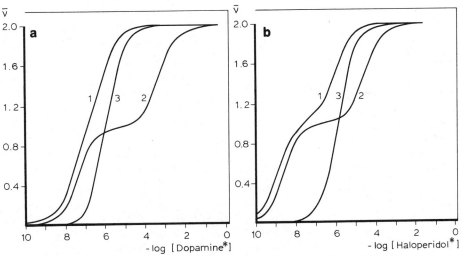

Fig. 28. Calculated log concentration curves for the sum of the fractional occupations of two classes of binding sites for (a) dopamine* and for (b) haloperidol* based on the binding constants given in the text. The figures give a binding curve for dopamine* [(a)1] and haloperidol* [(b)1] alone and such curves in the presence of dopamine at 670 nM [(a)3 and (b)2] and haloperidol at 920 nM [(a)2 and (b)3], concentrations in which the agents saturate the binding site with the lowest binding constant for 50%. Note: In the lower concentration ranges, dopamine more effectively than haloperidol displaces dopamine* (a) and haloperidol more effectively than dopamine displaces haloperidol* (b). At the higher dose range the reverse situation exists.

Remarkably, the partial agonists take an intermediate position in their affinities for both the preferential agonist binding sites and for the preferential antagonist binding sites. This may reflect that partial agonists not only biologically but also physicochemically take an intermediate position between agonists and the corresponding antagonists.

There is growing evidence, indeed, for the occurrence of two types of receptors for various neurotransmitters such as dopamine (Cools and Van Rossum, 1976; Cools, 1977) and noradrenaline (Doxey *et al.*, 1977), namely pre- and postsynaptic receptors. These two receptor populations can also be recognized on the basis of the existence of receptor stimulants and blockers which differentiate between the two receptor populations. Thus, for instance, for the dopaminergic receptors, dopamine and apomorphine act as stimulants and haloperidol and other neuroleptics as blockers on the postsynaptic (DA_e) receptors. On the other hand, DPI (3,4-dihydroxyphenylamino)-2-imidazoline acts as a stimulant and ergometrine as a blocker on the presynaptic (DA_i) receptors (Cools and Van Rossum, 1976; Cools, 1977).

20. Reflection

In this chapter our aim has been to show the simplest models for pharmacon–receptor–effector systems. One must be well aware that the simplest concept or model describing certain biological phenomena has priority over more complex concepts. The elaboration of more complex general models leads to the fields of theoretical biology or biomathematics. A disadvantage of the more complex models—and there is no limit to the complexity—is that, because of the large number of variables, rate constants, etc., involved, they usually will give a variety of answers in the interpretation of certain biological phenomena and thus do not show discrimination. The concept to be preferred for practical use is therefore one in which the number of operational variables is limited to the smallest number. Simple models as a rule are special cases of more complex general models and have the advantage of elucidating and emphasizing, in a heuristic form, fundamental concepts.

One may not expect, however, that one universal concept or theory of drug action will account for all mechanisms of action of bioactive agents.

ACKNOWLEDGMENT. The authors are indebted to Mrs. E. J. Klok for making computerized calculations and drawings based on the models outlined and to Mrs. A. R. H. Wigmans for help with the preparation of the manuscript.

References

Albuquerque, E. X., Eldefrawi, A. T., Eldefrawi, M. E., Mansour, N. A., and Tsai, M. C., 1978, Amantadine: Neuromuscular blockade by suppression of ionic conductance of the acetylcholine receptor, *Science* **199**:788.
Ariëns, E. J., 1964, *Molecular Pharmacology*, Vol. I, Academic Press, New York.

Ariëns, E. J., 1966a, Receptor theory and structure–action relationships, in: *Advances in Drug Research*, Vol. III (N. J. Harper and A. B. Simmonds, eds.), pp. 235–285, Academic Press, New York.

Ariëns, E. J., 1966b, Eine Molekulargrundlage für die Wirkung von Pharmaka, *Arzneim.-Forsch.* **16**:1376.

Ariëns, E. J., 1967, The structure–activity relationships of β-adrenergic drugs and β-adrenergic blocking drugs, *Ann. NY Acad. Sci.* **139**:606.

Ariëns, E. J., 1971a, A general introduction to the field of drug design, in: *Drug Design*, Vol. I (E. J. Ariëns, ed.), pp. 1–263, Academic Press, New York.

Ariëns, E. J., 1971b, Modulation of pharmacokinetics by molecular manipulation, in *Drug Design*, Vol. II (E. J. Ariëns, ed.), pp. 2–127, Academic Press, New York.

Ariëns, E. J., 1974, Drug levels in the target tissue and effect, *Clin. Pharmacol. Ther.* **16**:155.

Ariëns, E. J., 1976, Stereochemie und pharmakologische Wirkung, *Schriftenr. Bundesapoth. Wiss. Fortbild.* **IV**:77.

Ariëns, E. J., and Beld, A. J., 1977, The receptor concept in evolution, *Biochem. Pharmacol.* **26**:913.

Ariëns, E. J., and Simonis, A. M., 1960, Autonomic drugs and their receptors, *Arch. Int. Pharmacodyn. Ther.* **127**:479.

Ariëns, E. J., and Simonis, A. M., 1964, A molecular basis for drug action, *J. Pharm. Pharmacol.* **16**:137.

Ariëns, E. J., and Simonis, A. M., 1967, Cholinergic and anticholinergic drugs, do they act on common receptors?, *Ann. NY Acad. Sci.* **144**:842.

Ariëns, E. J., and Simonis, A. M., 1974, Design of bioactive compounds, *Top. Current Chem.* **52**:3.

Ariëns, E. J., and Simonis, A. M., 1976, Receptors and receptor mechanisms, in: *Beta-Adrenoceptor Blocking Agents* (P. R. Saxena and R. P. Forsyth, eds.), pp. 3–27, North-Holland, Amsterdam.

Ariëns, E. J., and Simonis, A. M., 1977, Chemical structure and toxic action. Avoidance of toxicity by molecular manipulation, in: *Drug Design and Adverse Reactions* (H. Bundgaard, P. Juul, and H. Kofod, eds.), pp. 317–330, Munksgaard, Copenhagen.

Ariëns, E. J., and Van Rossum, J. M., 1957, Affinity, intrinsic activity and the all-or-none response, *Arch. Int. Pharmacodyn. Ther.* **113**:89.

Ariëns, E. J., Van Rossum, J. M., and Simonis, A. M., 1956a, A theoretical basis of molecular pharmacology, *Arzneim.-Forsch.* **6**:282.

Ariëns, E. J., Van Rossum, J. M., and Simonis, A. M., 1956b, A theoretical basis of molecular pharmacology, part II, *Arzneim.-Forsch.* **6**:611.

Ariëns, E. J., Van Rossum, J. M., and Koopman, P. C., 1960, Receptor reserve and threshold phenomena, I. Theory and experiments with autonomic drugs tested on isolated organs, *Arch. Int. Pharmacodyn. Ther.* **127**:459.

Ariëns, E. J., Simonis, A. M., and Van Rossum, J. M., 1964, Drug–receptor interaction: Interaction of one or more drugs with different receptor systems, in: *Molecular Pharmacology*, Vol. 1 (E. J. Ariëns, ed.), pp. 287–393, Academic Press, New York.

Aronstam, R. S., Hoss, W., and Abood, L. G., 1977, Conversion between configurational states of the muscarinic receptor in rat brain, *Eur. J. Pharmacol.* **46**:279.

Beld, A. J., and Ariëns, E. J., 1974, Stereospecific binding as a tool in attempts to localize and isolate muscarinic receptors, part II. Binding of [+]-benzetimide, [−]-benzetimide and atropine to a fraction from bovine tracheal smooth muscle and to bovine caudate nucleus, *Eur. J. Pharmacol.* **25**:203.

Beld, A. J., Van den Hoven, S., Wouterse, A. C., and Zegers, M. A. P., 1975, Are muscarinic receptors in the central and peripheral nervous system different?, *Eur. J. Pharmacol.* **30**:360.

Belleau, B., 1964, A molecular theory of drug action based on induced conformational perturbations of receptors, *J. Med. Chem.* **7**:776.

Bennett, J. P., and Snyder, S. H., 1976, Serotonin and lysergic acid diethylamide binding in rat brain membranes: Relationship to postsynaptic serotonin receptors, *Mol. Pharmacol.* **12**:373.

Berde, B., Cerletti, A., and Konzett, H., 1961a, The biological activity of a series of peptides related to oxytocin, in: *Oxytocin* (R. Cadeyro-Barcia and H. Heller, eds.), pp. 247–265, Pergamon Press, Elmsford, N.Y.

Berde, B., Weidmann, H., and Cerletti, A., 1961b, Uber Phenylalanin²-Lysin-Vasopressin, *Helv. Physiol. Pharmacol. Acta* **19**:285.

Bregestovski, P. D., Iljin, V. I., Jurchenko, O. P., Veprintsev, B. N., and Vulfius, C. A., 1977, Acetylcholine receptor conformational transition on excitation masks disulphide bonds against reduction, *Nature (London)* **270**:71.

Burgen, A. S. V., 1970, Receptor mechanisms, *Annu. Rev. Pharmacol.* **10**:7.

Burgen, A. S. V., Hiley, C. R., and Young, J. M., 1974, The binding of [^{3}H]-propylbenzilylcholine mustard by longitudinal muscle strips from guinea-pig small intestine, *Br. J. Pharmacol.* **50**:145.

Burt, D. R., Creese, I., and Snyder, S. H., 1976, Properties of [^{3}H]haloperidol and [^{3}H]dopamine binding associated with dopamine receptors in calf brain membranes, *Mol. Pharmacol.* **12**:800.

Carlsson, E., Åblad, B., Brändström, A., and Carlsson, B., 1972, Differentiated blockade of the chronotropic effects of various adrenergic stimuli in the cat heart, *Life Sci.* **11**:953.

Changeux, J. P., Thiéry, J., Tung, Y., and Kittel, C., 1967, On the cooperativity of biological membranes, *Proc. Natl. Acad. Sci. USA* **57**:335.

Clark, A. J., 1937, *General Pharmacology*, Handbuch der experimentellen Pharmakologie, Vol. 4, Springer, Berlin.

Colquhoun, D., 1973, The relation between classical and cooperative models for drug action, in: *Drug Receptors* (H. P. Rang, ed.), pp. 149–182, Macmillan, New York.

Cools, A. R., 1977, Two functionally and pharmacologically distinct dopamine receptors in the rat brain, *Adv. Biochem. Psychopharmacol.* **16**:215.

Cools, A. R., and Van Rossum, J. M., 1976, Excitation-mediating and inhibition-mediating dopamine-receptors. A new concept towards a better understanding of electrophysiological, biochemical, pharmacological, functional and clinical data, *Psychopharmacòlogia* **45**:243.

Creese, I., Burt, D. R., and Snyder, S. H., 1976, Dopamine receptor binding predicts clinical and pharmacological potencies of antischizophrenic drugs, *Science* **192**:481.

Cuatrecasas, P., 1974, Membrane receptors, *Annu. Rev. Biochem.* **43**:169.

Cuatrecasas, P., 1975, Hormone–receptor interactions and the plasma membrane, in: *Cell Membranes: Biochemistry, Cell Biology and Pathology* (G. Weissmann and R. Claiborne, eds.), pp. 177–184, H. P. Publishing, New York.

Cuatrecasas, P., Hollenberg, M. D., Chang, K. J., and Bennett, V., 1975, Hormone receptor complexes and their modulation of membrane function, *Recent Prog. Horm. Res.* **31**:37.

Del Castillo, J., and Katz, B., 1955, On the localization of acetylcholine receptors, *J. Physiol. (London)* **128**:157.

de Robertis, E., 1975a, *Synaptic Receptors—Isolation and Molecular Biology*, Dekker, New York.

de Robertis, E., 1975b, Synaptic receptor proteins; isolation and reconstitution in artificial membranes, *Rev. Physiol. Biochem. Pharmacol.* **73**:9.

de Robertis, E., 1976, Synaptic receptor proteins and chemical excitable membranes, in: *The Structural Basis of Membrane Function* (Y. Hatefi and L. Djavadi Ohaniance, eds.), pp. 339–361, Academic Press, New York.

Doxey, J. C., Smith, C. F. C., and Walker, J. M., 1977, Selectivity of blocking agents for pre- and post-synaptic α-adrenoceptors, *Br. J. Pharmacol.* **60**:91.

Driel, C. van, Houthoff, H. J., and Baljet, B., 1973, The innervation of the myometrium, some histo-chemical observations in man and rat, *Eur. J. Obstet. Gynecol. Reprod. Biol.* **3**:11.

Ellenbroek, B. W. J., 1964, Stereoisomerie en biologische aktiviteit, Ph.D. Thesis, University of Nijmegen, Nijmegen, The Netherlands.

Ellenbroek, B. W. J., Nivard, R. J. F., Van Rossum, J. M., and Ariëns, E. J., 1965, Absolute configuration and parasympathetic action: Pharmacodynamics of enantiomorphic and diastereoisomeric esters of β-methylcholine, *J. Pharm. Pharmacol.* **17**:393.

Furchgott, R. F., 1954, Dibenamine blockade in strips of rabbit aorta and its use in differentiating receptors, *J. Pharmacol. Exp. Ther.* **111**:265.

Gaddum, J. H., 1937, The quantitative effects of antagonistic drugs, *J. Physiol. (London)* **89**:7P.

Goldberg, N. D., 1975, Cyclic nucleotides and cell function, in: *Cell Membranes: Biochemistry, Cell Biology and Pathology* (G. Weissmann and R. Claiborne, eds.), pp. 185–202, H. P. Publishing, New York.

Gosselin, R. E., and Gosselin, R. S., 1973, Tachyphylaxis of guinea-pig ileum to histamine and furtre-thonium, *J. Pharmacol. Exp. Ther.* **184**:494.

Grana, E., Lucchelli, A., and Zonta, F., 1974, Selectivity of β-adrenergic compounds, III. Classification of β-receptors, *Il Farmaco-Educ. Sci.* **29**:786.

Greaves, H. F., 1977, Membrane receptor–adenylate cyclase relationships, *Nature (London)* **265**:681.

Greengard, P., 1976, Possible role for cyclic nucleotides and phosphorylated membrane proteins in postsynaptic actions of neurotransmitters, *Nature (London)* **260**:101.

Grob, D., 1976, Myasthenia gravis, *Ann. NY Acad. Sci.* **274**:1.

Harms, A. F., and Nauta, W. Th., 1960, The effects of alkyl substitution in drugs. I. Substituted dimethylaminoethyl benzhydryl ethers, *J. Med. Pharm. Chem.* **2**:57.

Haydon, D. A., Hendry, B. M., and Levison, S. R., 1977, The molecular mechanisms of anaesthesia, *Nature (London)* **268**:356.

Hechter, O., and Braun, Th., 1971, Peptide hormone–receptor interaction: an informational transaction, in: *Structure–Activity Relationships of Protein and Polypeptide Hormones* (M. Margoulies and F. C. Greenwood, eds.), pp. 212–227, Excerpta Medica, Amsterdam.

Heidelberger, Ch., 1973, Current trends in chemical carcinogenesis, *Fed. Proc.* **32**:2154.

Hitchings, G. H., 1964, Antimetabolites and chemotherapy: Integration of biochemistry and molecular manipulation, in: *Chemotherapy of Cancer* (P. A. Plattner, ed.), pp. 77–87, Elsevier, Amsterdam.

Hofmann, K., 1960, Preliminary observations relating structures and function in some pituitary hormones, *Brookhaven Symp. Biol.* **13**:184.

Jacobs, S., and Cuatrecasas, P., 1976, The mobile receptor hypothesis and "cooperativity" of hormone binding. Application to insulin, *Biochim. Biophys. Acta* **433**:482.

Kachadorian, W. A., Wade, J. B., and DiScala, V. A., 1975, Vasopressin: Induced structural change in toad bladder luminal membrane, *Science* **190**:67.

Karlin, A., 1967, On the application of "a plausible model" of allosteric proteins to the receptor for acetylcholine, *J. Theor. Biol.* **16**:306.

Lands, A. M., Luduena, F. P., and Buzzo, H. J., 1967, Differentiation of β-receptors responsive to isoproterenol. *Life Sci.* **6**:2241.

Lee, A. G., 1977, Local anesthesia: The interaction between phospholipids and chlorpromazine, propranolol, and practolol, *Mol. Pharmacol.* **13**:474.

Lefkowitz, R. J., Caron, M. G., Limbird, L., Mukherjee, C., and Williams, L. T., 1976, Membrane-bound hormone receptors, in: *The Enzymes of Biological Membranes*, Vol. 4 (A. Martonosi, ed.), pp. 283–310, Wiley, New York.

Lüllmann, H., and Ziegler, A., 1973, A transient state concept of drug receptor interaction, *Naunyn-Schmiedeberg's Arch. Pharmakol.* **280**:1.

Lüllmann, H., Preuner, J., and Schaube, H., 1974, A kinetic approach for an interpretation of the acetylcholine-*d*-tubocurarine interaction on chronically denervated skeletal muscle, *Naunyn-Schmiedeberg's Arch. Pharmakol.* **281**:415.

Maddy, A. H., and Dunn, M. J., 1976, The solubilization of membranes, in: *Biochemical Analysis of Membranes* (A. H. Maddy, ed.), pp. 177–196, Chapman & Hall, London.

Maugh, T. H., 1976, Hormone receptors: New clues to the cause of diabetes, *Science* **193**:220.

McGuire, W. L., Carbone, P. P., and Vollmer, E. P., 1975, *Estrogen Receptors in Human Breast Cancer*, Raven Press, New York.

Monod, J., Wyman, J., and Changeux, J. P., 1965, On the nature of allosteric transitions: A plausible model, *J. Mol. Biol.* **12**:88.

Nickerson, M., 1956, Receptor occupancy and tissue response, *Nature (London)* **178**:697.

Papadimitriou, A., and Worcel, M., 1974, Dose–response curves for angiotensin II and synthetic analogues in three types of smooth muscle: existence of different forms of receptor sites for angiotensin II, *Br. J. Pharmacol.* **50**:291.

Paton, W. O. M., 1961, A theory of drug action based on the rate of drug receptor combination, *Proc. Roy. Soc. London Ser. B* **154**:21.

Pert, C. B., 1976, The opiate receptor, in: *Cell Membrane Receptors for Viruses, Antigens and Antibodies, Polypeptide Hormones and Small Molecules* (R. F. Beers and E. G. Bassett, eds.), pp. 435–448, Raven Press, New York.

Pert, C. B., and Snyder, S. H., 1974, Opiate receptor binding of agonists and antagonists affected differentially by sodium, *Mol. Pharmacol.* **10**:868.

Pfeiffer, C. C., 1956, Optical isomerism and pharmacological action, a generalization, *Science* **124**:29.

Podleski, T. R., and Changeux, J. P., 1970, On the excitability and cooperativity of the electroplax membrane, in: *Fundamental Concepts in Drug–Receptor Interactions* (J. F. Danielli, J. F. Moran, and D. J. Triggle, eds.), pp. 93–119, Academic Press, New York.

Raftery, M. A., 1975, Transductive coupling between acetylcholine, its receptor, and the postsynaptic membrane, in: *Functional Linkage in Biomolecular Systems* (F. O. Schmitt, D. M. Schneider, and D. M. Crothers, eds.), pp. 215–223, Raven Press, New York.

Rang, H. P., and Ritter, J. M., 1970a, On the mechanism of desensitization at cholinergic receptors, *Mol. Pharmacol.* **6**:357.

Rang, H. P., and Ritter, J. M., 1970b, The relationship between desensitization and the metaphilic effect at cholinergic receptors, *Mol. Pharmacol.* **6**:383.

Raynaud, J. P., 1977a, A strategy for the design of potent hormones, in: *Proceedings of the Vth International Symposium on Medicinal Chemistry, Paris, 1976,* Elsevier, Amsterdam.

Raynaud, J. P., 1977b, Une stratégie de récherche pour les hormones de synthèse: leurs interactions avec les récepteurs hormonaux, in: *Actualités Pharmacologiques* (J. Cheymol, J. R. Boissier, and P. Lechat, eds.), pp. 49–64, Masson et Cie, Paris.

Reinhardt, D., and Wagner, J., 1974, β-Adrenolytic antiarrhythmic and local anaesthetic effects of phenylephrine, *Naunyn-Schmiedeberg's Arch. Pharmakol.* **284**:245.

Rekker, R. F., Timmerman, H., Harms, A. F., and Nauta, W. Th., 1971, The antihistaminic and anticholinergic activities of optically active diphenhydramine derivatives, *Arzneim.-Forsch.* **21**:688.

Roach, P. J., 1977, Functional significance of enzyme cascade systems, *Trends Biochem. Sci.* **2**:87.

Robison, G. A., Butcher, R. W., and Sutherland, E. W., 1971, *Cyclic AMP,* Academic Press, New York.

Rocha e Silva, M., 1970, A thermodynamic approach to problems of drug antagonism. II. A microphysical model of the phenomenon of recovery, *Physiol. Chem. Phys.* **2**:503.

Rocha e Silva, M., 1978, Kinetics of antagonist action, in: *Histamine II and Anti-Histaminics* (M. Rocha e Silva, ed.), Handbuch der experimentellen Pharmakologie, Vol. 18, No. 2, pp. 295–332, Springer, Berlin.

Rodrigues de Miranda, J. F., Eikelboom, T. D., and Van Os, G. A. J., 1976, The extent of hydrophobic binding area studied by fatty acid binding to albumin, *Mol. Pharmacol.* **12**:454.

Rosell, S., and Belfrage, E., 1975, Adrenergic receptors in adipose tissue and their relation to adrenergic innervation, *Nature (London)* **253**:738.

Ross, E. M., and Gilman, A. G., 1977, Reconstitution of catecholamine-sensitive adenylate cyclase activity: Interaction of solubilized components with receptor-replete membranes, *Proc. Natl. Acad. Sci. USA* **74**:3715.

Rudinger, J., Pliška, V., and Krejči, I., 1972, Oxytocin analogs in the analysis of some phases of hormone action, *Recent Prog. Horm. Res.* **28**:131.

Ruffolo, R. R., Turowski, B. S., and Patil, P. N., 1977, Lack of cross-desensitization between structurally dissimilar α-adrenoceptor agonists, *J. Pharm. Pharmacol.* **29**:378.

Scatchard, G., 1949, Attractions of proteins for small molecules and ions, *Ann. NY Acad. Sci.* **51**:660.

Schaeffer, H. J., and Johnson, R. N., 1968, Enzyme inhibitors. XX. Studies on the hydrophobic and hydroxyl binding regions of adenosine deaminase. 9-(2-m-Bromoacetamidophenethyl)adenine, a new irreversible inhibitor of adenosine deaminase, *J. Med. Chem.* **11**:21.

Schramm, M., Orly, J., Eimerl, S., and Korner, M., 1977, Coupling of hormone receptors to adenylate cyclase of different cells by cell fusion, *Nature (London)* **268**:310.

Schueler, F. W., 1960, *Chemobiodynamics and Drug Design,* McGraw-Hill, New York.

Schwyzer, R., 1970, Programmierte Molekeln, *Experientia* **26**:577.

Seelig, S., and Sayers, G., 1973, Isolated adrenal cortex cells: ACTH agonists, partial agonists, antagonists; Cyclic AMP and corticosterone production, *Arch. Biochem. Biophys.* **154**:230.

Segura, E. T., and Biscardi, A. M., 1967, Changes in brain epinephrine and norepinephrine induced by afferent electrical stimulation in the isolated toad head, *Life Sci.* **6**:1599.

Singer, S. J., and Nicolson, G. L., 1972, The fluid mosaic model of the structure of cell membranes, *Science* **175**:720.

Sobel, A., Heidmann, T., Hofler, J., and Changeux, J. P., 1978, Distinct protein components from *Torpedo marmorata* membranes carry the acetylcholine receptor site and the binding site for local anesthetics and histrionicotoxin, *Proc. Natl. Acad. Sci USA* **75**:510.

Steinhardt, J., and Reynolds, J. A., 1969, in: *Multiple Equilibrium in Proteins* (B. Horecker, N. O. Kaplan, J. Marmur, and H. A. Scheraga, eds.), Chap. II, p. 12, Academic Press, New York.

Struyker Boudier, H. A. J., de Boer, J., Smeets, G. M. W., Lien, E. J., and Van Rossum, J. M., 1975, Structure–activity relationships for central and peripheral α-adrenergic activity of imidazoline derivatives, *Life Sci.* **17**:377.

Thron, C. D., 1973, On the analysis of pharmacological experiments in terms of an allosteric receptor model, *Mol. Pharmacol.* **9**:1.

Triggle, D. J., 1965, *Chemical Aspects of the Autonomic Nervous System,* Academic Press, New York.

U'Prichard, D. C., Greenberg, D. A., and Snyder, S. H., 1977, Binding characteristics of a radiolabeled agonist and antagonist at central nervous system alpha noradrenergic receptors, *Mol. Pharmacol.* **13**:454.

Van den Brink, F. G., 1973, The model of functional interaction, I and II, *Eur. J. Pharmacol.* **22**:270 and 279.

Van Rossum, J. M., 1958, Pharmacodynamics of cholinomimetic and cholinolytic drugs, Ph.D. thesis, University of Nijmegen, Nijmegen, The Netherlands.

Van Rossum, J. M., 1966, Limitations of molecular pharmacology, in: *Advances in Drug Research,* Vol. III (N. J. Harper and A. B. Simmonds, eds.), pp. 189–234, Academic Press, New York.

Van Rossum, J. M., 1968, Drug receptor theories, in: *Recent Advances in Pharmacology* (J. M. Robson and R. S. Stacey, eds.), pp. 99–133, Churchill, London.

Van Rossum, J. M., 1977, *Kinetics of Drug Action,* Handbuch der experimentellen Pharmakologie, Vol. 47, Springer, Berlin.

Van Rossum, J. M., and Ariëns, E. J., 1962, Receptor reserve and threshold phenomena, II. Theories on drug-action and a quantitative approach to spare receptors and threshold values, *Arch. Int. Pharmacodyn. Ther.* **136**:385.

Waser, P. G., 1975, *Cholinergic Mechanisms,* Raven Press, New York.

Waud, D. R., 1968, Pharmacological receptors, *Pharmacol. Rev.* **20**:49.

Waud, D. R., 1974, Adsorption isotherm vs. ion-exchange models for the drug–receptor reaction, *J. Pharmacol. Exp. Ther.* **188**:520.

Waud, D. R., 1976, Analysis of dose–response relationships, in: *Advances in General and Cellular Pharmacology,* Vol. 1 (T. Narahashi and C. P. Bianchi, eds.), 145–178, Plenum Press, New York.

Wick, H., 1972, unpublished data, Research Institute C. H. Boehringer Sohn, Ingelheim am Rhein, Germany.

Wyman, J., 1967, Allosteric linkage, *J. Am. Chem. Soc.* **89**:2202.

3

The Link between Drug Binding and Response: Theories and Observations

D. Colquhoun

A long time has elapsed since Paul Ehrlich enunciated his famous principle, *"Corpora non agunt nisi fixata."* Since then quite a lot has been learned about the binding of drugs to tissues and about the subsequent response. But knowledge of the mechanisms connecting these two events is still very tenuous and speculative. It is the purpose of this chapter to discuss the current state of knowledge of these phenomena in the case of the acetylcholine (ACh) receptor. Some recent reviews on related topics are Changeux (1975), Colquhoun (1975a), Ginsborg and Jenkinson (1976), Neher and Stevens (1977), Rang (1975), and Steinbach and Stevens (1976). First, the response will be considered.

1. The Response to Acetylcholine-Like Drugs

1.1. Methods of Investigation of the Response

The response to activation of ACh receptors at the skeletal neuromuscular junction can be measured in a number of ways; obviously it will be preferable, when the aim is to investigate mechanisms, to measure the primary response. Forty years ago the only response that could be measured was the tension developed in a whole muscle. This is a long way from the primary response. The introduction of intracellular microelectrodes (Ling and Gerard, 1949) permitted the measurement of the potential difference across muscle cell membranes, and this method was used in the classical work of Fatt and Katz (1951) to investigate the response of the end-plate region of skeletal muscle to ACh. They postulated that "small quantities of acetylcholine alter the end-plate surface in such a way that other ions can be rapidly transferred across it." This is very close to the present idea of the primary response. But it was not possible, in 1951, to measure the end-plate conductance in a satisfactory way.

D. Colquhoun ● Department of Pharmacology, St. George's Hospital Medical School, London, England.

The next step was achieved by Takeuchi and Takeuchi (1959, 1960), who clamped the membrane potential at a fixed value by passing current (controlled by a feedback circuit) through a second microelectrode. The current flowing in response to ACh could be measured, and the membrane conductance change produced by the drug could hence be inferred. By this time it had come to be widely supposed that the increase in membrane conductance produced by ACh was the result of the opening of pore-like *ion channels* in the membrane, and this is the current view (see the reviews cited above). The beauty of the voltage-clamp method is that the conductance change that is measured at any moment should be directly proportional to the number of ion channels that are open at that moment, so the response measured is quite close to the primary effect of the drug. However, the proportionality constant still could not be found, so the actual number of open channels could not be estimated, and consequently the properties of single channels, such as their conductance, could not be estimated either.

Another big advance started with the work of Katz and Miledi (1970, 1972, 1973a), who measured not only the depolarization produced by ACh but also the *fluctuation* of the membrane potential about the average level of depolarization. They showed that these fluctuations (ACh noise) were the result of random, moment-to-moment variations in the number of ion channels that were open. When current fluctuations were measured under the voltage-clamp method (Anderson and Stevens, 1973) it became possible, for the first time, to estimate the actual number of open channels (from the size—i.e., variance—of the fluctuations) and hence the conductance of a single open channel. And from the frequency of the fluctuations an estimate could be made of the mean lifetime of the open channel (Katz and Miledi, 1972; Anderson and Stevens, 1973).

Information about the kinetic behavior of ion channels similar to that obtained from the frequency of fluctuations can be obtained by observing the rate at which the system approaches ("relaxes" toward) a new equilibrium after it has been perturbed. The form of perturbation that has been most successfully applied so far is the *voltage-jump* method (Adams, 1975a; Neher and Sakmann, 1975). This method takes advantage of the fact that the number of channels open at equilibrium, in the presence of a constant agonist concentration, increases somewhat when the membrane potential is made more negative (hyperpolarized). If a step potential change is applied to the membrane, the relaxation of the drug-induced current can be observed. It has not yet proved possible to impose temperature jumps on the end-plate system, but observations on relaxation following jumps in the agonist concentration have, in effect, been made by assuming that the concentration of ACh falls rapidly (effectively a step reduction to zero) during a spontaneous miniature end-plate current. This, of course, is only possible for ACh (Katz and Miledi, 1972; Anderson and Stevens, 1973) or for agonists that can be released as false transmitters (Colquhoun *et al.*, 1977). No method of external drug application is known that can produce a sufficiently fast concentration change, but recently it has proved possible to produce a step change in agonist concentration by the use of a light-activated agonist (Lester and Chang, 1977).

While considering the properties of drug-operated ion channels it will be helpful to bear in mind the properties of the electrically operated sodium and potassium ion channels of nerve and muscle membrane and the description of these properties given by Hodgkin and Huxley (1952). Some degree of analogy among the various types of membrane ion channel might reasonably be expected.

1.2. The Nature of the Response to Acetylcholine

Acetylcholine, in a number of tissues, appears to cause the opening of ion channels that are permeable to sodium and potassium but relatively impermeable to chloride; an increase in calcium permeability also frequently occurs [see reviews by, for example, Ginsborg (1973) and Rang (1975)]. This appears to be the primary effect of ACh in, for example, the skeletal muscle end plate, autonomic ganglion cells, smooth muscle cells, and the electric organ of the eel (*Electrophorus*). At present it appears most likely, at the neuromuscular junction at least, that both Na^+ and K^+ go through a single sort of channel. However, the properties of this channel are by no means understood. In some ways it behaves like a simple ohmic pore; the instantaneous current–voltage relationship appears to be linear, both at the muscle end plate (Magleby and Stevens, 1972b) and in *Electrophorus* electroplaques (Lester *et al.*, 1975), just as it is for the sodium and potassium channels of nerve (Hodgkin and Huxley, 1952). However, the effect of changing the concentration of Na^+ and K^+ on the reversal potential for ACh is curious (Takeuchi and Takeuchi, 1960). These authors' observations did not follow the Goldman equation but fitted a simple equivalent circuit model, though only if it were supposed that the relative conductance increase to Na^+ and K^+ was unaffected by changes in the concentrations of these ions in the bathing fluid. The entire effect of changing the ion concentrations on the reversal potential had to be ascribed to changes in the Na and K equilibrium potentials [see discussions by Ginsborg (1973) and Rang (1975)]. Recent work (Lewis, 1978; Linder and Quastel, 1978) has not yet finally settled this problem.

1.3. The Response–Concentration Curve at Equilibrium

The equilibrium dose–response curve has been considered for a long time to be an essential part of any investigation of mechanism (Hill, 1909; Clark, 1933). It is a chastening fact that, even for the acetylcholine receptor, it has still not proved possible to observe directly the whole response–concentration curves, though, very recently, some good attempts have been made (see below).

The main reason for this is the phenomenon of desensitization. The response to a constant agonist concentration does not remain constant but falls away after its peak to reach eventually a steady level which may be much smaller than the peak. This happens at the muscle end plate (Fatt, 1950; Katz and Thesleff, 1957) and in electroplaques (Larmie and Webb, 1973; Lester *et al.*, 1975). Desensitization rate increases with agonist concentration, so the problem is least severe at the bottom end of the concentration–response curve. The effect can also be minimized by rapid application of the agonist, e.g., by ionophoresis. But until very recently it has been impossible to measure the concentration of drug produced by ionophoretic application, so this method has not been useful.

The second major problem in determining accurate response–concentration curves arises from the fact that the response measured must be the current (and hence conductance) in a voltage-clamped preparation if it is to be useful in determining mechanisms (see above). At the neuromuscular junction it is found that a microelectrode that is small enough to impale the cell without causing excessive damage is too small to allow passage of sufficient current for the upper end of the concentration–response curve to be measured easily.

For the reasons given above, only the lower part of response–concentration curves can, at the moment, be observed satisfactorily. The main feature of interest of these curves is that they do not follow a simple Langmuir hyperbola but are, in general, sigmoid, or positively cooperative, in shape. The Langmuir equation (Hill, 1909; Langmuir, 1918) is

$$p_A = \frac{x_A}{x_A + K_A}$$

$$= \frac{c_A}{c_A + 1} \tag{1}$$

where p_A is the fraction of binding sites that are occupied, x_A is the drug concentration, K_A is the dissociation equilibrium constant, and

$$c_A = \frac{x_A}{K_A}$$

is defined as a dimensionless measure of concentration. This equation describes binding to a set of identical and independent binding sites. The sigmoid nature of the response–concentration curve means that increasing the agonist concentration by a given factor causes, in the low concentration range, an increase in response by a *greater* factor (rather than by the *same* factor as in the case of the Langmuir equation), a phenomenon referred to as positive cooperativity. One way in which this could happen would be for the binding of one agonist molecule to facilitate the binding of subsequent molecules. But this is not the only possible mechanism—the facilitation of further response when some response already exists could occur at a later stage than drug *binding*. Whatever the mechanism of the observed cooperativity, its extent may be measured, quite empirically, by means of the Hill coefficient, n_H. This is defined as the slope of Hill plot, that is,

$$n_H = \frac{d \log\{[y(x) - y(0)]/[y(\infty) - y(x)]\}}{d \log x} \tag{2}$$

where $y(x)$ is the ordinate (e.g., response) corresponding with concentration x. At the bottom end of the curve, when $y(x) \ll y(\infty)$, this approaches the slope of the double logarithmic plot; i.e.,

$$n_H \simeq \frac{d \log[y(x) - y(0)]}{d \log x} \tag{3}$$

The Hill plot will be straight (with slope n) if the results follow an equation of the form

$$y = \frac{y_{\max} x^n}{x^n + K^n} \tag{4}$$

where $y_{\max}$ and K are constants. However, this equation does not describe any known physical mechanism for binding or for generation of a response. All physically possible mechanisms for these processes give Hill plots that are, to a greater or lesser extent, not straight, so n_H is not a constant and is not, in general, an integer [see, for example, Colquhoun (1973)].

Table I summarizes some values that have been observed for the Hill coefficient of the response to low concentrations of acetylcholine and other cholinomimetic drugs. There is, at the moment, no firm evidence for or against cooperativity in the response of smooth muscle. Despite the fact that the Hill coefficient would not be expected to be constant, it has not, in practice, usually been possible to detect consistent deviation from linearity of the Hill plot. When nonlinearity is suspected it is usual to report the maximum value of the Hill coefficient in the observed range, though it is quite possible that this is not the true maximum value [see, for example, Colquhoun (1973)].

It is clear from Table I that relatively powerful agonists such as acetylcholine and carbachol give a Hill slope of around 2 in a variety of tissues.

The studies cited in Table I did not, on the whole, go beyond the measurement of the Hill coefficient in the low agonist concentration range. Adams (1975c) applied agonist in the tissue bath and found similar Hill coefficients. Very recent work by Dreyer *et al.* (1978) [see Dreyer and Peper (1975)] and Dionne *et al.* (1978) has attempted to overcome the problems of desensitization and microelectrode limitation [see Dionne (1976) and Secs. 3.3 and 3.4] by inferring the equilibrium dose–response relationship from quantitative ionophoretic experiments. The relative drug concentrations at different times and at different points on the end plate were inferred from diffusion calculations in both studies, the total response at a given time being inferred by integration of the putative function relating concentration to response over the whole end plate. The absolute drug concentrations were inferred by measurement of transport numbers in the former paper and by means of an ion-sensitive microelectrode in the latter. Hill coefficients similar to those in Table I were found. The inferences about mechanism that were made from these more ambitious studies are discussed later, and some binding constants inferred from them are given in Table IVb.

The only drugs known, by direct conductance measurement, to be partial agonists at the frog end plate are diethyldecamethonium (see Table I), for which the maximum current is only 35 nA (Rang, 1973), and choline, for which it is about 120 nA (Adams, 1975c) (see also Sec. 3.4). However, there is no real doubt that decamethonium is also a partial agonist, as shown by its ability to antagonize the effects of more powerful agonists (Castillo and Katz, 1957b; Katz and Miledi, 1972, 1973a; Rang, 1973; Adams, 1976a), a property that is expected (Stephenson, 1956; Colquhoun, 1973) for a drug that is relatively ineffective in causing channel opening. The observations in Table I suggest that partial agonists produce lower Hill slopes, as might be expected [see Colquhoun (1973) and Sec. 3.4].

Table I
Some Observed Values for the Hill Coefficient

Drug[a]	Preparation	Method	Temp. (°C)	n_H	Ref.[b]
ACh, CCh, C10, PhTMA	*Torpedo*	Depolarization	20	~1.0	1
CCh	*Electrophorus*	Depolarization	—	1.8–2.0	2
C10	*Electrophorus*	Depolarization	—	1.6	2
PhTMA	*Electrophorus*	Depolarization	—	1.7	2
CCh	*Electrophorus*	Conductance (voltage clamp)	20–22	1.9	3
C10	*Electrophorus*	Conductance (voltage clamp)	20–22	1.45	3
ACh	*Electrophorus*	Na$^+$ efflux from microsacs	22	1.6	4
CCh	*Electrophorus*	Na$^+$ efflux from microsacs	22	1.5–1.7	4
C10	*Electrophorus*	Na$^+$ efflux from microsacs	22	1.7	4
PhTMA	*Electrophorus*	Na$^+$ efflux from microsacs	22	1.7	4
AThCh	*Electrophorus*	Na$^+$ efflux from microsacs	22	1.0	4
ACh	Cultured chick muscle	Rate of Na$^+$ uptake	2	2.0	5
CCh	Cultured chick muscle	Rate of Na$^+$ uptake	2	1.4	5
CCh	Cultured chick muscle	Rate of Na$^+$ uptake	36	1.9	5
ACh	Rat myotubes	Two-barrel ionophoresis	19–21	1.7	6
ACh	Snake end plate	Conductance (voltage clamp)	19–24	1.8	7
CCh	Frog end plate	Conductance (Martin correction)	20	1.3–2.1	8
CCh	Frog end plate	Conductance (voltage clamp)	20	ca. 2.0	9
EtC10	Frog end plate	Conductance (voltage clamp)	20	1.5	9
ACh	Frog end plate	Conductance (voltage clamp)	8, 23	2.7	10
CCh	Frog end plate	Conductance (voltage clamp)	8, 23	2.2	10
SCh	Frog end plate	Conductance (voltage clamp)	8, 23	2.0	10
C10	Frog end plate	Conductance (voltage clamp)	8, 23	1.5	10
Nic	Frog end plate	Conductance (voltage clamp)	8, 23	1.7	10
ACh	Denervated frog muscle	Conductance (voltage clamp)	8, 23	2.0	10
ACh	Mouse end plate	Conductance (voltage clamp)	38	2.6	11
ACh	Mouse end plate	Conductance (voltage clamp)	ca. 20	2.3	11
ACh	Denervated mouse muscle	Conductance (voltage clamp)	ca. 20	1.9	11

[a] *Drugs*: ACh, acetylcholine; CCh, carbamylcholine; C10, decamethonium; PhTMA, phenyltrimethylammonium; AThCh, acetylthiocholine; EtC10, decamethylene bis(ethyldimethylammonium); SCh, succinylcholine; Nic, nicotine.

[b] *References*: (1) Moreau and Changeux (1976), (2) Changeux and Podleski (1968), (3) Lester *et al.* (1975), (4) Kasai and Changeux (1971a), (5) Catterall (1975), (6) Land *et al.* (1977), (7) Hartzell *et al.* (1975), (8) Jenkinson and Terrar (1973), (9) Rang (1971), (10) Peper *et al.* (1975), (11) Dreyer *et al.* (1976a).

It has been found that the Hill coefficient measured by quantitative ionophoresis depends on calcium concentration, pH, and the presence or absence of an anticholinesterase agent (Sterz *et al.*, 1976). It would be expected that anything that altered the affinities of agonist for open or shut states or perturbed the receptor protein might have this effect.

1.4. The Kinetics of the Response

The rate at which drugs interact with their binding sites is investigated essentially by observing the rate at which the system approaches ("relaxes toward") equilibrium after it has been perturbed. The perturbation may be (1) an *imposed* change in any variable that affects the position of equilibrium (i.e., that alters one or more rate constants), for example, drug concentration, membrane potential, or temperature, or (2) a *spontaneous* fluctuation away from the equilibrium state, as observed in *noise* analysis. Drug-induced current fluctuations are usually analyzed by calculating from the observed fluctuations either (1) an autocovariance function in which the current at zero time is correlated with that t sec later and the correlation plotted against t (if a channel stays open for τ sec on average, then the correlation will die out for t that are much greater than τ) or (2) a spectral density in which the *quantity* (variance) of fluctuations at frequency f is plotted against f [for details, see, for example, Colquhoun and Hawkes (1977)]. Both methods give the same information.

When a perturbation is imposed, the results are most simply analyzed if the perturbation is in the form of a step—an essentially instantaneous change of concentration, etc.

1.4.1. Relationship between Methods of Studying Kinetics

The results of kinetic studies depend on the rate constants for the reactions that link the various states in which, according to the theory being tested, the system can exist. Once a theory has been postulated and values for the rate constants assigned, the predicted outcome of an experiment can be calculated relatively simply [see, for example, Colquhoun and Hawkes (1977)] as long as the rate constants and agonist concentration do not vary with time. In general, under these conditions, any theory based on the law of mass action that postulates that the system can exist in k kinetically distinct states (see Sec. 3.5) will show a time course of relaxation of the membrane conductance that is described by the weighted sum of $k - 1$ exponential components, $\sum w_i \exp(-t/\tau_i)$, with weights w_i and time constants τ_i that are related to the rate constants in the theory, often in quite a complicated way. And the autocovariance function of conductance fluctuations will also be the sum of $k - 1$ exponential components, $\sum w_i' \exp(-t/\tau_i)$, or, equivalently, the spectral density of the current fluctuations will be the sum of $k - 1$ Lorentzian components, $4 \sum w_i' \tau_i / [1 + (2\pi f \tau_i)^2]$, where f is frequency, so the spectral density falls to half its value at zero frequency, for each component, at $f = 1/2\pi\tau$. The observed time constants τ_i depend on the rate constants of the theory in exactly the same way whether relaxation (after an applied perturbation) or spontaneous fluctuations are observed. Thus the time constants observed in various sorts of experiments should be exactly the same if the conditions of the experiments are such that the rate constants are exactly the same (i.e., if the drug concentration, membrane potential, temperature, etc., are the same, because the values of at least some of the rate constants will depend on these variables). For example, a fluctuation experiment in a tissue equilibrated with 10 μM of agonist at a membrane potential of -80 mV and $T = 20°C$ should give exactly the same set of $k - 1$ time constants as (1) a voltage-jump experiment in the presence of 10 μM of agonist at $T = 20°C$ in which the membrane potential is stepped from -60 mV (or any other value) to -80 mV at $t = 0$ [insofar as rate constants change "instantaneously" with voltage; see Magleby and Stevens (1972a)] and as (2) a concentration-jump experiment at -80 mV and 20°C in which the agonist concentration is stepped from zero (or any other value) to 10 μM at $t = 0$. However, the coefficients (i.e., relative weightings w_i and w_i') of the various components are *not* the same in relaxation and fluctuation experiments [e.g., Colquhoun and Hawkes (1977)], so the two approaches give, in principle, complementary rather than identical information. It follows from the preceding remarks that if a single exponential component is seen in a relaxation experiment, then a single component with the same time constant should be seen in fluctuation experiments as long as (1) the conditions (concentration, membrane potential, etc.) are the same and (2) there is genuinely only one component present (it is quite possible in principle for a second component to be too small to see in a relaxation experiment but predominant in a fluctuation experiment).

1.4.2. Concentration-Jump Studies

In early studies the perturbation took the form of a sudden change in drug concentration (addition or removal of drug). A major problem in these studies was that the slowness of diffusion meant that the drug concentration *at the site of action* would not change in a step fashion, and there was usually uncertainty about the extent of the errors resulting from this fact. There can now be no real doubt that changes in concentration of drug in the fluid bathing a tissue are far too slow to enable the kinetics of agonists to be studied. All the work cited here has shown that acetylcholine interacts very rapidly with the receptor, as it obviously must do to enable synapses to operate at the observed rates.

1.4.2.1. Studies with Smooth Muscle. Agonists act rapidly, but the factors controlling their rate are not yet fully understood (Bolton, 1976). Potent antagonists have a high affinity for receptors, especially in smooth muscle, and they equilibrate with isolated tissues much more slowly than agonists. There is therefore reason to hope that their rate of reaction with receptors might be measurable by following reequilibration after step changes in the antagonist concentration in the fluid bathing the tissue. There is evidence that this can indeed be done; the evidence has recently been reviewed by Rang (1975, pp. 330–334) and will not be repeated here. But even in the case of antagonists, many published results are probably distorted by diffusion (Bolton, 1977).

1.4.2.2. Diffusion as a Rate-Limiting Factor. Certainly diffusion does affect measured rates in many tissues, especially when the substance that is diffusing is also bound and the amount that is bound is substantial compared with the amount that is free to diffuse. In fact the problem of telling whether diffusion or interaction with receptors is rate limiting is aggravated by the fact that frequently it is the binding that causes diffusion to be slow. This problem will not be discussed further here as it has been considered in a number of recent articles (Rang, 1966, 1975; Waud, 1967; Thron and Waud, 1968; Colquhoun and Ritchie, 1972; Colquhoun *et al.*, 1972; Katz and Miledi, 1973b; Colquhoun and Orentlicher,

1974; Colquhoun, 1975a,b; Magleby and Terrar, 1975; Gage, 1976; Roberts and Stephenson, 1976a,b; Colquhoun *et al.*, 1977; Kordaš, 1977; Bolton, 1977).

1.4.2.3. Studies on the Voluntary Muscle End Plate. The antagonists of acetylcholine at the nicotinic receptor of the voluntary muscle end plate, such as tubocurarine, have affinities (see Table IV) that are several orders of magnitude lower than those of the most potent antagonists at muscarinic receptors. It is, therefore, not surprising that they seem to act much more quickly. There is, *a fortiori*, no chance that the rate of *agonist* interaction with the receptor can be studied by changing the drug concentration in the tissue bath. More rapid application of drug can be achieved by ionophoretic ejection from a micropipette with its tip close to the end plate. Castello and Katz (1957a) found that after ionophoretic application, the depolarizing effects of acetylcholine and carbachol had similar half-decay times, about 40 msec, whereas the inhibitory effect of tubocurarine had a much longer half-decay time, 1400 msec. As in smooth muscle, agonists act faster than antagonists. (In the same paper it was shown that tubocurarine affects neither the resistance nor the capacitance of the end plate and that it works only when applied to the outside, not the inside, of the cell.) These observed rates place upper limits on the dissociation rates of these drugs from their binding sites, but Waud (1967) showed that they are probably limited by the rate of access of the drugs to their binding sites; the binding of drugs has a considerable effect on their rate of diffusion in the synaptic cleft, as it does in the extracellular space in other isolated tissues (see the references in Sec. 1.4.2.2). The larger the affinity, the slower binding will be, so it would be expected that antagonists would diffuse more slowly than agonists because of their higher affinities [see Sec. 1.4.2.2 and the review by Colquhoun (1975a)], and it is also to be expected that choline, which is a very weak antagonist, would have an effect that decays quite rapidly—in fact it is almost as fast, by ionophoretic application, as acetylcholine (Castillo and Katz, 1957b). Another problem that arises when drugs are applied ionophoretically is illustrated by the work of Castillo and Katz (1957b), who showed that decamethonium alone produces depolarization much more slowly than carbachol but can nevertheless cause a rapid inhibition of carbachol depolarization when both drugs are given together. Adams (1976a) has shown how this result may arise from the fact that decamethonium is less potent than carbachol and will therefore be acting over a much wider area of the membrane than carbachol when the ionophoretic pulse is such as to give similar depolarizations; once again it appears doubtful whether such experiments give any useful information about the rates of the actual drug–receptor interaction.

However, the experiments above do give an upper limit for the actual rates, and it is also possible to place a rough lower limit on the rate of tubocurarine action, based on several (rather indirect) lines of evidence that tubocurarine cannot dissociate to any great extent during the action of acetylcholine in a normal end-plate potential (Paton and Waud, 1967; Kruckenberg and Bauer, 1971; Ceccarelli *et al.*, 1973; Katz and Miledi, 1973b; Colquhoun *et al.*, 1977). This suggests that the tubocurarine–receptor complex must last at least a millisecond or so [especially as the rate of equilibration with ACh should be faster than the tubocurarine dissociation rate insofar as the acetylcholine does not saturate receptors during the end-plate potential (Rang, 1966)]. Similar conclusions were reached by Ferry and Marshall (1973), who observed that hexamethonium, which on its own is a curare-like blocker, could relieve the block of transmission produced by tubocurarine in rat diaphragm. This phenomenon resembles that seen in smooth muscle (Stephenson and Ginsborg, 1969; Ginsborg and Stephenson, 1974; see above), and Ferry and Marshall follow these authors in explaining it on the basis that hexamethonium is fast enough to equilibrate with acetylcholine during a nerve-evoked end-plate potential but tubocurarine is too slow to do so. On the basis of inevitably very indirect arguments they suggested that the true time constant for dissociation of hexamethonium for receptors might be around 2 msec compared with about 40 msec for tubocurarine. In an extension of this study, Blackman *et al.* (1975) found evidence that tubocurarine dissociated even faster, with a time constant of the order of 1 msec. But caution is required because, as in smooth muscle, this phenomenon is explicable, at least qualitatively, on the basis of diffusion and binding even if the rates of receptor interaction were infinitely fast [see, for example, Colquhoun (1975a)].

One case in which rates have been successfully measured by concentration-jump measurements up to now seems to be the study by Lester and Chang (1977). The *trans* isomer of bis-Q[3,3'-bis-α-(trimethylammonium)methyl azobenzene] is a much more potent agonist than the *cis* isomer, and the isomers interconvert rapidly when exposed to light of appropriate wavelengths (Bartels *et al.*, 1971). Lester and Chang equilibrated the innervated face of a voltage-clamped eel (*Electrophorus*) electroplaque with a solution containing mainly the *cis* isomer in the dark and exposed the preparation to a flash of light to cause a rapid increase in concentration of the *trans* isomer. The relaxation of the current flowing through the ion channels opened by the *trans* isomer was predominantly a single exponential, and the time constant of this relaxation (for a given agonist concentration, temperature, and membrane potential) was close to that found from voltage-jump relaxation measurements (see below).

Apart from the case just described, the only known method of applying an agonist sufficiently rapidly to enable concentration-jump observations to be made is application from the nerve ending itself. It appears that rate at which ACh disappears from the synaptic cleft following the release of a single packet is quite rapid compared with rate of relaxation back toward zero of the conductance increase produced by the ACh (i.e., rapid compared with the decay phase of the miniature end-plate current) (Magleby and Stevens, 1972a,b; Anderson and Stevens, 1973). Thus, insofar as the ACh concentration is negligible during the decay phase, the miniature end-plate current can be treated as a concentration-jump experiment in which the agonist concentration is reduced rapidly to zero. Of course this sort of experiment can be done only with ACh or with false transmitters that disappear from the cleft with similar rapidity, such as acetylmonoethylcholine (Colquhoun *et al.*, 1977). Results based on this idea will be described in more detail in later sections.

1.4.3. Fluctuation Analysis

Most of our knowledge of the kinetics of the response to stimulation of ACh receptors at the neuromuscular junction has come since 1970 when Katz and

Miledi published their first study of ACh-induced voltage fluctuations at the frog end plate (Katz and Miledi, 1970, 1972, 1973a). It was shown that fluctuations around the mean response of the frog end plate to cholinomimetic agonists resulted from moment-to-moment fluctuations in the number of open ion channels. They interpreted the frequency of fluctuations, recorded extracellularly, in terms of the mean lifetime of an open channel (see Sec. 3.5.1 for a discussion of the interpretation of fluctuation results) and showed that the mean lifetime of the open channel varied over a considerable range for different agonists (Katz and Miledi, 1973a). Intracellular voltage recording showed that the elementary event, supposed to result from the opening of a single ion channel, was about 0.3 μV (in frog) for ACh and also varied considerably from one agonist to another; however, most of this variability was a result of the differences in channel lifetime between the agonists, all the lifetimes being substantially less than the time constant for passive charging of the membrane. When this factor was taken into account (Colquhoun, 1975a, Eq. 3) there was no evidence for any difference in the conductance of the channels opened by the various agonists (Rang, 1975, Fig. 10).

Much subsequent work has been done with measurement of current fluctuations in voltage-clamped end plates. Almost all agonists have been found to produce current fluctuations which have an autocovariance function that can be fitted, within experimental error, by a single exponential component (or, equivalently, a spectral density that can be fitted by a single Lorentzian component). Single-component curves allow only two parameters to be estimated; with certain assumptions (see Sec. 3.5) observations of this sort can give estimates of the conductance of a single open channel (γ) and the mean lifetime of the open state (τ). Some estimates of these quantities are shown in Table II. The results are in general agreement with the suggestions from the first experiments published by Katz and Miledi. The single-channel conductance is almost constant for most of the agonists shown in Table II, almost all values for normal muscle are in the range 18–28 pS, but the mean open-channel lifetime varies over a 10- to 17-fold range for different agonists. Table III shows, for comparison, some estimates of the conductance of electrically operated ion channels in nerve. They are of the same order as for ACh-operated channels. The fact that they seem to be a bit smaller is, perhaps, not surprising, as the axon channels are more selective.

The time constant τ from fluctuation analysis is close to, or identical with, the time constant for the exponential decay of miniature end-plate currents (MEPCs), both for normal (Anderson and Stevens, 1973; Katz and Miledi, 1973b; Dreyer *et al.*, 1976a) and false (Colquhoun *et al.*, 1977) transmitters. This is as predicted by the theory of Magleby and Stevens (1972b), according to which the transmitter concentration in the synaptic cleft has fallen virtually to zero during the decay phase (see also Sec. 1.4.2).

The extrajunctional receptors in denervated muscle are seen to have a longer channel lifetime than junctional receptors by a factor of 3 or 4, though the dependence of lifetime on the nature of the agonist and on membrane potential is similar to that of junctional receptors. The single-channel conductance also appears to be somewhat reduced in extrajunctional receptors, though this is not quite so certain because of the problems that arise because of the higher agonist

Table II
Some Values of γ and τ

Drug[a]	Preparation	Method	Temp. (°C)	E_h (mV)[b]	γ (pS)	τ (msec)	Ref.[c]
ACh	Frog end plate	Focal ectracellular voltage	ca. 20	—	—	1.0	1, 2
CCh	Frog end plate	Focal extracellular voltage	ca. 20	—	—	0.3–0.4	1, 2
C10	Frog end plate	Focal extracellular voltage	ca. 20	—	—	0.1	1, 2
AThCh	Frog end plate	Focal extracellular voltage	ca. 20	—	—	0.12	1, 2
SubCh	Frog end plate	Focal extracellular voltage	ca. 20	—	—	1.65	1, 2
ACh	Frog end plate	Noise (voltage clamp)	18	−80	32	3	3
ACh	Frog end plate	Noise (voltage clamp)	10–15	−60 to −80	25.0	3.2	4
SubCh	Frog end plate	Noise (voltage clamp)	10–15	−60 to −80	28.6	5.6	4
HPTMA	Frog end plate	Noise (voltage clamp)	10–15	−60 to −80	18.8	1.0	4
PPTMA	Frog end plate	Noise (voltage clamp)	10–15	−60 to −80	12.8	0.83	4
ACh	Frog end plate	Noise (voltage clamp)	20–25	−65 to −80	18.6	0.83	5
CCh	Frog end plate	Noise (voltage clamp)	20–25	−65 to −80	18.4	0.33	5
SCh	Frog end plate	Noise (voltage clamp)	20–25	−65 to −80	20.0	0.23	5
C10	Frog end plate	Noise (voltage clamp)	20–25	−65 to −80	15.1	0.16	5
Nic	Frog end plate	Noise (voltage clamp)	20–25	−65 to −80	17.6	0.22	5
ACh	Frog denervated muscle	Noise (voltage clamp)	20–25	−65 to −80	7.5	3.0	5
ACh	Frog end plate	Noise (voltage clamp)	20	−80	23	0.95	6
ACh	Frog end plate	Noise (voltage clamp)	8	−80	—	3.1	6
CCh	Frog end plate	Noise (voltage clamp)	8	−80	19	1.0	6
SubCh	Frog end plate	Noise (voltage clamp)	8	−80	20	5.3	6
ACh	Frog denervated muscle	Noise (voltage clamp)	8	−80	15	11	6
CCh	Frog denervated muscle	Noise (voltage clamp)	8	−80	14	3.9	6
SubCh	Frog denervated muscle	Noise (voltage clamp)	8	−80	14	19	6
ACh	Frog end plate	Voltage-jump relaxation	8	−80	—	3.6	6
SubCh	Frog end plate	Voltage-jump relaxation	8	−80	—	6.6	6
CCh	Frog denervated muscle	Voltage-jump relaxation	8	−80	—	3.5	6
SubCh	Frog denervated muscle	Voltage-jump relaxation	8	−80	—	21	6
ACh	Frog denervated muscle	Single-channel recording	8	−120	—	26	7
CCh	Frog denervated muscle	Single-channel recording	8	−120	—	11	7
SubCh	Frog denervated muscle	Single-channel recording	8	−120	22.4	45	7
SubCh	Frog denervated muscle	Single-channel recording	8	−80	22.4	28	7
ACh	Frog end plate	Voltage-jump relaxation	20	−100	—	1.4	8
CCh	Frog end plate	Voltage-jump relaxation	20	−100	—	0.32	8
ACh	Frog end plate	Voltage-jump relaxation	2–6	−100	—	6.2	8
SubCh	Frog end plate	Voltage-jump relaxation	6	−86	—	14	9
ACh	Mouse end plate	Noise (voltage clamp)	23	−65 to −80	22[d]	1.0	10
ACh	Mouse end plate	Noise (voltage clamp)	39	−65 to −80	46[d]	0.3	10
ACh	Mouse denervated muscle	Noise (voltage clamp)	24	−65 to −80	13.4	1.4	10
ACh	Rat end plate	Noise (voltage clamp)	20	−80	24.9	1.2	11
AMECh	Rat end plate	Noise (voltage clamp)	20	−80	26.7	0.67	11
ACh	Rat end plate	Noise (voltage clamp)	21–23	−70	34[e]	0.96	12
ACh	Rat denervated muscle	Noise (voltage clamp)	21–23	−70	25[e]	3.8	12
ACh	Embryonic chick myoballs	Noise (voltage clamp)	37	−70 to −99	25–45	2.5–4.5	13
ACh	Eel electroplaque	Voltage-jump relaxation	15	−175	—	10[f]	14
CCh	Eel electroplaque	Voltage-jump relaxation	15	−175	—	3.3[f]	14
C10	Eel electroplaque	Voltage-jump relaxation	15	−175	—	2[f]	14

[a] Drug abbreviations are as in Table I, and SubCh, suberylcholine; HPTMA, 3-(m-hydroxyphenyl)propyltrimethyl-ammonium; PPTMA, 3-phenylpropyltrimethylammonium; AMECh, acetylmonoethylcholine.
[b] E_h is the holding potential.
[c] References: (1) Katz and Miledi (1972), (2) Katz and Miledi (1973a), (3) Anderson and Stevens (1973), (4) Colquhoun *et al.* (1975), (5) Dreyer *et al.* (1976b), (6) Neher and Sakmann (1975, 1976b), (7) Neher and Sakmann (1976a), (8) Adams (1975a), (9) Adams (1977a), (10) Dreyer *et al.* (1976a), (11) Colquhoun *et al.* (1977), (12) Sakmann (1975), (13) Lass and Fischbach (1976), (14) Sheridan and Lester (1975).
[d] Sudden transition in γ (but not τ) at 25.5°C.
[e] Assumes equilibrium potential of 0 mV.
[f] Concentration dependent; value given is that for zero concentration (by extrapolation).

Table III
Single-Channel Conductances for Electrically Operated Channels

Preparation	Channel specificity	Method	γ (pS)	Ref.[a]
Squid axon	Sodium	Noise	4	1
		Gating	>2.5	2
	Potassium	Noise	12	1
		NTEA entry[b]	≥2–3	7
Frog node of Ranvier	Sodium	Noise	8	3
		Gating	>3	4
	Potassium	Noise	4	5
Frog muscle	Sodium	TTX binding	0.6–15	6

[a] References: (1) Conti *et al.* (1975), (2) Keynes and Rojas (1974), (3) Conti *et al.* (1976), (4) Nonner *et al.* (1975), (5) Begenisich and Stevens (1975), (6) Ritchie and Rogart (1977), (7) Armstrong (1975).
[b] Nonyltriethylammonium.

concentration that has to be used as a result of the low density of extrajunctional receptors. Earlier reports that the reversal potentials of agonists were different for junctional and extrajunctional receptors have now been shown to be incorrect (Mallart *et al.*, 1976).

It is clear from the results in Table II that there is remarkably little difference between the results obtained in frog, rat, and mouse. This is true also of the temperature and potential dependence of the mean open-channel lifetime τ. In frog and toad this has been found to decrease *e*-fold for every 100–180 mV of depolarization (Takeuchi and Takeuchi, 1959; Kordaš, 1969; Magleby and Stevens, 1972a; Anderson and Stevens, 1973; Gage and McBurney, 1975) and to decrease with temperature with a Q_{10} of 2.5–3.5 (Takeuchi and Takeuchi, 1959; Kordaš, 1972; Magleby and Stevens, 1972b; Gage and McBurney, 1975). The voltage dependence is greater at low temperatures. In rat diaphragm end plates similar values are found, an *e*-fold decrease in τ per 109–117 mV of depolarization and a Q_{10} of 3.3 (Colquhoun *et al.*, 1977). In the frog, the single-channel conductance γ is almost independent of membrane potential and of temperature in the range 8–18°C (Anderson and Stevens, 1973; Lewis, 1978). Up to about 24°C this is true also in the mouse (Dreyer *et al.*, 1976a). But at 25.5°C the value of γ in the mouse omohyoid muscle end plate undergoes a sudden increase and thereafter remains relatively constant up to 39°C (Dreyer *et al.*, 1976a). A similar but less sharp transition has been observed in embryonic chick myoballs in culture (Lass and Fischbach, 1976).

The mean open-channel lifetime, whether estimated from the decay rate of MEPCs, fluctuations, or voltage-jump relaxation, has a voltage dependence that is very similar to that of the *equilibrium* conductance change produced by the agonist, which also increases with hyperpolarization, resulting in an upward curvature of the equilibrium current–voltage curve (Rang, 1973; Dionne and Stevens, 1975; Neher and Sakmann, 1975; Adams, 1976b). The interpretation of this observation is discussed later.

In a beautiful piece of work, Neher and Sakmann (1976a) have recently succeeded in measuring the current that flows through a single acetylcholine-operated channel (extrajunctional channel in denervated muscle). They demonstrated that the opening of a channel indeed produces a rectangular pulse of current, as had been assumed in the analysis of earlier fluctuation experiments. The pulse resembles that observed in artificial membranes with antibiotics (Hladky and Haydon, 1970). Furthermore, the mean length of the pulse and its amplitude (i.e., the single-channel conductance) agreed with the values inferred from noise measurements on the same preparation.

1.4.4. Voltage-Jump Relaxation Studies

The current that flows following imposition of a sudden jump in membrane potential at the skeletal muscle end plate has been investigated by Adams (1975a, 1977a,b) and Neher and Sakmann (1975). As might be expected from the single component seen in fluctuation experiments, the relaxation has been found to follow exponential curves for the agonists that have been tested, and, as expected (see Sec. 1.4.1), this exponential has a time constant which is close to that inferred from fluctuation experiments (see Table II). Recently Sakmann and Adams (1976) have reported a decrease in the relaxation time constant as the agonist concentration is increased; this decrease took place in the concentration range 2–100 μM for ACh (cholinesterase inhibitor present), and in this range the ratio of the variance of current fluctuations to the mean current (i.e., the apparent single-channel conductance) also fell. Similar decreases in relaxation time constant were seen for carbachol (100–500 μM) and suberylcholine (1–6 μM). Similar observations have been made with carbachol by D. Colquhoun, F. Dreyer, and R. E. Sheridan (in preparation). The interpretation of these results is discussed in Sec. 3.

Work on the voltage-clamped electroplaques of *Electrophorus* has also shown single exponential relaxation of the agonist-induced membrane current following a step change in membrane potential (Sheridan and Lester, 1975). In this work a decrease of the time constant τ for the exponential relaxation with increasing agonist concentration was clearly measured. The relationship between rate constant $(1/\tau)$ and agonist concentration is linear within the measurable range; no flattening could be detected at the highest agonist concentrations used. Unfortunately, currents were too small to allow the rate constant to be measured in the low concentration range where the equilibrium dose–response curve is sigmoid. When the relationship between relaxation rate constant and acetylcholine concentration is extrapolated to zero concentration (see Table II), the relaxation rate obtained is close, in both absolute value and voltage dependence, to the nerve-evoked postsynaptic current, a result that is exactly like that seen at the neuromuscular junction. Other agonists have zero-concentration relaxation time constants τ with values that are similar (and have similar voltage dependence) to those from fluctuation measurements at the neuromuscular junction (see Table II). The obvious interpretation is that the zero-concentration relaxation rate constant measures the mean open-channel lifetime, i.e., the channel closing rate. The slope of the relaxation rate constant–agonist concentration relationship, on the other hand, is not very voltage sensitive, in agreement with the apparent relatively slight voltage dependence of the channel opening rate at the neuromuscular junction (Dionne and Stevens, 1975; Neher and Sakmann, 1975).

The slope, however, is reduced in the presence of tubocurarine and during desensitization, as might be expected if the slope reflected the channel opening rate.

1.5. Anesthetics, Local Anesthetics, and Channel Blocking

A large number of substances are known that modify the opening of ion channels by agonists. At least some of these are thought to work by blocking ion channels, an idea that originates mainly from the work of Blackman (1959, 1970) and Blackman and Purves (1968).

General anesthetics and related uncharged compounds have been investigated using, mainly, the rate of decay of MEPCs as a measure of channel lifetime, i.e., as a relaxation experiment at zero ACh concentration (see Secs. 1.4.2 and 1.4.3). This work (Gage *et al.*, 1975; Gage, 1976) has shown that short-chain aliphatic alcohols, from methanol to pentanol, and also high (fatal) concentrations of ether lengthen the decay time constant for the MEPC, which is the reason for the increase in amplitude of miniature end-plate *potentials* (MEPPs) produced by ethanol. Ethanol also lengthened the time constant from fluctuation measurements. The relationship between effectiveness and oil–water partition coefficient suggested that the alcohols worked in a hydrophobic phase— most likely the membrane lipid surrounding the receptor. The results could be accounted for by supposing that the alcohols produce a change of the dielectric constant of the environment in which the channel closing takes place. If the channel closing is accompanied by a dipole moment change, then the rate of the closing reaction, and hence the mean open-channel lifetime, should be affected by the dielectric constant of the environment. The change in dipole moment of about 50 D proposed by Magleby and Stevens (1972b) to account for the voltage sensitivity of the channel closing rate (see Sec. 1.4.3) suggested an effective radius of the gating molecule of 1–1.5 nm (Gage *et al.*, 1975), certainly a plausible value.

On the other hand, octanol and low (anesthetic) concentrations of ether, halothane, chloroform, and enflurane all shortened the time constant for decay of MEPC, and ether and octanol have also been shown to reduce the time constant from fluctuation measurements (Katz and Miledi, 1973c; Gage *et al.*, 1974; Gage and Hamill, 1976; Gage, 1976). This shortening of the mean open-channel lifetime accounted for the reduction in MEPP amplitude produced by these agents. Again effectiveness correlated well with lipid solubility, suggesting an action in the membrane rather than a specific effect on receptors. It was suggested that the increased rate of channel closure might result from an increase in fluidity of the membrane. High concentrations of halothane and enflurane produced biphasic decay of the MEPC, as do hexanol, methyprylone, and the local anesthetics (see below).

Barbiturates also shorten the time constant for decay of MEPC (Seyama and Narahashi, 1975; Adams, 1976c) and hence reduce the amplitude of MEPP. They also, in lower concentrations, inhibit the response to ionophoretically applied pulses of carbachol, especially the second of a pair of pulses; the latter effect disappears exponentially as the interpulse interval is increased with a time constant of 150–200 msec (Adams, 1976c). The observations suggest that the blocking effect develops *after* receptor activation, as found by Steinbach (1968a,b) with local anesthetics (see below); they could not easily be explained by effects on the lipid environment of the channel of the sort described above. A more likely explanation (Adams, 1976c) is that barbiturates block open channels, though it is not certain whether they simply plug the channel or cause it to revert to a low-conductance form in some other way. Barbiturates appear to act in the *uncharged* form, unlike local anesthetics. In squid axon the uncharged form of pentobarbital also appears to be the cause of action potential inhibition, and it is thought to work on the inside of the axon membrane (Narahashi *et al.*, 1971; Frazier *et al.*, 1975). Methyprylone (a barbiturate-like hypnotic) produces a diphasic decay of nerve-evoked end-plate currents (Adams, 1976c), like local anesthetics, and showed no interpulse interaction (see above); it was proposed that it worked in the same way as barbiturates except that it dissociates much more rapidly from the open channel so a slow tail results from the transient appearance of unblocked channels as it dissociates, this component being too small and slow to be seen with barbiturates.

A large amount of work has been done on the effects of local anesthetics on end plate [see the references in Colquhoun (1975a, 1978) and Steinbach and Stevens (1976)] as well as on axons, and a number of hypotheses have been proposed. Recent work on end plates (Katz and Miledi, 1975; Beam, 1976; Ruff, 1977; Adams, 1977b) has employed fluctuation and voltage-jump relaxation analysis as well as looking at changes in end-plate currents. All three kinetic methods show at least two components in the presence of local anesthetics. This work has, on the whole, confirmed the view of Steinbach (1968a,b) that the local anesthetics combine with some part of the channel mechanism to produce a low-conductance state. The possibility that part, at least, of their effect results from changes in the lipid environment of the channel (Gage, 1976) is difficult to rule out, but at the moment this sort of mechanism seems not to be the major one (Ruff, 1977); other substances that produce biphasic MEPC decay, such as hexanol and high halothane concentrations, may work in the membrane, however.

Neher and Steinbach (1978) have made elegant recordings of the current through single channels, opened by suberylcholine, in the presence of the quaternary local anesthetics QX222 and QX314. They confirmed directly the inference made from earlier experiments that a single opening is split up, in the presence of QX222, into a series of short openings (of normal conductance). The fact that the total length of a burst of openings was longer than a normal opening suggests that the blocker is selective for the open state and that the channel cannot revert (or can revert only slowly) to the shut conformation until the blocker dissociates from the channel. This also suggests that the agonist cannot easily dissociate while the channel is in the open conformation; however many times the channel blocks and unblocks it still behaves in a manner characteristic of the agonist that is present until it finally shuts. The work of Nass *et al.* (1978) on eel electroplaque also suggests that the channel shuts very rapidly if bound agonist molecules are inactivated by photoisomerization. On the other hand, Adams (1977b) in a voltage-jump study of the tertiary local anesthetic procaine found that the procaine appeared not to be selective but bound to shut channels as well as to open channels. This might perhaps be expected on the basis of the hypothesis (Hille 1977a,b) that ionized and nonionized species can act at the same site in an ion channel, blocking it, but that ionized species can enter only when the channel is open. However, evidence against the idea that ionized and nonionized local anesthetics work at the same site in axons has been presented by Mrose and Ritchie (1978). This problem is discussed in more detail in Colquhoun (1978).

In *Aplysia* neurons it appears that not only procaine but also tubocurarine and hexamethonium work mainly by selective block of open ion channels, rather than by competition at receptors (Ascher *et al.*, 1978; Marty, 1978). Indeed Manalis (1977) suggested that tubocurarine might have an action of this sort at the neuromuscular junction too. Voltage-clamp experiments (Katz and Miledi, personal communication; Colquhoun *et al.*, 1978) do show effects that are compatible with channel block in frog muscle as well as the (normally much more important) competitive receptor block. For example, substantial tubocurarine concentrations at hyperpolarized membrane potentials can produce a shortening of the time constant normally measured in voltage-jump and noise experiments. This is consistent with curare associating with open channels with a rate constant not unlike that found for QX222 by Neher and Steinbach (1978). This rate is surprisingly rapid for largish molecules, which, on the basis of the voltage dependence of the apparent association rate, appear to bind right inside the channel. Following a hyperpolarizing voltage step in the presence of tubocurarine, the agonist-induced inward currents rise rapidly at first, despite the tubocurarine, but instead of staying constant, the current slowly shuts off again (time constant of the order of a second), and one interpretation of this is that, as in *Aplysia*, channels accumulate in the blocked form because of a rather slow dissociation of the blocker from the open channel (Colquhoun *et al.*, 1978). As in *Aplysia*, tubocurarine produces little change in single-channel conductance, and this is what would be expected for a simple channel blocker [see the discussion in Colquhoun (1978)]. The substantial reduction of single-channel conductance by quaternary local anesthetics seen by Ruff (1977) remains puzzling.

Although there is really no evidence that distinguishes between (1) combination of the local anesthetic, or other putative channel blocker, with a binding site that allosterically shuts the channel and (2) plugging of the channel by the local anesthetic, the latter view tends to be favored at present, largely (1) on the grounds of parsimony and (2) by analogy with tetrodotoxin block of sodium channels. The local anesthetic must enter the channel from outside since the onset of action is very rapid (Castillo and Katz, 1957a), and it is the charged form of the molecule that is active (Steinbach, 1968a); in fact many of the experiments have been done with quaternary analogues of conventional local anesthetics.

The axon, on the other hand, is quite insensitive to quaternary analogues when they are applied to the outside of the membrane, but they work well when applied *inside* the axon. Conventional nonquaternary local anesthetics can penetrate axon membranes in their uncharged forms, but once inside it is their charged form that produces block (Ritchie and Greengard, 1966) [though the uncharged form may contribute; see Hille (1977a,b) and Mrose and Ritchie (1978)]. There is evidence that quaternary anesthetics enter open sodium channels from the inside and block them (Strichartz, 1973). This is very similar to the proposed mechanism for the end plate except it takes place from the *inside* of the membrane. On the other hand, the charged form of tetrodotoxin blocks axon sodium channels from the outside [see Hille (1970)] and acts quite independently of local anesthetics in axons (Benzer and Raftery, 1972; Colquhoun *et al.*, 1972; Henderson *et al.*, 1973; Wagner and Ulbricht, 1976). Axon potassium channels can also be blocked, again, it is thought, by plugging. Tetraethylammonium and nonyltriethylammonium will do this (but not the corresponding methyl compounds), and they appear to enter open potassium channels from the *inside* of the membrane (Armstrong, 1971, 1975).

2. The Binding of Drugs to Acetylcholine Receptors

Although there are considerable gaps and uncertainties in our knowledge of the response to cholinomimetic agonists, knowledge of binding of these drugs is in an even more fragmentary state. This state of affairs results partly from the many technical problems associated with binding studies, and it is aggravated by the frequency with which results, not always very complete, are presented in rapid-publication journals. The reader of the literature is also likely to be confused by the frequency with which results of apparently similar experiments from different laboratories are contradictory.

2.1. Methods for Investigation of Binding

Ligands that are bound nearly irreversibly to the ACh receptor are the simplest for experimental purposes, insofar as the specificity of their binding to the receptor can be relied upon. The snake venoms such as α-bungarotoxin are the best known examples [see the review by Lee (1972)]. The labeled toxin can be incubated with the receptor preparation and unbound ligand subsequently removed by washing (for whole tissues), centrifugation (for homogenates), or gel filtration or ion exchange (for soluble receptor preparations).

Most ligands that are of direct interest (agonists and antagonists) are reversibly bound, on the time scale of experiments. Their binding can be measured in whole tissues or homogenates by bioassay of the loss of ligand from solution produced by incubation of the tissue in it, or by uptake of labeled ligand; in either case, a correction for the amount of unbound ligand in the extracellular space is applied (Clark, 1933; Paton and Rang, 1965; Brocklehurst and Colquhoun, 1965; Moore *et al.*, 1967; Keynes *et al.*, 1971; Colquhoun *et al.*, 1972; Almers and Levinson, 1975; Ritchie *et al.*, 1976). A second approach is to measure binding in homogenates or soluble receptor preparations directly by equilibrium dialysis. There are two major problems in this method: (1) some sort of check on the specificity of binding is needed, and (2) accurate measurements are difficult at higher ligand concentrations where the amount of free ligand in the dialysis chamber is large compared with the amount bound and the results are sensitive to small errors in, for example, the Donnan correction. Another approach that has been used widely for measuring the binding of reversible ligands is to measure the extent to which they slow down the rate of binding of an irreversible ligand such as α-bungarotoxin. One problem with this method is that it is not always possible to be sure that the reversible ligand and toxin compete in a simple mutually exclusive fashion for a single site. Another problem is that, although the protective ligand is supposed to be in equilibrium with the receptor, the toxin is not. The *rate* of toxin binding must be measured, so it is necessary to use a preparation in which diffusion of the toxin (which is likely to be very slow because of the high binding affinity; see Sec. 1.4.2.2) does not distort the binding rate. This problem led to wrong conclusions in early studies that used this method.

A problem with almost all methods of measuring binding is that they are slow. This means that usually only equilibrium measurements, not rate measurements, are possible with reversible ligands, which usually bind rather rapidly. Furthermore, if the ligand is an agonist, it has plenty of time to produce desensitization, so binding to the desensitized tissue rather than the native tissue is measured (see Sec. 3.4).

2.2. Binding at Equilibrium

Tables IVa and IVb summarize some of the more recent results for binding to various receptor preparations from *Electrophorus* and *Torpedo* and from skeletal muscle. Earlier results have been reviewed by Rang (1975) and Changeux (1975). There is seen to be a good deal of similarity between different species, especially in the binding of antagonist ligands.

The binding of acetylcholine itself is obviously of basic importance. In both *Electrophorus* and *Torpedo*, binding of ACh is often found to be heterogeneous. The affinity constants can be roughly divided into high- and low-affinity groups. In *Electrophorus* membrane preparations most often have predominantly low affinity, of the order of $K = 1$ μM, for ACh, but purification commonly causes a high-affinity form to become more predominant (Meunier and Changeux, 1973; Meunier *et al.*, 1974). This is true for other agonists, too, but antagonists have similar affinities for membrane fragments and purified receptor (see Table IVa). In *Torpedo*, on the other hand, membrane fragments show almost entirely a very high affinity ($K = 4$–30 nM) for ACh (see Table IVa), but purification of the receptor and storage of various chemical treatments can cause the appearance of a substantial, or large, proportion of binding sites with a low affinity for ACh (O'Brien and Gibson, 1975; Sugiyama and Changeux, 1975; Chang and Neumann, 1976), which may be a denatured form (O'Brien and Gibson, 1975; Gibson, 1976). The affinity of ACh for muscle receptors is similar to that of the low-affinity binding found in electric tissue.

The results in Table IVb show some estimates of equilibrium constants for binding of agonists that have been made indirectly from physiological studies. Only studies in which a serious attempt has been made to distinguish binding from other events (either by use of null methods or by analysis of the dose–response curve in terms of a realistic theory) have been included.

It is seen that the affinities found in all of these studies are a good deal lower than even the "low-affinity sites" seen in direct binding measurements (Table IVa). In interpreting apparent discrepancies of this sort it must always be borne

Table IVa
Measurements of the Equilibrium Dissociation Constant (K) for Binding

Drug[a]	K (μM)	Species	Receptor preparation	Method[b]	Ref.[c]
ACh	1.5	*Electrophorus*	Membrane	R (*Naja* α-toxin)	1
	0.06	*Electrophorus*	Purified	R (*Naja* α-toxin)	2
	0.046	*Electrophorus*	Triton extract	D ([³H]-ACh)	3
	0.5	*Electrophorus*	Triton extract	D ([³H]-ACh)	3
	0.045	*Electrophorus*	Purified	D ([³H]-ACh)	3
	0.44	*Electrophorus*	Purified	D ([³H]-ACh)	3
	0.008	*Torpedo*	Homogenate	D ([³H-ACh])	4
	0.068	*Torpedo*	Homogenate	D ([³H-ACh])	4
	0.020	*Torpedo*	Purified	D ([³H]-ACh)	5
	2.0	*Torpedo*	Purified	D ([³H]-ACh)	5
	0.012 (K_R)	*Torpedo*	Triton extract	D ([³H]-ACh)	6
	0.2 (K_T)	*Torpedo*	Triton extract	D ([³H]-ACh)	6
	0.014	*Torpedo*	Membrane	D ([³H]-ACh)	7
	0.004	*Torpedo*	Triton extract	D ([³H]-ACh)	7
	0.51	*Torpedo*	Triton extract	D ([³H]-ACh)	7
	0.008	*Torpedo*	Membrane	R (*Naja* α-toxin)	1
	0.008	*Torpedo*	Membrane	D ([³H]-ACh)	1
	0.03	*Torpedo*	Membrane	D ([³H]-ACh)	8
	~2.0	*Torpedo*	Purified	D ([³H]-ACh)	8
	0.47	Rat diaphragm (denervated)	Homogenate	R (α-BuTX)	9
	0.42	Rat diaphragm (denervated)	Triton extract	R (α-BuTX)	9
	0.2	Cat leg muscle (denervated)	Membrane	R (α-BuTX)	10
CCh	24	*Electrophorus*	Membrane	R (α-BuTX)	11
	20	*Electrophorus*	Membrane	I ([³H]-C10)	12
	40	*Electrophorus*	Membrane	R (*Naja* α-toxin)	1
	22	*Electrophorus*	Membrane	I ([³H]-C10)	13
	1.9	*Electrophorus*	Purified	I ([³H]-C10)	14
	0.5	*Torpedo*	Membrane	R (*Naja* α-toxin)	1
	0.4	*Torpedo*	Membrane	F (quinacrine)	15
	3.5	Rat diaphragm (denervated)	Homogenate	R (α-BuTX)	9
	6.3	Rat diaphragm (denervated)	Triton extract	R (α-BuTX)	9
	2.3	Rat diaphragm	Homogenate	R (α-BuTX)	9
	3.0	Cat leg muscle (denervated)	Membrane	R (α-BuTX)	10
C10	8	*Electrophorus*	Membrane	R (α-BuTX)	11
	0.3	*Electrophorus*	Membrane	D ([³H]-C10)	12
	0.8	*Electrophorus*	Membrane	R (*Naja* α-toxin)	1
	1.3	*Electrophorus*	Membrane	D ([³H]-C10)	13
	0.021	*Electrophorus*	Purified	D ([³H]-C10)	14
	0.8	*Torpedo*	Membrane	R (*Naja* α-toxin)	1
	0.74	*Torpedo*	Membrane	D ([³H]-C10)	1
	0.6	*Torpedo*	Membrane	F (quinacrine)	15
	2.1	Rat diaphragm (denervated)	Homogenate	R (α-BuTX)	9
	8.6	Rat diaphragm (denervated)	Triton extract	R (α-BuTX)	9
	3.0	Cat leg muscle (denervated)	Membrane	R (α-BuTX)	10

(continued)

Table IVa (continued)

Drug[a]	$K(\mu M)$	Species	Receptor preparation	Method[b]	Ref.[c]
C6	61	*Electrophorus*	Membrane	R (*Naja* α-toxin)	1
	62	*Electrophorus*	Purified	I ([^{3}H]-C10)	14
	40	*Torpedo*	Membrane	R (*Naja* α-toxin)	1
	60	*Torpedo*	Membrane	F (quinacrine)	15
	118	Rat diaphragm (denervated)	Homogenate	R (α-BuTX)	9
	89	Rat diaphragm (denervated)	Triton extract	R (α-BuTX)	9
	40	Cat leg muscle (denervated)	Membrane	R (α-BuTX)	10
TC	0.2	*Electrophorus*	Membrane	R (α-BuTX)	11
	0.17	*Electrophorus*	Membrane	R (*Naja* α-toxin)	1
	0.2	*Electrophorus*	Membrane	I (C10)	13
	0.39	*Electrophorus*	Purified	I (C10)	14
	0.17	*Torpedo*	Membrane	R (*Naja* α-toxin)	1
	0.17	*Torpedo*	Membrane	I (C10)	1
	0.23	*Torpedo*	Membrane	I (ACh)	1
	0.035, 0.1 (K_R)	*Torpedo*[d]	Membrane	I (ACh)	6
	0.4, 1.3 (K_T)	*Torpedo*	Membrane	I (ACh)	6
	0.4	Rat diaphragm (denervated)	Homogenate	R (α-BuTX)	16
	0.22	Rat diaphragm (denervated)	Homogenate	R (α-BuTX)	9
	0.38	Rat diaphragm (denervated)	Triton extract	R (α-BuTX)	9
	0.24	Rat diaphragm	Homogenate	R (α-BuTX)	9
	0.04	Cat leg muscle (denervated)	Membrane	R (α-BuTX)	10
	0.045[e]	Rat diaphragm	Purified (junctional)	R (α-BuTX)	17
	0.55	Rat diaphragm	Purified (extra-junctional)	R (α-BuTX)	17

[a] Drug abbreviations are as in Tables I and II, and C6, hexamethonium; TC, tubocurarine; α-BuTX, α-bungarotoxin.
[b] Methods: R, retardation of binding of the specified toxin; D, direct measurement of binding of specified ligand; I, inhibition of equilibrium binding of specified ligand; F, fluorescence of specified agent.
[c] References: (1) Weber and Changeux (1974), (2) Meunier *et al.* (1974), (3) Chang and Neumann (1976), (4) Eldefrawi *et al.* (1971), (5) Eldefrawi and Eldefrawi (1973b), (6) Gibson (1976), (7) O'Brien and Gibson (1975), (8) Sugiyama and Changeux (1975), (9) Colquhoun and Rang (1976), (10) Barnard *et al.* (1977), (11) Bulger and Hess (1973), (12) Fu *et al.* (1974), (13) Kasai and Changeux (1971b), (14) Meunier and Changeux (1973), (15) Grünhagen and Changeux (1976), (16) Colquhoun, *et al.* (1974), (17) Brockes and Hall (1975).
[d] Values given are for allosteric site (first number for α chain, second for β chain).
[e] See text.

in mind that the numbers cited may not be estimates of the same thing (see also Sec. 3). For example, the *comparison* and *interaction* null methods (Table IVb) estimate approximately the equilibrium constant K_T for binding to the shut conformation in a range of theories (Colquhoun, 1973); the values given by Dreyer *et al.* (1978) (Table IVb) are estimates of the effective equilibrium binding constant in the Hill equation [Eq. (4)], so it is a sort of mean equilibrium constant (and also the concentration for the 50% maximum *response*); the values given by Dionne *et al.* (1978) (Table IVb) are estimates of microscopic equilibrium constants for first and second agonist bindings to the shut conformation in a scheme

Table IVb

Some Values for Equilibrium Dissociation Constants for Binding of Agonists and Partial Agonists to the Shut Conformation That Have Been Inferred Indirectly from Physiological Studies[a]

Drug	Preparation	Method	K (μM)	Ref.[b]
MeN$^+$Et$_3$	Frog rectus	Interaction (Me$_4$N$^+$)	4070	1
	Frog rectus	Comparison (Me$_4$N$^+$)	2750	1
C10	Frog n.m.j. (voltage-clamp)	Interaction (CCh)	54	2
	Frog n.m.j. (voltage-clamp)	Comparison (CCh)	48.7	2
	Frog n.m.j. (voltage-clamp)	Irrev. (DNM)	39.4	2
Heptyl TMA	Guinea pig ileum	Comparison (CCh)	71	3
	Guinea pig ileum	Irrev. (DBN)	105	3
CCh	Frog n.m.j.	Dose–response	60–80	4
	Frog n.m.j.	Dose–response	336	5
	Frog n.m.j.	Dose–response	600, 80[c]	6
	Frog n.m.j.	Dose–response	300	7
ACh	Frog n.m.j.	Dose–response	30	7
	Frog n.m.j.	Dose–response	27	5

[a] Abbreviations: Drugs as in previous tables, and n.m.j., neuromuscular junction; DBN, dibenamine; DNM, dinaphthyldecamethonium mustard. The terms interaction, comparison, and irrev. refer to null methods for estimating the binding affinity of partial agonists [reviewed in Colquhoun (1973)]. Dose–response refers to analysis of the equilibrium dose–response curve; values are given only for cases where an attempt to estimate a pure *binding* constant (rather than ED50, for example) has been made.

[b] References: (1) Barlow *et al.* (1967), (2) Rang [unpublished; cited in Colquhoun (1973)], (3) Waud (1969), (4) Adams (1975c), (5) Dreyer *et al.* (1978), (6) Dionne *et al.* (1978), (7) Adams and Sakmann (1978c).

[c] Values for binding of first and second carbachol molecules, respectively.

like (**G**) (Sec. 3.2) except that both bindings are not assumed to have the same affinity; and the values given by Adams and Sakmann (1978c) (Table IVb) are estimates of the microscopic equilibrium constant K_T in scheme (**E**) (Sec. 3.2). Sheridan and Lester (1977) give values for ACh in eel electroplaque which are similar to those in Table IVb, though their values are not separated from the conformational equilibrium constant. On the other hand, the "equilibrium constants" from direct binding studies in Table IVa are mainly concentrations for half-saturation of binding rather than estimates of a particular equilibrium constant in a well-defined theory, so discrepancies between different estimates of agonist affinities are not surprising, though the possibility of changes from the native state remains a problem in many binding studies.

All of the methods make measurements at equilibrium except those involving measurement of retardation of the initial rate of binding of toxins. For the latter method it is clear that measurements must be done with a preparation in which the rate of toxin binding is not distorted by diffusion, and precautions must be taken to prevent rebinding of toxin that remains in the extracellular space after its removal from solution. Early studies in muscle frequently found that protection against toxin binding by various cholinergic ligands was incomplete (Miledi and Potter, 1971; Berg *et al.*, 1972; Albuquerque *et al.*, 1973; Chiu *et al.*, 1974; Libelius *et al.*, 1975). However, when suitable precautions are taken,

90% or more of toxin binding sites are protectable by both agonists and antagonists (Colquhoun *et al.*, 1974; Dolly and Barnard, 1974; Brockes and Hall, 1975; Colquhoun and Rang, 1976; Barnard *et al.*, 1977).

2.2.1. Cooperativity in Binding

It is crucial for knowledge of the mechanism of channel opening to know whether or not the binding of ligands to the receptor is cooperative or not. Unfortunately it has not yet been possible in most cases to investigate binding with sufficient precision for a definite answer to be given. The most precise studies have been on the high-affinity binding to unpurified *Torpedo* receptor. In this case cooperativity has been detected in some studies with Hill coefficients of 1.0–1.5 (Weber and Changeux, 1974) or 1.3–2.0 (Eldefrawi and Eldefrawi, 1973a; O'Brien and Gibson, 1975; Gibson, 1976), but cooperativity is not seen with low-affinity preparations. Cooperativity has not yet been detected with ligands other than ACh or in species other than *Torpedo*, though no direct binding studies to the muscle receptor have been possible yet.

2.2.2. Is There a Single Sort of Binding Site?

The vast majority of studies have assumed, and have been consistent with, the idea that toxins, agonists, and antagonists all compete for, and exclude each other from, the same site that acetylcholine combines with [see the reviews by Changeux (1975), Colquhoun (1975a), and Rang (1975)]. Nevertheless, there have, over a number of years, been periodical reports that appear to favor the idea that not all ligands combine with the same site but that sites either partially overlap or are completely separate but interact allosterically. The best documented sort of site other than the receptor itself is the site, putatively within the ion channel itself, responsible for the *channel-blocking* behavior of many substances (see Secs. 1.5 and 3.4). This *local anesthetic* binding site may well bind other antagonists (and even agonists), too. Many mechanisms can mimic competitive interaction over limited concentration ranges so they cannot be excluded, though frequently the evidence for separate sites comes from experimental evidence that is simply not in agreement with that of other workers. For example, short reports from Bulger and Hess (1973) and Fu *et al.* (1974) suggest that antagonists (bungarotoxin, tubocurarine) occupy sites that overlap incompletely with those for activators (carbachol, decamethonium) in *Electrophorus*. In the first of these, the association rate constant of bungarotoxin was an order of magnitude slower than seen by most workers, and an initial reversible stage of binding was seen which is more prominent than in most studies. In the report by Fu *et al.* (1974), it remains a possibility that the results were affected by decamethonium binding to acetylcholinesterase (Kasai and Changeux, 1971b), and their findings suggest that decamethonium should not provide complete protection against bungarotoxin binding, which, in most studies in both electric tissue and muscle, it does.

The most ambitious analysis so far of a binding study is that of Gibson (1976), who used *Torpedo* membrane fragments dissolved in Triton X-100. He found that his results could be fitted best by a four-subunit theory of the sort described by Monod *et al.* (1965) but with two pairs of nonidentical (α and β) subunits. Furthermore, each subunit was supposed to possess an ACh binding site and a second binding site which interacts allosterically with the ACh binding site. Nicotine appeared to act as a competitive inhibitor of ACh. But tubocurarine appeared to bind exclusively to the allosteric site and by doing so stabilized a conformation in which affinity of the ACh site for ACh on the β subunits was greatly reduced, producing *half-of-sites* ACh binding. Decamethonium was thought to act on both ACh and allosteric binding sites. Time will tell how this complex scheme fares. The theory certainly fits the data, but there are a large number of parameters to be estimated in the fitting of such complex theories, and the methods of curve fitting used provide no means of telling how well determined the parameters are. The binding constants inferred for tubocurarine in this study imply that it should bind cooperatively and should be a weak agonist. The evidence for these postulates, while not nonexistent (Gibson, 1976), is certainly slim [Katz and Miledi (1977) found no evidence for agonist activity at the end plate in a careful study]. Furthermore, the binding curves for ACh varied considerably from one receptor preparation to another; it is interesting that these variations could be fitted by varying only L_0, the equilibrium constant between active and inactive forms in the absence of drug, without changing any of the ACh binding parameters, but this variability obviously adds further uncertainty to the analysis.

In summary, the existence of more than one sort of binding site remains a possibility, but the evidence up to now is far from compelling except for the channel-blocking site discussed in Secs. 1.5 and 3.4. It should perhaps be borne in mind that a number of cases of ligands that appear to produce half-of-sites binding have been extensively investigated (and have given rise to some controversy) in the enzyme field (Seydoux *et al.*, 1974).

2.2.3. Binding to Junctional and Extrajunctional Receptors in Muscle

There are far fewer binding studies on receptors from muscle than on receptors from electric tissues. The results in Table IVa show, on the whole, tolerable agreement between the affinities found by different groups. All ligands appear to give a high degree of protection against toxin binding when the toxin binding rate constant is measured properly (see Sec. 2.1). In the study of Colquhoun and Rang (1976), both agonists and antagonists behaved like simple competitive inhibitors of α-bungarotoxin binding. Although the affinity found for ACh is similar to that of low-affinity sites in electric tissue, it is probably nevertheless greater than that of the native receptor for ACh (see below and Secs. 2.2 and 3.7).

The only serious discrepancy among laboratories is in the case of tubocurarine. The vast majority of estimates of the dissociation constant for tubocurarine in muscle by physiological methods (null methods usually) give values of about 0.1–0.5 μM, in a wide range of species [see the references in Colquhoun (1975a) and Ginsborg and Jenkinson (1976)], as do binding studies in electric tissue (see Table IVa). Colquhoun and Rang (1976) found similar values by retardation of bungarotoxin binding in homogenized and solubilized rat diaphragm, both in normal muscle and on the extrajunctional receptors of denervated muscle, in essential agreement with the classical study of Jenkinson (1960). Brockes and Hall (1975) found a similar value for purified extrajunctional receptor from denervated rat diaphragm but claim a much lower value for junctional receptors. However, inspection of their data shows that for both sorts of receptor 50% inhibition of the toxin binding rate occurs at a similar tubocurarine concentration (about 0.5 μM). The only difference between junctional and extrajunctional receptors is a slightly greater inhibition at very low tubocurarine concentrations for the former. There is no overall shift of the binding curve. Barnard *et al.* (1977), however, find a single binding site for tubocurarine of high affinity (0.04 μM) in denervated cat leg muscle membrane fragments. No explanation for this discrepancy is available at the moment. There have been some reports that low concentrations of tubocurarine (0.03–0.04 μM) will produce 50% inhibition of the nerve-evoked end-plate potential, whereas at extrajunctional receptors larger concentrations (0.2–0.8 μM) are needed for 50% inhibition of iontophoretic ACh potentials (Beranek and Vyskočil, 1967; Dolly *et al.*, 1977), but the uncertainties presented by the shape of the dose–response curve and the different rates of application [Beranek and Vyskočil (1967), however, did provide a control for this factor] make it dubious whether these values can be taken as estimates of equilibrium constants.

2.3. The Kinetics of Acetylcholine Binding

It is essential for any proper understanding of mechanism to know about the rate of drug binding as well as the rate of channel opening. But, unfortunately, hardly anything is known about the rate of the binding process. A number of indirect inferences have been made (see, for example, the discussion of the rate of tubocurarine action, Sec. 1.4.2.3). But, so far, there has been only one real attempt to measure binding rates directly. Chang and Neumann (1976) and Neumann and Chang (1976) found that ACh binding to purified receptor protein (low-affinity form) from *Torpedo* caused loss of several bound calcium ions. They performed temperature-jump experiments on the receptor in the presence of calcium, ACh, and murexide as a calcium indicator. The relaxation spectrum suggested that the binding of ACh proceeds in at least two steps, the association rate constant for the initial binding being about 2.4×10^7 M^{-1} sec^{-1}, certainly a plausible value in the light of enzyme studies where association rate constants between 10^7 and 10^8 M^{-1} sec^{-1} are usual [e.g., Gutfreund (1972)]. This is similar to the value suggested by Sheridan and Lester (1977). The other rate constants inferred from this study appeared, however, to be a good deal too slow to account for physiological events; the authors suggest that they could be related to desensitization.

It was found by Colquhoun and Rang (1976) that antagonist drugs bound much too rapidly for the rate to be measured by their retardation of bungarotoxin binding. On the other hand, agonists, which would certainly be expected to bind more rapidly than antagonists, took a measurable time (a time constant of several minutes) to produce their full inhibition of the bungarotoxin binding rate to denervated rat diaphragm homogenates. This rate is comparable with that of desensitization, so it was suggested, following Miledi and Potter (1971) and Lester (1972), that desensitization was a cause of the decreased rate of toxin binding produced by agonists. Technical limitations prevented any test of whether (1) the putative desensitized conformation of the receptor reacted at a lower rate with toxin than the active conformation or (2) the putative increased agonist affinity of the desensitized conformation caused an increased occupancy of binding sites by agonist, thus excluding the toxin to an increasing extent as desensitization took place. Weber *et al.* (1975) observed a very similar phenomenon in *Torpedo* membrane fragments and provided evidence for the latter explanation. On the other hand, Barnard *et al.* (1977) found no evidence for time dependence of the protective effect of agonists (or antagonists) in denervated cat leg muscle membrane preparations. If any such effect exists in their preparation, it is either too rapid to see (complete in 30 sec or so) or else very slow.

3. The Link between Drug Binding and Response

3.1. What Should a Mechanism Explain?

It will be clear from the preceding two sections, in which response to and binding of ligands were discussed, that there are still big gaps in our knowledge, despite the advances made in the last 10 years. Perhaps the biggest of these gaps is the uncertainty about the rate, and often even the extent, of binding of agonists in living tissue.

Needless to say, there is no lack of theories. Some of the questions that any proposed mechanism must explain are as follows.

1. By what mechanism does the binding of an agonist cause an ion channel to open?

2. How can the relationship between concentration and (a) binding and (b) response at equilibrium be explained? Why does this relationship appear to be cooperative? How can the high affinities for ACh observed in electric tissue be reconciled with the rapidity of drug action?

3. Why do some agonists produce a bigger maximum response than others? What is the physical basis of efficacy and partial agonism?

4. To what extent do actions of drugs at sites other than the ACh binding site matter? Are there important effects on allosteric sites? Do local anesthetic-like actions or channel-blocking actions contribute to the normal action of drugs or to desensitization?

5. What controls the *rate* of drug action? Is binding rapid relative to the subsequent conformation change or vice versa?

6. Why do fluctuation and relaxation spectra usually have only one predominant component (which suggests that the system can exist in only two kinetically distinguishable states)? What is the nature of these states? Are the assumptions made in the analysis of kinetic experiments valid?

7. Why does the apparent mean open-channel lifetime inferred from fluctua-

tion measurements appear to be much more strongly dependent on the nature of the agonist than the apparent mean single-channel conductance?

8. How is the concentration dependence of fluctuation and relaxation measurements to be accounted for?

9. How is the voltage dependence of (a) the equilibrium conductance change and (b) the rates measured in kinetic experiments to be explained?

3.2. Some Mechanisms

In this section some of the mechanisms that have been considered for drug action in recent work will be listed. The extent to which they can explain the above questions will then be discussed. Throughout the discussion, the macromolecular conformation of a protomer associated with the active (conducting) state of the ion channel will be denoted by R, and the conformation associated with the resting (nonconducting) state will be denoted by T (following Monod *et al.*, 1965). There may, of course, really be more than two conformations as well as the putative desensitized conformation(s), but there is no evidence that compels one to believe this at the moment.

The mechanisms under consideration can be classified roughly into (1) those that arbitrarily assign equilibrium constants, on the basis of experimental results, in such a way that states to which agonist is bound are more likely to be open than unliganded states without explaining why they are more likely to be open [for example, schemes (A), (D), and (F) below] and (2) those that postulate a physical mechanism for channel opening by agonists and for the cooperativity observed for this process [such as schemes (G)–(J)]. The limited amount of information that is usually available in the drug–receptor field means that it is normally proper, and necessary, to use the former, more empirical approach when trying to interpret experiments. But obviously the physical mechanism of the latter class of theories must be the ultimate aim, and even if the elucidation of such mechanisms needs more information than we have, the fact that some success has already been achieved in applying them to enzyme mechanisms makes it appropriate to bear them in mind while interpreting drug action. The equations for the state and binding functions at equilibrium and for the mean open-channel lifetime for the theories below are given in Table V for reference.

The simplest plausible mechanism is that proposed by Castillo and Katz (1957b). Rate constants are given as lowercase Greek symbols and equilibrium constants as uppercase Roman symbols. Binding equilibrium constants are defined as dissociation constants, and conformation equilibrium constants as (T-form)/(R-form):

$$
\begin{array}{c}
\mathrm{T} \\
K_{\mathrm{T}} \; \beta_{01} \Big\uparrow\!\!\Big\downarrow \beta_{10} \\
\mathrm{AT} \underset{\alpha_1}{\overset{\beta_1}{\rightleftharpoons}} \mathrm{AR} \\
L_1
\end{array}
\qquad (\mathbf{A})
$$

Table V

Equilibrium State and Binding Functions for Various Schemes[a]

Scheme	$p_R(0)$	$p_R(\infty)$	$p_R(c)$	$p_A(c)$	K_{eff}	m_R
(A)	0	$\dfrac{1}{1+L_1}$	$\dfrac{p_R(\infty)x}{x+K_{eff}}$	$\dfrac{x}{x+K_{eff}}$	$K_T\left(\dfrac{L_1}{1+L_1}\right)$	$\dfrac{1}{\alpha_1}$
(B)	$\dfrac{1}{1+L_0}$	1	$\dfrac{1+c_R}{1+c_R+L_0}$	$\dfrac{x}{x+K_{eff}}$	$K_R(1+L_0)$	$\dfrac{1+c_R}{\alpha_0}$
(H)	$\dfrac{1}{1+L_0}$	$\dfrac{1}{1+L_0M}$	$p_R(0)+\dfrac{[p_R(\infty)-p_R(0)]x}{x+K_{eff}}$	$\dfrac{x}{x+K_{eff}}$	$K_R\left(\dfrac{1+L_0}{1+L_0M}\right)$	$\dfrac{1+c_R}{\alpha_0+\alpha_1 c_R}$
(D)	0	$\dfrac{1}{1+L_2}$	$\dfrac{x^2}{dK_{1T}K_{2T}L_2}$	$\dfrac{(x/K_{1T})+(x^2/K_{1T}K_{2T})[(1+L_2)/L_2]}{d}$	—	$\dfrac{1}{\alpha_2}$
(E)	0	$\dfrac{1}{1+L_2}$	$\dfrac{1}{1+L_2[(1+c_T)/c_T]^2}$	$\dfrac{L_2c_T(1+c_T)+c_T^2}{L_2(1+c_T)^2+c_T^2}$	—	$\dfrac{1}{\alpha_2}$
(F)	0	$\dfrac{1}{1+L_2}$	$\dfrac{1}{1+L_1\{(1+c_T)^2/[2c_T+c_T^2(L_2/L_1)]\}}$	$\dfrac{c_T(1+c_TL_2/L_1)+L_1c_T(1+c_T)}{2c_T+c_T^2L_2/L_1+L_1(1+c_T)^2}$	—	$\dfrac{2c_T+c_T^2(L_2/L_1)}{2c_T\alpha_1+c_T^2(L_2/L_1)\alpha_2}$
(G)	0	$\dfrac{1}{1+L_2}$	$\dfrac{1}{1+(L_1/M)[(1+c_T)^2/(2c_R+c_R^2)]}$	$\dfrac{c_R(1+c_R)+(L_1/M)c_T(1+c_T)}{2c_R+c_R^2+(L_1/M)c_T(1+c_T)}$	—	$\dfrac{2c_R+c_R^2}{2c_R\alpha_1+c_R^2\alpha_2}$
(I)	$\dfrac{1}{1+L_0}$	$\dfrac{1}{1+L_0M^n}$	$\dfrac{1}{1+L_0[(1+c_T)/(1+c_R)]^n}$	$\dfrac{c_R(1+c_R)^{n-1}+L_0c_T(1+c_T)^{n-1}}{(1+c_R)^n+L_0(1+c_T)^n}$	—	$\dfrac{(1+c_R)^n}{\sum_{i=0}^{n}\binom{n}{i}c_R^i\alpha_i}$
(J)	0	$\left(\dfrac{1}{1+L_1}\right)^n$	$\left(\dfrac{1}{(1+L_1)}\dfrac{x}{(x+K_{eff})}\right)^n$	$\dfrac{x}{x+K_{eff}}$	$K_T\left(\dfrac{L_1}{1+L_1}\right)$	$\dfrac{1}{n\alpha_1}$

[a] The ligand concentration is denoted by x. Normalized concentrations are defined as $c_R = x/K_R$ and $c_T = x/K_T$, and the relative affinity for shut and open conformations as $M = K_R/K_T$. The fraction of channels that are open at concentration c is denoted by $p_R(c)$ [for scheme (J) this is the fraction of oligomers with all n subunits in the R conformation]. The fraction of binding sites occupied is denoted by $p_A(c)$. In scheme (D), d is defined as $1 + 2x/K_{1T} + x^2/K_{1T}K_{2T} + x^2/K_{1T}K_{2T}L_2$. The mean open lifetime is denoted by m_R. Notice that if all the α_i were identical, α say, then the mean lifetime of the open state in schemes (F), (G), (H), and (I) would reduce to $1/\alpha$. Otherwise m_R^{-1} is a weighted mean of the α_i.

An agonist molecule A combines with a shut channel, which may then, for reasons unspecified, open. In principle, of course, the channel must to some extent be capable of opening in the absence of agonists, so the predominant reaction pathway *could* be

$$
\begin{array}{c}
L_0 \\
T \underset{\alpha_0}{\overset{\beta_0}{\rightleftharpoons}} R \\
\alpha_{01} \Big\Vert \alpha_{10} \quad K_R \\
AR
\end{array}
\qquad \textbf{(B)}
$$

In view of the evidence that more than one agonist molecule may be involved in channel opening, something a bit more complicated is evidently needed.

Scheme (**C**): The simplest possible extension (Anderson and Stevens, 1973) of scheme (**A**), which is valid only at low agonist concentrations, is based on the observation that the conductance change is approximately proportional to x^n when the concentration x is low. The equilibrium equations are the same as those given for scheme (**A**) (Table V) but with x^n replacing x, and the predictions for kinetic experiments at constant concentration are the same as for scheme (**A**) (Anderson and Stevens, 1973). It may be noted that n, in this scheme, is essentially the experimentally observed Hill coefficient, so scheme (**C**) bears no necessary relationship with the postulate that n agonist molecules must bind in order to open a channel.

To get further it is necessary to make predictions for high as well as low concentrations of agonist. A simple extension of scheme (**A**) is

$$
\begin{array}{c}
T_2 \\
K_{1T} \quad \beta_{01} \Big\Vert \beta_{10} \\
AT_2 \\
K_{2T} \quad \beta_{12} \Big\Vert \beta_{21} \\
A_2 T_2 \underset{\alpha_2}{\overset{\beta_2}{\rightleftharpoons}} A_2 R_2 \\
L_2
\end{array}
\qquad \textbf{(D)}
$$

Note that both subunits, or protomers, have been assumed to undergo a concerted conformation change to the open state (so species such as TR or RT do not exist). The rate constants for binding are defined as microscopic rate constants so the macroscopic equilibrium constants for the first and second bindings would be $K_{1T}/2$ and $2K_{2T}$, respectively.

It is obviously necessary, if such mechanisms are to be tested, to have as few parameters in them as possible. One way to do this is to assume, until such time as there is evidence to the contrary, that the microscopic rate constants (β_{01}, β_{10}) for binding to a subunit in the T-conformation are the same for each binding, as is usual in the Monod–Wyman–Changeux mechanism, scheme (**I**) below. In other words, the binding steps *to a given form* (R or T) are assumed independent, though the overall binding function (Table V) is still cooperative (unless L_2 is very large). This assumption has also been made in the other schemes presented below but is not necessarily true. The predicted response is essentially the square of that predicted by scheme (**A**). In this case scheme (**D**) simplifies to:

$$
\begin{array}{c}
\mathrm{T_2} \\[2pt]
K_\mathrm{T} \quad \beta_{01}\big\|\beta_{10} \\[2pt]
\mathrm{AT_2} \\[2pt]
K_\mathrm{T} \quad \beta_{01}\big\|\beta_{10} \\[2pt]
\mathrm{A_2T_2} \underset{\alpha_2}{\overset{\beta_2}{\rightleftharpoons}} \mathrm{A_2R_2} \\[2pt]
L_2
\end{array}
\qquad (\mathbf{E})
$$

The number of parameters to be estimated is four in scheme (**E**) rather than six in scheme (**D**).

If binding is supposed, for whatever reason, to increase the probability of channel opening, it could be that one ligand can also, to some extent, cause opening as follows:

$$
\begin{array}{c}
\mathrm{T_2} \\[2pt]
K_\mathrm{T} \quad \beta_{01}\big\|\beta_{10} \quad L_1 \\[2pt]
\mathrm{AT_2} \underset{\alpha_1}{\overset{\beta_1}{\rightleftharpoons}} \mathrm{AR_2} \\[2pt]
K_\mathrm{T} \quad \beta_{01}\big\|\beta_{10} \\[2pt]
\mathrm{A_2T_2} \underset{\alpha_2}{\overset{\beta_2}{\rightleftharpoons}} \mathrm{A_2R_2} \\[2pt]
L_2
\end{array}
\qquad (\mathbf{F})
$$

This scheme, like schemes (**A**), (**C**), (**D**), and (**E**), assumes that the affinity of the agonist for the R-state is very high. If we allow that it may not be, we get

$$
\begin{array}{c}
\mathrm{T_2} \\[2pt]
K_\mathrm{T} \quad \beta_{01}\big\|\beta_{10} \quad L_1 \\[2pt]
\mathrm{AT_2} \underset{\alpha_1}{\overset{\beta_1}{\rightleftharpoons}} \mathrm{AR_2} \\[2pt]
K_\mathrm{T} \quad \beta_{01}\big\|\beta_{10} \qquad \alpha_{01}\big\|\alpha_{10} \quad K_\mathrm{R} \\[2pt]
\mathrm{A_2T_2} \underset{\alpha_2}{\overset{\beta_2}{\rightleftharpoons}} \mathrm{A_2R_2} \\[2pt]
L_2
\end{array}
\qquad (\mathbf{G})
$$

This has only one more rate constant to be estimated than scheme (**F**), because the introduction of the cyclic mechanism implies (because of the principle of microscopic reversibility) that $L_2 = L_1 M$, where $M = K_\mathrm{R}/K_\mathrm{T}$ is the relative affinity of the agonist for shut and open conformations. The opening of channels depends on this relative affinity. Apart from this, (**F**) and (**G**) are the same (Table V).

These schemes could, of course, easily be extended to the binding of more than two agonist molecules if necessary (see Sec. 3.3 for discussion). They can all be regarded as degenerate forms of the theory of Monod *et al.* (1965) (MWC theory), which postulates that only two conformations exist and that all subunits undergo a concerted transition between these two states [scheme **(I)**]. If schemes **(A)** and **(B)** are combined so either pathway is allowed, we get the following:

$$
\begin{array}{c}
L_0 \\[2pt]
\mathrm{T} \underset{\alpha_0}{\overset{\beta_0}{\rightleftharpoons}} \mathrm{R} \\[4pt]
K_{\mathrm{T}} \quad \beta_{01} \Big\| \beta_{10} \qquad \alpha_{01} \Big\| \alpha_{10} \quad K_{\mathrm{R}} \\[4pt]
\mathrm{AT} \underset{\alpha_1}{\overset{\beta_1}{\rightleftharpoons}} \mathrm{AR} \\[2pt]
L_1
\end{array}
\tag{H}
$$

In this scheme, unlike **(A)** and **(B)**, the equilibrium constants are, as in **(G)**, not all independent of each other. The principle of microscopic reversibility dictates that $L_1 = L_0 M$, where $M = K_{\mathrm{R}}/K_{\mathrm{T}}$, the relative affinities of the ligand for shut and open states, so the equilibrium constant for opening is now explicitly related to ligand binding. If more than one agonist molecule takes part, scheme **(H)** can be elaborated in either of two simple ways if we restrict ourselves to two conformations only. First, several protomers, n in general, undergo a concerted conformation change (Monod *et al.*, 1965; Karlin, 1967; Edelstein, 1972; Colquhoun, 1973; Thron, 1973):

$$
\begin{array}{c}
L_0 \\[2pt]
\mathrm{T}_n \underset{\alpha_0}{\overset{\beta_0}{\rightleftharpoons}} \mathrm{R}_n \\[4pt]
K_{\mathrm{T}} \quad \beta_{01} \Big\| \beta_{10} \qquad \alpha_{01} \Big\| \alpha_{10} \quad K_{\mathrm{R}} \\[4pt]
\mathrm{AT}_n \underset{\alpha_1}{\overset{\beta_1}{\rightleftharpoons}} \mathrm{AR}_n \\[2pt]
L_1 \\[8pt]
K_{\mathrm{T}} \Big\| \qquad\qquad \Big\| K_{\mathrm{R}} \\[6pt]
\mathrm{A}_n\mathrm{T}_n \underset{\alpha_n}{\overset{\beta_n}{\rightleftharpoons}} \mathrm{A}_n\mathrm{R}_n \\[2pt]
L_n
\end{array}
\tag{I}
$$

The relationship between binding and conformation change given for schemes **(G)** and **(H)** generalizes to $L_i = L_0 M^i$ (for $1 \leq i \leq n$).

Alternatively, by analogy with Hodgkin and Huxley (1952), a channel controlled by n *independent* protomers might be considered, the channel being open only when all n protomers are in the R-conformation. Each protomer might, for example, be as shown in scheme **(A)** or **(H)**. In the former case the mechanism may be represented as follows:

$$
\left[
\begin{array}{c}
\mathrm{T} \\[2pt]
K_{\mathrm{T}} \quad \beta_{01} \Big\| \beta_{10} \\[6pt]
\mathrm{AT} \underset{\alpha_1}{\overset{\beta_1}{\rightleftharpoons}} \mathrm{AR} \\[2pt]
L_1
\end{array}
\right]_n
\tag{J}
$$

Last, it may be considered that a whole continuous array of protomers could interact with each other so that the properties of any one would be influenced by the state of its neighbors (e.g., Changeux *et al.*, 1967). The dense packing of receptors on the subsynaptic membrane renders this sort of mechanism quite plausible, but Land *et al.* (1977) found that acetylcholine sensitivity was directly proportional to receptor density (measured with bungarotoxin) in cultured rat muscle, whereas the Hill coefficient was independent of receptor density. This suggests that interactions are restricted to the subunits of a single receptor–ionophore.

It is worth mentioning that all of the schemes outlined above, and many others, are entirely compatible with all the experimental results in the literature concerning the mutual interactions of full agonists, partial agonists, and antagonists. Although these experiments were analyzed in terms of the classical model $(A + T \rightleftharpoons AR)$ and were consistent with it, the fact that they were necessarily mostly null experiments (comparisons of concentrations at a constant response level) has been shown to imply that they can equally well be interpreted in terms of more detailed theories (Colquhoun, 1973; Thron, 1973).

Do Open but Unoccupied Channels Exist? There is no evidence that the species R which is included in schemes (**B**), (**H**), and (**I**) exists either in electric tissue or at the neuromuscular junction. Katz and Miledi (1977) showed that tubocurarine can produce a very small (40 μV) hyperpolarization at the frog end plate and discuss the possibility that the effect might depend on the presence of open channels in the absence of agonist. (Tubocurarine might suppress these, e.g., by having a greater affinity for T- than for R-conformations.) But the effect occurs only when cholinesterase is inhibited, so it was attributed to slow ACh leakage from nerve endings rather than spontaneous channel opening. Therefore the species R is usually, and properly, omitted from the analysis of experiments [see also Dionne *et al.* (1978), who suggest that $L_0 > 10^5$]. Another reason for doing this in kinetic experiments is because it is usual to subtract a control observation (of relaxation or fluctuations) from the observation made in the presence of drug. If any important number of channels were open in the absence of drug, most information about them would be lost in the process of subtraction (indeed the process of subtraction would be invalidated). The only reasons schemes such as (**B**), (**H**), and (**I**) are mentioned is that (1) it might be argued that some open but unoccupied channels *must* exist on thermodynamic grounds, (2) it maintains the analogy with enzyme work, and (3) the algebra is neater. Hardly compelling reasons.

3.3. The Concentration Dependence of Binding and Response at Equilibrium

The experiments described in Sec. 1 provide evidence for cooperativity in the equilibrium conductance–concentration curve. Schemes with a single binding site, such as (**A**), (**B**), and (**H**), all predict a hyperbolic response curve, whereas the other schemes can account for cooperativity. It should, perhaps, not be forgotten that there is still no real evidence, apart from the sigmoid nature of concentration–response curves, for the involvement of more than one agonist molecule in channel opening and that sigmoid curves could conceivably result from other causes (e.g., Nimmo and Bauermeister, 1976; Whitehead, 1970, p. 358). Dionne *et al.* (1975, 1978) have suggested that a scheme such as (**G**), but with different (higher) affinity for the second binding step, is a good, but not unique, description of their equilibrium results (see also Secs. 1.3, 2.2, and 3.7 and Table IVb). Dreyer *et al.* (1978) also found that cooperative binding, described by a Hill equation [Eq. (4), which can be regarded as a special case of scheme (**I**); see, for example, Colquhoun (1973)], fitted equilibrium data better than independent binding [scheme (**J**)] (see also Secs. 1.3 and 2.2 and Table IVb). Adams (1975c,d) suggested that his results at the frog end plate were best fitted by simple independent binding [scheme (**J**) with $n = 2$ subunits], but more recently Adams and Sakmann (1978c) have suggested that scheme (**E**) is preferable. The latter scheme should, unlike (**J**), show cooperativity in the binding function as well as in the response function (unless L_2 is very large; see Table V). Another reason for doubting the adequacy of scheme (**J**) is that it has been pointed out (Adams, 1977a; Colquhoun and Hawkes, 1977) that if simple versions (with rapid drug binding) of the n independent subunit hypothesis were correct, the time constants found in voltage-jump experiments (with small enough perturbations) should be n times slower than the time constant from fluctuation experiments (with low enough concentrations), whereas, in the range measured up to now, these time constants are very similar (Neher and Sakmann, 1975; Colquhoun *et al.*, 1978).

On the other hand, it is true that most binding experiments have failed to show the expected cooperativity in the binding of agonists; the most precise studies, on *Torpedo* receptor, have shown cooperativity, though only in the high-affinity component (see Sec. 2).

Although the equilibrium physiological studies cited above have, apart from Dreyer *et al.* (1978), assumed that there are two subunits that bind the agonist, it is generally agreed that this is a minimum number (in fact it is *too* small to be compatible with the Hill coefficients in Table I that are greater than 2). It would not be at all surprising if there were more subunits involved (e.g., Colquhoun, 1973; Gibson 1976), but it does not seem very likely that the number will be found unambiguously from electrophysiological studies. A substantial amount of recent work, both biochemical and morphological, has suggested that four to eight subunits may be present (Chang *et al.*, 1977; Cartaud *et al.*, 1978), though this work has not told us much about the function or interactions of the subunits (it is still not known whether the channel gating mechanism is made up of the same subunits that bind agonists). It is certainly common in other fields of biochemistry for the maximum Hill coefficient to be substantially less than the number of subunits (ligand binding sites); for example, (1) hemoglobin has four binding sites and a Hill coefficient of 2–3, depending on conditions; (2) NAD^+ binding to yeast glyceraldehyde-3-phosphate dehydrogenase (GPDH), which has four subunits, is cooperative with a Hill coefficient of about 2.3 (Kirschner *et al.*, 1971); and (3) the binding of the modifier analogue 5-bromocytidine triphosphate to six binding sites in aspartate transcarbamylase is very weakly cooperative (Eckfeldt *et al.*, 1970); the maximum Hill coefficient calculated from their data is only 1.4.

3.4. The Nature of Efficacy, Partial Agonists, and Desensitization

Partial agonists were originally defined as those that, even when they occupied all binding sites, were incapable of producing the full response that the tissue was capable of (Stephenson, 1956). At the molecular level the response of interest is the opening of ion channels, and a partial agonist would be one that is incapable of opening all ion channels, i.e., for which there is a relatively small probability that a completely liganded channel is open. It is not known whether *any* agonist is capable of opening most channels, but it is certainly known that some agonists can produce only a small conductance increase at the muscle end plate, e.g., the bis(ethyl dimethyl) analogue of decamethonium, choline, and decamethonium. In fact the first physical interpretation of partial agonists was proposed by Castillo and Katz (1957b) to explain the action of decamethonium. They used scheme (**A**) and suggested that the equilibrium between AT and AR was further to the left (i.e., $L_1 = \alpha_1/\beta_1$ is larger) for decamethonium than for, say, acetylcholine. And it is certainly true that fluctuation analysis, which is supposed to measure the mean channel lifetime $1/\alpha$ (see below and Table V), suggests that α_1 is larger for acetylcholine than for decamethonium (see Table II). But, until very recently, there has been little knowledge of β_1. However, despite the rather different schemes that were fitted to the results, three recent equilibrium studies (Dreyer *et al.*, 1978; Dionne *et al.*, 1978; Adams and Sakmann, 1978c) all suggest that the maximum fraction of channels that can be opened at the frog end plate by carbachol is about 50% (i.e., it is not a particularly efficacious agonist) and that this fraction is rather higher for acetylcholine.

The obvious extension of this approach is to suggest that an agonist will be weak if its binding is not very selective for the R-conformation, i.e., if $M = K_R/K_T$ is not very small. This approach mimics closely the classical behavior of partial agonists, including their interaction with more powerful agonists. In fact $(1/M) - 1$ is closely analogous to the *efficacy* of an agonist as defined by Stephenson (1956) [see Colquhoun (1973) and Thron (1973)]. It must, however, be stated that there is no proof at all of the validity of this approach, and M has never been determined for any agonist. Work in other biochemical fields might be expected to provide an analogy. There is certainly not a plethora of ligands that have been shown conclusively to show only marginal selectivity for one conformation. The binding of the modifier analogue 5-bromocytidine triphosphate to aspartate transcarbamylase is one candidate; the results of Eckfeldt *et al.* (1970) suggest that its affinity for one state is around 40% of its affinity for the other, and the maximum conformational change [from $p_R(\infty)$, Table V, scheme (**I**)] was about 80%. The original data on NAD^+ binding to yeast glyceraldehyde-3-phosphate dehydrogenase (Kirschner *et al.*, 1966) suggested a maximal shift of about 70% toward the R-state at high NAD^+ concentrations, but later work suggests that NAD^+ is quite a "strong agonist" (Kirschner *et al.*, 1971). In the study of Blangy *et al.*, (1968), the binding of a number of ligands to phosphofractokinase showed a high degree of selectivity, except for ATP, which showed no selectivity.

Of course most biochemical experiments concentrate on biologically important ligands, so the paucity of ligands with low selectivity may merely be testimony to the efficiency of nature. But it seems prudent to ask if there may be other explanations for the weakness of weak agonists than that outlined above.

The only other explanation that comes to mind is that weak agonists are weak because of some action on sites other than ligand binding sites. There are several reasons to expect that cationic substances might affect channels other than by action on the agonist binding sites. Local anesthetic cations are thought to block end-plate ion channels from the outside, and various charged substances can block axon sodium and potassium channels from both inside and outside; these have been discussed already (Sec. 1.5). Only rather small organic ions can penetrate *through* axon sodium and potassium channels [see the reviews by Landowne *et al.* (1975) and Armstrong (1975)]. However, there is some reason to believe that much larger compounds can penetrate the ion channels of muscle end plate, despite the fact that their conductance is not much bigger than that of axon ion channels, as shown in Tables II and III [and there are also reasons to believe that the permeability properties for inorganic ions differ from axon ion channels; see Ginsborg (1973) and Rang (1975, pp. 286–295)]. For example, it has been known for some time that decamethonium and carbachol can enter rat diaphragm muscle fibers in the end-plate region (Creese *et al.*, 1963; Creese and Maclagan, 1970, 1976; Creese and England, 1970; Creese *et al.*, 1971). The uptake appears to be saturable and is blocked by a low concentration of tubocurarine. It most probably enters through the ion channels that decamethonium itself opens. Furthermore, several organic cations can substitute for sodium: ammonium, methylammonium, ethylammonium, hydrazinium, and, to a lesser extent, tetramethylammonium, trimethylethylammonium, choline, and dimethyl-diethanolammonium (Furukawa *et al.*, 1956; Furukawa and Furukawa, 1959; Koketsu and Nishi, 1959; Nastuk, 1959). It was found by Colquhoun *et al.* (1975) that two agonists, phenylpropyltrimethylammonium and its *meta* hydroxy derivative, gave, on their own, fluctuation spectra that looked superficially like those produced by other agonists in the presence of local anesthetics, and, as in the latter case (Ruff, 1977), they gave a lower than usual single-channel conductance (Table II). [The reason for this is uncertain; see Colquhoun (1978).] They might well have local anesthetic-like side effects, but this has not yet been rigorously tested. However, there is good evidence (Adams and Sakmann, 1978b) that decamethonium can have a local anesthetic-like effect on frog muscle end plates. It remains to be seen whether a theory of this sort has wide validity and whether it can adequately explain the existing results on the interaction of partial and full agonists [see the review by Colquhoun (1973)].

Desensitization. The most common view of desensitization is that it can be explained by the existence of a particular (nonconducting) conformation of the receptor–channel protein, in addition to the resting and active conformations (Katz and Thesleff, 1957; Rang and Ritter, 1970). This approach fits many of the facts [see the reviews by Magazanik and Vyskočil (1973), Magazanik (1976), and Ginsborg and Jenkinson (1976)]. However, there are still a number of observations that do not seem to fit comfortably with this hypothesis. An indication of this is, perhaps, that even now no one has even attempted to postulate a complete kinetic scheme that includes resting, activated, *and* desensitized receptor conformations and also accounts for cooperativity. Another problem is to account for the wide range of substances that appear to be able to modify desensitization rates [see Magazanik (1976)]. It has been suspected for a long time [see Magazanik (1976)] that factors other than a special receptor conformation might contribute to desensitization (and to the equally mysterious phase II block of transmission that can be produced by drugs such as decamethonium); the most common proposal is that these phenomena may be linked, in some way, with the presence of agonist inside the cell (Creese *et al.*, 1963; Taylor and Nedergaard, 1965; Adams, 1975b). The most specific (but not yet quantitative) hypothesis of this sort has been advanced by Adams (1975b), who suggests that desensitization in frog muscle results from ion channel blockage of the sort discussed above; the slow component of recovery from desensitization, it is proposed, might result from closure of the channel, trapping the drug, or from slow loss of intracellular agonist. A further complication is the apparent involvement of the sodium pump in mammalian muscle (Creese *et al.*, 1976) but not frog muscle (Terrar, 1974).

3.5. Kinetics and Mechanism

It is a truism in the enzyme field that kinetic as well as equilibrium studies are necessary for investigation of mechanism. The relaxation and fluctuation studies reviewed in Sec. 1 showed that most agonists behave in a remarkably simple way. Only one component (one time constant) was usually visible in the measurable range.

3.5.1. What Does the Observation of a Single Time Constant Imply?

The first thing to remember is that observations are only possible over a limited range; in particular, no process faster than, very roughly, 100 μsec can be resolved at the moment. The observation of a single component suggests that the system can exist in two states only, i.e., that other states that may exist either (1) equilibrate rapidly compared with the fastest observations that are technically possible or (2) are present in small quantities only so that they may not be resolvable. (Although, for example, an equilibrium constant of 0.01 may be measurable, a component contributing 1% to a relaxation curve is not likely to be noticed.)

A number of rather general statements can be made about two-state systems (Colquhoun and Hawkes, 1977). For example, if we denote the mean lifetimes of the two states as m_1 and m_2, then the observed time constant τ (or rate constant, $\lambda = 1/\tau$) is related to the lifetimes very simply by

$$\lambda = \frac{1}{\tau} = m_1^{-1} + m_2^{-2} \tag{5}$$

and therefore

$$\tau = m_1 p_2 \tag{6}$$

or, equivalently,

$$\tau = m_2 p_1 \tag{7}$$

where p_1 and p_2 are the equilibrium fraction of the system in state 1 and state 2, respectively.

The main question now becomes, What is the nature of the two states? If binding of the drug is a rapid process, compared with the conformation change, we can say that state 1 = open state and state 2 = shut state (i.e., if there are several rapidly equilibrating open forms, they are all included in state 1, which is linked, by a relatively slow conformation change, to state 2, which consists only of shut channels). In this case the observed time constant is, from Eq. (6),

$$\tau = m_R p_T \tag{8}$$

so measurement of τ gives directly the mean lifetime of the open state m_R as long as only a small fraction of the channels is open (so $p_R \ll 1$ and $p_T = 1 - p_R \simeq 1$). It should, however, be noted that it is the mechanism, and the values of the *rate constants* in the mechanism, that is of primary interest, rather than the mean open lifetime itself.

In the case where state 1 consists only of open forms and state 2 consists only of closed forms, the single-channel conductance γ will be given by the well-known result [see, for example, Stevens (1972) and Colquhoun and Hawkes (1977)]

$$\gamma = \frac{\text{var}(I)}{\mu_1(V - V_{eq}) p_T} \tag{9}$$

where μ_1 is the mean drug-induced current observed to flow at potential V, V_{eq} is the equilibrium potential for the agonist, and var (I) is the variance of the current fluctuations about the mean. Low values for γ would obviously result if some of the variance were at frequencies too high to be registered by the recording apparatus, and this equation would be misleading if state 1 included closed as well as open forms (see below). However, it is valid even when there is more than one open form in state 1 [e.g., singly and doubly occupied forms in scheme (**F**)] as long as all open forms have the same conductance γ. If they have not, a mean conductance would be found that depended on the proportions of the various open forms that were present and hence probably on agonist concentration. So far no concentration dependence of γ, other than that which can be attributed to the decrease of p_T with concentration, has been observed for most agonists (see Sec. 1).

If the binding of agonist were slower than the conformation change, we could still have, effectively, two states linked by a relatively slow binding step. For example, all the forms in state 1 could be occupied by agonist molecules [for example, AT and AR in scheme (**A**)], and all those in state 2, not occupied [e.g., T in scheme (**A**)]. In this case, from the general relationship, Eq. (6), the observed time constant would be $\tau = m_1 p_2$, i.e., the mean lifetime of the *occupied* state multiplied by the fraction of unoccupied channels. But the occupied state is not open all of the time, so the channel conductance calculated from $\gamma = \text{var}(I)/\mu_1(V - V_{eq})$ would give the *mean* conductance of the occupied state (at low concentration so $p_2 \simeq 1$), i.e., the true open-channel conductance multiplied by the fraction of time for which the occupied channel is open; e.g.,

$$\gamma = \gamma_{\text{true}} \frac{\beta_1}{\alpha_1 + \beta_1} \tag{10}$$

for scheme (**A**).

It is worth emphasizing that for the fast binding approximation to hold it is not sufficient for the drug simply to dissociate rapidly [e.g., large β_{10} in scheme (**A**)]. This may be illustrated by a theoretical example [Example 1 in Colquhoun and Hawkes (1977)]. In that example, based on scheme (**A**), the lifetime of the open state, $1/\alpha_1$, was 1 msec, and $1/\beta_{10}$ was 0.1 msec. So if an agonist became bound and then dissociated again with no channel opening, the mean lifetime of the occupancy would be only 0.1 msec. However, if the channel opens, it will stay occupied longer. The other rate constants were such that the mean number of openings per occupancy, β_1/β_{10}, was 1.9. The mean lifetime of the occupied state, $m_{\text{occ}} = (\alpha_1 + \beta_1)/\alpha_1\beta_{10}$, was not 0.1 msec but 2.0 msec. Notice that this depends on the length of the mean *open* lifetime. It can be written as follows: m_{occ} = mean life of occupancy if there are no openings + (mean number of openings per occupancy × mean open lifetime). In this example, despite the fact that it is intermediate between the fast binding and fast conformation change extremes, only a single component is visible with a lowish agonist concentration. It has a time constant of about 2.8 msec, much more than the mean open lifetime. This value is accounted for by the fact that *if* a channel opens at all during an occupancy, it opens, on average, $1 + \beta_1/\beta_{10} = 2.9$ times in fairly rapid succession before the agonist dissociates again, each opening, on average, being 1 msec long. This example illustrates some of the pitfalls that might potentially be encountered in interpreting kinetic experiments. It emphasizes the relief that greeted the work (see Sec. 1.4.3) of Neher and Sakmann (1976a) which showed that multiple openings could not be seen at low concentrations of several agonists (though fast flickering between open and shut would not have been resolved) and that the directly observed open-channel lifetime and channel conductance agreed well with the values inferred from fluctuation measurements.

3.5.2. What Is the Rate-Limiting Step?

There is no conclusive, or even persuasive, evidence on the problem of whether the observed rates are controlled by the rate of drug binding or by the rate of conformation change (opening and shutting). Both propositions fit the available facts, though if binding were rate limiting it might be expected [from Eq. (10)] that the single-channel conductance would appear to vary from one agonist to another, whereas in fact it varies very little (see Table II). In most work so far, the working hypothesis has been adopted that a conformation change step (shut $\leftrightharpoons$ open) is rate limiting and that drug binding is relatively rapid in skeletal muscle [e.g., Magleby and Stevens (1972b), Katz and Miledi (1972), Anderson and Stevens (1973), and Dionne and Stevens (1975)], though Sheridan and Lester (1977) have suggested that binding may be rate limiting in eel electroplaque. The rapid-binding hypothesis has been based primarily on the putative analogy between the receptor mechanism and enzyme reactions; in the latter case binding has usually been found to be faster than conformation change [see Janin (1973) and the references in Dionne and Stevens (1975)]. On the other hand, Adams

(1977a) has advanced some indirect arguments in favor of a relatively slow binding step. It is clear that studies at low concentrations, far from saturation, cannot distinguish the two possibilities. There are great technical difficulties with electrophysiological studies at higher concentrations (see Sec. 1), but it seems likely that such studies will give essential clues. In particular, it is of great interest to know how the time constant(s) from kinetic studies depend on concentration, so the light that such measurements can cast on this problem, and on others, will be discussed next.

3.5.3. Concentration Dependence of Time Constants from Kinetic Studies

Most studies show only one time constant (see above and Sec. 1). It will be convenient to define state 1 as that favored by the agonist which contains all (or most) of the open states (and possibly some shut ones too) and state 2 as that containing only (or almost only) shut states.

According to Magleby and Stevens (1972b), the ACh concentration is negligible during most of the decay phase of the MEPC, so the decay phase is effectively relaxation with zero agonist concentration. The fact that the time constant τ for this process is close to that found in most fluctuation and voltage-jump studies (see Sec. 1) at the muscle end plate shows that the constant ACh concentration used in these studies is low in the sense that p_1 is negligible, so $p_2 \simeq 1$ and Eq. (6) gives $\tau \simeq m_1$, the mean lifetime of state 1. This seems to be at least approximately true for the false transmitter acetylmonoethylcholine as well (Colquhoun et al., 1977). [Those who find it a little odd that a relaxation study should give a mean open lifetime, on the grounds that channels may have been open some time before the timing is started, will find an explanation in Colquhoun (1970, 1971, pp. 84, 380).]

When higher agonist concentrations are used, τ becomes smaller, i.e., faster, than its zero-concentration value (see Sec. 1.4.4). Now if there are really only two kinetically distinguishable states, $\tau = m_1 p_2$, from Eq. (6), and p_2 must obviously decrease with concentration as more of the system changes to the conducting state. The observed decrease of τ with concentration may or may not be entirely accounted for by this factor; it is not yet known whether the mean lifetime of state 1 is concentration dependent.

If state 1 consisted only of open states (conformation change rate limiting) and m_1 were independent of concentration, it would be expected that the relation between τ and concentration would soon plateau for partial agonists if they work in the commonly supposed way; e.g., if a partial agonist could open only, at most, 5% of the channels, then p_2, and hence τ, would fall with concentration from $1.0m_R$ to $0.95m_R$. No such plateau has been observed so far (see Sec. 1). It could be, of course, that m_R is not independent of concentration (Table V shows that in general this is the case). For example, scheme (**B**) predicts (see Table V) an increase in τ with concentration, contrary to observation, when the binding step is rapid. It is known that this does happen for NAD^+ binding to yeast GPDH (Kirschner et al., 1971). In this system, quite unlike ACh receptors, the time constant related to the rate-limiting conformation change *increases* steeply with ligand concentration despite the decrease of p_2 ($= p_T$), because m_1 ($= m_R$), the mean lifetime of the R-conformation, increases with concentration.

This system is described by MWC theory [scheme (I)], so α_i/β_i must decrease with concentration. If all the α_i were constant (α say), i.e., $\beta_i = \alpha/L_0 M^i$, then m_R (Table V) reduces to $1/\alpha$, a constant, independent of concentration, i.e., independent of the number of ligands bound. In this case τ would decrease with concentration solely because of the decrease of p_T. Notice also that if all the α_i were the same, but not otherwise, it would make no difference how fast binding and dissociation to the R-conformation took place or whether it took place at all [schemes (D), (E), and (F)]; if binding were fast to the T-conformation, only one time constant would be observed as long as all open forms had the same conductance.

In the case of NAD$^+$, however, the evidence favored the opposite extreme case: All the β_i were roughly constant (β say), so $\alpha_i = \beta L_0 M^i$, and in this case m_R increases with concentration (the mean life of $A_i R_n$ being $1/\alpha_i$, which increases with the number of ligands, i); this is the reason the observed $\tau = m_R p_T$ increases with concentration despite the decrease in p_T. Indeed, in this case the mean lifetime of the *shut* state, m_T, is constant (viz. $1/\beta$), so, from Eq. (7), it is more helpful to write $\tau = m_T p_R$—the time constant increases in proportion to the fraction of open channels.

Insofar as the observed τ for ACh-like drugs continues to decrease with concentration, we can say that *either* (1) the agonist is not a partial agonist in the usual sense (see Sec. 3.4), *or* (2) m_R decreases with concentration throughout the range tested (in the MWC case, but not necessarily in general, α_i/β_i decreases with the number of ligands, i, so α_i *could* increase, unlikely as it may seem), *or* (3) the approximation that binding is fast relative to the conformation change is not good enough (e.g., as in the theoretical example in Sec. 3.5.1), *or* (4) states 1 and 2 are not as specified above at all.

What is expected if states 1 and 2 are not simply open and shut, respectively? For example, suppose an agonist binding step were rate limiting and the conformation change rapid. In most cases this would result in τ falling toward zero as concentration increases, without reaching a plateau, both for full and for "partial" (in the sense used above) agonists. Simple *general* predictions are difficult. One simple case is scheme (D), extended so that n agonist molecules must be bound before opening is possible. If the ith binding step were rate limiting (so state $2 = T_n, AT_n, ..., A_{i-1}T_n$), then m_1 would increase with the concentration x (proportional to x^{n-i} at high concentration), whereas p_2 would, as always, fall with concentration (proportional to $1/x^{n-i+1}$ at high concentration). Thus the time constant τ will decrease, and $\lambda = 1/\tau$ will increase, though not necessarily monotonically, with increasing concentration. And at high concentrations λ will increase linearly with concentration *whichever* binding step is rate limiting, the slope being the rate constant (M^{-1} sec^{-1}) for transition from state 2 to state 1. At zero concentration τ is the mean lifetime of state 1, as expected. Adams (1977a) suggests that the most attractive mechanism is scheme (E) with a *rapid* conformation change and $L_2 \ll 1$ (i.e., full agonist). In this case there is effectively only one time constant at *low* concentrations, $2\beta_{10}L_2$ (the other, β_{10}, being much faster), but this would not be so for a partial agonist (L_2 not very small), and furthermore, since state 1 is open only for a fraction $1/(1 + L_2)$ of the time, partial agonists should give a low estimate of mean channel conductance (see Sec. 3.5.1), which has so far not been seen (see Sec. 1); most agonists have very similar channel conductances. As Adams points out, there are two possible ways around these objections: (1) partial agonist current fluctuations could arise directly from opening and shutting during a long occupancy, the binding process being too slow to be seen; (2) partial agonists are actually full agonists in the usual sense ($L_2 \ll 1$), and there is some other explanation for their weakness (see Sec. 3.4 for further discussion).

3.6. What Is the Origin of Voltage Dependence?

It was pointed out in Sec. 1 that both the equilibrium response and the time constant τ from kinetic studies have very similar potential dependence; both increase roughly e-fold for each 100 mV of hyperpolarization. The usual interpretation of this observation is that the channel closing rate [α in scheme (A) or its equivalent in more complex schemes], which is supposed to be rate limiting, is also voltage dependent because of a change of dipole moment associated with the conformation change, whereas β has little or no voltage dependence (Magleby and Stevens, 1972b; Dionne and Stevens, 1975; Neher and Sakmann, 1975; Adams, 1976b). However, all workers agree that the evidence is inconclusive. Adams (1976b) has suggested that the binding step is voltage dependent as well as rate limiting. But Adams (1977a) points out that it is unlikely, in simple physical systems, for α to be voltage dependent while the rate constant for the reverse reaction, β, is not, so he therefore favored a scheme (E) with rate-limiting binding but voltage-dependent conformation change. However, in a recent abstract (Adams and Sakmann, 1976), it has been found that partial agonists (choline and decamethonium) lose the voltage dependence of the equilibrium response when applied in high concentrations; this is the predicted result for voltage-dependent binding but not for voltage-dependent conformation change. [In schemes such as (G), (H), and (I) this is true only if K_R and K_T have the same voltage dependence, so that efficacy is not itself voltage dependent.]

One factor that complicates this problem is that binding and conformation change are not independent in any of the schemes in Table V; a change in the *conformational* equilibrium constant(s), e.g., L_0, toward the R-conformation will usually result in an increase in *binding* (and, consequently, a slowing of diffusion in the synaptic cleft). For example, K_{eff} decreases, and hence binding increases, when either L_1 or K_T is reduced [Table V, scheme (A)]. This sort of phenomenon is presumed to underlie the initially unexpected lack of effect of anticholinesterase treatment on the voltage dependence of MEPC decay rate (Magleby and Stevens, 1972a; Gage and McBurney, 1975; Gage, 1976; Kordaš, 1977) and on the relative decay rates of normal and false MEPCs (Colquhoun *et al.*, 1977).

3.7. High Affinity Versus High Speed

It has been pointed out a number of times that the high affinities ($K_{eff} = $ 4–30 nM) reported for binding of ACh in some preparations (see Sec. 2, Table IVa) seem incompatible with the speed of synaptic action [e.g., Rang (1975, p. 355)]. For example, an equilibrium constant of 10 nM implies, even if the association rate constant were near the theoretical maximum [see Gutfreund (1972, p. 158)], that the dissociation rate constant would have to be very slow, 20 sec^{-1} at the most. A number of explanations for this have been suggested, not all of them valid. It should be noted that none of the schemes in Table V predict that high- and low-affinity sites (K_R and K_T) will be separately observable in equilibrium experiments. Furthermore, the concentrations needed for (1) 50% of maximal binding and (2) 50% of maximal conformational change never differ from each other by a large factor for a wide range of parameter values [for schemes (A) and (H) they are identical, viz. K_{eff}], though they both may be widely different from either K_R or K_T. It has been suggested (Meunier and Changeux, 1973; Gibson, 1976) that the receptor, in detergent solution, may adopt a conformation like the open (R) form, which has a high affinity for agonist, and Gibson's results could be fitted by changing only L_0 in scheme (I). This could certainly explain the observations, though we would be lucky if this were really the only change following removal of receptor from its membrane environment. There is, of course, no objection to K_R being very small (as in Gibson's study); in many simple schemes, such as (A), (D), (E), and (F), it is supposed to be zero. But most studies do not measure K_R but rather K_{eff}, the concentration for 50% of maximum binding.

References

Adams, P. R., 1975a, Kinetics of agonist conductance changes during hyperpolarization at frog endplate, *Br. J. Pharmacol.* **53**:308.

Adams, P. R., 1975b, A study of desensitization using voltage clamp, *Pflügers Arch.* **360**:135.

Adams, P. R., 1975c, An analysis of the dose–response curve at voltage-clamped frog endplate, *Pflügers Arch.* **360**:145.

Adams, P. R., 1975d, Drug interactions at the motor endplate, *Pflügers Arch.* **360**:155.

Adams, P. R., 1976a, A comparison of the time course of excitation and inhibition by iontophoretic decamethonium in frog endplate, *Br. J. Pharmacol.* **57**:59.

Adams, P. R., 1976b, Voltage dependence of agonist responses at voltage-clamped frog endplates, *Pflügers Arch.* **361**:145.

Adams, P. R., 1976c, Drug blockade of open endplate channels, *J. Physiol.* **260**:531.

Adams, P. R., 1977a, Relaxation experiments using bath-applied suberyldicholine, *J. Physiol. (London)* **268**:271.

Adams, P. R., 1977b, Voltage jump analysis of procaine action at frog endplate, *J. Physiol. (London)* **268**:291.

Adams, P. R., and Sakmann, B., 1976, Frog endplate current–voltage relations during bath application of full and partial agonists, *Pflügers Arch.* **365**:R146.

Adams, P. R., and Sakmann, B. 1978a, A comparison of current–voltage relations for full and partial agonists, *J. Physiol. (London)*, in press.

Adams, P. R., and Sakmann, B., 1978b, Decamethonium both opens and blocks endplate channels, *Proc. Natl. Acad. Sci. USA* **75**:2994.

Adams, P. R., and Sakmann, B., 1978c, Agonist-triggered endplate channel opening, *Biophys. J.* **21**:53a.

Albuquerque, E. X., Barnard, E. A., Chiu, T. H., Lapa, A. J., Dolly, J. O., Jansson, S., Daly, J., and Witkop, B., 1973, Acetylcholine receptor and ion conductance modulator sites at the murine neuromuscular junction: Evidence from specific toxin reactions, *Proc. Natl. Acad. Sci. USA* **70**:949.

Almers, W., and Levinson, S. R., 1975, Tetrodotoxin binding to normal and depolarized frog muscle and the conductance of a single sodium channel, *J. Physiol. (London)* **247**:482.

Anderson, C. R., and Stevens, C. F., 1973, Voltage clamp analysis of acetylcholine produced endplate current fluctuations at frog neuromuscular junction, *J. Physiol. (London)* **235**:655.

Armstrong, C. M., 1971, Interaction of tetraethylammonium ion derivatives with the potassium channels of giant axons, *J. Gen. Physiol.* **58**:413.

Armstrong, C. M., 1975, Ionic pores, gates and gating currents, *Q. Rev. Biophys.* **7**:179.

Ascher, P., Marty, A., and Neild, T. O., 1978, The mode of action of antagonists of the excitatory response to acetylcholine in *Aplysia* neurones, *J. Physiol. (London)* **278**:207.

Barlow, R. B., Scott, N. C., and Stephenson, R. P., 1967, The affinity and efficacy of onium salts on the frog rectus abdominis, *Br. J. Pharmacol.* **31**:188.

Barnard, E. A., Coates, V., Dolly, J. O., and Mallick, B., 1977, Binding of α-bungarotoxin and cholinergic ligands to acetylcholine receptors in the membrane of skeletal muscle, *Cell Biol. Int. Rep.* **1**:99.

Bartels, E., Wasserman, N. H., and Erlanger, B. F., 1971, Photochromic activators of the acetylcholine receptor, *Proc. Natl. Acad. Sci. USA* **68**:1820.

Beam, K. G., 1976, A quantitative description of end-plate currents in the presence of two lidocaine derivatives, *J. Physiol. (London)* **258**:301.

Begenisich, T., and Stevens, C. F., 1975, How many conductance states do potassium channels have?, *Biophys. J.* **15**:843.

Benzer, T. I., and Raftery, M. A., 1972, Partial characterization of a tetrodotoxin binding component from nerve membrane, *Proc. Natl. Acad. Sci. USA* **69**:3634.

Beranek, R., and Vyskočil, F., 1967, The action of tubocurarine and atropine on the normal and denervated rat diaphragm, *J. Physiol. (London)* **188**:53.

Berg, D. K., Kelly, R. B., Sargent, P. B., Williamson, P., and Hall, Z. W., 1972, Binding of α-bungarotoxin to acetylcholine receptors in mammalian muscle, *Proc. Natl. Acad. Sci. USA* **69**:147.

Blackman, J. G., 1959, The pharmacology of depressor bases, Ph.D. thesis, University of New Zealand.

Blackman, J. G., 1970, Dependence on membrane potential of the blocking action of hexamethonium at a sympathetic ganglion synapse, *Proc. Univ. Otago Med. Sch.* **48**:4.

Blackman, J. G., and Purves, R. D., 1968, Ganglionic transmission in the autonomic nervous system, *NZ Med. J.* **67**:376.

Blackman, J. G., Gauldie, R. W., and Milne, R. J., 1975, Interaction of competitive antagonists: The anti-curare action of hexamethonium and other antagonists at the skeletal neuromuscular junction, *Br. J. Pharmacol.* **54**:91.

Blangy, D., Buc, H., and Monod, J., 1968, Kinetics of the allosteric interactions of phosphofructokinase from *Escherichia coli, J. Mol. Biol.* **31**:13.

Bolton, T. B., 1976, On the latency and form of the membrane responses of smooth muscle to the iontophoretic application of acetylcholine or carbachol, *Proc. Roy. Soc. London Ser. B* **194**:99.

Bolton, T. B., 1977, Rate of offset of action of slow-acting muscarinic antagonists is fast, *Nature (London)* **270**:354.

Brockes, J. P., and Hall, Z. W., 1975, Acetylcholine receptors in normal and denervated rat diaphragm muscle. II. Comparison of junctional and extrajunctional receptors, *Biochemistry* **14**:2100.

Brocklehurst, W. E., and Colquhoun, D., 1965, Adsorption and diffusion of γ-globulin during passive sensitization of chopped guinea-pig lung, *J. Physiol. (London)* **181**:760.

Bulger, J. E., and Hess, G. P., 1973, Evidence for separate initiation and inhibitory sites in the regulation of membrane potential of electroplax. I. Kinetic studies with α-bungarotoxin, *Biochem. Biophys. Res. Commun.* **54**:677.

Cartaud, J., Benedetti, E. C., Sobel, A., and Changeux, J. P., 1978, A morphological study of the cholinergic receptor protein from *Torpedo marmorata* in its membrane environment and in its detergent-extracted purified form, *J. Cell. Sci.* **29**:313.

Castillo, J. del, and Katz, B., 1957a, A study of curare action with an electrical micro-method, *Proc. Roy. Soc. London Ser. B* **146**:339.

Castillo, J. del, and Katz, B., 1957b, Interaction at end-plate receptors between different choline derivatives, *Proc. Roy. Soc. London Ser. B* **146**:369.

Catterall, W. A., 1975, Sodium transport by the acetylcholine receptor of cultured muscle cells, *J. Biol. Chem.* **250**:1776.

Ceccarelli, D. J., Hurlbut, W. P., and Mauro, A., 1973, Turnover of transmitter and synaptic vesicles at the frog neuromuscular junction, *J. Cell. Biol.* **57**:499.

Chang, H. W., and Neumann, E., 1976, Dynamic properties of isolated acetylcholine receptor proteins: Release of calcium ions caused by acetylcholine binding, *Proc. Natl. Acad. Sci. USA* **73**:3364.

Chang, R. S. L., Potter, L. T., and Smith, D. S., 1977, Postsynaptic membranes in the electric tissue of *Narcine:* IV. Isolation and characterization of the nicotinic receptor protein, *Tissue Cell* **9**:623.

Changeux, J. P., 1975, The cholinergic receptor protein from fish electric organ, in: *Handbook of Psychopharmacology*, Vol. 6 (L. L. Iversen, S. D. Iversen, and S. H. Snyder, eds.), pp. 235–301, Plenum Press, New York.

Changeux, J. P., and Podleski, T. R., 1968, On the excitability and cooperativity of the electroplax membrane, *Proc. Natl. Acad. Sci. USA* **59**:944.

Changeux, J. P., Thiery, J., Tung, Y., and Kittel, C., 1967, On the cooperativity of biological membranes, *Proc. Natl. Acad. Sci. USA* **57**:335.

Chiu, T. H., Lapa, A. J., Barnard, E. A., and Albuquerque, E. X., 1974, Binding of *d*-tubocurarine and α-bungarotoxin in normal and denervated mouse muscles, *Exp. Neurol.* **43**:399.

Clark, A. J., 1933, *The Mode of Action of Drugs on Cells*, Edward Arnold, London; Williams & Wilkins, Baltimore.

Colquhoun, D., 1970, How long does a molecule stay on the receptor? Explanation of a paradox, *Br. J. Pharmacol.* **39**:215P.

Colquhoun, D., 1971, *Lectures on Biostatistics*, Clarendon Press, London.

Colquhoun, D., 1973, The relation between classical and cooperative models for drug action, in: *Drug Receptors* (H. P. Rang, ed.), pp. 149–182, Macmillan, New York.

Colquhoun, D., 1975a, Mechanisms of drug action at the voluntary muscle endplate, *Annu. Rev. Pharmacol.* **15**:307.

Colquhoun, D., 1975b, Receptor interaction and diffusion as factors limiting the rate of drug action, in: *Pharmacokinetics*, Symposium Proceedings No. 7, pp. 85–115, Institute of Mathematics and Its Applications, London.

Colquhoun, D., 1978, Noise: a tool for drug receptor investigation, in: *Cell Membrane Receptors for Drugs and Hormones* (L. Bolis and R. W. Straub, eds.), Raven Press, New York.

Colquhoun, D., and Hawkes, A. G., 1977, Relaxation and fluctuations of membrane currents that flow through drug-operated ion channels, *Proc. Roy. Soc. London Ser. B* **199**:231.

Colquhoun, D., and Orentlicher, M., 1974, Diffusion with binding in a cylinder, *J. Gen. Physiol.* **63**:182.

Colquhoun, D., and Rang, H. P., 1976, Effects of inhibitors on the binding of iodinated α-bungarotoxin to acetylcholine receptors in rat muscle, *Mol. Pharmacol.* **12**:519.

Colquhoun, D., and Ritchie, J. M., 1972, The kinetics of the interaction between tetrodotoxin and mammalian nonmyelinated nerve fibres, *Mol. Pharmacol.* **8**:285.

Colquhoun, D., Henderson, R., and Ritchie, J. M., 1972, The binding of labelled tetrodotoxin to non-myelinated nerve fibres, *J. Physiol. (London)* **227**:95.

Colquhoun, D., Rang, H. P., and Ritchie, J. M., 1974, The binding of tetrodotoxin and α-bungarotoxin to normal and denervated mammalian muscle, *J. Physiol. (London)* **240**:199.

Colquhoun, D., Dionne, V. E., Steinbach, J. H., and Stevens, C. F., 1975, Conductance of channels opened by acetylcholine-like drugs in muscle end-plate, *Nature (London)* **253**:204.

Colquhoun, D., Large, W. A., and Rang, H. P., 1977, An analysis of the action of a false transmitter at the neuromuscular junction, *J. Physiol. (London)* **266**:361.

Colquhoun, D., Dreyer, F., and Sheridan, R. E., 1978, The action of tubocurarine at the neuromuscular junction, *J. Physiol. (London)*, in press.

Conti, F., DeFelice, L. J., and Wanke, E., 1975, Potassium and sodium ion current noise in the membrane of the squid giant axon, *J. Physiol. (London)* **248**:45.

Conti, F., Hille, B., Neumcke, B., Nonner, W., and Stämpfli, R., 1976, Measurement of the conductance of the sodium channel from current fluctuations at the node of Ranvier, *J. Physiol. (London)* **262**:699.

Creese, R., and England, J. M., 1970, Decamethonium in depolarized muscle and the effects of tubocurarine, *J. Physiol. (London)* **210**:345.

Creese, R., and Maclagan, J., 1970, Entry of decamethonium in rat muscle studied by autoradiography. *J. Physiol. (London)* **210**:363.

Creese, R., and Maclagan, M., 1976, Labelled decamethonium in cat muscle, *Br. J. Pharmacol.* **58**:141.

Creese, R., Taylor, D. B., and Tilton, B., 1963, The influence of curare on the uptake and release of a neuromuscular blocking agent labelled with radioactive iodine, *J. Pharmacol. Exp. Ther.* **139**:8.

Creese, R., Taylor, D. B., and Case, R., 1971, Labelled decamethonium in denervated skeletal muscle, *J. Pharmacol. Exp. Ther.* **176**:418.

Creese, R., Franklin, G. I., and Mitchell, L. D., 1976, Two mechanisms for spontaneous recovery from depolarising drugs in rat muscle, *Nature (London)* **261**:416.

Dionne, V. E., 1976, Characterization of drug iontophoresis with a fast microassay technique, *Biophys. J.* **16**:705.

Dionne, V. E., and Stevens, C. F., 1975, Voltage dependence of agonist effectiveness at the frog neuro-muscular junction: Resolution of a paradox, *J. Physiol. (London)* **251**:245.

Dionne, V. E., Steinbach, J. H., and Stevens, C. F., 1975, Dose–response relationship at the frog neuromuscular junction, *Biophys. J.* **15**:266a.

Dionne, V. E., Steinbach, J. H., and Stevens, C. F., 1978, An analysis of the dose–response relation-ship at voltage-clamped frog neuromuscular junctions, *J. Physiol. (London)* **281**:421.

Dolly, J. O., and Barnard, E. A., 1974, Affinity of cholinergic ligands for the partially purified acetyl-choline receptor from mammalian skeletal muscle, *FEBS Lett.* **46**:145.

Dolly, J. O., Albuquerque, E. X., Sarvry, J. M., Mallick, B., and Barnard, E. A., 1977, Binding of perhydro-histrionicotoxin to the postsynaptic membrane of skeletal muscle in relation to its blockade of acetylcholine-induced depolarization, *Mol. Pharmacol.* **13**:1.

Dreyer, F., and Peper, K., 1975, Density and dose–response curve of acetylcholine receptors in frog neuromuscular junction, *Nature (London)* **253**:641.

Dreyer, F., Müller, K. D., Peper, K., and Sterz, R., 1976a, The *M. omohyoideus* of the mouse as a convenient mammalian muscle preparation, *Pflügers Arch.* **367**:115.

Dreyer, F., Walther, C., and Peper, K., 1976b, Junctional and extrajunctional acetylcholine receptors in normal and denervated frog muscle fibres: Noise analysis experiments with different agonists, *Pflügers Arch.* **366**:1.

Dreyer, F., Peper, K., and Sterz, R., 1978, Determination of dose–response curves by quantitative ionophoresis at the frog neuromuscular junction, *J. Physiol. (London)* **281**:395.

Eckfeldt, J., Hammes, G. G., Mohr, S. C., and Wu, C. W., 1970, Relaxation spectra of aspartate transcarbamylase. I. Interaction of 5-bromocytidine triphosphate with native enzyme and regulatory subunit, *Biochemistry,* **9**:3353.

Edelstein, S., 1972, An allosteric mechanism for the acetylcholine receptor, *Biochem. Biophys. Res. Commun.* **48**:1160.

Eldefrawi, M. E., and Eldefrawi, A. T., 1973a, Cooperativities in the binding of acetylcholine to its receptor, *Biochem. Pharmacol.* **22**:3145.

Eldefrawi, M. E., and Eldefrawi, A. T., 1973b, Purification and molecular properties of the acetyl-choline receptor from *Torpedo* electroplax, *Arch. Biochem. Biophys.* **159**:362.

Eldefrawi, M. E., Britten, A. G., and Eldefrawi, A. T., 1971, Acetylcholine binding to *Torpedo* electroplax: Relationship to acetylcholine receptors, *Science* **173**:338.

Fatt, P., 1950, The electromotive action of acetylcholine at the motor end-plate, *J. Physiol. (London)* **111**:408.

Fatt, P., and Katz, B., 1951, An analysis of the end-plate potential recorded with an intracellular electrode, *J. Physiol. (London)* **115**:320.

Ferry, C. B., and Marshall, A. R., 1973, An anti-curare effect of hexamethonium at the mammalian neuromuscular junction, *Br. J. Pharmacol.* **47**:353.

Frazier, D. T., Murayama, K., Abbott, N. J., and Narahashi, T., 1975, Comparison of the action of different barbiturates on squid axon membranes, *Eur. J. Pharmacol.* **32**:102.

Fu, J. L., Donner, D. B., and Hess, G. P., 1974, Half-of-the-sites reactivity of the membrane-bound *Electrophorus electricus* acetylcholine receptor, *Biochem. Biophys. Res. Commun.* **60**:1072.

Furukawa, T., and Furukawa, A., 1959, Effects of methyl- and ethyl-derivatives of NH_4^+ on the neuro-muscular junction, *Jpn. J. Physiol.* **9**:130.

Furukawa, T., Takagi, T., and Sugihara, T., 1956, Depolarization of end-plates by acetylcholine externally applied, *Jpn. J. Physiol.* **6**:98.

Gage, P. W., 1976, Generation of end-plate potentials, *Physiol. Rev.* **56**:177.

Gage, P. W., and Hamill, O. P., 1976, Effects of several inhalation anaesthetics on the kinetics of postsynaptic conductance changes in mouse diaphragm, *Br. J. Pharmacol.* **57**:263.

Gage, P. W., and McBurney, R. N., 1975, Effect of membrane potential, temperature and neostigmine on the conductance change caused by a quantum of acetylcholine at the toad neuromuscular junction, *J. Physiol. (London)* **244**:385.

Gage, P. W., McBurney, R. N., and van Helden, D., 1974, End-plate currents are shortened by octanol: Possible role of membrane lipid, *Life Sci.* **14**:2277.

Gage, P. W., McBurney, R. N., and Schneider, G. T., 1975, Effects of some aliphatic alcohols on the conductance change caused by a quantum of acetylcholine at the toad end-plate, *J. Physiol. (London)* 244:409.

Gibson, R. E., 1976, Ligand interactions with the acetylcholine receptor from *Torpedo californica*. Extensions of the allosteric model for cooperativity to half-of-site activity, *Biochemistry* **15**:3890.

Ginsborg, B. L., 1973, Electrical changes in the membrane in junctional transmission, *Biochim. Biophys. Acta* **300**:289.

Ginsborg, B. L., and Jenkinson, D. H., 1976, Transmission of impulses from nerve to muscle, in: *Handbuch der Experimentellen Pharmakologie,* Vol. 42 (E. Zaimis, ed.), pp. 229–364, Springer, Berlin.

Ginsborg, B. L., and Stephenson, R. P., 1974, On the simultaneous action of two competitive antagonists, *Br. J. Pharmacol.* **51**:287.

Grünhagen, H.-H. and Changeux, J. P., 1976, Studies on the electrogenic action of acetylcholine with *Torpedo marmorata* electric organ IV, *J. Mol. Biol.* **106**:497.

Gutfreund, H., 1972, *Enzymes: Physical Principles,* Wiley, New York.

Harrington, L., 1973, A linear dose–response curve at the motor end-plate, *J. Gen. Physiol.* **62**:58.

Hartzell, H. C., Kuffler, S. W., and Yoshikami, D., 1975, Postsynaptic potentiation: Interaction between quanta of acetylcholine at the skeletal neuromuscular synapse, *J. Physiol. (London)* **251**:427.

Henderson, R., Ritchie, J. M., and Strichartz, G. R., 1973, The binding of labelled saxitoxin to the sodium channels in nerve membranes, *J. Physiol. (London)* **235**:783.

Hill, A. V., 1909, The mode of action of nicotine and curari, determined by the form of the contraction curve and the method of temperature coefficients, *J. Physiol. (London)* **39**:361.

Hille, B., 1970, Ionic channels in nerve membranes, *Prog. Biophys. Mol. Biol.* **21**:1.

Hille, B., 1977a, The pH-dependent rate of action of local anesthetics on the node of Ranvier, *J. Gen. Physiol.* **69**:475.

Hille, B., 1977b, Local anesthetics: Hydrophilic and hydrophobic pathways for the drug–receptor interaction, *J. Gen. Physiol.* **69**:497.

Hladky, S. B., and Haydon, D. A., 1970, Discreteness of conductance change in bimolecular lipid membranes in the presence of certain antibiotics, *Nature (London)* **225**:451.

Hodgkin, A. L., and Huxley, A. F., 1952, A quantitative description of membrane current and its application to conduction and excitation in nerve, *J. Physiol. (London)* **117**:500.

Janin, J., 1973, The study of allosteric proteins, *Prog. Biophys.* **27**:79.

Jenkinson, D. H., 1960, The antagonism between tubocurarine and substances which depolarize the motor end-plate, *J. Physiol. (London)* **152**:309.

Jenkinson, D. H., and Terrar, D. A., 1973, Influence of chloride ions on changes in membrane potential during prolonged application of carbachol to frog skeletal muscle. *Br. J. Pharmacol.* **47**:363.

Karlin, A., 1967, On the application of "a plausible model" of allosteric proteins to the receptor for acetylcholine, *J. Theor. Biol.* **16**:306.

Kasai, M., and Changeux, J. P., 1971a, *In vitro* excitation of purified membrane fragments by cholinergic agonists, I, *J. Membr. Biol.* **6**:1.

Kasai, M., and Changeux, J. P., 1971b, *In vitro* excitation of purified membrane fragments by cholinergic agonists, III, *J. Membr. Biol.* **6**:58.

Katz, B., and Miledi, R., 1970, Membrane noise produced by acetylcholine, *Nature (London)* **226**:962.

Katz, B., and Miledi, R., 1972, The statistical nature of the acetylcholine potential and its molecular components, *J. Physiol. (London)* **224**:665.

Katz, B., and Miledi, R., 1973a, The characteristics of "end-plate noise" produced by different depolarizing drugs, *J. Physiol. (London)* **230**:707.

Katz, B., and Miledi, R., 1973b, The binding of acetylcholine to receptors and its removal from the synaptic cleft, *J. Physiol. (London)* **231**:549.

Katz, B., and Miledi, R., 1973c, The effect of atropine on acetylcholine action at the neuromuscular junction, *Proc. Roy. Soc. London Ser. B* **184**:221.

Katz, B., and Miledi, R., 1975, The effect of procaine on the action of acetylcholine at the neuromuscular junction, *J. Physiol. (London)* **249**:269.

Katz, B., and Miledi, R., 1977, Transmitter leakage from motor nerve endings, *Proc. Roy. Soc. London Ser. B* **196**:59.

Katz, B., and Thesleff, S., 1957, A study of the "desensitization" produced by acetylcholine at the motor end-plate, *J. Physiol. (London)* **138**:63.

Keynes, R. D., and Rojas, E., 1974, Kinetics and steady-state properties of the charged system controlling sodium conductance in the squid giant axon, *J. Physiol. (London)* **239**:393.

Keynes, R. D., Ritchie, J. M., and Rojas, E., 1971, The binding of tetrodotoxin to nerve membranes, *J. Physiol. (London)* **213**:235.

Kirschner, K., Eigen, M., Bittman, R., and Voigt, B., 1966, The binding of nicotinamide-adenine dinucleotide to yeast D-glyceraldehyde-3-phosphate dehydrogenase: Temperature jump relaxation studies on the mechanism of an allosteric enzyme, *Proc. Natl. Acad. Sci. USA* **56**:1661.

Kirschner, K., Gallego, E., Schuster, I., and Goodall, D., 1971, Cooperative binding of nicotinamide-adenine dinucleotide to yeast glyceraldehyde-3-phosphate dehydrogenase I. Equilibrium and temperature jump studies at pH 8.5 and 40°C, *J. Mol. Biol.* **58**:29.

Koketsu, K., and Nishi, S., 1959, Restoration of neuromuscular transmission in sodium-free hydrazinium solution, *J. Physiol. (London)* **147**:239.

Kordaš, M., 1969, The effect of membrane polarization on the time course of the end-plate current in frog sartorius muscle, *J. Physiol. (London)* **204**:493.

Kordaš, M., 1972, An attempt at an analysis of the factors determining the time course of the end-plate current, *J. Physiol. (London)* **224**:333.

Kordaš, M., 1977, On the role of junctional cholinesterase in determining the time course of the end-plate current, with an appendix by P. Jakopin and M. Kordaš, *J. Physiol. (London)* **270**:133.

Kruckenberg, P., and Bauer, H., 1971, Die Dissoziationkonstante zwischen Curare and dem Acetylcholin-Receptor, *Pflügers Arch.* **326**:184.

Land, B. R., Podleski, T. R., Salpeter, E. E., and Salpeter, M. M., 1977, Acetylcholine receptor distribution on myotubes in culture correlated to acetylcholine sensitivity, *J. Physiol. (London)* **269**:155.

Landowne, D., Potter, L. T., and Terrar, D. A., 1975, Structure–function relationships in excitable membranes, *Annu. Rev. Physiol.* **37**:485.

Langmuir, I., 1918, The adsorption of gases on plane surfaces of glass, mica and platinum, *J. Am. Chem. Soc.* **40**:1361.

Larmie, E. T., and Webb, G. D., 1973, Desensitization in the electroplax, *J. Gen. Physiol.* **61**:263.

Lass, Y., and Fischbach, G. D., 1976, A discontinuous relationship between the acetylcholine-activated channel conductance and temperature, *Nature (London)* **263**:150.

Lee, C. Y., 1972, Chemistry and pharmacology of polypeptide toxins in snake venoms, *Annu. Rev. Pharmacol.* **12**:265.

Lester, H., 1972, Vulnerability of desensitized or curare-treated acetylcholine receptors to irreversible blockade by cobra toxin, *Mol. Pharmacol.* **8**:632.

Lester, H. A., and Chang, H. W., 1977, Response of acetylcholine receptors to rapid photochemically produced increases in agonist concentration, *Nature (London)* **266**:373.

Lester, H. A., Changeux, J. P., and Sheridan, R. E., 1975, Conductance increases produced by bath application of cholinergic agonists to *Electrophorus* electroplaques, *J. Gen. Physiol.* **65**:797.

Lewis, C. A., 1978, The ion-concentration dependence of the reversal potential and the single channel conductance of ion channels at the frog neuromuscular junction, *J. Physiol. (London)*, in press.

Libelius, R., Eaker, D., and Karlsson, E., 1975, Further studies on the binding properties of cobra neurotoxin to cholinergic receptors in mouse skeletal muscle, *J. Neural Transm.* **37**:165.

Linder, T. M., and Quastel, D. M. J., 1978, A voltage-clamp study of the permeability change induced by quanta of transmitter at the mouse end-plate, *J. Physiol. (London)* **281**:535.

Ling, G., and Gerard, R. W., 1949, The normal membrane potential of frog sartorius fibers, *J. Cell. Comp. Physiol.* **34**:383.

Magazanik, L. G., 1976, Functional properties of postjunctional membrane, *Annu. Rev. Pharmacol.* **16**:161.

Magazanik, L. G., and Vyskočil, F., 1973, Desensitization at the motor end-plate, in: *Drug Receptors* (H. P. Rang, ed.), pp. 105–119, Macmillan, New York.

Magleby, K. L., and Stevens, C. F., 1972a, The effect of voltage on the time course of end-plate currents, *J. Physiol. (London)* **223**:151.

Magleby, K. L., and Stevens, C. F., 1972b, A quantitative description of end-plate currents, *J. Physiol. (London)* **223**:173.

Magleby, K. L., and Terrar, D. A., 1975, Factors affecting the time course of decay of end-plate currents: A possible cooperative action of acetylcholine on receptors at the frog neuromuscular junction, *J. Physiol. (London)* **244**:467.

Mallart, A., Dreyer, F., and Peper, K., 1976, Current–voltage relation and reversal potential at junctional and extrajunctional ACh-receptors of the frog neuromuscular junction, *Pflügers Arch.* **362**:43.

Manalis, R. S., 1977, Voltage-dependent effect of curare at the frog neuromuscular junction, *Nature (London)* **267**:366.

Marty, A., 1978, Noise and relaxation studies of ACh induced currents in the presence of procaine, *J. Physiol. (London)* **278**:237.

Meunier, J.-C., and Changeux, J.-P., 1973, Comparison between the affinities for reversible cholinergic ligands of a purified and membrane bound state of the acetylcholine receptor protein from *Electrophorus electricus*, *FEBS Lett.* **32**:143.

Meunier, J.-C., Sealock, R., Olsen, R., and Changeux, J.-P., 1974, Purification and properties of the cholinergic receptor from *Electrophorus electricus* electric tissue, *Eur. J. Biochem.* **43**:371.

Miledi, R., and Potter, L. T., 1971, Acetylcholine receptors in muscle fibres, *Nature (London)* **233**:599.

Monod, J., Wyman, J., and Changeux, J.-P., 1965, On the nature of allosteric transitions: A plausible model, *J. Mol. Biol.* **12**:88.

Moore, J. W., Narahashi, T., and Shaw, T. I., 1967, An upper limit to the number of sodium channels in nerve membrane? *J. Physiol. (London)* **188**:99.

Moreau, M., and Changeux, J.-P., 1976, Studies on the electrogenic action of acetylcholine with *Torpedo marmorata* electric organ, I. *J. Mol. Biol.* **106**:457.

Mrose, H. E., and Ritchie, J. M., 1978, Local anesthetics: Do benzocaine and lidocaine act at the same site?, *J. Gen. Physiol.* **71**:223.

Narahashi, T., Frazier, D. T., Deguchi, T., Cleaves, C. A., and Ernau, M. C., 1971, The active form of pentobarbital in squid giant axons, *J. Pharmacol. Exp. Ther.* **177**:25.

Nass, M. M., Lester, H. A., and Krouse, M. E., 1978, Response of acetylcholine receptors to photo-isomerizations of bound agonist molecules, *Biophys. J.*, in press.

Nastuk, W. L., 1959, Some ionic factors that influence the action of acetylcholine at the muscle end-plate membrane, *Ann. NY Acad. Sci.* **81**:317.

Neher, E., and Sakmann, B., 1975, Voltage dependence of drug induced conductance in frog neuromuscular junction, *Proc. Natl. Acad. Sci. USA* **72**:2140.

Neher, E., and Sakmann, B., 1976a, Single-channel currents recorded from membrane of denervated frog muscle fibres, *Nature (London)* **260**:799.

Neher, E., and Sakmann, B., 1976b, Noise analysis of drug-induced voltage clamp currents in denervated frog muscle fibres, *J. Physiol. (London)* **258**:705.

Neher, E., and Steinbach, J. H., 1978, Local anaesthetics transiently block currents through single acetylcholine-receptor channels, *J. Physiol. (London)* **277**:153.

Neher, E., and Stevens, C. F., 1977, Conductance fluctuations and ionic pores in membranes, *Annu. Rev. Biophys. Bioeng.* **6**:345.

Neumann, E., and Chang, H. W., 1976, Dynamic properties of isolated acetylcholine receptor protein: Kinetics of the binding of acetylcholine and Ca ions, *Proc. Natl. Acad. Sci. USA* **73**:3994.

Nimmo, I. A., and Bauermeister, A., 1976, Cooperativity in the cyclic model for drug–receptor interaction, *Biochem. Pharmacol.* **25**:1903.

Nonner, W., Rojas, E., and Stämpfli, R., 1975, Displacement currents in the node of Ranvier. Voltage and time dependence, *Pflügers Arch.* **354**:1.

O'Brien, R. D., and Gibson, R. E., 1975, Conversion of high affinity acetylcholine receptors from *Torpedo californica* electroplax to an altered form, *Arch. Biochem. Biophys.* **169**:458.

Paton, W. D. M., and Rang, H. P., 1965, The uptake of atropine and related drugs by intestinal smooth muscle of the guinea pig in relation to acetylcholine receptors, *Proc. Roy. Soc. London. Ser. B* **163**:1.

Paton, W. D. M., and Waud, D. R., 1967, The margin of safety in neuromuscular transmission, *J. Physiol. (London)* **191**:59.

Peper, K., Dreyer, F., and Müller, K.-D., 1975, Analysis of cooperativity of drug–receptor interaction by quantitative iontophoresis at frog motor endplates, *Cold Spring Harbor Symp. Quant. Biol.* **40**:187.

Rang, H. P., 1966, The kinetics of action of acetylcholine antagonist in smooth muscle, *Proc. Roy. Soc. London Ser. B* **164**:488.

Rang, H. P., 1971, Drug receptors and their function, *Nature (London)* **231**:91.

Rang, H. P., 1973, in: *Receptor biochemistry and biophysics, Neurosci. Res. Program Bull.* **11**:220.

Rang, H. P., 1975, Acetylcholine receptors, *Q. Rev. Biophys.* **7**:283.

Rang, H. P., and Ritter, J. M., 1970, On the mechanism of desensitization at cholinergic receptors, *Mol. Pharmacol.* **6**:357.

Ritchie, J. M., and Greengard, P., 1966, On the mode of action of local anesthetics, *Annu. Rev. Pharmacol.* **6**:405.

Ritchie, J. M., and Rogart, R. B., 1977, The binding of labelled saxitoxin to the sodium channels in normal and denervated mammalian muscle, and in amphibian muscle, *J. Physiol. (London)* **269**:341.

Ritchie, J. M., Rogart, R., and Strichartz, G. R., 1976, A new method of labelling saxitoxin and its binding to non-myelinated fibres of the rabbit vagus, lobster walking leg, and garfish olfactory nerves. *J. Physiol. (London)* **261**:477.

Roberts, F., and Stephenson, R. P., 1976a, The use of different agonists in antagonist affinity constant estimations, *Br. J. Pharmacol.* **57**:395.

Roberts, F., and Stephenson, R. P., 1976b, The kinetics of competitive antagonists on guinea-pig ileum, *Br. J. Pharmacol.* **58**:57.

Ruff, R. L., 1977, A quantitative analysis of local anaesthetic alteration of miniature end-plate currents and end-plate current fluctuations, *J. Physiol. (London)* **264**:89.

Sakmann, B., 1975, Noise analysis of acetylcholine induced currents in normal and denervated rat muscle fibres, *Pflügers Arch.* **359**:R89.

Sakmann, B., and Adams, P. R., 1976, Fluctuation and relaxation analysis of end-plate currents of frog neuromuscular junction induced by different concentrations of bath applied agonists, *Pflügers Arch.* **365**:R145.

Seyama, I., and Narahashi, T., 1975, Mechanism of blockade of neuromuscular transmission by pentobarbital, *J. Pharmacol. Exp. Ther.* **192**:95.

Seydoux, F., Malhotra, O. P., and Bernard, S. A., 1974, Half-site reactivity, *CRC Crit. Rev. Biochem.* **2**:227.

Sheridan, R. E., and Lester, H. A., 1975, Relaxation measurements on the acetylcholine receptor, *Proc. Natl. Acad. Sci. USA* **72**:3496.

Sheridan, R. E., and Lester, H. A., 1977, Rates and equilibria at the acetylcholine receptor of *Electrophorus* electroplaques, *J. Gen. Physiol.* **70**:187.

Steinbach, A. B., 1968a, Alteration by xylocaine (Lidocaine) and its derivatives of the time course of the end-plate potential, *J. Gen. Physiol.* **52**:144.

Steinbach, A. B., 1968b, A kinetic model for the action of xylocaine on the receptors for acetylcholine, *J. Gen. Physiol.* **52**:162.

Steinbach, J. H., and Stevens, C. F., 1976, Neuromuscular transmission, in: *Neurobiology of the Frog* (R. Llinas and W. Prect, eds.), Springer, Berlin.

Stephenson, R. P., 1956, A modification of receptor theory, *Br. J. Pharmacol.* **11**:379.

Stephenson, R. P., and Ginsborg, B. L., 1969, Potentiation by antagonist, *Nature (London)* **222**:790.

Sterz, R., Dreyer, F., and Peper, K., 1976, ACh-receptors: Dependence of the Hill-coefficient on pH and drugs, *Pflügers Arch.* **362**:R30.

Stevens, C. F., 1972, Inferences about membrane properties from electrical noise measurements, *Biophys. J.* **12**:1028.

Strichartz, G. R., 1973, The inhibition of sodium currents in myelinated nerve by quaternary derivatives of lidocaine, *J. Gen. Physiol.* **62**:37.

Sugiyama, H., and Changeux, J.-P., 1975, Interconversions between different states of affinity for acetylcholine of the cholinergic receptor protein from *Torpedo marmorata*, *Eur. J. Biochem.* **55**:505.

Takeuchi, A., and Takeuchi, N., 1959, Active-phase of frog's end-plate potential, *J. Neurophysiol.* **22**:395.

Takeuchi, A., and Takeuchi, N., 1960, On the permeability of end-plate membrane during the action of transmitter, *J. Physiol. (London)* **154**:52.

Taylor, D. B., and Nedergaard, O. A., 1965, Relation between structure and action of quaternary ammonium neuromuscular blocking agents, *Physiol. Rev.* **45**:523.

Terrar, D. A., 1974, Influence of SKF525A congeners, strophanthidin and tissue culture media on desensitization in frog skeletal muscle, *Br. J. Pharmacol.* **51**:259.

Thron, C. D., 1973, On the analysis of pharmacological experiments in terms of an allosteric receptor model, *Mol. Pharmacol.* **9**:1.

Thron, C. D., and Waud, D. R., 1968, The rate of action of atropine, *J. Pharmacol. Exp. Ther.* **160**:91.

Wagner, H.-H., and Ulbricht, W., 1976, Saxitoxin and procaine act independently on separate sites of the sodium channel, *Pflügers Arch.* **364**:65.

Waud, D. R., 1967, The rate of action of competitive neuromuscular blocking agents, *J. Pharmacol. Exp. Ther.* **158**:99.

Waud, D. R., 1969, On the measurement of the affinity of partial agonists for receptors, *J. Pharmacol.* **170**:117.

Weber, M., and Changeux, J.-P., 1974, Binding of *Naja nigricollis* ^{3}H–α-toxin to membrane fragments from *Electrophorus* and *Torpedo* electric organs. 2. Effect of cholinergic agonists and antagonists on the binding of the tritiated α-neurotoxin, *Mol. Pharmacol.* **10**:15.

Weber, M., David-Pfeuty, T., and Changeux, J.-P., 1975, Regulation of binding properties of the nicotinic receptor protein by cholinergic ligands in membrane fragments from *Torpedo marmorata*, *Proc. Natl. Acad. Sci. USA* **72**:3443.

Whitehead, E., 1970, The regulation of enzyme activity and allosteric transmission, *Prog. Biophys. Mol. Biol.* **21**:321.

Kinetics of Cooperative Binding

A. De Lean and D. Rodbard

1. Overview

Numerous models have been developed to describe *cooperative* phenomena in binding and enzyme systems. De Meyts has proposed a general model for cooperative binding in which the equilibrium constant of dissociation (K_d) is a linear function of receptor occupancy (De Meyts and Roth, 1975). This model has been applied to a wide variety of systems, ranging from insulin, catecholamines, lectins, and small peptides to drugs such as the opiates. The existence of such a system is inferred from the kinetics of binding; equilibrium studies are insufficient. However, to date, a detailed mathematical treatment of the nonsteady state has not been available. We shall consider a special important case where the dissociation rate constant k' (or k_{off}) changes linearly from k'_e to k'_f as fractional receptor occupancy changes from 0 to 1 while the association rate constant k (or k_{on}) is invariant. The simultaneous differential equations for labeled and unlabeled ligand binding to a single class of homogeneous cooperative sites are formulated and the exact analytical solutions obtained. Properties of the model are examined in several coordinate systems, including the titration curve, Hill plot, and Scatchard plot. For the most extreme case of negative cooperativity ($k'_e/k'_f \rightarrow 0$), the Hill coefficient tends toward a lower limit of 0.5. For the most extreme case of positive cooperativity, the midpoint of the titration curve shifts to the left by a factor of 2, the Scatchard plot becomes horizontal, and the Michaelis–Menten plot remains concave to the right. Using binding parameters representative of the insulin–receptor system, we have obtained computer simulations of association, dissociation, and equilibrium binding curves. Association curves for a large range of ligand concentrations in the presence of either positive or negative cooperativity were almost indistinguishable from those expected in the absence of cooperativity. Using low tracer concentration for association, perfect washing to remove unbound tracer, and a 100-fold dilution, the model predicts first-order

A. De Lean and D. Rodbard ● Endocrinology and Reproduction Research Branch, National Institute of Child Health and Human Development, National Institutes of Health, Bethesda, Maryland. ADL is a Postdoctoral Fellow of the Medical Research Council of Canada.

dissociation with minimal rebinding. Under these conditions, in the presence of negative cooperativity, increasing concentrations of unlabeled ligand progressively enhance the apparent dissociation rate. The converse is observed for positive cooperativity.

De Meyts has generated considerable experimental support for the applicability of this model to the IM-9 lymphocyte insulin receptor (De Meyts et al., 1973; De Meyts et al., 1976; De Meyts, 1976b). However, others (Pollet et al., 1977) have challenged the applicability of this model to the insulin system on the basis of findings that (1) the association kinetics appear compatible with a simple, noncooperative, bimolecular reaction; (2) initial dissociation rates were apparently unaffected by the concentration of labeled ligand used during association (and, presumably, occupancy of receptors); and (3) late in dissociation, the curves for high concentration of labeled ligand showed, paradoxically, a slower rate of dissociation than the curves for low ligand concentration. In an analysis of this controversy, we shall examine the dissociation experiment in more detail, with special consideration of the separate and combined effects of three potentially important sources of systematic errors: incomplete washing, presence of nonspecific binding, and failure to obtain "instantaneous" mixing and measurement of the "true" initial binding at the onset of dissociation. In the ideal case of perfect washing and addition of unlabeled insulin, the dissociation rate is positively correlated with changes in receptor occupancy. However, the relationship between receptor occupancy and the *apparent* dissociation rate can be radically altered in the presence of even as low as 3–5% residual free hormone at the onset of dissociation, either alone or together with a "lag" period of a few minutes between the onset of dissociation and measurement of "initial" binding. Thus, the present model and computer simulation studies suggest ways in which one might reconcile the conflicting observations.

In addition, we have made a preliminary study of the basic properties of a second model, where the association rate constant (k_{on}) is a simple linear function of receptor occupancy. This model also allows for either positive or negative cooperativity, but the properties of the system are quite different from the model in which k_{off} is a variable.

The present mathematical analyses should provide a basis for improved experimental design to test the ability of this class of cooperative models to describe ligand binding systems. The relationship of these models to classical models of cooperativity is discussed.

2. General Introduction

Cooperative phenomena play an important role in several biochemical systems, such as oxygen binding to hemoglobin and binding of substrates, inhibitors, and allosteric modifiers to regulatory enzymes. In 1965, Monod, Wyman, and Changeux speculated that the effects of hormones may involve cooperative phenomena (Monod et al., 1965). Cooperative behavior has recently been claimed for the binding of steroid and peptide hormones, drugs, and neurotransmitters to their receptors (De Meyts et al., 1973; Frazier et al., 1974; Limbird et al., 1975; Limbird and Lefkowitz, 1976).

Table I

Nomenclature

α	De Meyts' cooperativity factor $= K_f/K_e = k'_e/k'_f$
δ	Interaction factor ($= 1/\alpha = K_e/K_f = k'_f/k'_e$)
ε	Interaction factor for model II
μ	Scatchard interaction factor (4,31)
a, b, c	Coefficients of quadratic equation
a, b, c, d	Parameters of four-parameter logistic equation
B or $[H^*R]$	Concentration of labeled ligand bound
B_0	B at $t = 0$ (or after selected time lag)
$\% B_0$	"Normalized" counts bound $= B/B_0 \times 100$
F	Free (unbound) ligand concentration
$F_{0.5}$	Free ligand concentration resulting in $Y = 0.5$
H	Generic for unlabeled ligand
H^*	Generic for labeled ligand (or ligand in general)
h	Total ligand concentration (unlabeled)
h^*	Total ligand concentration (labeled)
K	Equilibrium constant of association (function of Y)
K_e or K_0	Initial K for $Y = 0$
K_f	Final K for $Y = 1$
k	Association rate constant (a function of Y in model II)
k_0	Initial k when $Y = 0$
k'	Dissociation rate constant (a function of Y in model I)
k'_e	Initial k' for $Y = 0$
k'_f	Final k' for $Y = 1$
R	Generic for receptor or unoccupied receptor
r	Total binding site concentration
t	Time
U	Total concentration of occupied receptors
U_0	U at $t = 0$
U_1, U_2	Roots of quadratic equation
U_1	Equilibrium value of U
Y	Fractional occupancy of receptors ($= U/r$)

Numerous models have been proposed for protein–ligand interactions which involve cooperativity, allosteric effects, of site–site interactions (Hill, 1910; Adair, 1925; Pauling, 1935; Scatchard, 1949; Atkinson et al., 1965; Monod et al., 1965; Koshland et al., 1966; Levitzki and Koshland, 1969). These models have been used extensively to describe binding of oxygen to hemoglobin, enzyme kinetics, and neurotransmitter systems (Magar, 1972; Colquhoun, 1973). These models have been extensively reviewed and discussed in the literature (Fletcher et al., 1970; Magar, 1972; Colquhoun, 1973; Whitehead, 1973; Herzfeld and Stanley, 1974). Several authors have pointed out the interrelationships and fundamental interconvertibility of many of these models. Even after more than 60 years of study of oxygen binding to hemoglobin, it appears impossible to demonstrate definitive superiority for any one model. In the field of enzymology, the selection among models has often led to serious, even acrimonious debate. It appears impossible to establish the "uniqueness" of a model for a given biochemical system either in theory or practice. Nevertheless, these models have been helpful in guiding experimental design and as a basis for synthesizing vast amounts of data. In the case of hemoglobin, independent physical chemical data (e.g., X-ray diffraction) supplement kinetic data to provide a picture of the molecular mechanisms involved in the cooperative effects. However, in many enzyme–substrate systems and virtually all hormone–receptor systems, kinetic studies must proceed in the absence of any information regarding intramolecular events. While kinetic analyses may provide clues regarding the nature of such mechanisms, it cannot establish them.

Models for cooperative binding may be classified into two categories: those based on a specified hypothetical underlying molecular mechanism (Monod *et al.*, 1965; Koshland *et al.*, 1966; Pauling, 1935; Changeux *et al.*, 1967), i.e., those based on putative conformational change, and those which are *mechanism-free*, i.e., those which make no assumptions regarding molecular geometry (Adair, 1925; Hill, 1910; Scatchard, 1949; De Meyts, 1976a). In view of our ignorance of the mechanisms involved in most systems, we shall confine our attention to the latter group of models.

The prototype of the *mechanism-free* or *mechanism-unspecified* models is the Adair or *stepwise* model (Adair, 1925; Fletcher *et al.*, 1970), shown in reaction scheme (**I**):

$$H + R \rightleftharpoons HR \qquad K_1$$

$$H + HR \rightleftharpoons H_2R \qquad K_2$$

$$H + H_2R \rightleftharpoons H_3R \qquad K_3 \qquad\qquad (\mathbf{I})$$

$$\vdots$$

$$H + H_{n-1}R \rightleftharpoons H_nR \qquad K_n$$

The constants K in scheme (**I**) are stepwise association constants and differ from equilibrium affinity constants used throughout the chapter (Fletcher *et al.*, 1970; Magar, 1972). Although this scheme has the superficial appearance of a "mechanism," we prefer to regard it as a mechanism-free model, since it makes no explicit assumptions about how or why the intrinsic affinity constants might change with binding of successive molecules of ligand. Further, this model is sufficiently general to include either the *concerted* or *induced-fit* type of molecular mechanisms (Fletcher *et al.*, 1970 ; Magar, 1972) : In effect, only the saturating stoichiometry is specified.

The *Hill equation* (Hill, 1910; Atkinson *et al.*, 1965) or the use of a linear segment for the *Hill plot*, i.e., $\log[B/(B_{\max} - B)]$ vs. $\log(\text{free})$, may also be regarded as a general, empirical description of a (cooperative) binding system, without making any assumptions regarding the molecular mechanisms. The properties of the Hill equation are frequently examined in the case of specified molecular models, e.g., tri- or tetramolecular collisions, but these are special, limiting cases. In general, a Hill plot with noninteger slope does not correspond to and is not restricted to any specified molecular mechanism.

Scatchard (1949) considered the case of noncooperative binding to multiple independent classes of sites, and (often overlooked in the same paper) he also discussed the case where the free energy of binding, $\Delta F = -RT \ln(K)$, changes linearly with the fractional occupancy of the binding protein (Scatchard, 1949; Hunston, 1975). This model has been successfully applied to the binding of ions, dyes, and other small molecules to proteins. In this model, the equilibrium constant of association (K) or its reciprocal, the dissociation rate constant (K_d), is an exponential function of occupancy (Fig. 1, dotted line). This model is plausible from a thermodynamic point of view but makes no assumptions regarding the nature of conformational changes (or other structural events) which result in the progressive change of K and ΔF.

De Meyts and Roth (1975) recently proposed another simple mathematical model: for small or even moderate degrees of cooperativity (e.g., a sixfold change observed for the binding of insulin to its receptors on mononuclear cells. De Meyts considered the case where K_d is linearly related to fractional receptor occupancy (De Meyts, 1976a). This model is closely related to Scatchard's *exponential* model: for small or even moderate degrees of cooperativity (e.g., a sixfold change in K_d as the receptor becomes saturated), the exponential and linear relationships are similar, and it would be nearly impossible to discriminate between them even when experimental errors are small. In such cases, De Meyts' model may be regarded as an approximation to Scatchard's. This interconvertibility is useful, for, as will be seen below, the simpler models facilitate mathematical treatment of the kinetics of association and dissociation.

In many applications, the use of rapid *perturbation* kinetics has provided many additional insights into the nature of molecular mechanisms (Castellan, 1963; Hammes and Schimmel, 1967; Craig *et al.*, 1971; Lancet and Pecht, 1976; Neuman and Chang, 1976). These methods require the availability of both ligand and binder in highly purified form and at high concentrations. Accordingly, at the present time, they are not applicable to the study of most hormone, drug, and neurotransmitter receptors, which are not available in sufficient purity, quantity, and stability. However, it is still possible to study the kinetics of association and dissociation under a wide variety of conditions (time, temperature,

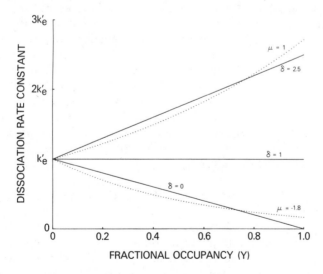

Fig. 1. Relationship between K_d and/or k_{off} (ordinates) and receptor occupancy Y (abscissa). The linear relationship suggested by De Meyts is shown (solid line) together with the exponential relationship discussed by Scatchard. For moderate ratios of k'_f/k'_e, the linear and exponential relationships would be experimentally indistinguishable.

pH, ionic milieu, ligand concentrations, and receptor occupancy). These *kinetic* studies can potentially provide much more information than equilibrium studies alone: it is well known that it is impossible to discriminate between multiplicity (heterogeneity) of binding sites and negative cooperativity on the basis of equilibrium studies alone (Fletcher *et al.*, 1970; Magar, 1972). At least potentially, both the design and analysis of such kinetic experiments could be facilitated by a mathematical model to describe the system. Appropriate kinetic studies should enable us to determine whether changes in k_{on} or k_{off} are predominantly responsible for the changes in the equilibrium constant of association (or whether it would be necessary to invoke changes in both k_{on} and k_{off} or to consider still more complex reaction mechanisms).

In this chapter, we shall attempt to (1) describe the basic properties of an extension of De Meyts' model (where k_{off} is a *linear* function of occupancy), both at equilibrium and for the kinetics of association and dissociation, with explicit treatment of both labeled and unlabeled ligand; (2) indicate how computer simulation studies may be helpful in resolving current controversy regarding the nature of the insulin–receptor system; (3) develop the mathematical properties of binding systems in which the association rate constant is a simple linear function of receptor occupancy and describe appropriate experiments to identify such a system; and (4) discuss the above models in the context of still more complicated reaction schemes.

3. Model I: k_{off} as a Linear Function of Occupancy

De Meyts *et al.* have postulated that the nonlinear Scatchard plot obtained for the insulin–lymphocyte system (De Meyts *et al.*, 1973) could be explained by the presence of destabilizing site–site interactions (negative cooperativity) rather than the existence of multiple independent classes of binding sites. They proposed a new and remarkably simple model (De Meyts and Roth, 1975) involving a minimum of assumptions and parameters. This model is quantitatively compatible with most available equilibrium data for the interaction of insulin with its receptors (De Meyts *et al.*, 1976; De Meyts, 1976b) and is at least qualitatively compatible with most non-steady-state data. De Meyts' model makes the fundamental assumption that the reciprocal of the equilibrium constant of association is a linear function of receptor occupancy (De Meyts, 1976a) or that the equilibrium constant of association (K) is given by

$$K = \frac{K_e}{1 + Y(1/\alpha - 1)} \tag{1}$$

where K_e is the average equilibrium constant at infinitesimal receptor occupancy, K_f is the mean affinity as Y approaches unity, Y is fractional receptor occupancy, and α is an interaction factor equal to K_f/K_e. Weber (1975), Koren and Hammes (1976), and Triggle and Triggle (1976) have suggested that changes in K are primarily, if not exclusively, effected by changes in the dissociation rate constant (k') rather than changes in the association rate constant (k), since the latter is presumably diffusion limited in most examples involving binding of hormones or neurotransmitters to isolated cells, membranes, or solubilized receptors.

3.1. Assumptions

Model I is based on four major assumptions:

1. Ligand(s) and receptor are homogeneous.
2. The association rate constant is independent of receptor occupancy.
3. Site–site interactions resulting in changes of affinity (by virtue of changes in the dissociation rate constant) are "instantaneous" relative to changes in occupancy of binding sites. Further, the parameters K and k' are descriptive of the ensemble (i.e., they are average macroscopic values).
4. Ligands are univalent; the valency of the receptor is unspecified.

Reaction scheme (**II**) shows the simple bimolecular reaction of two distinguishable but kinetically identical ligands (H*) and (H) with binding molecule (R) according to the mass action law:

$$H^* + R \underset{k'}{\overset{k}{\rightleftharpoons}} H^*R$$
$$H + R \underset{k'}{\overset{k}{\rightleftharpoons}} HR \qquad \textbf{(II)}$$

In the following, we adopt the arbitrary convention that H* represents the labeled and H the unlabeled ligand. Both H* and H have the same affinity K_e and the same cooperativity factor α. In reaction scheme (**II**), k is the association rate constant, and k' is the average dissociation rate "constant" (perhaps better designated *coefficient*), which is a linear function of total receptor occupancy Y, changing from k'_e to k'_f as Y changes from 0 to 1. Thus, we have

$$k' = k'_e[1 + (\delta - 1) Y] \qquad (2)$$

where

$$Y = \frac{U}{r} \qquad (2a)$$

$$U = [H^*R] + [HR] \qquad (2b)$$

$$\delta = \frac{k'_f}{k'_e} = \frac{1}{\alpha} = \frac{K_e}{K_f} \qquad (2c)$$

Here U is the total concentration of occupied receptors and r is total receptor concentration. Equation (2) is directly related to Eq. (1). The interaction factor δ is simply the reciprocal of De Meyts' α; use of δ leads to slightly simpler and more compact mathematical expressions. Assumptions 1 and 3 imply that changes in occupancy and in the dissociation rate constant are processes occurring randomly (uniformly) throughout the system. Otherwise, we make no restrictive assumptions whatever regarding the geometrical or physical–chemical nature of the cooperative effect. The formulation of the family of differential equations corresponding to reaction scheme (**II**), together with its analytical solution, is given in Appendix A.

3.2. Properties of the Model

3.2.1. Equilibrium

The saturation curves for a wide spectrum of values of the cooperativity factor δ are shown in Fig. 2(a). In the absence of cooperativity ($\delta = 1$) half-maximal receptor occupancy is obtained at a *free* ligand concentration, $F_{0.5}$, equal to the reciprocal of the equilibrium association constant K_e. In the case of negative cooperativity, as δ increases, the binding curve is progressively shifted to the right. For positive cooperativity ($\delta < 1$), the curves shift progressively to the left until a limiting case is reached; as δ approaches 0, $F_{0.5}$ approaches $0.5/K_e$. In general, $F_{0.5}$ is related to δ and K_e by

$$F_{0.5} = \frac{\delta + 1}{2K_e} \tag{3}$$

Figure 2(b) shows the Scatchard plots corresponding to the binding iso-therms of Fig. 2(a). The Scatchard plot is now perhaps the most frequently used method for linearization of noncooperative binding isotherms and is particularly useful for detection of nonlinearity, i.e., heterogeneity of sites or cooperativity. For negative cooperativity, the Scatchard plots are concave up; for positive

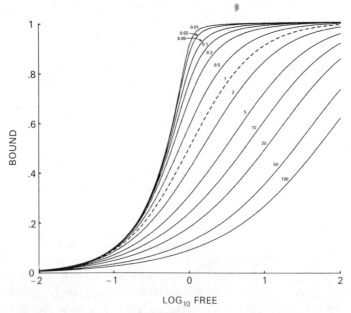

Fig. 2(a). Binding curves at equilibrium for increasing degrees of both negative ($\delta > 1$) and positive ($\delta < 1$) cooperativity when $K_e = 1$ and $r = 1$. For negative cooperativity ($\delta = 2, 5, 10, 20, 50, 100$), the titration curves are shifted progressively to the right of the noncooperative curve (dashed line, $\delta = 1$), while for positive cooperativity ($\delta = 0.5, 0.2, 0.1, 0.05, 0.02, 0.01$) the curves shift to the left up to a limiting factor of 2 in terms of $F_{0.5}$.

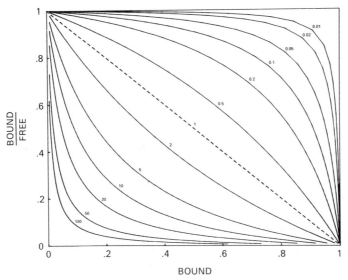

BOUND

Fig. 2(b). Scatchard plots corresponding to the curves in Fig. 2(a). The plot is linear for the nonco-operative case (dashed line, $\delta = 1$), while it is concave upward for negative cooperativity ($\delta = 2 - 100$) and concave downward for positive cooperativity ($\delta = 0.5 - 0.01$). For reciprocal values of δ (e.g., 5 and 0.2), the curves are symmetrical relative to the line for $\delta = 1$. The initial slope of the Scatchard plot for positive cooperativity can never be positive, so there is never a "hook" in the initial part of the curve, for this model.

cooperativity ($\delta < 1$), the curves are concave down. Even a small departure of δ from unity results in significant curvature, indicating the sensitivity of the Scatchard plot for the detection of cooperativity (Boeynaems and Dumont, 1975). Reciprocal values of δ (e.g., 5 and $\frac{1}{5}$) give rise to symmetrical curves relative to the straight line for $\delta = 1$ (dashed line). Even for the most extreme case of positive cooperativity, the initial slope can never become positive. Thus, the present model can never give rise to the *low-dose hook* often thought to be a *sine qua non* of positive cooperativity. Accordingly, the present model cannot give rise to a sigmoidal Michaelis–Menten plot: a plot of bound versus free would always be concave to the right, without an inflection point.

Figure 2(c) shows the corresponding Hill plots. The dashed line shows the Hill plot for $\delta = 1$. For negative cooperativity (e.g., $\delta = 2 - 100$) the Hill plots show minimal departure from linearity in the range customarily examined experimentally (ordinate between -1 and 1). The minimum slopes of these Hill plots (n_{Hmin}) occur when $F = 1/K_e$ and decrease progressively as δ increases. However, n_H evaluated when $F = 1/K_e$ has a limiting lower value of 0.5. For positive cooperativity, the Hill plots display severe nonlinearity, which increases progressively as $\delta \to 0$. The maximum slope of the Hill plots, n_{Hmax}, also occurs when $F = 1/K_e$. As in Fig. 2(a), the curves cannot shift to the left beyond a limiting (asymptotic) case. Figure 2(d) shows the local slopes of the Hill plot or apparent $n_H = d \ln [B/(r - B)]/d \ln (F)$, plotted as a function of $\log_{10}F$. Notice that

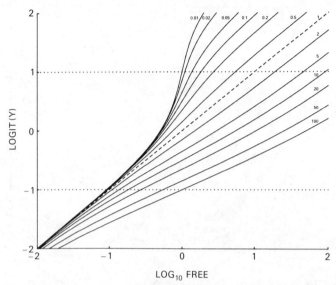

Fig. 2(c). Hill plot of the binding curves shown in Fig. 2(a). For the noncooperative case one has a straight line in the central region, but for all other cases the curves are nonlinear. For negative cooperativity ($\delta = 2 - 100$) the curves show minimal departure from linearity within the usual experimental range (delineated by the dotted lines), but for positive cooperativity there is a severe curvature within the readily observable range. Note: logit $(Y) = \log_e[Y/(1 - Y)]$.

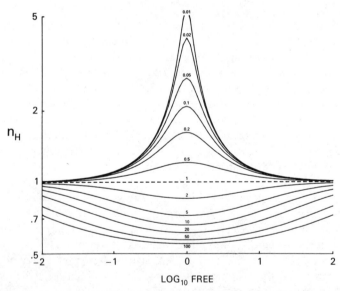

Fig. 2(d). Slope of the Hill plots of Fig. 2(c) as a function of $\log_{10}$ of free ligand concentration. The logarithmic scale used for the ordinate is used to display the effects of both negative and positive cooperativity. For $\delta < 1$, the local slope exhibits a sharp peak for free ligand concentrations near $1/K_e$. For $\delta > 1$, the slope is close to its minimum value for a broad range of ligand concentrations around $F = 1/K_e$.

Table II
Relationship between n_H and δ when $F = 1/K_e$

δ	n_H	δ	n_H
10	0.658	$\frac{1}{2}$	1.207
9	0.667	$\frac{1}{3}$	1.366
8	0.677	$\frac{1}{4}$	1.500
7	0.689	$\frac{1}{5}$	1.618
6	0.704	$\frac{1}{6}$	1.725
5	0.724	$\frac{1}{7}$	1.823
4	0.750	$\frac{1}{8}$	1.914
3	0.789	$\frac{1}{9}$	2.000
2	0.854	$\frac{1}{10}$	2.081
1	1.000		

for positive cooperativity the slopes can become very large and have a sharp, nearly symmetrical peak when F is in the vicinity of $1/K_e$. However, for negative cooperativity, the local n_H has a broad plateau and is relatively insensitive to the arbitrary choice of F for calculating $n_{H_{min}}$. In general, when $F = 1/K_e$, the slope of the Hill plot, n_H, is given by

$$n_H = \frac{\sqrt{\delta} + 1}{2\sqrt{\delta}} \tag{4}$$

Selected numerical values for this relationship between δ and $n_{H_{min}}$ (or $n_{H_{max}}$) are given in Table II. These values may be useful as initial estimates for iterative parameter fitting (see Sec. 3.3).

3.2.2. Association Curves

The association curves for reaction scheme (**II**) (labeled ligand only) are shown in Fig. 3(a) in the case of negative cooperativity, with a representative set of parameters. In this example, the association curves for varying concentrations of labeled hormone do not display any obvious peculiarity which would indicate the presence of cooperativity. When these curves were fitted using a noncooperative model (i.e., the present model but with δ constrained equal to unity), there were only minimal differences between the predicted and "fitted" curves. However, we did not find any identifiable characteristic of the discrepancy between the true curves for the cooperative model and those fitted with $\delta = 1$, since these discrepancies were affected by the arbitrary choice of ligand concentrations, time periods, and the magnitude of simulated random errors. For real experimental data, these subtle differences are likely to be obscured by experimental error, unless the cooperative effect is of extreme magnitude. Figure 3(b) shows an attempt to linearize the association curves of Fig. 3(a). In the case of a noncooperative homogeneous system, under conditions where the ratio of bound/total ligand is insignificant, the simplified equation for the kinetics of association (Waud, 1968) should give rise to a family of straight lines by plotting

$\log_e[B_{eq}/(B_{eq} - B)]$ vs. time, provided that the amount of ligand bound at equilibrium (B_{eq}) has been accurately evaluated. The linearity of this plot has been considered as evidence for applicability of a noncooperative model (Bockaert et al., 1973; Lissitsky et al., 1975; Pollet et al., 1977). Careful testing would reveal nonlinearity only in the absence of any experimental errors. For ligand concentrations much larger than those of binding sites $(h > r)$, the slope in Fig. 3(b) is approximately given by $kh^* + k'$, while for the converse case with massive excess of binding sites, the slope asymptotically approaches $kr + k'$. Use of linear regression of the slopes of the curves shown in Fig. 3(b) vs. h^* would provide the following estimates of the rate constants: $k = 0.01160$ nM^{-1} and $k' = 0.09951$ min^{-1}. Accordingly, the association rate constant can be nearly perfectly estimated by this approach, while the dissociation rate constant obtained was close to the arithmetic average of k'_e and k'_f (0.0231 and 0.1386 min^{-1} in the present case). However, this method is fraught with difficulties in statistical curve fitting.*

Figures 3(c) and 3(d) show the case of positive cooperativity when δ is $\frac{1}{6}$ and all other binding parameters are identical to those used for Figs. 3(a) and 3(b). Again, the association curves, Fig. 3(c), do not reveal any obvious departure from those expected in the absence of cooperativity. However, Fig. 3(d) indicates an inversion of the position of two curves, since the line corresponding to a total ligand concentration of 2 nM has a lower slope than those for lower concentrations (0.2 and 0.02 nM). This reversal might easily be missed in an experimental system which does not provide accurate measurement of association at extremely low receptor occupancy.

Figures 3(e) and 3(f) show the Scatchard plots of non-steady-state binding in the cases of negative ($\delta = 6$) and positive ($\delta = \frac{1}{6}$) cooperativity. For a noncooperative system, Scatchard plots obtained prior to equilibrium are nearly linear and lead to an underestimate of the affinity constant (Rodbard et al., 1971). For negative cooperativity, the non-steady-state Scatchard plot is linear only very early [Fig. 3(e), at 5 min] during association. The curve bends progressively and shows a typical upward concavity well before achievement of equilibrium [compare Figs. 3(e) and 3(a)]. As expected, the region of the right-hand curve corresponding to large concentrations of ligand and nearly complete receptor occupancy reaches the equilibrium value well before the left-hand end of the curve for low ligand concentrations. In the case of positive cooperativity [Fig. 3(f)], the Scatchard plot remains nearly linear over the experimentally accesible range, even relatively late during the association process [compare with Fig. 3(c)]. Moreover, the non-steady-state curves show steep downward curvature in the region near complete occupancy. Ignoring this nearly inaccessible region could lead to a severe overestimate of the binding capacity and to the erroneous interpretation that the Scatchard plots were linear. Thus, the presence of positive cooperativity would be easily missed.

---

*In our simulation studies for association we varied the ligand concentration over four orders of magnitude in the absence of experimental errors. In practice, the range of hormone concentration for association is restricted to one or two orders of magnitude. Failure to include points at high fractional occupancy (≈ 0.95) of receptors can lead to severely biased estimates of parameters K, k, k', and r.

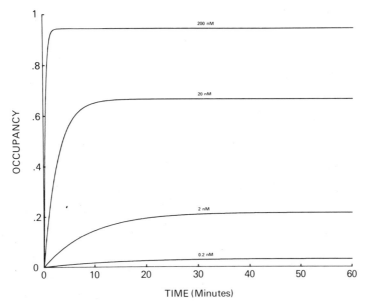

Fig. 3(a). Kinetics of association for increasing concentrations of labeled ligand (0.02, 2, 20, 200 nM). Binding parameters: $r = 4$ nM, $k = 0.01155$ nM^{-1} min^{-1}, $k'_e = 0.0231$ nM^{-1} min^{-1}, $\delta = 6$. These curves do not make the presence of negative cooperativity obvious.

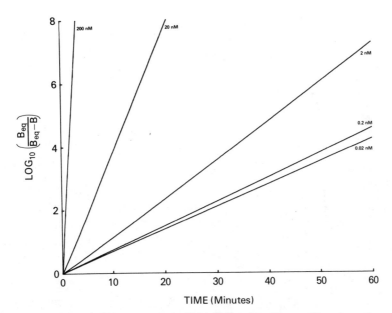

Fig. 3(b). Linear transform of the curves of Fig. 2(a) using $\log_e[B_{eq}/(B_{eq} - B)]$ on the ordinate. These curves do not show any obvious curvature. Linear regression of the average slope of these lines as a function of ligand concentration (total or free) provides a good estimate of the association rate constant k (Waud, 1968) when the range of ligand concentrations used is large, as in this example. However, the estimate of the dissociation rate constant obtained lies between k'_e and k'_f and is subject to large errors.

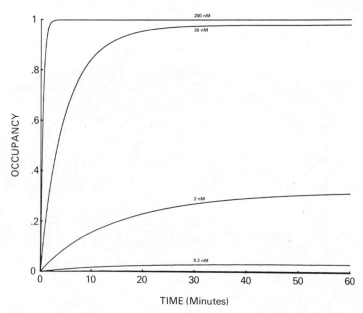

Fig. 3(c). Association kinetics for various concentrations of labeled ligand in the case of positive co-operativity [parameters as in Fig. 3(a) except $\delta = \frac{1}{6}$]. Any one curve would not reveal the presence of positive cooperativity. However, the relative position of the curves differs from Fig. 3(a): The curve for 20 nM has moved toward that for 200 nM. This shift is subtle and might easily be overlooked.

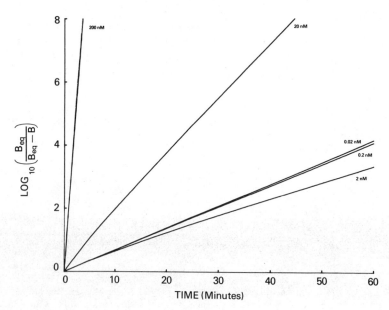

Fig. 3(d). Transform of association curves from Fig. 3(c) using $\log_e[B_{eq}/(B_{eq} - B)]$. In the presence of positive cooperativity there is a reversal in the expected relative position of the curves for ligand concentrations of 0.2 and 2 nM.

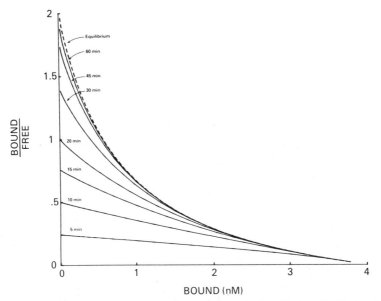

Fig. 3(e). Scatchard plots of binding curves prior to attainment of equilibrium (5–60 min of association) compared with the curve at equilibrium (dashed line) for the case of negative cooperativity. Parameters as in Fig. 3(a). For very brief incubations (e.g., 5–10 min), the curve is nearly linear, but it bends and shows a typical upward concavity well before reaching equilibrium. The right-hand region of the curve reaches equilibrium first, as expected from the mass–action law.

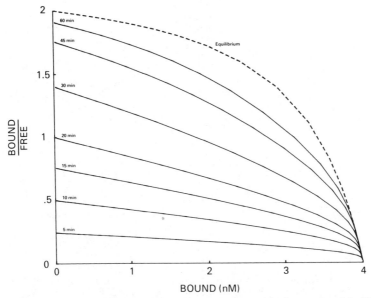

Fig. 3(f). Equilibrium Scatchard plots in the case of positive cooperativity [parameters as in Fig. 3(c)]. The curves are nearly linear for most of their experimentally accessible segment much later than in the case of negative cooperativity [Fig. 3(e)]. Even at 60 min, when the usual association curves [Fig. 3(c)] suggest "achievement" of equilibrium, the Scatchard plot is still far from the curve at true equilibrium (dashed line).

3.2.3. Dissociation Curves

Dissociation studies should be the most sensitive approach to validate a cooperative model in which k' is a function of occupancy (De Meyts et al., 1973; De Meyts et al., 1976; De Meyts, 1976a,b). Indeed, the applicability of the negative cooperative model to the insulin–receptor system and other systems has been claimed (De Meyts et al., 1973; Frazier et al., 1974; Limbird et al., 1975; Limbird and Lefkowitz, 1976) and disputed (Pollet et al., 1977; Jacobs and Cuatrecasas, 1976; Cuatrecasas and Hollenberg, 1975) on the basis of these kinds of studies. In general, the labeled hormone, at low concentration, is allowed to react with receptor for a sufficiently long period to approach (but not necessarily attain) equilibrium. At this point, one can separate the occupied and unoccupied receptors on the membranes or cells from the medium (containing unbound H*), e.g., by means of filtration or centrifugation. Ideally, with perfect washing, this could be done without perturbing the concentrations of H*, R, and H*R. One may then introduce several perturbations, e.g., dilution, addition of unlabeled ligand (H), addition of fresh receptor, addition of soluble or insoluble adsorbants to reduce [H*] to negligible levels, or combinations of the above. Numerous other perturbations are possible (e.g., change of temperature), but the above are likely to be (and have been) most informative and readily performed. Figure 4(a) shows the effect of various degrees of dilution for the case of negative

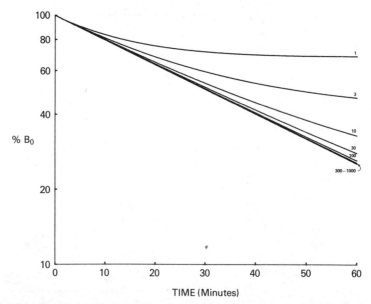

Fig. 4(a). Effect of increasing dilution on dissociation of ligand–receptor complex following instantaneous and perfect washing of free ligand after association (60 min). Binding parameters: $k = 0.01155$ nM^{-1} min^{-1}, $k_e = 0.0231$ min^{-1}, $\delta = 6$, $r = 4$ nM, $h = 0.02$ nM. Under these conditions, washing only or washing and low dilution ($\times$ 3, 10, 30) result in nonlinear dissociation curves due to rebinding of dissociated ligand, leading to a new equilibrium at lower fractional occupancy of binding sites. A 100-fold dilution results in a virtually complete prevention of rebinding; larger dilutions ($\times$ 300, 1000) do not further improve the linearity of the dissociation curves.

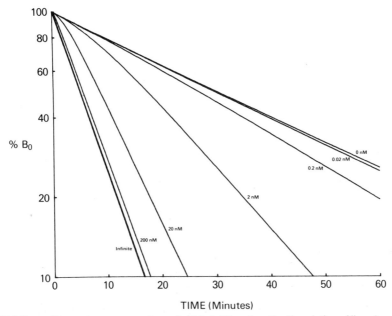

Fig. 4(b). Effects of increasing concentrations of unlabeled ligand on the dissociation of ligand–receptor complex after association (60 min) followed by perfect washing and 100-fold dilution for negative cooperativity [$\delta = 6$, with other parameters as in Fig. 3(a)]. Addition of unlabeled ligand at concentrations above 0.2 nM results in a significant increase in the dissociation rate. For massive excess of unlabeled ligand (200–2000 nM), the dissociation curve becomes first order. For intermediate concentrations (0.2, 2, and 20 nM), the dissociation curve exhibits a characteristic initial downward curvature.

cooperativity. For practical purposes, a dilution of 100-fold is virtually identical with *infinite* dilution for this case where the labeled ligand concentration is infinitesimal. Notice that $\log(B/B_0)$ is a linear function of time (i.e., B/B_0 is a simple negative exponential function of time) only in the case of effectively infinite dilution; otherwise, the curves are ultimately concave up and have a horizontal plateau corresponding to a new equilibrium. In the case of positive cooperativity, the effect of increasing dilution and washing is nearly identical to that shown in Fig. 4(a), since the initial occupancy is negligible in both cases, so k' remains equal to k'_e. Figure 4(b) shows the dissociation curves for the case of 100-fold dilution combined with instantaneous addition of various concentrations of unlabeled ligand in the case of negative cooperativity. The curve for dilution only has a slope which corresponds to k'_e, since we begin with infinitesimal occupancy for this *tracer* concentration of labeled ligand. With increasing concentration of unlabeled ligand the slope becomes progressively steeper, until it reaches its maximal possible value, corresponding to k'_f. In the extreme cases, the relationship between $\log(B/B_0)$ and t is perfectly linear. However, for all intermediate concentrations (e.g., the five shown here), the curve has a characteristic downward concavity, which is a hallmark of negative cooperativity in the ideal case of perfect washing, absence of nonspecific binding, and infinite dilution, where rebinding of labeled ligand is impossible. This curvature arises because the total

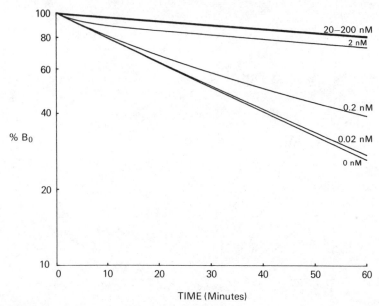

Fig. 4(c). Effect of increasing concentrations of unlabeled ligand on the dissociation of labeled ligand under conditions identical to Fig. 4(b) but with positive cooperativity ($\delta = \frac{1}{6}$). Here the presence of unlabeled ligand at concentrations above 0.02 nM decelerates dissociation. At intermediate concentrations (0.2, 2 nM) there is an initial upward curvature of the dissociation curve. A similar upward curvature could arise due to rebinding of dissociated ligand [Fig 4(a)] or due to heterogeneity of receptor.

occupancy of the receptors is increasing with time (as the unlabeled ligand H binds to the receptor), and hence the instantaneous dissociation rate (the derivative of the curve) increases progressively to its limiting value. In contrast, in the case of positive cooperativity [$\delta = \frac{1}{6}$, Fig. 4(c)] we obtain a family of curves for various concentrations of unlabeled ligand which are concave upward for intermediate ligand concentration, as the dissociation rate for the receptor changes from the relatively fast k'_e to the relatively slow k'_f. However, this pattern would not be diagnostic of positive cooperativity: the same would be expected in the case of heterogeneity of binding sites with the presence of several independent classes or orders of sites. The distinction between these two mechanisms of upward concavity might be made on the basis of the curvature of the equilibrium Scatchard plots (provided both mechanisms were not present simultaneously).

3.3. Discussion

We have described the general properties of a cooperative model based on the linear relationship between the dissociation rate "constant" and the fractional occupancy of the receptor. The model predicts that, in the presence of negative cooperativity, the minimum Hill slope n_H cannot be lower than 0.5. Hence if a binding system were to exhibit a Hill coefficient lower than 0.5, the present model could not apply. In such a case, the low n_H would be suggestive of receptor heterogeneity operating alone or superimposed on a cooperative system. Also, the present model cannot account for a hook effect, i.e., a paradoxical rise in the bound-to-total (B/T) or bound-to-free (B/F) ratios for ligand as ligand concentration increases [cf. Fig. 2(b)]. In the most extreme case of positive cooperativity attainable with this model, the B/F ratio remains constant at $K_e r$ until occupancy reaches 100%, at which point there is a discontinuity and B/F plummets to zero.

The fact that the kinetics of *association* of model I can be distinguished from those of a noncooperative system only with extreme difficulty clearly indicates that association studies are of little value in discriminating between cooperative and noncooperative models when only changes of the dissociation rate coefficient are involved. At most, association studies may confirm the assumption that the association rate constant is (reasonably) independent of receptor occupancy.

Study of nonequilibrium Scatchard plots [Fig. 3(f)] indicates that the presence of positive cooperativity is likely to be missed if binding data are not obtained at the ultimate equilibrium.

Curve Fitting : The availability of an exact analytical solution for the differential equations describing model I vastly facilitates numerical curve fitting by obviating the need for numerical integration. Indeed, the curve-fitting routines become only a little more complicated than in the noncooperative case (Rodbard and Weiss, 1973). However, we now have four fitted parameters, viz. r, $K = k/k'_e$, k, and δ. As in the noncooperative case, it is often advantageous to reparameterize in terms of the logarithms of these parameters, to provide better stability, better "normality" of the distributions of the errors in the parameters, and to prevent the parameters from becoming negative. In the case of δ, the logarithmic transformation appears especially desirable in view of the exquisite sensitivity of δ to curvature of the Scatchard plot (or to small changes in n_H). As usual, weighting is advisable unless preliminary analysis indicates the exceptional uniformity of variance for the dependent variable.* Preliminary analysis indicates that there will usually be very strong correlation in the errors in δ, r, and k'_e, and similarly there is a strong negative correlation between errors in $K = k/k'_e$ and r. One may use Monte Carlo methods to evaluate the precision (or even feasibility) of parameter estimation for any given experimental design for any given magnitude of experimental errors. Such an approach permits rational experimental design.

One intriguing approach to curve fitting in the case of negative cooperativity —which may be useful for making initial estimates of parameters—is to exploit the relationship between δ and n_H [given by Eq. (4) and illustrated in Table II] and the apparent linearity of the Hill plots [Fig. 2(c)]. Thus, the Hill equation may be used to describe the binding isotherm over a considerable range. The procedure is simply to plot the bound ligand concentration on the ordinate vs. the log of the free (or total, if this approximates free) ligand concentration on the

*Computer programs for weighted least-squares nonlinear curve fitting for the cooperative and noncooperative cases of model I and for the four-parameter logistic function are available on request.

abscissa, corresponding to Fig. 2(a). One then estimates the parameters of the four-parameter logistic equation (Rodbard and Frazier, 1975):

$$y = \frac{a - d}{1 + (X/c)^b} + d \tag{5}$$

where y = concentration of bound ligand, X = free (or total) ligand concentration, a = fitted parameter corresponding to concentration of bound ligand when the free ligand concentration is zero (ideally zero), b corresponds to the mean slope of the Hill plot (i.e., n_H), c corresponds to $F_{0.5}$ or $1/K_e$ if X represents free or $(1/K_e + r/2)$ if X represents total ligand, and the difference $d - a$ corresponds to the binding capacity r (or B_{max}). Fitting of Eq. (5) is relatively straightforward, since the parameters are usually well conditioned with fairly small interactions. Computer programs are available for this purpose from the authors (see the footnote on p. 161). The parameter δ can then be approximated from n_H or b using Eq. (4) or Table II. By permitting a to assume nonzero values, one can partially compensate for the fact that a straight line on the Hill plot does not correspond exactly to the true binding isotherm in this model and also correct for nonspecific binding and/or background counts.

The present model retains the attractive features of De Meyts' equilibrium model, requiring a minimum number of assumptions and using a minimum number of parameters. Nothing is expressed or implied regarding the underlying biochemical or thermodynamic mechanisms involved. As such, it is more general than explicit mechanistic models (Monod *et al.*, 1965; Koshland *et al.*, 1966) but less specific in this predictions. Nevertheless, it makes a sufficient number of predictions to permit meaningful testing of the model and to optimize experimental design to maximize the efficiency of such testing.

4. Application to the Insulin–Receptor System

4.1. The Controversy

The insulin–receptor system does not behave like a simple bimolecular reaction between hormone and a homogeneous class of independent binding sites. Instead, it displays a nonlinear Scatchard plot which is concave upward and an enhanced rate of dissociation of labeled (tracer) insulin in the presence of high concentrations of unlabeled insulin which increases receptor occupancy (De Meyts *et al.*, 1973). These findings led De Meyts to suggest the presence of negative cooperativity, and he proposed a model in which the equilibrium constant of dissociation K_d increases linearly with occupancy (De Meyts, 1976a). Further, De Meyts found that increasing occupancy during association by increasing the concentration of labeled ligand resulted in a diminished ability of unlabeled insulin to accelerate dissociation (De Meyts *et al.*, 1976). Thus, occupancy by either labeled or unlabeled ligand appeared to have equivalent effects, consistent with the above model. Recently, Pollet *et al.* (1977) used an almost identical experimental system to show that (1) the association kinetics of this system appear consistent with a simple noncooperative bimolecular model and (2) variation of "labeled" ligand (actually a mixture of labeled and unlabeled hormone) over a 1000-fold range had virtually no effect on the early dissociation kinetics in the presence of an intermediate concentration of unlabeled ligand; the dissociation curves appeared virtually identical whether occupancy was increasing, decreasing, or remaining constant. While these findings appear to be inconsistent with De Meyts' model, Pollet did confirm the nonlinearity of the Scatchard plot for insulin binding and the enhancement of dissociation of labeled hormone by the addition of unlabeled hormone at all concentrations of labeled and unlabeled ligand tested.

In Sec. 3, we examined the properties of model I under ideal conditions, in which the factors promoting the dissociation, e.g., dilution, washing, and addition of unlabeled ligand (hereafter termed the *perturbation*), occur instantaneously, with immediate measurement of the true binding at zero time, B_0. We shall now consider the dissociation studies in more detail when there is (1) incomplete washing; (2) the presence of a small contribution from nonspecific, rapidly dissociating low-affinity sites; and/or (3) a delay between perturbation and the measurement of initial binding B_0.

4.2. Experimental Design

The following experimental simulations were performed using the parameters shown in Table III. These parameters were based on available data for the insulin–receptor system and represent a compromise between the values reported by De Meyts and Pollet, which were generally in reasonably good agreement. The concentrations of labeled and unlabeled ligand and the incubation times were selected to simulate Pollet's experimental conditions (Pollet *et al.*, 1977).

1. *Perfect washing* ("ideal case" or 0% contamination). Association of labeled ligand at widely varying concentrations was followed by dissociation in the presence or the absence of an intermediate concentration of unlabeled ligand. Labeled ligand concentrations ranged from 0.02 to 200 nM ($0.01/K_e$ to $100/K_e$). These concentrations were designed to provide occupancy of receptors ranging from 0.003 to 0.94 at the end of the 60-min association period. Dissociation was initiated by *washing*, to remove all unbound ligand, and by a 100-fold dilution. By use of an intermediate unlabeled ligand concentration of 2 nM ($1/K_e$) during dissociation, we have simulated cases where occupancy was either increasing, decreasing, or remaining approximately constant during the dissociation period, depending on whether the concentration of labeled hormone during association was smaller, equal to, or larger than the concentration of unlabeled hormone during dissociation.

2. *No washing* (100% contamination). The above experiment was repeated, but the washing step was entirely omitted. Hence, there was 100% carry-over of both the bound and unbound labeled ligand. Dilution and addition of unlabeled ligand were as above.

Table III

Parameters Used for Simulations of Insulin–Receptor Binding

Association rate (k)	$=$	$0.01155 \, \text{nM}^{-1} \, \text{min}^{-1}$
Association time	$=$	$60 \, \text{min}$
Initial affinity constant (K_e)	$=$	$0.5 \, \text{nM}^{-1}$
Initial dissociation rate (k'_e)	$=$	$0.0231 \, \text{min}^{-1}$
Maximum dissociation rate (k'_f)	$=$	$0.1386 \, \text{min}^{-1}$
Cooperativity factor $(\delta = k'_f/k'_e)$	$=$	6
Receptor concentration	$=$	$4 \, \text{nM}$
Labeled hormone concentration	$=$	$0.02, 2, \text{and } 200 \, \text{nM}$
Unlabeled hormone concentration	$=$	$2 \, \text{nM}$
Dilution	$=$	100-fold

3. *Time lag*. The simulated experiment was repeated with the introduction of a 10-min lag period between the "true" onset of dissociation and the measurement of initial binding (B_0). The time lag was introduced by using the amount of labeled hormone bound at 10 min, instead of the true value immediately after the onset of dissociation, as "B_0" for normalizing the dissociation curves. The dissociation curves were plotted using the true duration of dissociation on the time scale, so that the time lag is evident on the graphs.

4. *Time lag and incomplete washing combined*. A lag period before the "initial" measurement of binding during dissociation was introduced together with contamination by unbound labeled hormone due to incomplete washing. The initial measurement B_0 was delayed 10 min while washing efficiency was set at 95%, leading to 5% residual free labeled hormone at the onset of dissociation.

4.3. Simulation Results

1. *Perfect washing* ("ideal case"). When the dissociation experiment was performed with perfect washing, we obtained the results shown in Fig. 5. The dissociation curves are expressed in terms of $\% \, B_0$ (on a log scale) vs. time, where $\% \, B_0$ is the amount of labeled hormone bound divided by B_0, the amount bound at the onset of dissociation $(t = 0)$. In the case of a single class of noncooperative bindings sites, the dissociation curve should be linear with a slope of $k'/2.303$ under conditions which prevent rebinding of dissociated hormone. In the presence of cooperativity, the dissociation curves will be nonlinear. Other factors, such as multiplicity of classes of receptors and/or rebinding of labeled hormone after dissociation, can also give rise to nonlinear dissociation curves which are concave up.

As expected, the apparent dissociation rate is slower for low concentrations of labeled hormone and faster for the high concentrations. The initial rates of dissociation range between the limits corresponding to k'_f and k'_e. Thirty minutes after the onset of the dissociation, all dissociation curves tend to be parallel. By 120 min, all dissociation curves reach a new equilibrium plateau, with a very low occupancy relative to the initial conditions. The initial downward curvature of the dissociation curve for low concentrations of labeled hormone and an intermediate

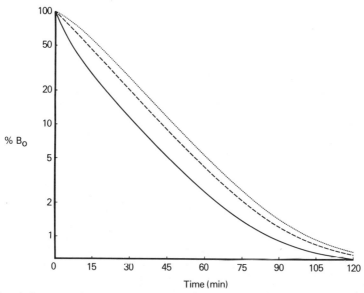

Fig. 5. Dissociation curves for increasing concentrations of labeled hormone. Association (60 min) is performed with increasing concentrations of labeled hormone (0.02, 2, and 200 nM). Dissociation curves for low (0.02 nM), intermediate (2 nM), and high (200 nM) labeled hormone concentration are shown, respectively, by dotted, dashed, and solid lines. Association is stopped and dissociation initiated by perfect washing of unbound labeled hormone (100% removal of unbound labeled ligand) and by a 100-fold dilution in the presence of three fixed concentrations of unlabeled hormone. Parameters are given in Table III. The dissociation curves for the labeled hormone are plotted as % B_0 = $B/B_0 \times 100$ (on a log scale) vs. time, where B_0 is the concentration of labeled hormone bound at the onset of dissociation.

concentration of unlabeled hormone is not apparent in the experimental curves reported to date (De Meyts et al., 1973; De Meyts et al., 1976; De Meyts, 1976a; Pollet et al., 1977), suggesting either that the model is inadequate or that the downward curvature was missed, due to "nonideal" experimental conditions.

2. *No washing*. In the case of 100% carry-over of unbound labeled hormone, the initial part of the dissociation curves (Fig. 6) is similar to that observed following perfect washing (Fig. 5). The initial instantaneous dissociation rate is still related to receptor occupancy. For low concentrations of labeled hormone, we still observe a downward concavity of the dissociation curve. As dissociation progresses, the curves for high labeled ligand concentrations cross over those for lower concentrations, resulting in an inversion of their late positions. According to the cooperative model, one would have expected that the dissociation curve for the larger concentrations of labeled hormone would always be steeper and positioned below those for lower concentrations of labeled hormone. Such an inversion of the expected order of the normalized dissociation curves is similar to that observed experimentally by Pollet.

3. *Lag period*. The effect of a time lag between perturbation and measurement of initial binding (experiment 3) is shown in Fig. 7. A delay of 10 min could result in obliteration of the initial downward concavity and a reduction of the

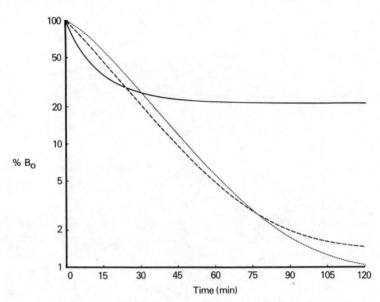

Fig. 6. Dissociation curves for increasing concentration of labeled hormone in the case of no washing (100% contamination). Dissociation curves for low (0.02 nM), intermediate (2 nM), and high (200 nM) labeled hormone concentration are shown, respectively, by dotted, dashed, and solid lines.

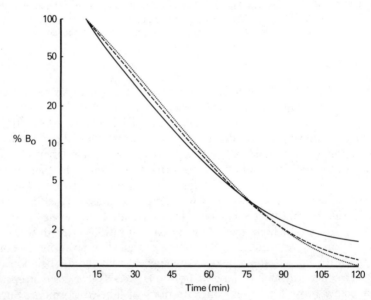

Fig. 7. Effect of a time lag on the shape of the dissociation curves. The time lag between the onset of dissociation and the initial measurement of B_0 was introduced by use of the amount of hormone bound after 10 min of dissociation as an apparent "B_0." Dissociation curves for low (0.02 nM), intermediate (2 nM), and high (200 nM) labeled hormone concentration are shown, respectively, by dotted, dashed and solid lines. The true time scale is used on the abscissa; hence, the curve begins when $t = 10$. Other experimental conditions are as described in Fig. 5.

curvature of the initial part of the dissociation curve. Further, this time lag also results in a reversal of the relative position of the curves late in dissociation, similar to that observed in experiment 2. Thus, the paradoxical inversion of the dissociation curves can be simulated both by incomplete washing of labeled hormone and by the time lag. However, only the time lag has been found to obliterate the initial downward curvature (Fig. 7). This does not exclude the possibility that some other source of error, i.e., temperature perturbations, may explain the disappearance of this downward curvature.

4. *Lag period and contamination.* The combined effects of carry-over of labeled hormone and introduction of an artifactual time lag are shown in Fig. 8. When a 10% contamination is combined with a 10-min delay, the initial downward curvature is no longer present, and the concave upward curves appear to be "stacked" in the reverse order compared with the ideal case (Fig. 5). Even with only 3–5% contamination and a 5–10 min delay (data not shown), obliteration of the downward curvature and the reversal of the position of the curves are still observed. Thus, when the contribution of residual free labeled hormone and the delay between association and dissociation are present simultaneously, model I can reproduce the major findings of the studies of both De Meyts *et al.* (1976) and Pollet *et al.* (1977). Although one might ordinarily regard a 3–5% carry-over of labeled ligand and a 5–10 min delay in manipulation as acceptable (e.g., for equilibrium studies), these two factors can have surprisingly large effects on the properties of the dissociation curves.

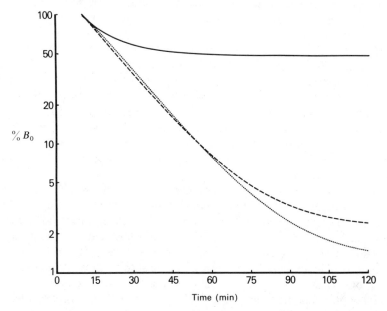

Fig. 8. Combined effect of a 10-min time lag with incomplete washing (10% residual free hormone) on the shape of the dissociation curves. The carry-over of labeled hormone was introduced by setting the concentration of free labeled hormone at 10% of its value immediately after 100-fold dilution. Other experimental conditions are as in Fig. 5. Dissociation curves for low (0.02 nM), intermediate (2 nM), and high (200 nM) labeled hormone concentration are shown, respectively, by dotted, dashed, and solid lines.

4.4. Discussion

The present computer simulation studies indicate that the predictions of our mathematical model for cooperative binding can reproduce the experimental data of neither De Meyts nor Pollet unless we consider the contributions of incomplete washing (or, equivalently for present purposes, rapidly dissociating nonspecific binding) and delays before obtaining measurements of binding after perturbation. These practical experimental problems are likely to be present at least to some extent in the insulin–lymphocyte system, as in most other binding systems. Nonspecific binding in this system has been reported to represent approximately 3% of the total hormone (De Meyts, 1976a). Moreover, this nonspecific binding component would be expected to dissociate most rapidly during the first few minutes of the dissociation process (De Meyts et al., 1976). Thus, even if the binding data were "corrected" for nonspecific binding, the rapid mutation of the nonspecific compartment into free hormone would be likely to result in unexpectedly serious contamination of the binding system. In view of the importance of nonspecific binding, it would be desirable for experimental results to be presented in terms of *raw counts bound* rather than in terms of binding above nonspecific counts or % B_0, i.e., counts bound relative to an arbitrary time origin. Only in this manner can one evaluate the effects of nonspecific counts, incomplete washing, and time lags.

The effect of temperature on the dissociation rate constant k'_f in the presence of an excess of unlabeled hormone has been studied for the insulin receptor (De Meyts et al., 1976) and the β-adrenergic receptor (Limbird and Lefkowitz, 1976). As expected, k'_f increases with temperature. Thus, lowering of temperature from 15 to 4°C during centrifugation at the end of the association period (De Meyts et al., 1976; Pollet et al., 1977) might be expected to "freeze" the binding process and to minimize the effect of the delay required for centrifugation. However, the effects of temperature on the other binding parameters are not known for these systems. Moreover, the temperature jumps might also introduce unpredictable conformational changes in the binding sites. The reversal of these changes in the binding properties of the receptors may not follow immediately the shift from 4°C back to 15°C, and an altered state may conceivably be present during the early phase of dissociation. Thus, temperature jumps should be considered as perturbations, but their simulation with the present model would require a more complete knowledge of the temperature dependence of the binding parameters. In the absence of a complete understanding of these effects, it would be advisable to avoid such temperature changes. It is conceivable that thorough studies of temperature jumps might contribute to a better understanding of hormone–receptor interactions, in analogy to relaxation kinetic studies already applied to chemical reactions (Castellan, 1963), enzyme reactions (Hammes and Schimmel, 1967), antibody–hapten binding (Haselkorn et al., 1974), and drug–receptor interactions (Neuman and Chang, 1976).

The absence of the *hallmark* of negative cooperativity, i.e., the downward curvature of log(% B_0) vs. time (De Meyts et al., 1973; De Meyts et al., 1976; De Meyts, 1976b; Pollet et al., 1977), can be attributed either to failure of the model or to a combination of systematic experimental artifacts. We favor the latter explanation, but a definitive resolution of this question will require improved

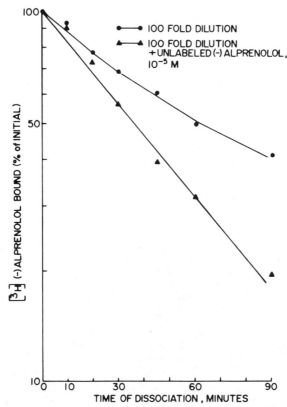

Fig. 9. Enhancement of the dissociation rate of alprenolol from frog erythrocytes (Limbird *et al.,* 1975, Fig. 2) in the presence of excess unlabeled alprenolol (reproduced with permission of L. E. Limbird and R. L. Lefkowitz). In the presence of excess unlabeled ligand (triangles), the initial part of the dissociation curve appears to show the downward curvature expected when negative cooperativity is present in the "ideal" case (Fig. 5). A similar pattern was observed in several, though not in all, experiments.

methods for washing, addition of unlabeled ligand, mixing, and initial sampling which would minimize artifacts. The downward curvature of the dissociation curve would not be expected if the concentration of unlabeled hormone were in large excess, so intermediate concentrations ($\sim 1/K$) should be used for this purpose.

To examine early time points, it is likely that separation methods based on filtration would be superior to methods based on relatively slow sedimentation. As an example, Fig. 9 shows the "De Meyts" dissociation experiment for the case of the β-adrenergic receptor in frog erythrocytes (Limbird *et al.,* 1975; Limbird and Lefkowitz, 1976). The experimental data strongly suggest an initial downward curvature. These results were obtained under conditions which tend to minimize lag periods by use of filtration for separating bound and free hormone. For the β-adrenergic system, nonspecific binding appears to be less rapidly dissociating than in the case of the insulin receptor, corresponding to "improved washing."

Table IV
Comparison of Predicted and Observed Properties of the Insulin–Receptor System

	Properties	Predicted	Observed
1.	Scatchard plot concave up and symmetrical	Yes	Yes
2.	Apparently noncooperative association curves	Yes	Yes
3.	Enhanced dissociation rate with increasing occupancy for "dilution + excess unlabeled ligand"	Yes	Yes
4.	Downward curvature of dissociation curves for "dilution + intermediate unlabeled ligand"	Yes	No
5.	Linear dissociation curves for high dilution with or without large excess of unlabeled ligand	Yes	No
6.	Apparent independence of dissociation rate and initial occupancy under selected nonideal conditions	Yes	??

The simulations shown in Sec. 4.3 were performed with the "standard" set of parameters shown in Table III. However, the major effects of the experimental variables remain qualitatively and nearly quantitatively the same despite considerable changes in parameters over a two- or threefold range. We have repeated these simulations using several different sets of parameters. Decrease of k' (due to change in k'_e or in δ) effectively expands the time scale, whereas increasing k' compresses it. Likewise, changes in K result in a corresponding change in the concentrations of all reactants if the reaction patterns are to remain the same (a tenfold increase in K corresponds to a tenfold decrease in h^*, h, and r). Thus, the major properties of the model are quite insensitive to changes in the parameters over a wide enough range, so that the major conclusions of the present study should be valid, even if the parameters of Table III were modified in light of additional studies.

Mathematical modeling of the kinetics of hormone–receptor binding should help in the interpretation of available (and seemingly conflicting) experimental results and in the design of more refined and definitive experiments. In the case of the insulin receptor, the model can resolve at least part of some current controversy regarding the existence of negative cooperativity. Table IV summarizes the comparison between the model and the experimental data. Some discrepancies between the model and the experiments remain to be explained, especially the observation that, under all experimental conditions, the dissociation curves are nonlinear and concave up when portrayed on a logarithmic ordinate.

5. Model II: k_{on} as a Linear Function of Occupancy

5.1. Introduction

It is generally agreed that in systems where the association rate constant is mainly determined by diffusion factors, differences in affinity among analogues of the same hormone reflect differences in the dissociation rate constant, the association rate constant being invariant (Weber, 1975; Triggle and Triggle, 1976; Koren and Hammes, 1976). However, several exceptions have been observed (Taylor *et al.*, 1970; Yoda, 1972). In most cases, ligand interactions are probably complex processes involving multiple steps between the initial interaction or nucleation and the fully interacting state (Richards *et al.*, 1975; De Lisi and Metzger, 1976; Engel and Winklmair, 1972; Ariëns and Simonis, 1976; Triggle and Triggle, 1976; De Lean *et al.*, 1979). Any cooperative process which would alter one or more of these binding steps would result in a change in the apparent association rate "constant."

While specific mechanistic models could be considered, a general non-mechanistic cooperative model considering changes in the association rate constant should provide a basis for studying the kinetics of association and dissociation of cooperative binding. In analogy with the previous model, the exponential relationship between the rate constant and fractional occupancy corresponding to the equilibrium model of Scatchard (Fig. 10) might well be approximated by a linear relationship, provided that the interactions are not too strong.

This model is the conceptual counterpoint of model I. Its properties are different and require separate study. Moreover, a complete study of the cooperative properties of a binding system should include consideration of changes in both association and dissociation rate constants.

As expected, equilibrium binding data are insufficient to characterize the model. Dissociation experiments would not be informative if the association rate constant were the only changing rate constant. Specific experimental tests may reveal positive cooperative effects; these consist of two-stage association experiments to examine the effect of increasing receptor occupancy on the initial association rate of labeled hormone.

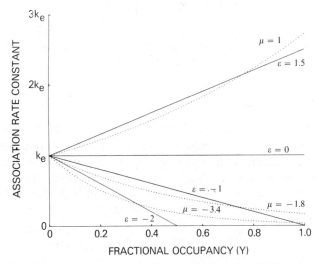

Fig. 10. Model II: relationships between k_{on} and fractional occupancy. The linear relationship implied by model II is shown (solid line) and compared to Scatchard's exponential model (dotted lines). In the case of severe negative cooperativity ($\varepsilon < -1$), model II would result in an apparent reduction of binding capacity, since k_{on} becomes zero and the reaction would stop before all sites initially present become saturated.

5.2. The Model

In analogy with model I (Sec. 3), we consider reaction scheme (**III**):

$$H + R \underset{k'}{\overset{k}{\rightleftharpoons}} HR \tag{III}$$

where H is the hormone, R is the receptor, and HR is the hormone–receptor complex. Here k is the *variable* association rate constant (coefficient) and k' is the *invariant* dissociation rate constant. The association rate constant k is related to the fractional occupancy Y of the receptor sites according to

$$k = k_0(1 + \varepsilon Y) \tag{6}$$

where k_0 is the initial association rate constant when all the binding sites are initially empty and ε is an interaction factor analogous to δ. When $\varepsilon = 0$, there is no cooperative effect, and k is invariant ($k = k_0$). When $\varepsilon \neq 0$, the binding process will be positively or negatively cooperative depending on whether ε is positive or negative, respectively. The analytical solution to the differential equation corresponding to reaction scheme (**III**) is given in Appendix C.

If both labeled (H*) and unlabeled (H) hormone are present, the reaction can be summarized as

$$H^* + \underset{k'}{\overset{k}{\rightleftharpoons}} H^*R$$
$$H\ \ + \underset{k'}{\overset{k}{\rightleftharpoons}} HR \tag{IV}$$

where

$$k = k_0(1 + \varepsilon Y) \tag{7}$$

$$Y = \frac{U}{r} \tag{8}$$

$$U = [H^*R] + [HR] \tag{9}$$

The corresponding differential equations are shown in Appendix D. There is no general analytical solution to these equations, but some approximations could be obtained under special conditions (Appendix D).

5.3. Properties of the Model

5.3.1. Equilibrium

The binding isotherms for various degrees of positive ($\varepsilon > 0$) and negative ($\varepsilon < 0$) cooperativity are shown in Fig. 11. In the case of positive cooperativity, the curve is shifted toward the lower concentration range and becomes steeper as ε increases. The free hormone concentration $F_{0.5}$ corresponding to 50% maximal occupancy when $\varepsilon > -1$ is given by the following relationship:

$$F_{0.5} = \frac{2}{K_0(\varepsilon + 2)} \tag{10}$$

where K_0 is the ratio k_0/k'. In the case of negative cooperativity, there is a decrease in the apparent binding capacity when $\varepsilon < -1$. The maximum binding capacity then becomes

$$B_{max} = -\frac{r}{\varepsilon} \tag{11}$$

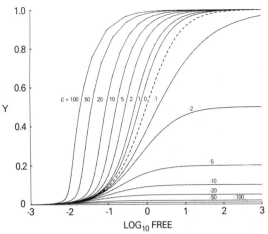

Fig. 11. Model II. Equilibrium binding isotherms for various degrees of either negative ($\varepsilon < 0$) or positive ($\varepsilon > 0$) cooperativity. $K_0 = 1$ and $r = 1$. For positive cooperativity, the curves shift progressively to the left. For negative cooperativity, the maximum binding capacity is progressively diminished.

This apparent decrease results from the fact that when $\varepsilon < -1$ the association rate constant will become negligible before all the binding sites are filled.

The corresponding Hill plots are shown in Figs. 12(a) and 12(b). In the case of positive cooperativity, the Hill plot is not linear. It can be shown that the free hormone concentration for which the Hill plot is steepest is

$$\hat{F} = \frac{1}{K_0(\varepsilon + 1)} \tag{12}$$

and does not coincide with $F_{0.5}$. In the case of negative cooperativity, the Hill plot is clearly nonlinear only for values of $\varepsilon > -2$. In the presence of a higher degree of negative cooperativity, the Hill plot is a straight line with a slope approaching unity.

The corresponding Scatchard plots are shown in Fig. 13. In the case of positive cooperativity (Fig. 13, left panel), the Scatchard plot is bell-shaped with a maximum positioned near the center of the plot, provided $\varepsilon > 1$. In the case of negative cooperativity (Fig. 13, right panel), the Scatchard plot is curvilinear with weak interactions ($0 > \varepsilon > -1$). When interactions are stronger, the plot again becomes nearly linear with a severe underestimate of the total number of receptor sites.

A low-dose hook is observed for positively cooperative interactions (provided $\varepsilon > 1$) in the case of displacement curves of the labeled hormone by the unlabeled species [Fig. 14(a)]. This low-dose hook effect is most pronounced when both the labeled ligand and the receptor concentrations are very small ($0.001/K_0$) relative to the initial affinity constant [Fig. 14(b)]. When both the labeled hormone and the receptor concentrations are high ($>1/K_0$), the hook flattens and tends to disappear [Fig. 14(c)]. For practical reasons, displacement curves are usually constructed experimentally with both the labeled hormone and the receptor concentration near the reciprocal of the affinity constant ($0.1/K_0 - 1/K_0$), and thus

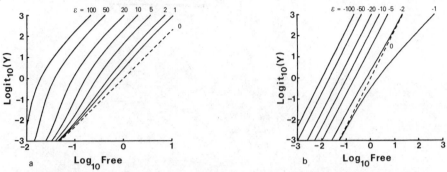

Fig. 12(a). Hill plots of the binding isotherms shown in Fig. 11 in the case of positive cooperativity ($\varepsilon > 0$). These plots are nonlinear even in the central range, which is readily accessible experimentally. (b). Hill plots of the binding curves shown in Fig. 11 for the case of negative cooperativity. The logit transform was calculated using the apparent B_{max} [Eq. (11)]. For extreme values of negative cooperativity ($\varepsilon < -2$), the curves become essentially linear. Note: $\text{Logit}(Y) = \log_e[Y/(1 - Y)]$.

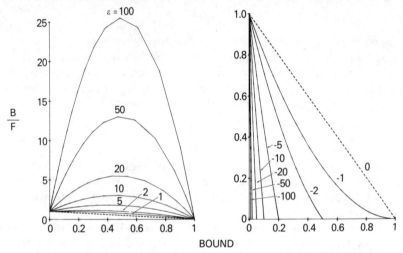

Fig. 13. Scatchard plot of binding equilibria for the curves shown in Fig. 11. In the case of positive cooperativity (left panel), the curves are bell-shaped (biphasic). In the case of negative cooperativity (right panel), the plots are curvilinear for low degrees of interactions ($\varepsilon \simeq -1$) but approach linearity in the case of strong negative interactions.

such hooks may easily be missed. In the case of radioimmunoassays, the high specific activity of the iodinated hormone permits the use of low concentrations of the ligand, and such low-dose hook effects have been observed (Matsukura et al., 1971; Weintraub et al., 1973).

5.3.2. Association Curves

Since the cooperative properties of the present model are based on changes in the association rate constant, one would expect that the cooperative properties

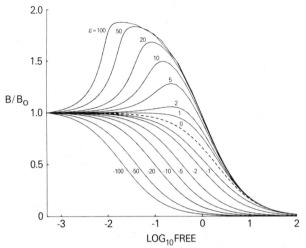

Fig. 14(a). Displacement curves using a constant infinitesimal concentration of labeled ligand ($h^* = 0.001/K_0$) and increasing concentrations of unlabeled ligand (abscissa) for various extents of positive or negative cooperativity. Ordinate: $[H^*R]/[H^*R]_0$ or B/B_0. $K_0 = 1$, $r = 1$. For positive cooperativity, there is a "hook effect" in the low-dose region if $\varepsilon > 1$. For negative cooperativity, the curves shift progressively to the left and downward, due to the progressive loss of effective affinity and binding capacity.

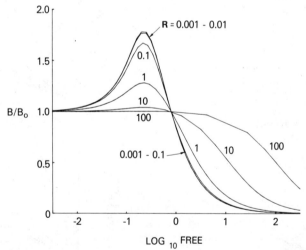

Fig. 14(b). Effect of varying receptor concentrations on the low-dose hook effect seen in Fig. 14(a) when the labeled hormone is maintained at a low value ($h^* = 0.001/K_0$). For $r > 1/K_0$, the hook effect is progressively diminished. $K_0 = 1$.

should be experimentally observable with association experiments. Figure 15(a) shows the time course of association of intermediate concentrations of labeled hormone for varying degrees of positive and negative cooperativity. An early S-shaped association curve can be observed when there is a high degree of positive cooperativity [Fig. 15(b)]. For a high degree of negative cooperativity, the association curve plateaus early at low receptor occupancy.

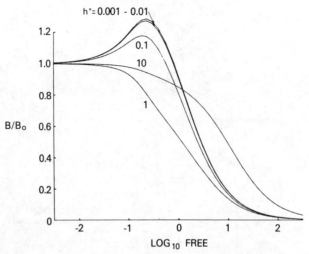

Fig. 14(c). Effect of concentration of the labeled hormone on the magnitude of the hook effect. Here the receptor concentration is held constant ($r = 1/K_0$). The hook is apparent only for low values of labeled hormone concentration ($h^* < 1/K_0$). $K_0 = 1$.

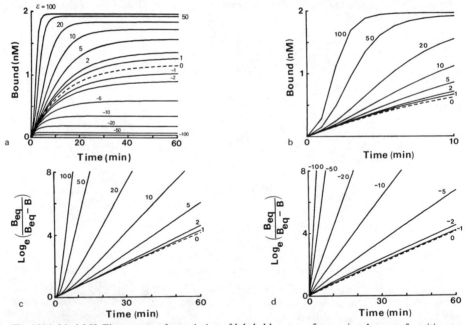

Fig. 15(a). Model II. Time course of association of labeled hormone for varying degrees of positive or negative cooperativity. Rate constants have been chosen to be similar to those used in Table III. $k_0 = 0.01155$ nM^{-1} min^{-1}; $k' = 0.0231$ min^{-1}; $r = 4$ nM, $h^* = 2$ nM. ε varies from -100 to $+100$. (b). Expanded view of the early part of Fig. 15(a) showing the sigmoidal (S-shaped) initial segment of the association curve for high degrees of positive cooperativity. (c). Transformation to linearize first-order or pseudo-first-order reactions, applied to association curves of Fig. 15(a), in the case of positive cooperativity. Note the upward curvature for early time points, corresponding to the *autocatalytic* or positive cooperative effect. (d). Transformation of association curves as in Fig. 15(c) in the case of negative cooperativity. Departure from linearity would be difficult to detect, at least for this set of parameter values.

When the association curves are transformed using the pseudo-first-order linear transform, $\log_e[B_{eq}/(B_{eq} - B)]$, the simulated curves clearly depart from linearity in the case of positive cooperativity [Fig. 15(c)]. However, this non-linearity may occur so early that it would be very easily missed experimentally. In the case of negative cooperativity [Fig. 15(d)], the transform leads to nearly straight lines. Thus, the analysis of experimentally obtained individual association curves may fail to reveal any significant clue to the presence of cooperative interactions. However, simultaneous analysis of a family of association (and dissociation) curves should be more indicative of site–site interactions.

5.3.3. Dissociation Curves

The dissociation experiments described in Sec. 3.2.3 for model I were informative of changes in the dissociation rate constant. However, for model II, the dissociation curves following "perfect" washing and "infinite" dilution are superimposable with the curves obtained after washing, dilution, and addition of excess unlabeled hormone, as expected. In the presence of rebinding (inadequate dilution or incomplete washing), addition of excess unlabeled hormone may have a detectable effect.

5.4. Testing for Positive Cooperativity

In the case of positive cooperativity due to increase of the association rate constant with receptor occupancy, a diagnostic experiment based on a two-stage association experiment could be performed. The receptor sites are first incubated with unlabeled hormone to increase occupancy so that the association rate constant is increased. Then a tracer amount of labeled hormone is added, and the resulting association curves are compared with the control curve obtained when the binding sites were not preincubated with unlabeled hormone (Fig. 16). If the concentration of unlabeled hormone is properly chosen, the increase in the association rate constant will counterbalance the reduction of free binding sites so that preincubation with unlabeled hormone will result in a paradoxical increase in the apparent association rate constant. Appendix E provides the details for calculating the optimum concentration of unlabeled hormone to be used during the preincubation. Figure 16 shows the effect of increasing the concentration of unlabeled hormone during preincubation for this experiment. Use of concentrations of unlabeled hormone larger than the optimum value results in the expected slowing down of association instead of the paradoxical enhancement of tracer binding. If the labeled hormone concentration is too large, then the total fractional occupancy will increase during the second incubation period and move away from its optimal value, resulting in a loss of the enhancing effect. Similarly, the unbound unlabeled hormone should not be washed out at the end of the preincubation period; otherwise fractional occupancy will drop from its optimum value, also resulting in a loss of the desired effect.

A similar paradoxical enhancement of tracer binding following preincubation with unlabeled hormone might be observed in a system where the receptor sites are being degraded during preincubation, provided the presence of unlabeled hormone would prevent such a degradation by occupying and stabilizing the

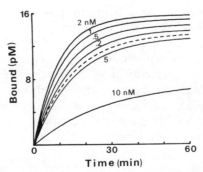

Fig. 16. Effect of increasing the concentration of unlabeled hormone during the preincubation period. Parameters are similar to those in Table III: $k_0 = 0.01155$ nM^{-1} min^{-1}, $k' = 0.0231$ min^{-1}, $\varepsilon = +5$, $r = 4$ nM, $h^* = 0.02$ nM. Duration of the preincubation = 60 min. Here, the optimal value of h is 2.04 nM. Choice of h is critical. Dashed line: control, $h = 0$.

sites. Such a phenomenon is common for cytosol steroid receptors (Chamness and McGuire, 1975). This alternate explanation for enhanced binding could be ruled out in separate control experiments, checking for permanent loss of binding capacity following preincubation in the presence of ligand.

Another feasible test would be to preincubate the labeled hormone first and then add unlabeled hormone ("hot first, then cold," Fig. 17). The optimum conditions for this test are the same as those for the former. The first incubation with hormone at low concentration should approach equilibrium. The second incubation should be initiated by addition of an optimum concentration of unlabeled hormone. This approach has the advantage that receptor occupancy can be monitored during the first incubation, so that receptor degradation might be detectable.

Figure 17 shows the effect of varying concentrations of unlabeled hormone during the second incubation period. Concentrations of unlabeled hormone in excess of the optimum value produce a transient increase in labeled hormone binding, followed by a sustained decrease. The paradoxical increase in binding in these kinetic studies will occur when there is a hook in the Scatchard plot, i.e., when $\varepsilon > 1$.

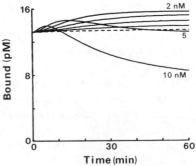

Fig. 17. Effect of variation of the concentration of the unlabeled ligand during the second stage of incubation when the labeled ligand has been introduced in a preincubation. Parameters are as in Fig. 16, except for the order of addition of labeled and unlabeled ligands. Note the definite optimum value for h and the narrow region surrounding this which permits observation of *enhanced* binding. Use of preincubation with labeled ligand facilitates controls for degradation of ligand or receptor.

6. General Discussion

In this chapter, we have concentrated most of our attention on the two special restricted cases where either the dissociation constant or the association constant changes as a simple function of occupancy. In view of the dozens of models available for hemoglobin, enzymes, and the acetylcholine receptor, why confine our attention to these extensions of the Scatchard and De Meyts models?

First, the properties and applications of the other models have already been discussed in the literature, *in extenso*. Second, the present group of models has the primary virtue of simplicity. While it would be controversial to argue that the present models are more simple than the Adair model, they rival even this classical model in this regard. For a divalent receptor, the Adair model would involve a total of five parameters (four rate constants and the receptor concentration), rather than four parameters as in the present models. However, models I and II provide predictions for the kinetics of binding, while the properties of Adair's model have been developed only for equilibrium. The sequence of binding steps is arbitrary and may not correspond to the real sequence of events during association and dissociation. The kinetic properties of the Adair model may not be so simple as its properties at equilibrium, and many additional assumptions may be required.

Third, the present models vastly facilitate the study of reaction kinetics under a wide variety of conditions and types of experimental design, since analytical solutions are available for the differential equations and because the number of parameters is still manageable. Most treatments of cooperativity in terms of *enzyme kinetics* deal with the *steady-state kinetics*, i.e., the catalytic velocity of the enzyme for given concentrations of substrate and allosteric effectors. However, these enzyme kinetics correspond to the *equilibrium* case for ligand binding. We note that the Monod–Wyman–Changeux (1965), Changeux *et al.* (1967), and Koshland–Nemethy–Filmer (1966) models have been developed in the context of the equilibrium case, and these authors have not considered the corresponding non-steady-state kinetics. Only recently have Herzfeld and Stanley (1974, Figs. 3–8) considered the transient-state kinetics of oxygen binding to hemoglobin both experimentally and theoretically.

After having explored models I and II in exhaustive detail, the experimentalist may wish to examine more complicated models, e.g., those involving changes in both k_{on} and k_{off} and those involving changes in conformational state, viz.

$$H + R \rightleftharpoons HR \rightleftharpoons HR' \tag{V}$$

Here, HR' represents a ligand–receptor complex differing in its properties (conformation, activity) from HR. This reaction scheme is not cooperative *per se*: If the four rate constants were truly constant, one should have a linear Scatchard plot. The mathematical properties of this model have been considered (Ross *et al.*, 1977). This model, well known in the enzymology literature (Kirsch, 1973), is quite relevant here. While the Scatchard plot is linear, the dissociation curves will be biphasic (i.e., the sum of two negative exponential terms). Thus, this scheme, combined with the premise that one or more of the rate constants is a (linear) function of occupancy, might provide a further generalization of the cooperative models I and II, encompassing both the cooperative properties and the hitherto unexplained apparent heterogeneity in the dissociation kinetics of insulin.

We cannot overemphasize that all of the analyses in this chapter have been highly oversimplified. While we have considered the effects of incomplete washing and the effects of errors in the time of measurement of B_0, we have ignored a host of other experimental practicalities, e.g., the problems of separation of bound and free ligands. It is incumbent on the experimentalist to indicate that his method(s) for separation of bound and free does not significantly perturb the *partitioning* previously established (Rodbard and Catt, 1972).

Degradation or instability of either or both hormone and receptor is another major source of experimental artifact (Ketelslegers *et al.*, 1975). It is possible that the rapid rate of insulin degradation may account for the failure of Gliemann *et al.* (1975) to observe the cooperative effect in fat cells. Undoubtedly, more complex models will be progressively explored as the available data permit and require.

Perhaps most disturbing is the problem of *mixing* and the unstirred layer. De Meyts has provided an impressive rebuttal to the argument that the enhanced dissociation observed upon addition of unlabeled ligand is due to an unstirred layer surrounding the cell (De Meyts *et al.*, 1973; De Meyts, 1976b). Perhaps the *structural specificity* arguments and the demonstration of ligands which bind but do not induce cooperativity are most convincing. Change of viscosity (e.g., by use of sucrose, glycerol) is one approach to examine whether the *enhanced dissociation* depends on the diffusion coefficient of the ligand in the medium (De Lisi and Metzger, personal communication), as would be expected for the unstirred layer mechanism. According to the unstirred layer hypothesis, the retention effect should be proportional to the local concentration of binding sites in the membrane (De Lisi and Metzger, 1976). Increasing or decreasing the local concentration of sites should proportionally alter the apparent enhancement of dissociation rate with unlabeled hormone (De Lisi, personal communication). These experiments, readily applicable to enzymes or binding proteins in free solution, may be problematic in interpretation when applied to living cells or even membranes.

Pollet has interpreted his results as indicative of heterogeneity of the receptor, with one predominant form over the physiological range of concentrations. In view of the nonlinear Scatchard plot, it would be plausible to assume that the class of sites with the lower affinity (K) would have the higher k_{off}. In such cases, as the degree of total occupancy of the receptor increases, we would expect an increase in the fraction in the rapidly dissociating pool. Thus, the slope of the log–dissociation curve, $d \ln (B/B_0)/dt$, should increase as occupancy increases. If one were effectively at *infinite dilution* and in the absence of an unstirred layer, then addition of unlabeled ligand at the onset of dissociation should have no effect. Both of these predicitons are at variance with Pollet's findings: The wide variation in the mass of tracer during association gave rise to no change in the initial rate velocity of dissociation, whereas addition of unlabeled ligand at the onset of dissociation still had a major effect. Hence, it is diffult to reconcile the available data with a simple model of receptor heterogeneity. This would be plausible only if the lower-affinity class of sites were to have the same dissociation rate constant as the high-affinity class of sites. This would imply that the second class of sites, present in larger number than the first (based on "curve peeling" of the Scatchard plots), would have a slower association rate, but no evidence was found for this on the basis of the association experiment.

It may be difficult to distinguish between the kinetics of a heterogeneous receptor, as proposed by Pollet *et al.* (1977) and other workers, and the combination of a sequential (stepwise) reaction of the type shown in reaction scheme (**V**). Thus, it becomes helpful to perform computer simulations using both models in an attempt to find optimal conditions to discriminate between these two possibilities.

Another alternative explanation for Pollet's results might be that the labeled and unlabeled ligands display different properties with respect to rate constants and/or to the cooperative effect. Since some analogues of insulin fail to enhance the dissociation rate, it is conceivable that the process of iodination results in damage, fragmentation, or other modification of the insulin molecule such that its ability to induce the change in k_{off} is diminished or lost. This might explain the failure to observe an enhanced rate of dissociation in the case of a variable amount of labeled ligand during association but the retention of the effect of unlabed insulin added at the onset of the dissociation period. There are several arguments against this possibility. First, Pollet presumably used the same iodination procedure and insulin preparations as De Meyts; yet De Meyts *et al.* (1976) did observe an enhanced rate of dissociation as the mass of labeled ligand was varied. Second, although Pollet refers to a change in the mass of tracer, it should be noted that the specific activity of the labeled reagent was not constant. To avoid too drastic an increase in the number of counts, he progressively reduced the specific activity of the labeled insulin by addition of unlabeled ligand. Hence, even if the labeled reagent were unable to induce the putative *conformation change* (or whatever is needed to enhance k_{off}), the progressively increasing mass of unlabeled ligand should have been able to do so. Ideally, the experimentalist should indicate that the results obtained are independent of the specific activity of tracer and/or independent of the specific activity or method of labeling.

Several workers have proposed alternative hypotheses which might explain the enhanced dissociation in the presence of unlabeled ligand during dissociation. J. Wolfe, B. Mullin, and A. Minton (personal communications) have suggested that one might be dealing with a *concerted substitution* reaction, e.g., as follows:

$$H + H^*R \rightleftharpoons [H^*RH] \rightleftharpoons HR + H^* \qquad (VI)$$

Hence, the unlabeled ligand (H) would result in an enhanced rate of dissociation if the net velocity of reaction scheme (**VI**) were faster than the intrinsic rate of dissociation of H* from R. Minton has shown quantitatively that the above kind of reaction mechanism, when combined with heterogeneity of sites, would be compatible with some published results; the dissociation rate becomes (nearly) independent of initial occupancy but markedly altered by the presence of unlabeled ligand during dissociation. However, the concerted substitution model would be incompatible with the findings of De Meyts *et al.* (1976) that the extent of enhanced dissociation does depend on the concentration of labeled ligand used during association. This model is perhaps the simplest case of what might be regarded as the *multisubsite* model. A detailed treatment of this important case is presented elsewhere (De Lean *et al.*, 1979).

Leaving aside the applicability of the model, we would note that the concerted substitution reaction, though regarded by some as a noncooperative model, may equally well be regarded as an extreme case of a cooperative reaction. Thus, we encounter semantic problems. While the concerted substitution reaction mechanism is distinct from the one proposed by De Meyts (and considered at length here), it may be regarded as the ultimate form of *negative cooperativity* if we regard the receptor as though it were divalent rather than monovalent. The molecular species [H*RH] and/or [HRH*] is a putative transitory intermediate. Thus, we have a *half-of-sites* reactivity, which is generally regarded as a cooperative mechanism. This reaction is similar to the bimolecular nucleophilic substitution mechanism (S_N2) described for small organic molecules. With noncovalent macromolecular binding, it would usually be regarded as a cooperative (though not necessarily an *allosteric*) model.

7. A Guide to the Experimentalist

Characterization of a binding system becomes a progressively more demanding task. How does one proceed to identify positive or negative cooperativity and evaluate whether this is due primarily to changes in k_{on} or k_{off}? A brief summary is given in Table V. For negative cooperativity involving k_{off}, the crucial experiment is the "De Meyts experiment," i.e., dissociation following dilution, with and without addition of moderate to large concentrations of unlabeled ligand. The downward curvature, if present, is confirmatory (Figs. 4(b), 9). The major competing hypotheses are the presence of *rebinding* of hormone after dissociation (i.e., insufficient dilution), the presence of an unstirred layer, and the concerted substitution mechanism. The same experiment should also be able to detect positive cooperativity, with a decrease in k_{off} with increasing occupancy. However, the shape of the dissociation curves would then be indistinguishable from that expected in the case of heterogeneity of sites. The second critical experiment to evaluate this model is dissociation after various degrees of occupancy have been achieved using different concentrations of labeled (and unlabeled) ligand during the association phase (De Meyts *et al.*, 1976; Pollet *et al.*, 1977).

The evaluation of changes in k_{on} is considerably more difficult. Indeed, in the case of negative cooperativity, it is perhaps impossible to design an experiment based on overall macroscopic association kinetics (as opposed to rapid perturbation or relaxation methods) which will distinguish between changes in k_{on} for a single molecule and simple heterogeneity of sites. For positive cooperativity, the situation is only slightly better. Here we can perform two types of experiments: either preincubation with unlabeled ligand, followed by addition of labeled ligand, or preincubation with labeled ligand, followed by addition of unlabeled material. In either case, we are looking for a paradoxical increase in binding of labeled ligand due to occupancy by the unlabeled material. Unfortunately, a

Table V
Summary of Tests for Cooperativity

Varying constant	Cooperativity type	Experiment	Characteristics	Artifacts
k_{off}	Negative	Dissociation	Downward curvature Enhanced dissociation rate with increasing occupancy	Unstirred layer Rebinding Heterogeneity of binding sites
	Positive	Dissociation	Decreased dissociation rate with increasing occupancy	Heterogeneity of binding sites Multistep binding
k_{on}	Positive	Association	Enhanced association rate after preincubation with unlabeled hormone	Degradation of binding sites

qualitatively similar pattern is expected in the relatively frequent case where there is degradation of receptor.

Obviously, the problem of establishing a model is vastly complicated when two or more mechanisms are operating simultaneously, e.g., heterogeneity of sites combined with cooperativity or combined with ligand or receptor degradation. With the relatively crude preparations of intact cells or membranes in common use today, this multiplicity of effects is a likely occurrence. With progressive purification of the systems, more reliable kinetic and descriptive biochemical data should become available. Nevertheless, we would argue that it is desirable to employ quantitative data analysis, i.e., computerized least-squares curve fitting, using a specified model, and simultaneous analysis of all available experimental data. For instance, analysis of association, equilibrium, or dissociation data alone is usually inadequate to permit verification of compatibility with any given model. Only by simultaneous curve fitting of all three types of experiments, with *constraints* such that the same parameters are used throughout, can we obtain enough information to discriminate among the multitudinous possibilities. This kind of simultaneous curve fitting is available on several general-purpose modeling programs. Among the first of these is SAAM (Berman and Weiss, 1971). Similar analyses are available on MODELAIDE (Shrager, 1970) and MLAB (Knott and Reece, 1972). In the past, this kind of curve fitting has required rather large programs, extensive computer facilities, and a fairly sophisticated user. Recently, we have demonstrated the feasibility of the use of much smaller, special-purpose programs on smaller machines in the BASIC language, with considerable ease of operation by persons with limited experience with computers (De Lean *et al.*, 1978). These programs make it possible to evaluate the *goodness of fit* of the model to the data using all of the data for all ligand and/ or receptor concentrations and all durations of the reaction simultaneously. This provides a much more informative analysis than fitting individual segments of the data [e.g., initial-reaction velocities, time necessary to reach half-maximal saturation ($t_{1/2}$), equilibrium, etc.]. Such global curve fitting is extremely sensitive to relatively small departures from the model but requires a large collection of experimental data with fairly low *noise*. A careful study of the deviations of the data from their predicted values should provide a critical test for the model and might suggest directions for further modifications and improvements.

Appendix A: Model I: Differential Equations and Solutions

The equations describing total binding U are

$$\frac{dU}{dt} = k(h^* + h - U)(r - U) - k'U \tag{A1}$$

$$h^* = [H^*] + [H^*R] \tag{A2}$$

$$r = [R] + [H^*R] + [HR] \tag{A3}$$

$$h = [H] + [HR] \tag{A4}$$

where h^* and h are the total concentrations of labeled and unlabeled ligands and t is time. Combining Eq. (A1) with statements of conservation of mass [Eqs. (A2)–(A4)], the definition of Y [Eq. (2a)], and Eq. (2), one obtains a Ricatti differential equation:

$$\frac{dU}{dt} = aU^2 - bU + c \tag{A5}$$

where

$$a = k\left(1 - \frac{\delta - 1}{rK_e}\right) \tag{A6}$$

$$b = k\left(h^* + h + r + \frac{1}{K_e}\right) \tag{A7}$$

$$c = k(h^* + h)r \tag{A8}$$

This equation has the exact analytical solution

$$U = \frac{U_1[(U_2 - U_0)/(U_1 - U_0)] - U_2 \exp[-a(U_2 - U_1)t]}{[(U_2 - U_0)/(U_1 - U_0)] - \exp[-a(U_2 - U_1)t]} \tag{A9}$$

where

$$U_0 = U(0) = [H^*R] + [HR] \qquad \text{when } t = 0 \tag{A10}$$

$$U_1 = \frac{b - (b^2 - 4ac)^{1/2}}{2a} \tag{A11}$$

$$U_2 = \frac{b + (b^2 - 4ac)^{1/2}}{2a} \tag{A12}$$

This solution is algebraically similar to that obtained in the simple non-cooperative case when $\delta = 1$ and $a = k$ [Eq. B7 of Rodbard and Weiss (1973), Eq. 1 of Vassent (1974)]. The present form is preferable computationally, since it avoids ratios of large positive exponentials as t becomes large. U_1 is the asymptotic equilibrium value of U as time becomes infinite. When only labeled hormone is present ($h = 0$), Eq. (A9) completely describes the kinetics of reaction scheme (**II**). However, when both labeled and unlabeled ligands are present but with different fractions bound, we must solve the following differential equation simultaneously with Eq. (A1) in order to describe the binding of labeled ligand H* :

$$\frac{dB}{dt} = k(r - U)(h^* - B) - k'B \tag{A13}$$

Here B is the concentration of H*R as a function of time and U is the function of time given by Eq. (A9). This linear differential equation can be rearranged as

$$\frac{dB}{dt} + (kr + k'_e - aU)B = k(r - U)h^* \tag{A14}$$

The solution to Eq. (A14) is then given by

$$B = \left(\frac{h^*}{h^* + h} \right) U$$
$$+ \left[B_0 - \left(\frac{h^*}{h^* + h} \right) U_0 \right] \left(\frac{U_2 - U}{U_2 - U_0} \right) \exp\left[-(kr + k'_e - aU_1)t \right]$$

$$\tag{A15}$$

where B_0 is the value of B when $t = 0$ and other symbols are as defined above.

Appendix B: Addition of Fresh (Empty) Receptors

When the same preparation containing the insulin receptors is present during both association and dissociation of the labeled hormone, the binding sites are regarded as homogeneous. However, when fresh free receptors are added during the course of the experiment, this leads to heterogeneity of the rate constants. In that case, reaction scheme (**II**) no longer adequately describes the binding system, and a more complex model is required:

$$\text{H*} + \text{R}_1 \underset{k'_1}{\overset{k}{\rightleftharpoons}} \text{HR}_1$$
$$\text{H*} + \text{R}_2 \underset{k'_2}{\overset{k}{\rightleftharpoons}} \text{HR}_2$$
$$\text{H} + \text{R}_1 \underset{k'_1}{\overset{k}{\rightleftharpoons}} \text{HR}_1 \tag{VII}$$
$$\text{H} + \text{R}_2 \underset{k'_2}{\overset{k}{\rightleftharpoons}} \text{HR}_2$$

where the symbols bear the same meaning as in reaction scheme (**II**). The subscripts 1 and 2 indicate whether the indexed quantities are related to old (1) or to fresh (2) receptors. In general, the dissociation rates will be different ($k'_1 \neq k'_2$)

except when the fractional occupancies for the old and the fresh receptors become equal. There is no general analytical solution for this case, but the following differential equations can be solved numerically, e.g., using the MLAB interactive system developed by Knott and Reece (1972):

$$\frac{dB_1}{dt} = k(h^* - B_1 - B_2)(r_1 - U_1) - k'_1 B_1 \tag{B1}$$

$$\frac{dB_2}{dt} = k(h^* - B_1 - B_2)(r_2 - U_2) - k'_2 B_1 \tag{B2}$$

$$\frac{dU_1}{dt} = k(h^* + h - U_1 - U_2)(r_1 - U_1) - k'_1 U_1 \tag{B3}$$

$$\frac{dU_2}{dt} = k(h^* + h - U_1 - U_2)(r_2 - U_2) - k'_2 U_2 \tag{B4}$$

where h^* = total concentration of labeled hormone; h = total concentration of unlabeled hormone; r_1, r_2 = total concentrations of old and fresh receptors; B_1, B_2 = concentration of labeled hormone bound to old and fresh receptors; U_1, U_2 = concentration of hormone bound (labeled + unlabeled) to old and fresh receptors, respectively; k = common association rate constant; k'_1, k'_2 = distinct dissociation rate coefficient for old and fresh receptors. The dissociation rate constants are linearly related to the fractional occupancy of their respective receptor, according to Eq. (2). In general, the solution to Eqs. (B1)–(B4) will differ from the approximate analytical solution which could be obtained by considering a common average occupancy $\overline{Y} = (U_1 + U_2)/(r_1 + r_2)$ and using reaction scheme (II) to describe the system after instantaneous mixing of the old and the fresh receptors. The latter method, although simpler (and with an available analytical solution), would not adequately describe the initial conditions after mixing when there is a marked disparity in the occupancies and dissociation rates for the old and fresh receptors. Due to the nonlinearity of the system, the properties of a system with "average" parameters are not identical to the "average" properties of the heterogeneous system.

Appendix C: Model II: Labeled Ligand Only

The time course of hormone–receptor complex B in the case of only one ligand is given by the following differential equation:

$$\frac{dB}{dt} = P(B) = p_0 B^3 + p_1 B^2 + p_2 B + p_3 \tag{C1}$$

where

$$p_0 = \frac{\varepsilon k_0}{r} \tag{C2}$$

$$p_1 = k_0 - \varepsilon k_0 - \frac{\varepsilon h k_0}{r} \tag{C3}$$

$$p_2 = hk_0\varepsilon - k_0h - k_0r - k' \tag{C4}$$

$$p_3 = hrk_0 \tag{C5}$$

$$h = \text{total hormone concentration}$$

$$r = \text{total receptor concentration}$$

If B_{eq} is the concentration of B at equilibrium, then Eq. (C1) can be written as

$$\frac{dB}{dt} = (B - B_{eq}) \, Q(B) \tag{C6}$$

where

$$Q(B) = aB^2 + bB + c \tag{C7}$$

$$a = p_0 \tag{C8}$$

$$b = p_1 + p_0 B_{eq} \tag{C9}$$

$$c = p_2 + p_1 B_{eq} + p_0 B_{eq}^2 \tag{C10}$$

The value B_{eq} can be obtained as the root of the cubic polynomial $P(B)$ which satisfies the condition $0 \le B \le r$ (or h if $h < r$). Equation (C6) can be be solved analytically, and the resulting equation is more conveniently expressed as a function of B when time is considered:

$$t = \frac{1}{2Q(B_{eq})} \log_e \left[\frac{(B - B_{eq})^2 \, Q(B_0)}{(B_0 - B_{eq})^2 \, Q(B)} \right] - (2aB_{eq} + b)I \tag{A11}$$

where B_0 is the concentration B at $t = 0$ and I is an integral whose representation depends on the discriminant $d = b^2 - 4ac$ of the quadratic polynomial $Q(B)$. If $d > 0$, then

$$I = \frac{1}{\sqrt{d}} \log_e \left[\left(\frac{2aB + b - \sqrt{d}}{2aB + b + \sqrt{d}} \right) \left(\frac{2aB_0 + b + \sqrt{d}}{2aB_0 + b - \sqrt{d}} \right) \right] \tag{C12}$$

If $d = 0$, then

$$I = \frac{2}{2aB_0 + b} - \frac{2}{2aB + b} \tag{C13}$$

If $d < 0$, then

$$I = \frac{2}{\sqrt{d}} \left[\arctan \left(\frac{2aB + b}{\sqrt{d}} \right) - \arctan \left(\frac{2aB_0 + b}{\sqrt{d}} \right) \right] \tag{C14}$$

To evaluate B as a function of a given time $\hat{t}$, we could use the implicit function

$$g(B) = \hat{t} - f(B) \tag{C15}$$

The value $\hat{B}$ such that $g(\hat{B}) = 0$ can be obtained by Newton's method. However, for limiting cases ($t \to 0$ or $t \to \infty$), this method is likely to diverge. To stabilize the process, the variable B may be reparameterized as

$$Z = \left(\frac{B - B_0}{B_{eq} - B} \right) \tag{C16}$$

The implicit function (C15) then becomes

$$g(Z) = \hat{t} - f\left(\frac{B_0 + ZB_{eq}}{Z + 1} \right) \tag{C17}$$

If Z is an "old" approximation of $\hat{Z}$ such that $g(\hat{Z}) = 0$, then

$$Z_{new} = Z + g(Z)(B_{eq} - B_0)\frac{P(B)}{(B_{eq} - B)^2} \tag{C18}$$

is a better approximation of $\hat{Z}$.

Appendix D: Model II: Labeled and Unlabeled Ligands

When both labeled (H*) and unlabeled (H) hormones are present, the reactions can be described by the following set of differential equations:

$$\frac{dU}{dt} = k(h^* + h - U)(r - U) - k'U \tag{D1}$$

$$\frac{dB}{dt} = k(h^* - B)(r - U) - k'B \tag{D2}$$

where

$$B = [H^*R] \tag{D3}$$

$$U = [H^*R] + [HR] \tag{D4}$$

$$k = k_0\left(1 + \frac{\varepsilon U}{r} \right) \tag{D5}$$

h^* = total labeled hormone concentration

h = total unlabeled hormone concentration

r = total receptor concentration

These differential equations may be written as

$$\frac{dU}{dt} = P(U) = p_0 U^3 + p_1 U^2 + p_2 U + p_3 \tag{D6}$$

where

$$p_0 = \frac{\varepsilon k_0}{r} \tag{D7}$$

$$p_1 = k_0 - \varepsilon k_0 - \frac{\varepsilon(h^* + h)k_0}{r} \tag{D8}$$

$$p_2 = \varepsilon k_0(h^* + h) - k_0(h^* + h + r) - k' \tag{D9}$$

$$p_3 = k_0 r(h^* + h) \tag{D10}$$

and

$$\frac{dB}{dt} + [G(U) + k']B = h^*G(U) \tag{D11}$$

where

$$G(U) = k_0 r + k_0(\varepsilon - 1)U - \frac{k_0 \varepsilon U^2}{r} \tag{D12}$$

Equation (D6) can be solved analytically using the method described in Appendix C. Equation (D11) does not admit a general analytical solution, but both Eqs. (D6) and (D11) could be integrated numerically. In the special case where the total hormone–receptor complex concentration is nearly constant, Eq. (D11) admits the following analytical solution:

$$B = \frac{1}{k' + G(U)}\left(h^*G(U) + [B_0 k' + B_0 G(U) - h^*G(U)]\exp\{-[k' + G(U)]t\}\right) \tag{D13}$$

where B_0 is the initial value for B at $t = 0$. This result will be helpful in optimizing experimental conditions (Appendix E).

Appendix E: Optimization of Testing for Model II with ε > 0

We wish to find optimal conditions for the test for positive cooperativity due to changes in the association rate constant. In the first kind of experiment (Fig. 16), the receptors are preincubated with unlabeled hormone until equilibrium is reached. One expects that addition of a tracer amount of labeled hormone during a second incubation will result in enhancement of the initial association rate of the labeled hormone when compared to a control experiment with no hormone during preincubation.

We shall consider the case where the labeled hormone concentration is low ($h^* \ll 1/K_0$), so that the fractional occupancy of the binding sites (U) does not change significantly during the second incubation. The initial association rate at the onset of the second incubation is

$$\frac{dB}{dt} = k_0\left(1 + \frac{\varepsilon U}{r}\right)(r - U)h^* \tag{E1}$$

where U is nearly constant but $B \ll U$ is increasing. The paradoxical enhancement of the association rate following preincubation with hormone results from the fact that the positive contribution to the factor $(1 + \varepsilon U/r)$ overcomes the negative influence of the reduction of the free receptors [R].

The optimum fractional occupancy U maximizes the initial rate of association dB/dt and satisfies the requirement that $\partial(dB/dt)/\partial U = 0$. Thus optimum occupancy $\hat{U}$ is

$$\hat{U} = \frac{r(\varepsilon - 1)}{2\varepsilon}$$

(E2)

Since U is nearly constant,

$$\frac{dU}{dt} = k_0 \left(1 + \frac{\varepsilon U}{r} \right)(r - U)(h - U) - k'U = 0$$

(E3)

and thus

$$\hat{h} = \hat{U} + \frac{2(\varepsilon - 1)}{K_0(\varepsilon + 1)^2}$$

(E4)

Optimum conditions can be found only if $\varepsilon > 1$. Interestingly, this condition is also required for the appearance of a hook in the Scatchard plot.

ACKNOWLEDGMENTS. P. De Meyts provided numerous stimulating discussions and access to unpublished manuscripts and showed unequivocal positive cooperativity. R. Pollet provided access to unpublished data and a critical review of the manuscript. P. J. Munson provided a helpful review of the manuscript and assistance with the derivations. G. H. Weiss reviewed the mathematical derivations. Dr. L. Limbird and R. J. Lefkowitz kindly provided the data shown in Fig. 9.

References

Adair, G. S., 1925, The hemoglobin system. VI. The oxygen dissociation curve of hemoglobin, *J. Biol. Chem.* **63**:529.

Ariëns, E. J., and Simonis, A. M., 1976, Receptors and receptor mechanisms, in: *Beta-Adrenoreceptor Blocking Agents* (P. R. Saxena and R. P. Foreyth, eds.), Chap. 1, North-Holland, Amsterdam.

Atkinson, D. E., Hathaway, J. A., and Smith, E. C., 1965, Kinetics of regulatory enzymes, *J. Biol. Chem.* **240**:2682.

Berman, M., and Weiss, M. P., 1971, *User's Manual for SAAM, US DHEW PHS, NIH, NIAMD, Version SAAM 25,* U.S. Government Printing Office, Washington, D.C.

Bockaert, J., Roy, C., Rajerison, R., and Jard, S., 1973, Specific binding of ^{3}H lysine-vasopressin to pig kidney plasma membranes. Relationship of receptor occupancy to adenylate cyclase activation, *J. Biol. Chem.* **248**:5922.

Boeynaems, J. M., and Dumont, J. E., 1975, Quantitative analysis of the binding of ligands to their receptors, *J. Cyclic Nucleotide Res.* **1**:123.

Castellan, G. W., 1963, Calculation of the spectrum of chemical relaxation for a general reaction mechanism, *Ber. Bunsenges. Phys. Chem.* **67**:898.

Chamness, G. C., and McGuire, W. L., 1975, Scatchard plots: Common errors in correction and interpretation, *Steroids* **26**:538.

Changeux, J. P., Thiery, J., Tung, Y., and Kittel, C., 1967, On the cooperativity of biological membranes, *Proc. Natl. Acad. Sci. USA* **57**:335.

Colquhoun, D., 1973, The relation between classical and cooperative models for drug action, in: *Drug Receptors* (H. P. Rang, ed.), Chap. 11, University Park Press, Baltimore.

Craig, M. E., Crothers, D. M., and Doty, P., 1971, Relaxation kinetics of dimer formation by self-complementary oligonucleotides, *J. Mol. Biol.* **62**:383.

Cuatrecasas, P., and Hollenberg, M. D., 1975, Binding of insulin and other hormones by nonreceptor materials: Saturability, specificity and apparent "negative cooperativity," *Biochem. Biophys. Res. Commun.* **62**:31.

De Lean, A., Munson, P. J., and Rodbard, D., 1978, Simultaneous analysis of families of sigmoid curves. Application to bioassay, radioligand assay, and physiological dose–response curves, *Am. J. Physiol.* **235**:E97.

De Lean, A., Munson, P. J., and Rodbard, D., 1979, Multivalent ligand binding to multi-subsite receptors: Application to hormone–receptor interactions, *Mol. Pharmacol.,* in press.

De Lisi, C., and Metzger, H., 1976, Some physical chemical aspects of receptor-ligand interactions, *Immunol. Commun.* **5**:417.

De Meyts, P., 1976a, The negative cooperativity of the insulin receptor: Mathematical linking of affinity to receptor and subunit structure, Abstract 24, Endocrine Society Meeting, San Francisco.

De Meyts, P., 1976b, Cooperative properties of hormone receptors in cell membranes, *J. Supramol. Struct.* **4**:241.

De Meyts, P., and Roth, J., 1975, Cooperativity in ligand binding: A new graphic analysis, *Biochem. Biophys. Res. Commun.* **66**:1118.

De Meyts, P., Roth, J., Neville, D. M., Jr., Gavin, J. R., and Lesniak, M. A., 1973, Insulin interactions with its receptors: Experimental evidence for negative cooperativity, *Biochem. Biophys. Res. Commun.* **55**:154.

De Meyts, P., Bianco, A. R., and Roth, J., 1976, Site–site interactions among insulin receptors: Characterisation of the negative cooperativity, *J. Biol. Chem.* **251**:1877.

Engel, J., and Winklmair, D., 1972, Equilibrium and kinetics of cooperative binding and cooperative association, in: *Enzymes, Structure and Function,* VIII FEBS Meeting (J. Drenth, R. A. Oosterbaan, and C. Veeger, eds.), pp. 29–44, American Elsevier, New York.

Fletcher, J. E., Spector, A. A., and Ashbrook, J. D., 1970, Analysis of macromolecule–ligand binding by determination of stepwise equilibrium constants, *Biochemistry* **9**:2580.

Frazier, W. A., Boyd, L. F., Pulliam, M. W., Szutowicz, A., and Bradshaw, R. A., 1974, Properties of binding sites for ^{125}I-nerve growth factor in embryonic heart and brain, *J. Biol. Chem.* **249**:5918.

Gliemann, J., Gammeltoft, S., and Vinten, J., 1975, Time course of insulin-receptor binding and insulin-induced lipogenesis in isolated rat fat cells, *J. Biol. Chem.* **250**:3368.

Hammes, G. G., and Schimmel, P. R., 1967, Relaxation spectra of enzymatic reactions, *J. Phys. Chem.* **71**:917.

Haselkorn, D., Friedman, S., Givol, D., and Pecht, I., 1974, Kinetic mapping of the antibody combining site by chemical relaxation spectrometry, *Biochemistry* **13**:2210.

Herzfeld, J., and Stanley, H. E., 1974, A general approach to co-operativity and its application to the oxygen equilibrium of hemoglobin and its effectors, *J. Mol. Biol.* **82**:231.

Hill, A. V., 1910, The possible effects of the aggregation of the molecules of haemoglobin on its dissociation curves, *J. Physiol. (London)* **40**:190.

Hunston, D. L., 1975, Two techniques for evaluating small molecule–macromolecule binding in complex systems, *Anal. Biochem.* **63**:99.

Jacobs, S., and Cuatrecasas, P., 1976, The mobile receptor hypothesis and cooperativity of hormone binding. Application to insulin, *Biochem. Biophys. Acta* **433**:482.

Ketelslegers, J-M., Knott, G. D., and Catt, K. J., 1975, Kinetics of gonadotropin binding by receptors of the rat testis. Analysis by a nonlinear curve fitting method, *Biochemistry* **14**:3075.

Kirsch, J. F., 1973, Mechanism of enzyme action, *Annu. Rev. Biochem.* **42**:205.

Knott, G. D., and Reece, D., 1972, MLAB: a civilized curve-fitting system, in: *Proceedings of the ON LINE' 72 International Conference,* Vol. 1, pp. 497–526, Brunel University, England.

Koren, R., and Hammes, G. G., 1976, A kinetic study of protein–protein interactions, *Biochemistry* **15**:1165.

Koshland, D. E., Jr., Nemethy, G., and Filmer, D., 1966, Comparison of experimental binding data and theoretical models in proteins containing subunits, *Biochemistry* **5**:365.

Lancet, D., and Pecht, I., 1976, Kinetic evidence for hapten-induced conformational transition in immunoglobulin, *Proc. Natl. Acad. Sci. USA* **73**:3549.

Levitzki, A., and Koshland, D. E., Jr., 1969, Negative cooperativity in regulatory enzymes. *Proc. Natl. Acad. Sci. USA* **62**:1121.

Limbird, Lee E., and Lefkowitz, R. J., 1976, Negative cooperativity among beta-adrenergic receptors in frog erythrocyte membranes, *J. Biol. Chem.* **251**:5007.

Limbird, L. E., De Meyts, P., and Lefkowitz, R. J., 1975, Beta-adrenergic receptors: Evidence for negative cooperativity, *Biochem. Biophys. Res. Commun.* **64**:1160.

Lissitsky, S., Fayet, G., and Verrier, B., 1975, Thyrotropin–receptor interaction and cyclic AMP-mediated effects in thyroid cells, *Adv. Cyclic Nucleotide Res.* **5**:133.

Magar, M. E., 1972, *Data Analysis in Biochemistry and Biophysics,* Academic Press, New York, Chap. 13, pp. 401–428.

Matsukura, S., West, C. D., Ichikawa, Y., Jubig, W., Harada, G., and Tyler, F. H., 1971, A new phenomenon of usefulness in the radioimmunoassay of plasma adrenocorticotropin, *J. Lab. Clin. Med.* **77**:490.

Monod, J., Wyman, J., and Changeux, J. P., 1965, On the nature of allosteric transitions: A plausible model, *J. Mol. Biol.* **2**:88.

Neuman, E., and Chang, H. W., 1976, Dynamic properties of isolated acetylcholine receptor protein: Kinetics of the binding of acetylcholine and calcium ions, *Proc. Natl. Acad. Sci. USA* **73**:3994.

Pauling, L., 1935, The oxygen equilibrium of hemoglobin and its structural interpretation, *Proc. Natl. Acad. Sci. USA* **1**:186.

Pollet, R. J., Standaert, M. L., and Haase, B. A., 1977, Insulin binding to the human lymphocyte receptor, *J. Biol. Chem.* **252**:5828.

Richards, W. G., Aschman, D. G., and Hammond, J., 1975, Conformational flexibility in physiologically active amines, *J. Theor. Biol.* **52**:223.

Rodbard, D., and Catt, K. J., 1972, Mathematical theory of radioligand assays: The kinetics of separation of bound and free, *J. Steroid Biochem.* **3**:255.

Rodbard, D., and Frazier, G. R., 1975, Statistical analysis of radioligand assay data, *Methods Enzymol.* **37(B)**:3.

Rodbard, D., and Weiss, G. H., 1973, Mathematical theory of immunoradiometric (labeled antibody) assays, *Anal. Biochem.* **52**:10.

Rodbard, D., Ruder, H. J., Vaikaitis, J., and Jacobs, H. S., 1971, Mathematical analysis of kinetics of radioligand assays: Improved sensitivity obtained by delayed addition of labeled ligand, *J. Clin. Endocrinol.* **33**:343.

Ross, E. M., Maguire, M. E., Sturgill, T. W., Biltonen, R. L., and Gilman, A. G., 1977, Relationship between the beta-adrenergic receptor and adenylate cyclase, *J. Biol. Chem.* **252**:5761.

Scatchard, G., 1949, The attractions of proteins for small molecules and ions, *Ann. NY Acad. Sci.* **51**:660.

Shrager, R. I., 1970, MODELAIDE: a computer graphics program for the evaluation of mathematical models, *Technical Report No. 5,* Division of Computer Research and Technology, National Institutes of Health, Department of Health, Education and Welfare, Bethesda.

Taylor, P. W., King, R. W., and Burgen, A. S. V., 1970, Kinetics of complex formation between human carbonic anhydrases and aromatic sulfonamides, *Biochemistry* **9**:2638.

Triggle, D. J., and Triggle, C. R., 1976, *Chemical Pharmacology of the Synapse,* Chap. 2, Academic Press, New York.

Vassent, G., 1974, Multi-random binding molecular interactions: a general kinetic model, *J. Theor. Biol.* **44**:241.

Waud, D. R., 1968, Pharmacological receptors, *Pharmacol. Rev.* **20**:49.

Weber, G., 1975, Energetics of ligand binding to proteins, *Adv. Protein Chem.* **29**:1.

Weintraub, B. D., Rosen, S. W., McCammon, A., and Perlman, R. L., 1973, Apparent cooperativity in radioimmunoassay of human chorionic gonadotropin, *Endrocrinology* **92**:1250.

Whitehead, E. P., 1973, The mathematical formalism and the physical understanding of allosteric interactions in proteins, *Acta Biol. Med. Ger.* **3127**:2.

Yoda, A., 1972, Structure–activity relationship of cardiotonic steriods for the inhibition of sodium- and potassium-dependent adenosine triphosphate, *Mol. Pharmacol.* **9**:51.

5

Distinction of Receptor from Nonreceptor Interactions in Binding Studies

Morley D. Hollenberg and Pedro Cuatrecasas

1. Defining a Pharmacologic Receptor

Studies of the binding of ligands to various putative pharmacologic receptors are entering a new phase. Over the past decade, sufficient expertise has been developed to measure with confidence the binding of highly radioactively labeled probes to quantitatively small numbers of high-affinity binding sites, which in most cases represent pharmacologic receptors of interest. Now, these techniques are being employed to study *receptors* for an ever-burgeoning number of pharmacologically active agents [Cuatrecasas and Hollenberg (1976); collection of reviews edited by Blecher (1976)]. Furthermore, it is becoming apparent that the task of assessing receptor characteristics by ligand binding studies (e.g., receptor changes related to desensitization phenomena) may require a more sophisticated approach than may have been initially considered adequate. It thus appears appropriate to develop an overview for such studies, particularly from the point of view of distinguishing ligand binding sites which typify "true" receptor characteristics (therefore *specific* binding sites) from those cellular binding sites which do not reflect receptor properties (therefore *nonspecific* or *nonreceptor* binding sites).

Critical to an interpretation of ligand binding data is an understanding of the receptor concept, which has been so productive in analyzing, from a pharmacologic point of view, the action of endogenous or foreign ligands (e.g., hormones, drugs) at the molecular, cellular, or even whole-organism level. The main themes can be traced principally from the work of Langley (1906), Ehrlich (1956), and Clark (1926a,b). So astute were the observations of these investigators

Morley D. Hollenberg ● Division of Clinical Pharmacology, Departments of Medicine and of Pharmacology and Experimental Therapeutics, The Johns Hopkins University School of Medicine, Baltimore, Maryland. *Pedro Cuatrecasas* ● The Wellcome Research Laboratories, Burroughs Wellcome Company, Research Triangle Park, North Carolina.

that it is worth quoting, at least in part, the germ cells of the concept as they perceived it.

In 1908, Ehrlich (1956), in recounting his work with tetanus toxin, wrote:

> I assumed accordingly that tetanus poison, for example, must unite with certain chemical groups of the protoplasm of cells and of the motor ganglion cells in particular, and that this chemical union is the prerequisite for and the cause of the disease. I have, therefore, described these groups in short, as poison receptors, or simply as receptors. . . .

Working at the same time on the action of nicotine and curare on muscle, Langley (1906) arrived at conclusions analogous to those of Ehrlich:

> The mutual antagonism of nicotine and curari on muscle can only satisfactorily be explained by supposing that both combine with the same radicle of the muscle, so that nicotine–muscle compounds and curari–muscle compounds are formed. Which compound is formed depends upon the mass of each poison present and the relative chemical affinities for the muscle radicle. . . . Since neither curari nor nicotine, even in large doses, prevents direct stimulation of muscle from causing contraction, it is obvious that the muscle substance which combines with nicotine or curari is not identical with the substance which contracts. It is convenient to have a term for the specially excitable constituent and I have called it the receptive substance. It receives the stimulus and, by transmitting it, causes contraction.

It was left to the elegant work of Clark (1926a,b) to put the observations of acetylcholine and atropine action on a quantitative basis, so as to indicate that vanishingly small quantities of drug (Clark estimated 10^{-14} mol or about 20,000 molecules per cell) were required to produce a cellular response. It was, therefore, appreciated early that only very small amounts of receptors were present on responsive cells.

A further dimension of the receptor concept came from the work by Ehrlich (1956) with certain classes of trypanocidal dyes. It was his assumption that the compounds worked by binding to receptors in the protoplasm, and he utilized Fischer's *lock-and-key* concept to analyze the development of drug resistance. He observed that drug resistance was characterized by a strict chemical specificity, whereby organisms resistant to one compound in a series of chemical analogues, e.g., fuchsin, were also resistant to a large series of related compounds, i.e., triphenylmethane dyes. In contrast, a strain resistant to one series of compounds was still sensitive to another series of compounds, presumably due to the action on a separate receptor. The strict stereochemical requirements of a variety of receptors for many classes of pharmacologically active compounds in a wide variety of organisms are now well established.

It is important to note that the concept of a receptor for a given ligand is inextricably tied to the biological responsiveness of a particular system (enzyme, cell, or organism) to the ligand of interest. Most often, the term *receptor* is used to identify that (probably rate limiting) cellular component which recognizes the ligand of interest so as to set in motion the series of events leading to a cellular response. Ligand binding studies to date have accentuated the recognition function of the receptor; more recently a great deal of attention is being focused on receptor interactions consequent to the initial binding events. In all studies, however, the final analysis of the data describing a receptor must be placed in the context of cellular responsiveness. It is relevant to point out, in this regard, that while the action of some agents (e.g., ethanol as an inebriant or argon as a general anesthetic) may possibly be dealt with in terms of the receptor concept, the practicalities of isolating the relevant *rate-limiting* recognition factor may be insurmountable.

The binding studies to which this chapter will direct attention are, in view of the above considerations, limited to those ligands which interact with high specificity and affinity with biological systems. In summary, the receptor can be thought of as that highly selective chemical recognition factor, present in or on cells in very small numbers, which, in interacting with a certain chemical series of compounds, alters in a (probably) rate-limiting way some critical cellular biochemical process so as to lead ultimately to a generalized cellular response. In some instances, the receptor may turn out to be a membrane-associated macromolecule, e.g., insulin or acetylcholine receptor; alternatively the receptor may represent an important enzyme (e.g., dihydrofolate reductase) in an important metabolic pathway. In the subsequent sections of this chapter, an approach to the analysis of binding data related to receptor function is presented. While no single formula or approach may suffice for studies with all ligands of interest, a consideration of the binding data in the light of the following discussions may serve to distinguish receptor from nonreceptor interactions.

2. Criteria for Receptor Interactions

Given the challenge of indentifying the recognition macromolecule responsible for the biological action of a certain agent, one must decide upon the criteria which will be useful in evaluating the binding characteristics of cellular isolates. The analysis of biological responsiveness provides a most useful key, such that binding studies for each ligand studied must be interpreted in the context of available biological data. For instance, it has long been recognized that insulin acts at extremely low concentrations (serum levels less than 10^{-9} M) in a saturable, reversible, dose-dependent manner in recognized *target* tissues to cause a biological effect. Furthermore, structural analogues of insulin have been characterized with established relative potencies in causing a biological response, e.g., in potency, insulin = desalanine insulin > proinsulin > desoctapeptide insulin. It is, therefore, reasonable to assume that the binding of insulin to a putative recognition macromolecule will exhibit (1) a high affinity in keeping with the low concentration range over which insulin acts, (2) saturability, (3) reversibility, (4) a tissue distribution consistent with those organs on which insulin is known to act, (5) a relative binding affinity of analogues in keeping with the relative biological activities, and (6) a binding isotherm readily related to the concentration range over which insulin is known to act. When such criteria *are* met for a binding substance, it is reasonable to assume that the macromolecule isolated does indeed represent the insulin *receptor*. However, when the binding data conflict with information derived from studies of insulin action, then it is necessary to question very seriously the interpretation of such data; the possibility that a nonreceptor interaction may be under study cannot, in such circumstances, be ruled out.

The criteria, as outlined above, while extremely useful and while perhaps representative of the "best one can do," short of reconstitution experiments with receptor isolates, are necessary but may not be sufficient proof in certain circumstances, that a binding substance does in fact represent "the receptor." In the sections that follow we shall deal with some of the difficulties and successes in using such criteria to evaluate binding data in connection with studies of pharmacologic receptors.

3. The Problem of Relating Binding to Biological Responsiveness

Since, as discussed above, the term *receptor* unavoidably implies biological responsiveness, it is reasonable to speculate as to the kind of information responsiveness *per se* can yield concerning the nature of the initial ligand–recognition site interaction. The situation of evaluating concentration–efficacy relationships for drugs or hormones is closely analogous to the evaluation of substrate–velocity relationships for enzymes; in the latter case there is an enormous advantage in (usually) having available a large amount of chemical information concerning interacting species (enzyme and substrate) and the product of the reaction. For receptor interactions, one is in the difficult position of having chemical information usually only about one of the reacting species, namely the drug. The end product of the drug–receptor interaction is not a ligand metabolite (indeed in a majority of circumstances, effects resulting from hormone–receptor interactions are caused without metabolic conversion of the hormone) but rather an ultimate change in the biological system (e.g., muscle contraction, acid secretion, change in color, sleep) which can be quantified so as to represent the familiar *velocity* term of enzyme kinetics. Because of the many uncertainties concerning the coupling of receptor occupation to *response*, it is the situation that only limited information can be gained from dose–response curves concerning the affinity and mechanism of interaction of stimulatory ligands with receptors.

As a first approximation, the concentration of ligand producing a half-maximal biological effect (ED_{50}) can be taken to be representative of the affinity of the ligand–recognition site. Additionally, the relative ED_{50}s of a series of congeners can be expected to be indicative of the relative affinities of compounds for the recognition site. In the idealized case at low receptor concentration, where response is likely to be linearly related to receptor occupation and when a maximum response is achieved when all receptors are occupied, the ED_{50} represents the equilibrium dissociation constant, K_D. It is, however, likely that such a situation is the exception rather than the rule, such that, depending on the nature of coupling of receptor occupation to response, the ED_{50} may be less than, equal to, or may exceed the true K_D for the ligand–receptor interaction. The essence of the above discussion is that while a number of theoretical models for drug action have been proposed and reviewed at length (Barlow and Stephenson, 1970; Changeux *et al.*, 1970; Colquhoun, 1973; Cuatrecasas and Hollenberg, 1976; Hall 1972; Paton, 1961; Paton and Rang 1965; Rodbard, 1973; Stephenson, 1956; Waud, 1968), it is clear that the shapes of dose–response curves cannot confirm or disprove any postulated mode of interaction of an agonist ligand with the receptor. Rather, dose–response curves are useful in comparing relative affinities of ligands with the receptor and in determining whether the ligand of interest in a series of compounds can cause the same maximal response (indicative of a closely similar interaction with the recognition site) as other analogues. Importantly, dose–response curves also provide invaluable comparative data for studies with chemical and/or enzymatic probes of receptor characteristics. Perturbations which effect responsiveness to a ligand in a characteristic way (e.g., shift of dose–response curve to the right, indicative of a lower ligand affinity) would be expected to affect the ligand–recognition site interaction in a similar way.

Surprisingly, it is the inhibition of biological responsiveness by antagonists rather than responsiveness *per se* which reveals accurate information about a receptor recognition site. The derivation of the equations relating the biological action of competitive antagonists to ligand–receptor affinity rests on the powerful principle of the *null* hypothesis, whereby it is assumed that when the response to one concentration of agonist in the absence of inhibitor is the same as the response to a higher concentration of agonist in the presence of competitive inhibitor, then the amount of agonist reaching the receptor in the two situations is identical. It is thus possible by biossay to measure the concentration dependence of an antagonist's ability to prevent agonist access to the receptor site, so as to determine the antagonist equilibrium dissociation constant (Schild, 1949; Arunlakshana and Schild, 1959). No assumption need be made about the coupling mechanism between agonist and response. Receptor antagonists can thus serve to characterize receptors in a variety of tissues according to biological criteria, which can then be compared with the characteristics of a receptor present in a subcellular isolate. The approach using antagonists has been particularly useful for ligand binding studies of both nicotinic and muscarinic cholinergic receptors, where a remarkably close agreement is observed between affinities measured by binding and affinities estimated by biossay (Paton and Rang, 1965; Cohen and Changeux, 1975; Birdsall and Hulme, 1976).

It is important to note that antagonists which combine irreversibly with receptors may provide a biological means of assessing agonist affinities [reviewed by Waud (1968)] by effectively blocking a fixed proportion of receptor from agonist access. Such an approach has not yet found as widespread use as has the one with reversible antagonists in providing accurate ligand affinities to be correlated with binding studies. Part of the problem in the use of such irreversible blocking agents stems from a lack of absolute specificity of such agents, which, like benzilylcholine mustard (Gill and Rang, 1966), may alkylate sites other than the receptor (Gupta *et al.*, 1976) and for which adducts tend to hydrolyze. Nonetheless, the use of irreversible antagonists to assess agonist affinities by biological criteria may aid in the interpretation of binding studies which detect multiple binding sites for an active ligand of interest.

In summary, it is evident that the analysis of biological responsiveness provides an important but limited amount of information concerning agonist–recognition and antagonist–recognition site interactions. The data are, however, of critical importance in interpreting ligand binding studies. For certain ligands there is a wealth of information concerning the actions of many analogues [e.g., for cholinergic agents, Horn (1975)] in which context measurements of ligand binding can be thoroughly evaluated. In many cases, the biological data can be evaluated in a simple responsive system (e.g., isolated adipocyte) amenable to disruption and subsequent study of ligand binding. In other cases (e.g., the central nervous system) a convenient isolated responsive system may not be possible such that, for interpreting binding data, reliance must be placed upon experimental evidence (structure–action, relative potencies) derived from studies in the whole animal. Enormous success with this approach has been achieved in studies of receptors for opioids (Snyder *et al.*, 1975).

4. Nonspecific Binding: Definition and Examples of Complications of Binding Data Analysis

The saturation of ligand binding sites for potent pharmacologic agents occurs over concentration ranges which are remarkably low (usually lower than 10^{-7} M; more often than not, lower than 10^{-9} M) relative to the higher, more familiar concentration ranges (about $10^{-6} - 10^{-3}$ M) over which many cellular enzymes interact with substrates. Even at the low concentrations of pharmacologic interest, it can be reasonably assumed that a given ligand will bind to *many* cellular structures, including the specific receptor: The main characteristics which will distinguish receptor from nonreceptor binding will be the chemical specificity and ligand affinity, in addition to the other properties alluded to in the above discussion. There are two principal types of nonreceptor binding which can be detected. The first kind, termed *strictly nonspecific* binding, represents the adsorption of radiolabeled ligand in a manner which is not competitive with the binding of the parent ligand. A second kind of binding is termed *specific nonreceptor* binding. In such cases, the ligand may bind in a chemically specific manner to a nonreceptor macromolecule. It is often a problem to distinguish between the two major types of nonspecific binding from the binding of ligands to receptor macromolecules.

There are two principal approaches which are currently in use to evaluate the specificity of ligand binding. One method simply measures the entire binding isotherm, as has been done for the binding of ^{3}H-labeled atropine to smooth muscle (Paton and Rang, 1965); the data are subsequently analyzed mathematically, so as to dissect the isotherm into two or more saturable (or nonsaturable) components, one of which may be related quantitatively to the biological data for the ligand of interest. For instance, in the study of atropine binding, it was possible to detect a saturable component with an equilibrium dissociation constant of 1.1×10^{-9} M, which agreed exactly with the estimate of atropine affinity based on bioassay measurements [either by dose ratio or from biologically estimated rate constants (Paton and Rang, 1965)]. By definition, that part of the binding isotherm which can be numerically correlated to the biological action of the ligand is considered *specific*.

A second approach, now more widely used than the above method, employs essentially the principle of isotopic dilution. In practice, the amount of radioactive ligand bound to a receptor preparation (either soluble or particulate) is measured in both the presence and absence of an excess (usually greater than 50-fold) of unlabeled compound. It is presupposed that the radioactively labeled compound competes effectively at the receptor site for the unlabeled compound. Thus, in the presence of a sufficient excess of unlabeled ligand, the amount of radioactive ligand bound will be substantially reduced. For instance, should the concentration of radioactive ligand be sufficient to saturate a putative binding site (hence no more ligand can be bound at higher ligand concentrations), then the addition of a 50-fold excess of unlabeled ligand would reduce the amount of radioactivity bound 50-fold. In such an instance, it would be evident that all of the radioactively labeled ligand is bound in a specific manner. It is, however, usually the rule rather than the exception in such binding studies that there is an amount of radioactivity adsorbed to the receptor preparation for which the unlabeled compound cannot compete, even at very high concentrations. For practical purposes, the amount of radioactive ligand *not* displaced is considered to be bound in a *strictly nonspecific* manner, as opposed to the specific nonreceptor binding to be discussed below.

The strictly nonspecific binding may in some cases exhibit a high affinity, comparable to or perhaps even greater than that of the receptor-related binding, so as to constitute a substantial amount of the binding even at low ligand concentrations. The exact nature of this nonspecific binding is a matter for speculation. The nonspecific radioactive uptake may represent binding of degradation products of the labeled compound, transpeptidation of an iodinated tyrosine residue, dissociation and diffusion of the radioactive atom (e.g., ^{3}H or ^{125}I) from the labeled compound into a cell preparation, or some other interfering process. Empirically, it is observed that the nonspecific binding, while potentially of high affinity, usually exhibits such a large capacity that it is difficult, if not impossible, to demonstrate saturability of binding. Additionally, it is often the case (see, however, the example of *talc* binding below) that nonspecific binding exhibits a rapid time course (seconds to minutes) which for high-affinity receptors such as that for insulin contrasts with the comparatively longer time course (20–60 min for equilibrium at 24°C) of receptor binding. Thus in some cases, despite the presence of receptor-specific binding, its detection may be rendered impossible because of obscuring levels of *background* nonspecific binding. As a rule, the criteria of saturability, chemical and target tissue specificity, appropriate affinity, and reversibility can serve to distinguish the strictly nonspecific binding interactions from those which are receptor specific.

Binding competition data using unlabeled analogues of the radioactively labeled parent compound prove most useful in evaluating the characteristics of a receptor preparation. For example, in studies of receptors for cholinergic agents, catecholamines, insulin, adrenocorticotrophic hormone, and opiates, the use of numerous chemical analogues has given tremendous support to the hypothesis that the binding measurements reflect the relevant pharmacologic receptor. Regrettably, studies with many ligands are hampered for lack of suitable active analogues. It is important to point out, however, that even when relatively large numbers of analogues are available, some aspects of the binding-competition data may be difficult to interpret.

In some studies, the competition data may reveal more than one binding site. For example, in a study of the binding characteristics of dopamine receptors in membranes from calf caudate, it is observed that ^{3}H-labeled dopamine binds to a single site, whereas the ^{3}H-labeled antagonist, haloperidol, binds to multiple saturable sites of which the higher-affinity component is influenced by dopamine and its antagonists (Creese et al., 1975). In contrast, in studies with cholinergic receptors in rat cerebral cortex a single binding site is observed for antagonists (Yamamura and Snyder, 1974; Birdsall and Hulme, 1976), whereas studies with radioactive cholinergic agonists display at least two binding sites (Birdsall and Hulme, 1976), only one of which agonist binding sites approximates the apparent affinity for agonists determined from dose–response curves. The existence of such multiple-affinity receptor sites can be rationalized in terms of certain models such as a *two-state* model of receptor function (Changeux, 1966; Karlin, 1967; Changeux and Podleski, 1968; Changeux et al., 1970; Edelstein, 1972; Snyder, 1975) or as a consequence of the *mobile receptor* hypothesis (Cuatrecasas, 1974; Cuatrecasas and Hollenberg, 1976; Jacobs and Cuatrecasas, 1976). Nonetheless, infrequently, if ever, are the *low-affinity* ligand binding sites detected in various studies evaluated with the same thoroughness as are the *high-affinity* sites. Since most studies describing multiple ligand sites deal with membrane preparations derived from heterogeneous cell types and since certain ligands such as atropine can block receptors for more than one agonist (atropine antagonizes histamine with a K_i of about 10^{-5} M as well as acetylcholine, with a K_i of about 10^{-9} M), the interpretation of observed multiple binding sites in terms of a *single* pharmacologic receptor is fraught with difficulty. A detailed analysis over the entire concentration range of binding, using concentrations of labeled ligand which may occupy all binding sites, along with a series of competitive analogues may be required to resolve the question as to whether a particular binding site among several is or is not representative of the pharmacologic receptor.

In the above discussion, mention was made of *strictly nonspecific binding*, meaning that amount of binding of radioactive ligand for which unlabeled ligand cannot compete. This type of nonspecific binding is to be clearly distinguished from the competitive binding of structural analogues (also mentioned above) at a site which may *not* represent a pharmacologic receptor. A convenient paradigm to think of, in this context, concerns the binding by human serum albumin (HSA) of families of compounds at two qualitatively distinct sites (Sudlow et al., 1976). For instance, the structurally related coumarin analogues warfarin, phenprocoumon, and acenocoumarin all interact with site I on HSA with similar but quantitatively distinguishable effects on the conformation of HSA; significantly, the interaction of R(+)warfarin can be readily distinguished from that of S(−)-warfarin. While it is clear that HSA does not represent a pharmacologic receptor, the described interaction of the coumarin drugs with HSA in other circumstances could be so interpreted; the binding to the same HSA site of drugs pharmacologically unrelated to the coumarins (e.g., phenylbutazone, sulfinpyrazone) should dispel the suspicion that the HSA binding site might represent a "receptor" related to the anticoagulant action of warfarin.* It is evident that the binding of

---

*The paradigm is particularly useful, since, *in vivo*, phenylbutazone is known to potentiate warfarin action, so as to imply in the absence of other information a possible interaction at a warfarin "receptor" site. A consideration of the data presented allows the reader to understand the nonreceptor nature of this particular drug interaction.

warfarin analogues to albumin (other examples will be discussed below) is certainly specific in that special chemical configurations are required for binding. This type of binding, which undoubtedly does occur for many pharmacologically active agents, including polypeptide hormones, is termed *specific nonreceptor binding* in distinction from the *strictly nonspecific* binding previously discussed.

The importance of distinguishing both strictly nonspecific and specific nonreceptor binding is of the utmost importance for the mathematical analysis of binding data, e.g., by the method of Scatchard (1949). In one sense, the incorrect assessment of the total amount of nonreceptor binding present in a receptor preparation is equivalent to an underestimate of the *enzyme blank* in the assay of enzymatic activity. The underestimate of nonspecific binding would lead to a downward curvature of a Scatchard plot (Rodbard, 1973; McGhee and von Hippel, 1974). In addition to the publications of these authors, including an excellent analysis of binding data related to hormone action (Rodbard, 1973), there are a number of discussions of the complex analysis of Scatchard plots [for a discussion of and references to binding studies in particular, see the review by Cuatrecasas and Hollenberg (1976)] which indicate that nonlinearity in such plots may not represent another set of receptor-specific binding sites but may result from (in addition to underestimates of nonspecific binding) (1) positive ligand–ligand cooperativity (concave down), (2) simultaneous binding of a ligand at two or more sites (concave up), (3) differences in the affinity of labeled ligand (concave up), (4) inaccurate determination of the proportion of bound and free ligand (concave up), (5) negative ligand–ligand cooperativity (concave up), and (6) lack of complete binding equilibrium.

As also discussed by Buller *et al.* (1976), special considerations may be necessary in the case of steroid hormone (S) binding studies, where at low concentrations of cytoplasmic receptor–hormone complex (SR) the rate of SR association with nuclear acceptor N (a reaction of considerable interest) may be as slow or slower than the dissociation of SR to S + R with a consequent nonspecific association of the steroid S with nuclear proteins to form the complex SN (instead of SRN). The above considerations should serve to indicate how difficult it may be to interpret the binding data in a manner relevant to the pharmacologic receptor of interest. Nonetheless, forewarned of such complications, one may be pleasantly surprised when data analysis proves straightforward and consistent with a simple interpretation.

5. Estimating the Affinity of the Unlabeled Ligand

As mentioned previously, one important source of error in interpreting binding data may arise if the affinity of the unlabeled ligand differs from that of the labeled derivative. A direct approach to this source of error involves the determination by bioassay of the potency of the labeled ligand, as has been done for ^{125}I-labeled derivatives of insulin (Cuatrecasas, 1971; Freychet *et al.*, 1971a,b; Gliemann *et al.*, 1975), epidermal growth factor [both *in vivo* (Carpenter *et al.*, 1975) and *in vitro* (Hollenberg, 1975)], cholera toxin (Cuatrecasas 1973; Bennett and Cuatrecasas, 1976), and other labeled ligands. If the bioassay is sufficiently sensitive to detect small changes in ligand potency (ED_{50}) and if a sufficiently

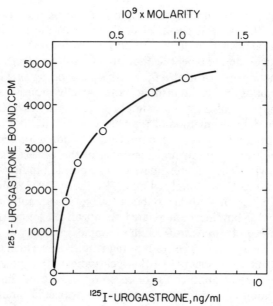

Fig. 1. Binding of ^{125}I-labeled urogastrone to human fibroblast monolayers. The specific binding at increasing concentrations of ^{125}I-labeled human urogastrone (124 cpm/pg) to confluent human skin-derived monolayers (35.6 μg of protein per monolayer; approximately 120,000 cells) was measured by previously described methods (Cuatrecasas and Hollenberg, 1976). The receptor concentration can be calculated to be approximately 14 pM, so as to yield a quotient, [receptor]/K_D, of 0.09, which is appropriate for *zone A* behavior (see the text). [Data from Hollenberg and Gregory (1978).]

large proportion of the labeled preparation contains the radioisotope,* then it is a reasonable assumption that the *unlabeled* derivative possessing the same potency as the labeled compound binds to the receptor with the same affinity; to postulate otherwise would be in defiance of Occam's razor.

It is often the case that the entire binding isotherm may be determined with the radiolabeled ligand, as for the interaction of human epidermal growth factor–urogastrone (EGF–URO) with cultured human fibroblasts (Fig. 1). Since the labeled and unlabeled derivatives possess the same biological activity, it can be concluded that the ligand affinity measured for the radioactively labeled ligand (half-maximal saturation of binding at 10^{-10} M) reflects the binding affinity of the unlabeled compound.

It should be pointed out, however, that there are a number of unavoidable sources of experimental error in assigning an accurate specific radioactivity to a ^{125}I-labeled polypeptide such as insulin or EGF–URO. In addition, because of technical considerations (high nonspecific binding), it may not be feasible (especially for relatively low-affinity binding sites) to determine the entire binding isotherm. It is thus advantageous to make use of the unlabeled ligand itself, for which accurate spectrophotometric estimates of concentration can be made, to estimate ligand affinity by the following procedures.

---

* As indicated in the following discussion, it is *not* necessary for binding studies that a preparation be homogeneously labeled, e.g., monoiodo derivatives of peptides.

The approach involved in the examples to be described (Hollenberg and Gregory, 1978) stems directly from the analysis of the kinetics of enzyme reactions in the presence of competitive inhibitors [for comprehensive reviews, see Webb (1963) and Segel (1975)]. In essence, the amount of radioactive ligand bound at a given concentration takes the place of the velocity of an enzymatic reaction; the competition for binding by unlabeled ligand is considered as a reduction in reaction velocity in the presence of enzyme inhibitor. It is evident that all the mathematical derivations for enzyme kinetics will apply; the affinity of the labeled compound may or may not be required for the estimation of the affinity of the unlabeled ligand, depending on the experimental design. The examples chosen illustrate three approaches, with the added constraint that, as is the case with many studies with cultured cells, the number of samples available for measurement is assumed to be relatively small.

In a large number of published studies of ligand binding, a tracer amount of labeled compound is added, and the binding competition by the native ligand or its analogues is determined. A similar experiment is shown in Fig. 2, which demonstrates the competition for ^{125}I-labeled EGF–URO binding to human fibroblast monolayers by unlabeled ligand. Since, as calculated from the data in Fig. 1, the absolute receptor concentration relative to the K_D for EGF–URO can be shown to be negligible (i.e., [receptor]$/K_D \leq 0.1$), the concentration of unlabeled ligand at which 50% of the bound $[^{125}$I]-EGF–URO is displaced (IC$_{50}$) can be used to

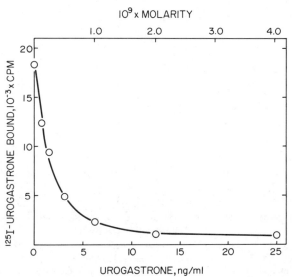

Fig. 2. Competition by unlabeled urogastrone for $[^{125}$I]urogastrone binding. The binding of ^{125}I-labeled urogastrone (2.4 ng/ml; 630 cpm/pg) to monolayers exposed to increasing concentrations of native urogastrone was measured by previously described methods (Cuatrecasas and Hollenberg, 1976). The total amount of labeled urogastrone bound is depicted; from the IC$_{50}$ determined in this experiment (1.3 ng/ml) and an estimate of the K_D for $[^{125}$I]urogastrone from the data in Fig. 1 (1.1 ng/ml), a value of 66 pM for the K_D of unlabeled urogastrone can be calculated according to Eq. (1). [Data from Hollenberg and Gregory (1978).]

yield the equilibrium dissociation constant of the unlabeled ligand (K_L) according to the equation (Cheng and Prusoff, 1973)

$$K_L = \frac{IC_{50}}{1 + (L^*/K_{L^*})} \qquad (1)$$

where L^* represents the absolute concentration of labeled ligand and K_{L^*} the dissociation constant of the labeled ligand, as estimated from the data in Fig. 1.

For the above calculation, the absolute concentration of receptor is low enough so that the very important effects of high values of the quotient, [receptor]/K_D (zone B or zone C behavior), described for enzyme inhibitor studies by Straus and Goldstein (1943), may be neglected. In many instances, as alluded to in the following paragraph, the effect of receptor concentration relative to K_D cannot be neglected, and corrections to Eq. (1) must be applied (Jacobs et al., 1975; Rodbard, 1973).

A striking example of the effect of receptor concentration on binding competition curves is shown in Fig. 3. It is evident that increasing the concentration of membrane (EGF-receptor) markedly increased the concentration of unlabeled EGF required to inhibit by 50% the binding of ^{125}I-labeled EGF. This situation is entirely analogous to the effects of increased enzyme concentration on the concentration of inhibitor required to reduce enzymatic activity by 50%, as discussed by Straus and Goldstein (1943) in terms of the *zone behavior* of enzymes. A factor of utmost importance is the *specific concentration* of enzyme given by the quotient E/K, where E is the concentration of enzyme and K is the equilibrium dissociation constant for the enzyme–inhibitor interaction. Only

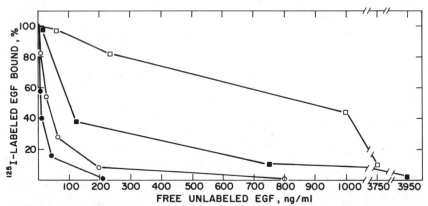

Fig. 3. Competitive displacement of ^{125}I-labeled epidermal growth factor (EGF) from placenta membranes by native EGF. Varying concentrations (23 µg/ml, ● ; 70 µg/ml, ○ ; 388 µg/ml, ■ ; 2330 µg/ml, □) of placenta membranes were incubated in 0.2 ml of Krebs–Ringer–bicarbonate buffer, 0.1% albumin, pH 7.4, at 24°C for 10 min with various concentrations of unlabeled EGF. ^{125}I-labeled EGF (0.29 ng/ml, ● , or 1.8 ng/ml, ○ , □ , ■) was added, and the incubation was continued for an additional 30 min at 24°C. The suspensions were filtered through a Millipore filter to determine binding. Binding in the presence of 10^{-6} M EGF (nonspecific binding) has been subtracted. The concentration of free, unlabeled EGF was calculated as follows: (cpm of ^{125}I-labeled EGF added minus total cpm bound) × (concentration of unlabeled EGF added)/(cpm of ^{125}I-labeled EGF added). [Data from Jacobs et al. (1975).]

in the cases where $E/K \leq 0.1$ (zone A) does the fractional inhibition become independent of enzyme concentration such that the concentration of inhibitor causing 50% inhibition is equivalent to the inhibitor's Michaelis constant. By analogy, in studies with receptors, calculations of affinity constants based on binding-competition data are readily done only if zone A behavior is observed; unfavorable values of receptor concentration, relative to the dissociation constant (i.e., $[R]/K_D \geq 1$), may obtain in experiments where radioligands of relatively low specific activity (e.g., ≤ 40 Ci/mmol) are used such that inappropriately high concentrations of membrane receptor may be required to yield measurable binding. In such instances (i.e., zone B or zone C conditions) the uncorrected binding-competition data may yield spurious estimates of the receptor affinity such that a distinction of specific from nonspecific binding based on affinities of analogues may prove impossible. The tests to determine the zone behavior of enzyme systems have been described (Straus and Goldstein, 1943); the analogous approach for receptor systems of interest is recommended to avoid this complication in the interpretation of binding data. The importance of receptor concentration in determining the *relative* potencies of a series of competitive inhibitors of the binding of a labeled ligand has also been pointed out by Rodbard (1973).

Another way in which binding inhibition data can be seen to be analogous to enzyme–velocity relationships is demonstrated by use of the *dose-ratio* method, originally employed in pharmacologic studies by Arunlakshana and Schild (1959) in connection with inhibitor studies (discussed above) and extended for the analysis of binding data by Chang *et al.* (1975) (see also Cuatrecasas and Hollenberg, 1976). The application of this method for the analysis of binding to relatively small numbers of fibroblast monolayers is demonstrated in Fig. 4. As indicated, the binding of radioactive ligand with increasing concentration is measured both in the absence and in the presence of several fixed concentrations of unlabeled ligand. From the binding isotherms generated for the radioactive ligand, the ratio of the concentration of *labeled* ligand necessary to achieve a given amount of binding (e.g., 3000 cpm, Fig. 4) in the *presence* of unlabeled ligand (X_1, Fig. 3) relative to the concentration of labeled ligand necessary to achieve the *same* amount of binding in the *absence* of unlabeled ligand (X_2) is measured so as to yield the dose ratio (DR $= X_1/X_2$). As discussed by Chang *et al.* (1975), a plot of 1/DR vs. $[\text{EGF–URO}]/(\text{DR} - 1)$ yields, as the ordinate intercept, the equilibrium dissociation constant. In practice, as is necessary because of the limited number of samples usually available for studies of cultured cells and as is routinely done for the measurements of dose ratios for inhibitor action in biologically responsive systems, results of two or more independent experiments can be pooled, as indicated in the upper panel of Fig. 4. It is evident that assumptions as to the *absolute* concentration of labeled ligand and the affinity of the labeled ligand do not enter into the estimate of ligand affinity by this technique; the value obtained reflects only the affinity of the unlabeled ligand.

One further way an estimate of the unlabeled ligand affinity may be determined, without necessitating a knowledge of the affinity or concentration of the labeled ligand, is to use the so-called *Dixon plot* for enzyme inhibition (Fig. 5). According to this approach, two (or if possible, more) binding-competition curves are measured for the unlabeled ligand at two or more concentrations of the la-

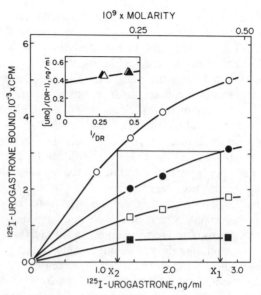

Fig. 4. Estimate of affinity of unlabeled urogastrone by the dose-ratio method. The specific binding at increasing concentrations of ^{125}I-labeled urogastrone (378 cpm/pg) to human fibroblast monolayers was determined in the absence (○) and presence of three fixed concentrations of unlabeled urogastrone: ●, 0.56 ng/ml; □, 1.12 ng/ml; ■, 2.24 ng/ml. From the three upper binding isotherms, dose ratios (DRs) were estimated, as indicated (DR $= X_1/X_2$), and a plot of [urogastrone]/(DR $-$ 1) vs. 1/DR was constructed, as shown in the inset, to yield, as the intercept, the K_D for unlabeled urogastrone (Chang *et al.*, 1975). The inset shows data calculated from two independent experiments. [Data from Hollenberg and Gregory (1978).]

beled ligand. Analogous to experiments with enzymes, where velocity is measured vs. substrate concentration for two or more concentrations of inhibitor, the plot of 1/cpm bound vs. ligand concentration yields, as the abscissa value corresponding to the intersection point, the magnitude of the dissociation constant for the unlabeled ligand. Since information about the affinity and concentration of the labeled ligand is not required, the accuracy of the measurements depends on the estimate of cpm bound and on the spectrophotometric estimate of the concentration of the stock sample of unlabeled ligand. This approach has the added advantage that the *same* data may be used to estimate the affinity of the unlabeled ligand from the IC_{50}, as discussed above; this second method of calculation does, however, require the values for the absolute concentration and affinity of the labeled ligand.

It should be evident that *any* of the mathematical approaches for the measurement of enzyme inhibitor properties will apply for the evaluation of binding characteristics; it is simply a matter of choosing the most appropriate experimental design and method of data analysis for the particular ligand and receptor system of interest. As indicated in Table I, there is reasonable agreement between the values estimated for the binding affinity of labeled and unlabeled EGF–URO so as to indicate that the unlabeled and labeled molecules interact with the receptor with essentially the same affinity. It is evident that for such an estimate it is unnecessary to isolate and characterize the labeled ligand used for study; all

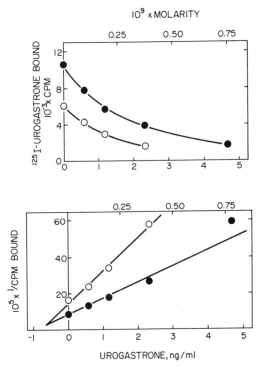

Fig. 5. Estimate of affinity of unlabeled urogastrone: Dixon plot of binding inhibition. Confluent human fibroblast monolayers were incubated with two concentrations of ^{125}I-labeled urogastrone (539 cpm/pg; ●, 1.9 ng/ml, and ○, 0.95 ng/ml) in the presence or absence of increasing concentrations of unlabeled urogastrone. The specific binding of radioactive urogastrone was determined as previously described (Cuatrecasas and Hollenberg, 1976), and the binding competition data (upper figure) were plotted according to Dixon (lower figure) so as to yield, as the intersection point, a value of 97 pM for the K_D; a value of 95 pM can be estimated from the IC_{50} values determined from the data in the upper figure. [Data from Hollenberg and Gregory (1978).]

Table I

Estimates of Receptor Affinity for Labeled and Unlabeled Urogastrone– EGF[a]

Method of analysis	Information about labeled peptide needed?	K_D ng/ml	pM
Binding isotherm, ^{125}I-labeled peptide	Yes	1.09 ± 0.01^b	177
Displacement curve, IC_{50}	Yes	0.50 ± 0.09^c	80
Dose ratio	No	0.37	60
Dixon plot	No	0.60	97

[a] Values in this table were estimated solely from the data presented in this chapter.
[b] Value represents the mean $\pm \frac{1}{2}$ range for two estimates of the K_D for *labeled* urogastrone–EGF based on Figs. 1 and 4.
[c] Value represents the mean $\pm$ S.D. for three estimates of the K_D of *unlabeled* urogastrone based on data in Figs. 2 and 5; values based on the dose ratio and Dixon plot also represent the affinity of the unlabeled peptide.

that is required is that the labeled and unlabeled compounds compete for binding at the same receptor site.

6. Examples of Receptor-Like Nonreceptor Interactions

In the example of albumin binding sites discussed above, it is evident that certain families of compounds may interact with a macromolecule so as to feign, in part, receptor binding. So difficult may be the distinction of receptor from nonreceptor binding in certain circumstances that it is worth describing in detail one or two "blind alleys" in studies with hormones, where the principles outlined in this chapter *have* succeeded in proving the nonreceptor nature of binding.

The most instructive example presently in the literature concerns the difficult task (now successful) of identifying the β-adrenergic recognition site responsible for catecholamine action [see the discussions in the reviews by Hollenberg and Cuatrecasas (1975), Cuatrecasas and Hollenberg (1976), and Lefkowitz *et al.* (1976)]. Initially, highly promising results were obtained with the use of ^{3}H-labeled norepinephrine, which binds with high affinity in a saturable manner to target tissues such as liver and fat cells. The affinity of binding (K_D approximately 10^{-6}–10^{-7} M) is closely consistent with the saturable dose–response curves (lipolysis or stimulation of adenylate cyclase) for catecholamines in these tissues. At first it thus appeared (according to the criteria discussed previously) that binding of the ^{3}H-labeled catecholamine derivatives did reflect a receptor interaction; so confident were some investigators of this fact that several publications appeared to support such a hypothesis. However, on review of the binding-competition data, it was apparent that the degree of ligand specificity did not correlate exactly with the previously documented biological potencies of certain agents. Importantly, the (+) isomers of catecholamines, known to possess very low biological potencies, were just as potent in competing for binding at the putative receptor site as were the very potent (−) isomers. Further experimentation revealed that compounds devoid of any adrenergic action whatsoever (pyrocatechol) were equally capable of competing for the ligand binding site. The key, however, to demonstrating the nonreceptor interaction of all of the catechol compounds stems from the understanding of the use of inhibitors as discussed above. Clearly, a compound which interacts at a receptor site but which possesses no biological activity *per se* must be a competitive antagonist for active compounds. The inability of pyrocatechol or (+)norepinephrine (Fig. 6) to act as a competitive catecholamine antagonist at concentrations which saturate the binding of the putative receptor sites proved beyond any doubt that in liver and fat cell membrane preparations the binding of ^{3}H-labeled catecholamines represents a specific nonreceptor site (Cuatrecasas *et al.*, 1974). Strikingly, the kind of approach used to establish the *nonreceptor* nature of binding of ^{3}H-labeled catecholamines to liver and fat cell membranes has been used successfully by U'Prichard and Snyder (1977) to measure α-adrenergic receptor binding in brain-derived membranes.

A second example of specific nonreceptor binding came to light in studies with insulin (Cuatrecasas and Hollenberg, 1975) and proves applicable to studies

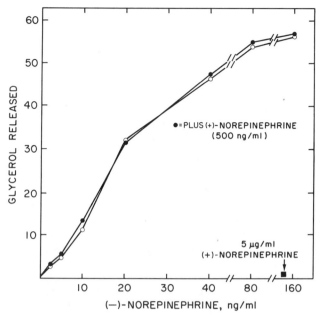

Fig. 6. Effect of (+)norepinephrine on the lipolytic response of fat cells to (−)norepinephrine. The lipolytic response after 80 min at 37° is expressed as micromoles of glycerol released in the medium per millimole of cell triglyceride. [From Cuatrecasas *et al.* (1974).]

with other peptide hormones. In one series of experiments, designed to measure the binding of insulin to lymphocytes, a *control* was included in which cells were omitted from the medium containing appropriate amounts of all other reagents. After incubation with vigorous agitation in glass tubes and upon filtration over Millipore filters, it was observed that more radioactivity was trapped from samples without native insulin than from samples including native insulin in the medium (Table II). Under other circumstances, this difference in radioactivity *bound* would

Table II
Specific Binding of Insulin in the Absence of Tissue by Vigorous Shaking of Assay Mixtures[a,b]

	Binding (cpm)			
	Glass tubes		Plastic tubes	
Addition	Shaking	No shaking	Shaking	No shaking
None	25,300 ± 2500	3260 ± 380	3230 ± 290	2880 ± 310
Native insulin, 5 μg/ml	14,800 ± 1500	2940 ± 760	3100 ± 110	2500 ± 200

[a]Incubation media (0.2 ml) consisting of Hanks' buffer containing 0.1 % albumin were incubated with and without vigorous agitation at 24°C for 40 min in the presence of [^{125}I]insulin (2×10^5 cpm, 1.1 Ci/μmol) with and without native insulin. Glass or plastic tubes (12×75 mm) were used as indicated, and binding (cpm $\pm \frac{1}{2}$ range of duplicates) was determined by filtration over EA cellulose acetate Millipore filters. Similar results are obtained if the buffer used is Krebs–Ringer–bicarbonate, 0.1 % albumin.
[b]Data from Cuatrecasas and Hollenberg (1975).

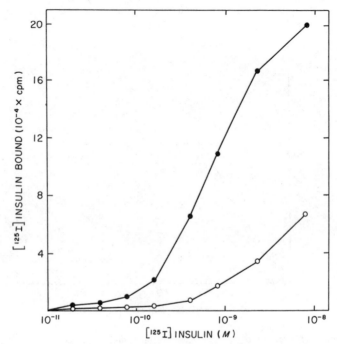

Fig. 7. Dependence on insulin concentration of specific binding of insulin during vigorously agitated assays in glass tubes in the absence of tissue. Incubations consisting of Hanks' buffered saline containing 0.1 % albumin (0.2 ml) were shaken vigorously for 50 min at 24°C in the presence of varying concentrations of ^{125}I-labeled insulin (1.6 Ci/μmol) and in the presence and absence of native insulin (50 μg/ml). Binding was determined by filtration over EG cellulose acetate Millipore filters. Specific binding (●) was calculated by subtracting the binding observed in the presence of native insulin (nonspecific binding, ○) from the total binding (not shown) in the absence of native insulin. [Data from Hollenberg and Cuatrecasas (1975).]

be interpreted as *specific* counts bound. Additionally, when the apparent affinity of the *binding* was estimated (Fig. 7) half-maximal saturation was observed at concentrations (about 10^{-9} M) uncomfortably close to the estimated K_D of a binding site for insulin reported in fat cells and other cells (e.g., lymphocytes) in a number of studies. Vigorous agitation in glass, but not plastic, tubes was found to be a prerequisite for this as yet unexplained artifactual binding (Table II).

The studies with glass tubes were extended to observations with small amounts of talc, which demonstrated not only an unsuspected long time course of the nonreceptor binding (Fig. 8) but also a greater potency of insulin compared to proinsulin in competing for the binding of [^{125}I]insulin to talc (Fig. 9). Additionally, native insulin was found to accelerate the off rate of [^{125}I]insulin bound to talc.

In many ways, the observations of the binding of insulin to inert materials fulfill the operational criteria thought to be characteristic of receptor interactions, i.e., high affinity, reversibility, saturability, some degree of ligand specificity, and even negative cooperativity [claimed by some to characterize insulin–receptor interactions (De Meyts, 1976)]. However, on evaluating all of the data (e.g., growth

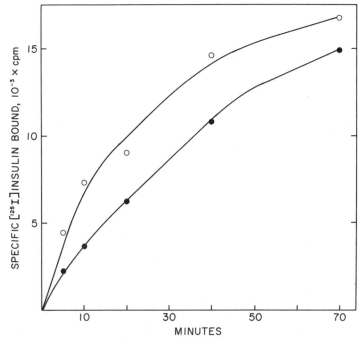

Fig. 8. Time course of binding of ^{125}I-labeled insulin to talc. Binding was measured at 37°C in the presence (●) and in the absence (○) of native insulin.

hormone displaces insulin from talc), it is evident that the talc interaction does not correlate with the known biological data for the action of insulin and insulin analogues.

Far from arguing against the use of binding studies for the study of receptor characteristics, the data for binding to inert materials were collected both to emphasize how potentially difficult it might be in some systems to distinguish receptor from nonreceptor binding and to illustrate the potential success of applying *in sufficient detail* the principles outlined in this chapter.

A number of other instances of specific nonreceptor binding deserve mention. ^{125}I-labeled transferrin can bind to polypropylene culture tubes in a manner that mimics published data purporting to measure cell binding of transferrin (Phillips, 1976). Furthermore, [125I]glucagon can bind in a displaceable manner to certain Millipore filters (Cuatrecasas *et al.*, 1975), opiate drugs bind to cerebroside sulfate stereospecifically and with a relative affinity reflecting *in vivo* potency (Loh *et al.*, 1974), [3H]naloxone binds stereospecifically to glass filters (S.H. Snyder, personal communication), and numerous drugs interact in a highly specific manner with naturally occurring macromolecules (Goldstein *et al.*, 1974). Additionally, except under special conditions of ionic strength, progesterone–receptor complex binds to multiple (probably nonreceptor) sites in isolated nuclei (Spelsberg *et al.*, 1976). In the absence of a biologically responsive system in which context such data can be placed, the distinction of receptor from nonreceptor binding becomes difficult, if not impossible.

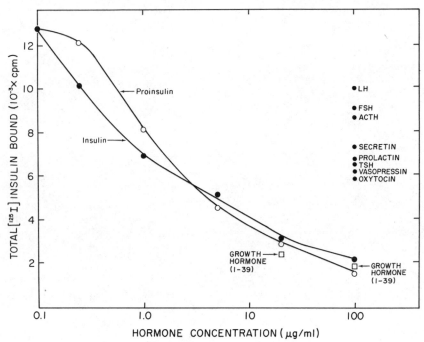

Fig. 9. Specificity of binding of insulin to talc. Samples (0.2 ml) containing 100 μg/ml talc were incubated at 24°C for 30 min with 2.4×10^5 cpm [^{125}I]insulin (1.1 Ci/μmol) in the presence of various unlabeled hormones. The latter were added 5 min before addition of [^{125}I]insulin. Binding was determined by Millipore (e.g., cellulose acetate) filtration.

7. Conclusion

It is evident from the discussions presented in this chapter that the distinction of receptor from nonreceptor interactions in binding studies may rarely be a simple task. The approach to such a distinction must be closely coupled with the receptor concept, as developed at the turn of the century, and must rest on the accumulated pharmacologic data relevant to the biological process of interest in which the particular receptor functions. From such a perspective, the criteria of (1) high and appropriate ligand affinity, (2) saturability, (3) reversibility, (4) ligand specificity, and (5) tissue specificity may be usefully employed, with the realization that even when all such criteria are satisfied, additional independent experimental data (e.g., an unexpected effect of enzymatic "receptor" perturbation on all five criteria for a receptor of interest) may indicate that a binding interaction first thought to represent a receptor may only closely feign such an interaction. It should be evident from the discussions in this chapter that no *single* formula or tool can be recommended to ensure unequivocally in all cases that a measured binding interaction truly represents a pharmacologic receptor. Rather, an understanding of the receptor concept as it has developed over the years, along with the judicious and thorough application of the principles outlined in this chapter, may minimize the risks of misinterpreting binding data.

ACKNOWLEDGMENTS. A portion of the work presented in this chapter was performed in the Alan Bernstein Memorial Laboratories of the Johns Hopkins University School of Medicine and was supported in part by a Basil O'Connor Research Starter Grant from the National Foundation–March of Dimes and by grants (75-22; 76-28) from the Maryland Division of the American Cancer Society. M.D.H. is an Investigator of the Howard Hughes Medical Institute.

References

Arunlakshana, O., and Schild, H. O., 1959, *Br. J. Pharmacol.* **14**:48.

Barlow, R. B., and Stephenson, R. P., 1970, in: *A Companion to Medical Studies*, Vol. 2 (R. Passmore and J. S. Robson, eds.), p. 31, Blackwell's, Oxford.

Bennett, V., and Cuatrecasas, P., 1976, in: *Methods in Receptor Research* (M. Blecher, ed.), Part I, p. 73, New York.

Birdsall, N. J. M., and Hulme, E. C., 1976, *J. Neurochem.* **27**:7.

Blecher, M. (ed.), 1976, *Methods in Receptor Research, Parts I and II*, Dekker, New York.

Buller, R. F., Schrader, W. T., and O'Malley, B., 1976, *J. Steroid Biochem.* **7**:321.

Carpenter, G., Lembach, K. J., Morrison, M. M., and Cohen, S., 1975, *J. Biol. Chem.* **250**:4297.

Chang, K. J., Jacobs, S., and Cuatrecasas, P., 1975, *Biochim. Biophys. Acta* **406**:294.

Changeux, J.-P., 1966, *Mol. Pharmacol.* **2**:369.

Changeux, J.-P., and Podleski, T. R., 1968, *Proc. Natl. Acad. Sci. USA* **59**:944.

Changeux, J.-P., Blumenthal, R., Kasai, M., and Podleski, T., 1970, in: *Molecular Properties of Drug Receptors* (R. Porter and M. O'Connor, eds.), p. 197, Churchill, London.

Cheng, Y.-C., and Prusoff, W. H., 1973, *Biochem. Pharmacol.* **22**:3099.

Clark, A. J., 1926a, *J. Physiol.* **61**:530.

Clark, A. J., 1926b, *J. Physiol.* **61**:547.

Cohen, J. B., and Changeux, J.-P., 1975, *Ann. Rev. Pharmacol.* **15**:83.

Colquhoun, D., 1973, in: *Drug Receptors* (H. P. Rang, ed.), p. 149, Butterworth's, London.

Creese, I., Burt, D. R., and Snyder, S. H., 1975, *Life Sci.* **17**:993.

Cuatrecasas, P., 1971, *Proc. Natl. Acad. Sci. USA* **68**:1264.

Cuatrecasas, P., 1973, *Biochemistry* **12**:3547.

Cuatrecasas, P., 1974, *Annu. Rev. Biochem.* **43**:169.

Cuatrecasas, P., and Hollenberg, M. D., 1975, *Biochem. Biophys. Res. Commun.* **62**:31.

Cuatrecasas, P., and Hollenberg, M. D., 1976, *Adv. Protein Chem.* **30**:251.

Cuatrecasas, P., Tell, G. P. E., Sica, V., Parikh, I., and Chang, K.-J., 1974, *Nature (London)* **247**:92.

Cuatrecasas, P., Hollenberg, M. D., Chang, K.-J., and Bennett, V., 1975, *Recent Prog. Horm. Res.* **31**:37.

De Meyts, P., 1976, in: *Methods in Receptor Research* (M. Blecher, ed.), Part I, Chap. 11, Dekker, New York.

Edelstein, S. J., 1972, *Biochem. Biophys. Res. Commun.* **48**:1160.

Ehrlich, P., 1956, Nobel lecture, 1908, in: *The Collected Papers of P. Ehrlich*, Vol. III (F. Himmelweit, M. Marquardt, and H. Dale, eds.), p. 183, Pergamon Press, Elmsford, N.Y.

Freychet, P., Roth, J., and Neville, D. M., 1971a, *Proc. Natl. Acad. Sci. USA* **68**:1833.

Freychet, P., Roth, J., and Neville, D. M., 1971b, *Biochem. Biophys. Res. Commun.* **43**:400.

Gill, E. W., and Rang, H. P., 1966, *Mol. Pharmacol.* **2**:284.

Gliemann, J., Gammeltoft, S., and Vinten, J., 1975, *J. Biol. Chem.* **250**:3368.

Goldstein, A., Aronow, L., and Kalman, S. M., 1974, *Principles of Drug Action*, Wiley, New York.

Gupta, S. K., Moran, J. F., and Triggle, D. J., 1976, *Mol. Pharmacol.* **13**:1019.

Hall, Z. W., 1972, *Annu. Rev. Biochem.* **41**:925.

Hollenberg, M. D., 1975, *Arch. Biochem. Biophys.* **171**:371.

Hollenberg, M. D., and Cuatrecasas, P., 1975, in: *Biochemical Actions of Hormones*, Vol. III (G. Litwack, ed.), p. 41, Academic Press, New York.

Hollenberg, M. D., and Gregory, H., 1978, *Mol. Pharmacol.*, submitted.

Horn, H. S., 1975, in: *Handbook of Psychopharmacology*, Vol. 2 (L. L. Iversen, S. D. Iversen, and S. H. Snyder, eds.), p. 129, Plenum Press, New York.

Jacobs, S., and Cuatrecasas, P., 1976, *Biochim. Biophys. Acta* **433**:482.

Jacobs, S., Chang, K.-J., and Cuatrecasas, P., 1975, *Biochem. Biophys. Res. Commun.* **66**:687.

Karlin, A., 1967, *J. Theor. Biol.* **16**:306.

Langley, J. N., 1906, *Proc. Roy. Soc. London Ser. B* **78**:170.

Lefkowitz, R. J., Limbird, L. E., Mukherjee, C., and Caron, M. G., 1976, *Biochim. Biophys. Acta* **457**:1.

Loh, H. H., Cho, T. M., Wu, Y. C., and Way, E. L., 1974, *Life Sci.* **14**:2231.

McGhee, J. D., and von Hippel, P. H., 1974, *J. Mol. Biol.* **86**:469.

Paton, W. D. M., 1961, *Proc. Roy. Soc. London Ser. B* **154**:21.

Paton, W. D. M., and Rang, H. P., 1965, *Proc. Roy. Soc. London Ser. B* **163**:1.

Phillips, J. L., 1976, *Biochem. Biophys. Res. Commun.* **71**:726.

Rodbard, D., 1973, in: *Receptors for Reproductive Hormones* (B. W. O'Malley, and A. R. Means, eds.), p. 289, Plenum Press, New York.

Scatchard, G., 1949, *Ann. NY Acad. Sci.* **51**:660.

Schild, H. O., 1949, *Br. J. Pharmacol.* **4**:277.

Segel, I. H., 1975, *Enzyme Kinetics,* Wiley, New York.

Snyder, S. H., 1975, *Biochem. Pharmacol.* **24**:1371.

Snyder, S. H., Pasternak, G. W., and Pert, C. B., 1975, in: *Handbook of Psychopharmacology,* Vol. 5 (L. L. Iversen, S. D. Iversen, and S. H. Snyder, eds.), p. 329, Plenum Press, New York.

Spelsberg, T. C., Pikler, G. M., and Webster, R. A., 1976, *Science* **194**:197.

Stephenson, R. P., 1956, *Br. J. Pharmacol.* **11**:379.

Straus, O. H., and Goldstein, A. J., 1943, *J. Gen. Physiol.* **26**:559.

Sudlow, G., Birkett, D. J., and Wade, D. N., 1976, *Mol. Pharmacol.* **12**:1052.

U'Prichard, D. C., and Snyder, S. H., 1977, *Life Sci.* **20**:527.

Waud, D. R., 1968, *Pharmacol. Rev.* **20**:49.

Webb, J. L., 1963, *Enzyme and Metabolic Inhibitors,* Vol. 1, Academic Press, New York.

Yamamura, H., and Snyder, S. H., 1974, *Proc. Natl. Acad. Sci. USA* **71**:1725.

Incorporation of Transport Molecules into Black Lipid Membranes

Robert Blumenthal and Adil E. Shamoo

1. Introduction

Receptors are cellular components which have the ability to selectively recognize external signals and which transduce those signals into a message that is meaningful for the cell. The transduction step is, in most cases, mediated by a membrane permeability change induced by a transport molecule associated with the receptor. This fact is immediately obvious in the case of receptors for neurotransmitters where the interaction directly causes a depolarization or hyperpolarization of the postsynaptic membrane due to ionic permeability change. Evidence is, however, mounting that the action of receptors for hormones (Rasmussen, 1975) or antigens (Lauf, 1975) is also associated with ionic permeability change, leading to Rasmussen's notion of metallic ions as *second messengers*. We shall not review the second messenger concept here, but we note it in view of the importance of transport molecules in receptor action.

On the other hand, our concepts about the action of active transport systems such as $(Na^+ + K^+)$–ATPase or Ca^{2+}–ATPase lead to the notion of a receptor site (or gate) for the metallic ion in the transport molecule. The gate undergoes an energy mediated transduction step, thereby negotiates the passage of the ion through a membrane channel and promotes active accumulation into or extrusion from the cell.

In view of the fact that transport molecules are so intimately involved in receptor action it is important to study their mechanism of action either as separate entities or, preferably, in conjunction with the recognition part of the system. Refined techniques have been developed to study the recognition sites of receptors in great detail (Kahn, 1976).

The assay for the transport molecules has been furthered by development of the bimolecular lipid membrane (BLM) technique (Mueller and Rudin, 1969b;

Robert Blumenthal ● Laboratory of Theoretical Biology, National Cancer Institute, National Institutes of Health, Bethesda, Maryland. *Adil E. Shamoo* ● Department of Radiation Biology and Biophysics, University of Rochester School of Medicine and Dentistry, Rochester, New York.

Mueller *et al.*, 1964) about 15 years ago. In view of the success in characterizing membrane permeability mechanisms in great detail in the case of ionophorous antibiotics using the BLM technique (Mueller and Rudin, 1969b), it seems hopeful to apply those methods to transport molecules associated with cell membrane receptor systems. The liposomal assay for transport function is discussed in Chapter 1 in this volume. Although transport function is more readily reconstituted using the liposomal system, the BLM allows a more detailed study of the mechanism of action of the reconstituted transport molecule.

Two general approaches have been applied to the reconstitution problem. The approach emphasized by Shamoo and Goldstein (1977) is to start with purified material from a well-established transport system such as the $(Na^+ + K^+)$–ATPase, the Ca^{2+}–ATPase, or the acetylcholine receptor and to test whether its BLM activity corresponds to the intact transport system in terms of its ion dependency and/or selectivity and the effect of specific inhibitors.

A different approach is to isolate and purify a given membrane protein (preferentially a receptor system such as rhodopsin) suspected to be associated with a transport molecule and to use the BLM to verify the putative transport properties associated with the receptors. Some of these efforts have been reviewed recently by Montal (1976). We shall discuss both approaches in Secs. 5 and 6. In the other sections we shall review the methodology of working with BLMs, the well-characterized methods to determine mechanisms of ion permeation mediated by carriers or channels, and the often annoying interaction of soluble proteins and detergents with BLMs.

2. Methodology

Methods for formation and characterization of BLMs have been described in two books (Jain, 1972; Tien, 1974) and in extensive reviews in methodology books (Mueller and Rudin, 1969a; Fettiplace *et al.*, 1975; Andreoli, 1974; Finkelstein, 1974b). Finkelstein (1974b) has some amusing references to the elusive stability problems of BLMs and recommends certain cleaning rituals and other incantations to overcome those difficulties. In this chapter we shall not give a detailed account of how to set up a BLM laboratory, but we shall discuss some salient aspects of methodology with respect to incorporation of transport molecules.

2.1. Formation and Composition of BLMs

Black lipid membranes are formed from an approximately 1% (w/v) solution of surfactant molecules (most commonly phospholipids) in an organic solvent (generally either an aliphatic hydrocarbon or a chloroform–methonal mixture). The membrane-forming solution is spread under water across a circular hole in a plastic or Teflon septum. Initially a thick film is formed which, when viewed at $10\times$ to $20\times$ magnification in reflected light, shows beautiful interference colors. With time, black spots appear in the colored film; the spots grow and coalesce until the entire film is black. The light reflected from the front interface between lipid and water undergoes a 180° phase shift and destructively interferes with light reflected from the rear face, since the film thickness is small compared to the wavelength of visible light (3800–7600 Å). A number of lines of evidence based on electron microscopy, capacitance measurements, variation of lengths of hydrocarbon chains, and water permeability suggest that the black film is a bilayer membrane.

The most popular membrane-forming solutions are egg lecithin plus cholesterol in decane, a mixture of ox-brain lipids in chloroform–methanol and oxidized cholesterol. They derive their popularity from the fact that they form the most stable films. The above-mentioned books and reviews list a variety of the membrane-forming solutions that have yielded satisfactory results in a variety of laboratories.

The question of how much solvent is present in a formed BLM has been subject to numerous investigations. The most direct studies use double labeling of solvent and lipid and the recovery of a sample withdrawn from the BLM after formation. Pagano *et al.* (1972), using an elegant mercury droplet technique to sample a formed BLM, found 37% solvent in a glyceryl monoleate–hexadecane BLM. The sampling technique might be of use to measure the amount of incorporated transport molecules associated with a BLM.

The indirect method for measuring black film composition developed in the laboratory of Haydon (Fettiplace *et al.*, 1971) is based on the application of the Gibbs equation, which yields a relationship between interfacial tension, activity of the lipid in a nonpolar solvent, and the surface excess of lipid in the interface. The method includes estimation of the thickness of the BLM from the capacitance and dielectric constant. Fettiplace *et al.* (1971) found that the amount of solvent retained in the BLM depended on its chain length. For instance, with glycerol monoleate films, the film thickness decreased from 46.5 to 32.2 Å, and the volume fraction of solvent decreased from 45% to 17% in the following series of hydrocarbon solvents: C_7, C_{10}, C_{14}, and C_{16}. The solvent content found by Pagano *et al.* (1972), which was higher than that deduced by Fettiplace *et al.* (1971), almost certainly reflects the presence of small lenses in glyceryl monoleate–hexadecane BLMs which are not picked up by the surface thermodynamic method. When using lecithin, Fettiplace *et al.* (1971) found that they could obtain essentially dry films (0% solvent) with hexadecane.

Alternative methods for the formation of BLMs include the spherical bilayer membranes (Pagano and Thompson, 1967), which are more suitable for tracer flux studies because of their large area, and the *dry* planar lipid bilayer (Montal and Mueller, 1972; Montal, 1974).

The Pagano and Thompson (1967) bilayers are formed by injecting a droplet in a density gradient in a glass cell containing aqueous electrolyte. A syringe microburette filled with electrolyte solution is attached to small-diameter polyethylene tubing, and the tip of the polyethylene tubing is coated with a membrane-forming solution. This tubing is inserted into the density gradient, and a droplet of electrolyte solution, coated with lipid, is injected from the syringe into the density gradient solution. The droplet, containing a spherical membrane surrounding an aqueous salt solution, comes to rest at isodense level. In their final state the droplets contain a spherical bilayer membrane with an area up to 1 cm² and a bulk lipid level at the top of the spherical bilayer occupying less than 10% of the surface of the sphere. For electrical measurements micropipettes have been used to gain access to the interior. The large surface area makes those bilayers very suitable for tracer flux studies. They have also been used by Yguerabide and Stryer (1971) in their study of the location, orientation, and rotational mobility of fluorescent chromophores embedded into the spherical bilayer. The advantage of the Pagano and Thompson bilayer for these studies is that different orientations of the chromophores are available when the membranes are illuminated by an appropriately sized and positioned beam of light.

The Montal and Mueller (1972) bilayers are formed by the apposition of the hydrocarbon chains of two lipid monolayers at an air–water interface. The advantages of these planar bilayer membranes over black lipid membranes are that they contain no hydrocarbon solvents, that asymmetric bilayers can be formed, and, consequently, that direct reconstitution of transport molecules can be carried out.

2.2. Electrical Properties of BLMs

For rapid screening of incorporation of transport molecules into BLMs ion permeability and selectivity are the simplest and most straightforward properties to measure. They are determined from the electrical resistance of the BLM and the diffusion potentials generated by a salt gradient. Transport molecules that do not mediate the passage of a charged ion or net current will of course be missed by the electrical techniques. For instance, a glucose carrier or a transport molecule carrying ions as a neutral complex will not carry any net current through the BLM. The properties of these transport molecules then has to be measured with isotope tracers by adding a radiolabeled species on one side of the BLM and measuring the appearance of the tracer on the other side.

Conductance is measured by applying a step voltage and recording the current response. The membrane is placed in series with an *ammeter* resistance and a voltage source, and current (I) is passed and measured by means of calomel or Ag/AgCl.

Capacitance can be obtained from the time constant of the current response to a voltage step or by applying a potential difference by means of a function generator across the membrane in the form of a triangular wave: at $V = 0$ the height of the current response is a measure of the capacitance (C) of the BLM, from the relation $I = C(dV/dt)$. More accurate capacitance measurements have been obtained with the AC bridge method (Fettiplace *et al.*, 1975).

To measure ion selectivity, ion concentration differences between the two compartments separated by the BLM are created by adding small aliquots of concentrated salt solution to one of the compartments. The voltage source is switched to open circuit, and membrane potentials (V_m) are measured with a high-input-impedance electrometer. Most commonly calomel electrodes or silver–silver chloride electrodes are used. The silver–silver chloride electrodes are favored by Eisenman and his co-workers (Szabo *et al.*, 1969) because there are no uninterpretable corrections for liquid junction potentials. Initially, salts of chloride at the same concentration are present on both sides of the BLM, and the membrane potential should be zero if the electrodes are completely reversible. Then concentrated salts of nitrate (NO_3^- and Cl^- have essentially identical free-solution mobilities) are added rather than salts of *chloride* because silver–silver chloride electrodes respond to changes in chloride concentration as well as to the membrane potential. Calomel electrodes are to be used if the total absence of chloride ion is required, but aside from the liquid junction potentials, which change with solution concentration, leakage of KCl from the junction poses cumbersome problems in the interpretation of these measurements.

An estimate of relative ionic permeabilities may be made by computing the ionic transference numbers from the expression

$$V_m = - \sum_{i=1}^{n} t_i E_i \tag{1}$$

E_i, the equilibrium potential of the ith ion, is given by the Nernst expression,

$$E_i = \frac{RT}{z_i F} \ln \frac{a_i^{\mathrm{I}}}{a_i^{\mathrm{II}}} \tag{2}$$

where R, T, and F are, respectively, the gas constant, absolute temperature, and Faraday's number; z_i is the valence of the ith ion; and a_i^{I} and a_i^{II} are the activities of the ith ion in the two aqueous phases bathing the membrane. The ionic transference number t_i is defined as

$$t_i = \frac{g_i}{g_m} \tag{3}$$

where g_i is the conductance of the ith ion and g_m is the total membrane conductance. It follows from this definition that

$$\sum_{i=1}^{n} t_i = 1 \tag{4}$$

If a membrane is completely ion selective for one ionic species ($t_i = 1$ for that species and $t_i = 0$ for all other species), the membrane potential is given by

$$V_m = - \frac{RT}{z_i F} \ln \frac{a_i^{\mathrm{I}}}{a_i^{\mathrm{II}}} = - \frac{2.3 RT}{z_i F} \log \frac{a_i^{\mathrm{I}}}{a_i^{\mathrm{II}}} \tag{5}$$

Since $2.3 RT/F = 58$ mV at room temperature, the membrane potential will change by -58 mV/decade concentration ratio of a monovalent cation-selective membrane, by 58 mV/decade for an anion-selective membrane, and by -29 mV/decade for a bivalent cation.

A more accurate relationship between ion permeabilities and the membrane potential is given by the Goldman–Hodgkin–Katz equation (Goldman, 1943; Hodgkin and Katz, 1949), which for an asymmetrical solution of sodium chloride (NaCl) yields

$$V_m = \frac{RT}{F} \ln \left[\frac{(P_{\mathrm{Na}}/P_{\mathrm{Cl}}) a_{\mathrm{Na}}^{\mathrm{I}} + a_{\mathrm{Cl}}^{\mathrm{II}}}{(P_{\mathrm{Na}}/P_{\mathrm{Cl}}) a_{\mathrm{Na}}^{\mathrm{II}} + a_{\mathrm{Cl}}^{\mathrm{I}}} \right] \tag{6}$$

where $P_{\mathrm{Na}}/P_{\mathrm{Cl}}$ is the permeability ratio of sodium and chloride. A detailed expression for the Goldman equation for more than one cationic species and for the case of divalent vs. monovalent cation selectivity can be found in the review by Shamoo and Goldstein (1977).

3. Mechanisms of Ion Permeability

The concept of carriers and channels was introduced to reconcile the general impermeability of lipid bilayers with the fact that many molecules which are virtually insoluble in lipids can cross membranes readily by a path which is saturable and highly specific for the substrate. Carriers are assumed to be lipid-soluble molecules that bind substrate on one side of the membrane, diffuse through the lipid bilayer, and dissociate on the other side of the membrane, releasing free substrate. The binding site accounts for specificity, and under the assumption of a very rapid association–dissociation of carrier–substrate complex, the rate-limiting step of diffusion of a limited number of carrier–substrate complexes through the lipid bilayer accounts for saturation. If a carrier has no particular orientation in the membrane and does not interact specifically with a particular membrane lipid, it is described as a truly mobile carrier (Shamoo, 1975; Wyssbrod et al., 1971).

Channels are aqueous filled pathways spanning the membrane. These channels may have high selectivity (Shamoo, 1975; Goldstein and Solomon, 1960; Hille, 1970). Channels may or may not have highly specialized gates (Armstrong, 1975). The gates could interact with the transporting molecule or with other controlling factors, such as voltage (Armstrong, 1975). It may be difficult to distinguish gated channels from carriers.

3.1. Carriers

For many years the carrier has been a most useful concept in explaining specificity of transport across membranes, although no one has actually demonstrated that it exists. For instance, the specificity between two such chemically similar ions as sodium and potassium could only be explained by invoking a membrane constituent with a highly sophisticated structure. The concept of a carrier was first proposed as a component of the membrane reversibly binding to electrolytes and enhancing their lipid solubility (Osterhout and Stanley, 1932).

Since lipids are the basic constituents of membranes, it was thought that certain lipids may serve as ion carriers. This led to membrane fractionation to extract specific lipids from cell membranes which will enhance the lipoid solubility of ions or substrates. Lipids that enhanced ion or substrate solubility into organic solvents were then studied as possible ion carriers.

Solomon and his colleagues (1956) and Kirschner (1964) found that a number of purified phospholipids (e.g., phosphatidylserine, sphingomyelin, and acetyl-phosphate), but not cholesterol, enhanced the lipoid solubility of K^+ 8 to 14 times more than Na^+ into the organic solvent phase. Schulman and his co-workers (Rosano et al., 1961) studied the mechanism of ion flux migration through non-aqueous liquid membranes. Their electrical potential measurements across nonaqueous liquid membranes showed that K^+ ions migrated more easily than Na^+ ions through the "oil"–water interface. This finding agrees with the activation energy value for migration of Na^+ and K^+ through "oil"–water. Stearic acid was also shown to act as a Na^+ carrier across an organic solvent phase (Moore and Schechter, 1969). Further evidence for phospholipids as possible ion carriers came from Schneider and Wolff (1965), who found that phospholipids bind iodide and other anions. These other anions displace bound iodide in a sequence which parallels their ability to displace iodide bound within slices of thyroid. Several other compounds have been claimed to have ionophoric activity using equilibrium extraction and bulk transport methods. These compounds include lecithin (Hyman, 1971), cephalin (Feinstein, 1964), phosphatidylserine (Solomon et al., 1956; Feinstein, 1964; Kirschner, 1964), cytochrome c (Margoliash et al., 1970), free fatty acids (Wojtczak, 1974), and prostaglandins (Kirtland and Baum, 1972; Carafoli and Crovetti, 1973). In a more recent report, Tyson et al. (1976) reported a survey of various phospholipids for ionophoric activity. By using the bulk transport assay method, it was found that cardiolipin and phosphatidic acid were the most active in transporting divalent and monovalent cations into an organic solvent phase. Neither of the compounds just mentioned has any known relationship to transport function in membranes.

LeFevre et al. (1964) showed that glucose and sorbitol (a nontransported sugar) were similarly solubilized into a chloroform phase when the phospholipid fraction from erythrocyte membranes was present. It was, however, found that erythrocyte phospholipids could not discriminate between the two glucose isomers and that ribose and inositol at 0.4 M did not compete with each other in erythrocyte phospholipid extract induced transport in an aqueous/chloroform/aqueous system. These experiments clearly showed that the phospholipids were poor candidates for a physiological carrier system (LeFevre et al., 1968).

The inhibitory effect of pronase on anion transport in red blood cells was taken as evidence against membrane lipid participation in anion transport (Passow, 1971). This finding added further credence to the concept that lipids cannot serve as selective carriers in cell membranes.

The ability of phospholipids to bind different ions and substrates cannot account for the specificity of various transport systems or the specificity of the individual transport system. Since all of these lipid or lipid-like materials extracted by organic solvent procedure have not produced results directly implicating them as the transport carriers, this approach was abandoned in favor of the search of a proteinaceous material as the transport carrier.

Pressman and his co-workers (Moore and Pressman, 1964) observed that the phenomenon of antibiotic-induced mitochondrial transport is related to the formation of an alkali ion complex by such antibiotic compounds. To describe the transport properties of these antibiotics generally, Pressman introduced the term *ionophorous* agents. The ionophores have served as valuable models for illustrating dynamic properties of membrane-bound carriers as well as the possible molecular basis of ion selectivity. They have also been employed extensively as tools for perturbing biological systems (Pressman, 1973). Recently Pressman and deGuzman (1974) have applied ionophores as pharmocological agents affecting cardiovascular systems.

Strictly speaking, ionophorous antibiotics are transport molecules incorporated into black lipid membranes. They have, however, not been found to bear any relationship to biological transport functions. They are actually designed to change permeabilities of cells, and they tend to kill them because the cell will not maintain its osmotic or ionic balance when permeated by those ionophores. An interesting question is why the bacteria that synthesize those antibiotics do not commit suicide.

The careful work on mechanisms of ion permeation carried out in various laboratories [for review articles, see Shamoo (1975)] provides a clear framework for investigating the mode of action of functional transport systems.

3.2. Channel Formers

Collander and Barlund (1933) collected a comprehensive set of accurate measurements on the plant cell *Chara ceratophylla*, assessing the variation of permeability constant with chemical structure of the penetrating species. The relatively good correlation between permeability constant and olive oil–water partition coefficient confirmed Overton's rule which established the lipid nature of the cell membrane. The fact, however, that a number of small molecules such as water, methanol, and formamide penetrate faster than would be predicted from their oil–water partition coefficients led to the notion of the lipid-porous sieve mosaic character of the cell membrane.

The detailed study of these pore structures in membranes came with the development of the BLM and the discovery by Mueller and Rudin (1963) that EIM (excitability-inducing material), a proteinaceous material of bacterial origin, interacts with the BLM to produce a voltage-dependent conductance system. Subsequently other antibiotics such as gramicidine, alamethicin, and monoazomycin and the polyenes nystatin and amphothericin B were found to induce voltage-dependent pores in BLMs (Haydon and Hladky, 1972; Finkelstein, 1974a). In the case of EIM, alamethicin, and monoazomycin, the induced conductance is voltage dependent.

A significant feature of channels in BLMs is that a single event brought about by the insertion of a channel into BLM or by the opening of an inserted channel can be measured. The conductance of a unit channel is given by

$$g = \frac{\pi r^2 \Lambda c}{d} \tag{7}$$

where r is the radius of the channel, d the length, Λ the equivalent conductivity of the ions in the channel, and c the concentration. According to a current model for the structure of the gramicidin channel (Urry et al., 1975), it has a radius of 2 Å and a length of 30 Å. The 2-Å radius is also consistent with Finkelstein's (1974a) measurements of water and urea permeability through gramicidin-doped BLMs. Assuming then that the ion concentration and mobility of the gramicidin channel are equal to that in free solution, we obtain a conductance of $5 \times 10^{-11} \Omega^{-1}$ in 0.1 M KCl. (The equivalent conductivity of 0.1 M KCl in free solution is 129 Ω^{-1} cm^{-2} equiv^{-1}; *Handbook of Chemistry and Physics 1973–1974*, 54th edition, Chemical Rubber Company Press, Cleveland, Ohio, p. D132). This is about the conductance measured from the discrete conductance steps measured in a BLM.

The unit conductance of channels in biological membranes which is obtained from the conductance fluctuation analysis is in the same "ball park." For instance, in the frog neuromuscular junction the acetylcholine-sensitive postsynaptic channels are thought to have a conductance of $3 \times 10^{-11} \, \Omega^{-1}$ (Katz and Miledi, 1971; Anderson and Stevens, 1973; Sachs and Lecar, 1973). In the BLM unit conductance events are readily measured because the background conductance of an unmodified BLM is in the range of $10^{-11} \, \Omega^{-1}$. Because biological membrane preparations, on the other hand, have much larger conductance, of the order of $1 \, \Omega^{-1} \, cm^{-2}$, increases due to single events cannot be measured directly but have to be deduced from the autocorrelation spectrum of current fluctuations.

The opening of a channel with a conductance of $10^{-11} \, \Omega$ in a 0.1-V gradient gives rise to a current of 10^{-12} A, according to Ohm's law. We can deduce how many ions move per second:

$$10^{-12} \, A \rightarrow 10^{-12} \, C/sec \xrightarrow[\text{faraday}]{\overset{\text{divide}}{\underset{\text{by}}{}}} 10^{-17} \, equiv/sec \xrightarrow[\substack{\text{Avogadro's} \\ \text{number}}]{\overset{\text{multiply}}{\underset{\text{by}}{}}} 6 \times 10^6 \, ions/sec$$

A carrier will generally move about 10^3 ions/sec, which would give rise to a unit of current fluctuation of about 10^{-16} A which is not picked up by the BLM measuring system. Using a measuring system which would allow observation of transient increases in conductance down to $5 \times 10^{-13} \, \Omega$, provided they lasted for about 100 msec, Hladky and Haydon (1970) did not observe step changes in a BLM doped with nonactin, which is a well-established carrier ionophore antibiotic.

Unit conductance steps have been observed in the case of the other channel formers EIM (Bean *et al.*, 1969; Ehrenstein *et al.*, 1970), alamethicin (Gordon and Haydon, 1972; Eisenberg *et al.*, 1973), gramicidin (Hladky and Haydon, 1970, 1972), and keyhole limpet hemocyanin (Alvarez *et al.*, 1975; Blumenthal, 1975). In the case of EIM the unit conductance is $4 \times 10^{-10} \, \Omega^{-1}$, which would be consistent with a radius of 4 Å.

It is very important to make the distinction whether discrete conductance changes correspond to single conductance events, since they could also be caused by spurious fluctuations in the BLM lipid matrix, as has been demonstrated by Yafuso *et al.* (1974). The following criteria have been applied to identifying discrete conductance changes with events related to channel formation.

1. The opening and closing of channels follow Poisson statistics (Haydon and Hladky, 1972; Ehrenstein *et al.*, 1974).

2. The kinetic behavior of single channels is similar to that of many channels (Ehrenstein *et al.*, 1974).

3. The ion selectivity of single and many channels is similar (Latorre *et al.*, 1972).

4. The unit conductance is independent of voltage and directly proportional to salt concentration and viscosity of the medium.

5. The rate of channel formation is proportional to membrane thickness (Hladky and Haydon, 1972).

6. The slope of a discrete current change seen when applying a voltage in the form of a triangular wave extrapolates back to zero current at the equilibrium potential (to $V = 0$ in equal salt concentrations) (Ehrenstein et al., 1970).

7. A formed channel can be destroyed by applying a proteolytic enzyme to the opposite side from where the channel is inserted (Bean et al., 1969). This is to distinguish a discrete step due to local disruption of the lipid structure from that of a bonafide pore.

8. The water permeability and nonelectrolyte permeability should be consistent with the notion of a water-filled pore in a BLM (Finkelstein, 1974a).

9. Temperature dependence: (a) Latorre et al. (1974) performed an elegant experiment to distinguish between a channel mechanism and a local disruption mechanism of EIM in oxidized cholesterol membranes. The EIM channel has two states in 0.1 M KCl: an open state of 0.4 nmho and a closed state of 0.08 nmho. In the open state the conductance increased with temperature as predicted for a wide electrolyte-filled pore, whereas in the closed state the conductance decreased with increasing temperature. The interpretation of the conductance decrease is that the greater fluidity of the lipids at the higher temperature will seal the "cracks" around the protein, which initially caused the local disruption. (b) With BLMs formed of glyceryl dipalmitate or glyceryl distearate, Krasne et al. (1971) showed that the carriers nonactin and valinomycin lost their effectiveness in mediating ion conductance at the temperature where the BLMs became solid, as judged visually. By contrast the effects of the channel former gramicidin appeared to be the same on solid and liquid membranes. The findings are interpreted as reduction of mobility of the carriers by freezing, with little effect on their solubility in the BLM. Channel-mediated transport does not depend on a mobile component; therefore it is not affected by the state of the lipid.

4. Models of Interactions of Proteins with BLMs

The BLM has been used extensively to study soluble protein–lipid interactions with different objectives in mind. Usually soluble proteins are added to BLMs as control experiments for the incorporation of transport molecules. In some cases it has been thought that the BLM could be used as a transducer system to monitor enzyme–substrates as antigen–antibody reactions (Del Castillo et al., 1966).

The liposomal model has, however, been used more successfully to study protein–lipid interactions since these interactions can be monitored by other techniques (e.g., differential scanning calorimetry, fluorescence, freeze fracture) in addition to studying permeability effects. Kimelberg (1976) has recently written an excellent review on protein–liposome interactions. One of the most striking conclusions derived from Kimelberg's review is that many soluble proteins such as hemoglobin and cytochrome c have the capability to interact hydrophobically with phospholipids. This fact certainly poses a caveat for interpretation of incorporation of transport molecules into BLMs. The initial electrostatic interactions are translated into hydrophobic interactions, which causes local disruption of the lipid matrix, resulting in a permeability change.

In the case of the BLM, Montal (1972) observed conductance changes induced by polypeptides with opposite charge to the surface charge of the BLM. The most striking example of an interaction of a soluble protein with a BLM to produce conductance changes is keyhole limpet hemocyanin, which is a respiratory protein of high molecular weight dissolved in the hemolymph of certain invertebrates.

Pant and Conran (1972) showed that associated hemocyanin produces electrical conductance increases of as much as six orders of magnitude when added to the aqueous solution after forming a BLM and that the conductance was voltage dependent and was abolished at pH 4 and 10. Discrete current fluctuations first observed by Alvarez *et al.* (1975) indicated that possibly discrete pores formed by associated hemocyanin in the bilayer membrane give rise to the conductance. Blumenthal (1975) observed that by dissociating hemocyanin at alkaline pH, the hemocyanin still induced conductance changes, but the discrete pores and voltage dependence were abolished.

The different ways in which proteins might interact with lipid bilayer membranes are summarized in Fig. 1. The dark area represents the hydrophobic region and the light area the hydrophilic region of the protein. The arrows represent possible ion pathways. Although the proteins are represented as globular

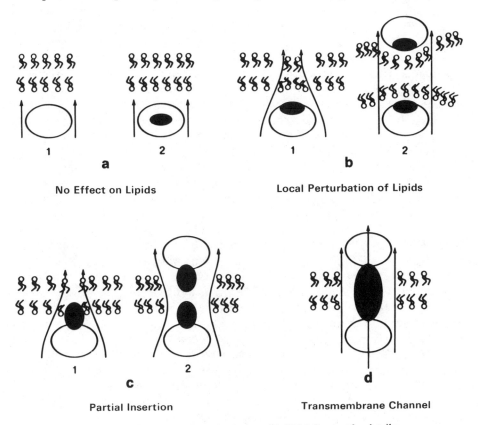

Fig. 1. Types of interaction of proteins with BLM. See text for details.

proteins, the discussion will certainly pertain to other conformations. Figure 1(a1), for instance, represents a protein without any hydrophobic region. In the situation of Fig. 1(a), binding of the protein to the surface will produce no effect on lipid bilayer permeability. Figure 1(b) represents a protein with hydrophobic region on the surface of the proteins which interacts with lipid bilayer. Two of these proteins may interact with both sides of the bilayer, as shown in Fig. 1(b2).

The protein might interact initially by charge interactions followed by increases in permeability by locally lowering the phase-transition temperature of the lipid. Papahadjopoulos *et al.* (1973) actually examined the temperature dependence of sodium diffusion through dipalmitoylphosphatidyl glycerol liposomes, particularly in the temperature region where the hydrocarbon chains undergo a phase transition from crystalline to liquid crystalline. Their results show that at temperatures below phase-transition temperature the liposomes were almost impermeable to sodium ions but that the diffusion increased with increasing temperature, reaching a maximum of about 100-fold increase in diffusion rate at a temperature coinciding with the midpoint of the lipid phase transition. They explained their observations by suggesting that boundary regions separating domains of phospholipid molecules in crystalline and liquid–crystalline configurations possess a high permeability to sodium ions. Proteins interacting hydrophobically with a lipid bilayer might create these boundary regions of relatively high permeability. Figure 1(c) represents a protein with a hydrophobic moiety which can penetrate one layer of the BLM. Two of these proteins may interact with both sides of the bilayer, as shown in Fig. 1(c2). The penetration results in a permeability change due to local disruption of the BLM structure. The situation diagrammed in Fig. 1(c) is representative of the mode of action of low concentrations of detergents on BLMs.

Ionic or nonionic detergents have been reported to increase BLM conductance (Shamoo *et al.*, 1976a,b; Seufert, 1965; Rosenberg and Pant, 1970; Minassian-Saraga and Wietzerbin, 1970; Van Zutphen *et al.*, 1972; Ksenzhek *et al.*, 1974). Anionic detergents may cause the BLM to become cation selective, and cation detergents may cause the BLM to become anion selective. Relatively large doses of detergents cause the BLM to become leaky to all ions, and eventually higher dosages will cause rupture (Shamoo *et al.*, 1976a,b).

Figure 1(d) represents protein with hydrophobic region spanning the membrane. Such interaction is most likely to produce little, if any, permeability changes due to disruption of the lipid bilayer, but the ions will pass through the channel itself. A transport molecule such as $(Na^+ + K^+)$–ATPase will clearly interact with the membrane as shown in Fig. 1(d), especially as it is known to interact with ligands from both sides of the cell membrane. Since detergent is required to remove the transport molecules from the membrane, it follows that hydrophobic bonding between protein and bilayer lipid is dominant.

It is likely that all the reported "incorporations" of transport molecules into BLMs follow the models presented in Figs. 1(b) and (c). Mostly, it is very simple to decide whether BLM conductance data fit models of the type shown in Figs. 1(b) and (c): Since the permeability change is caused by a local disruption of the lipid membrane matrix, adding more protein will lead to fast eventual disruption of the BLM. The consequence of this statement is not to throw the data of these interactions away, as has been done by numerous workers, but to exploit the results, as has been done in our laboratories (Shamoo and Albers, 1973; Shamoo *et al.*, 1974; Blumenthal and Shamoo, 1974; Shamoo, 1974; Shamoo and Ryan, 1975), to identify components of a transport system in terms of their selectivity and the effects of inhibitors.

The puzzling data on *incorporation selectivity*, i.e., the fact that the protein requires a certain ion (Na^+ or Ca^{2+}) to produce a conductance change but once the conductance change has been produced it loses its selectivity, can also be explained by the model. In the presence of Na^+ or Ca^{2+} the conformation of the protein is such that it exposes the part of the hydrophobic surface of the protein that can interact with the hydrophobic surface of the lipid. The ions still have to pass the permeability barrier in the vicinity of the protein, and thus the protein can still confer some ion specificity upon the BLM. The model in Fig. 1(c) could even give rise to discrete conductance fluctuations: Every time a local disruption occurs it gives rise to a conductance increase with the magnitude of a pore.

As mentioned in the previous section, these discrete fluctuations can easily be distinguished from "real" channel fluctuations. First, the local disruption bump will not give rise to a channel of uniform size (conductance magnitude). It can easily be abolished by increasing the temperature. The experiments of Bean *et al.* (1969) on digestion of EIM channels on the other side of the BLM and by Latorre *et al.* (1974) on the disappearance of the "closed" channel of EIM with temperature provide excellent criteria for deciding whether we are dealing with a real channel. In the case of hemocyanin it is not clear whether the discrete fluctuations are caused by a channel such as shown in Fig. 1(d).

In some cases the *bilateral* action of protein was felt to provide some evidence for pore formation in analogy with the half pores in the case of nystatin and gramicidin (Finkelstein, 1974a). However, a one-sided local disruption will not be sufficient to produce a conductance change, and addition to the protein to the other side will significantly enhance the effect.

In summary, even if one can demonstrate hydrophobic interactions in protein–BLM interactions, this is not a sufficient demonstration that a pore indeed has formed. The uniformity of channel size and the Bean *et al.* (1969) and Latorre *et al.* (1974) experiments could provide additional evidence that we have reconstituted the integral part into a BLM.

5. Ionophorous Properties in BLMs of Functional Transport Molecules

5.1. Ca^{2+}–ATPase: Dissection of a Transport System

$(Ca^{2+} + Mg^{2+})$–ATPase is an integral protein of the sarcoplasmic reticulum membrane involved in the active ATP-dependent uptake of Ca^{2+} from the myoplasm. The $(Ca^{2+} + Mg^{2+})$–ATPase has been purified extensively, and its ATP-dependent Ca^{2+} transport function has been reconstituted in sonicated phospholipid liposomes (see Chapter 1 in this volume).

The reconstitution data indicate that the primary *pump* machine for Ca^{2+}

resides in the $(Ca^{2+} + Mg^{2+})$-ATPase molecule. Therefore, it is reasonable to search for the Ca^{2+}-transporting site (Ca^{2+}-mediating site). The Ca^{2+}-transporting site may consist of either a gate, a channel, a carrier, or a gated channel. This led Shamoo and MacLennan (1974) to test the intact enzyme for Ca^{2+}-ionophoric activity in BLM assay. They demonstrated the existence of Ca^{2+}-dependent and -selective ionophoric activity associated with the intact sarcoplasmic reticulum $(Ca^{2+} + Mg^{2+})$-ATPase molecule employing the BLM conductance assay. Conductance increments were shown with the $(Ca^{2+} + Mg^{2+})$-ATPase by either (1) succinylation of the intact enzyme in order to solubilize the enzyme, (2) sonication of insoluble $(Ca^{2+} + Mg^{2+})$-ATPase into the BLM bathing fluid, or (3) tryptic digestion of $(Ca^{2+} + Mg^{2+})$-ATPase (Shamoo and MacLennan, 1974, 1975; Shamoo et al., 1976a,b; Shamoo and Ryan, 1975). The relative conductance change and the relative permeability exhibited by this material has the following sequence:

$$Ba^{2+} > Ca^{2+} > Sr^{2+} > Mg^{2+} > Zn^{2+}, Na^+, K^+, Cs^+, Li^+, \text{ and } Rb^+$$

In the presence of $(Ca^{2+} + Mg^{2+})$-ATPase the BLM conductance increase was dependent on the presence of Ca^{2+} when compared to other divalent cations, as was the BLM selectivity for Ca^{2+} when compared to other ions. The ionophoric activity sequence of $(Ca^{2+} + Mg^{2+})$-ATPase is consistent with the overall Ca^{2+} transport in sarcoplasmic reticulum and resembles the selectivity sequence reported for Ca^{2+} ionophore X-537A in phosphatidylcholine black films (Celis et al., 1974).

To further link the Ca^{2+}-ionophoric activity with the Ca^{2+} transport system, several common inhibitors were sought. It was found that Ca^{2+} conductance was inhibited by divalent cations such as Zn^{2+}, Mn^{2+}, and Hg^{2+}, and by La^{3+}. This inhibition is consistent with the assumption that these ions compete for the Ca^{2+}-ionophoric site. At least two sodium ions are required to inhibit the Ca^{2+} conductance (Shamoo and Ryan, 1975; Shamoo et al., 1976a). The inhibitory action of Na^+ on the Ca^{2+} ionophore among all other monovalent cations tested is consistent with the observation that Na^+ inhibits Ca^{2+} uptake into sarcoplasmic reticulum vesicles (Masuda and deMeis, 1974). As was indicated, $HgCl_2$ completely blocked the ionophoric activity in the presence of Ca^{2+}, while methylmercury (also a sulfhydryl group blocker) had no effect (Shamoo et al., 1976a). Therefore $HgCl_2$ inhibition of Ca^{2+} ionophore is not due to its binding to a sulfhydryl group but rather may be a replacement of Ca^{2+} at the Ca^{2+}-ionophoric site. Also, Hg^{2+} inhibition of Ca^{2+} conductance is similar to the above characteristics for other divalent cations, and the mode of inhibition is competition for the Ca^{2+}-ionophoric site. Furthermore, mercuric chloride inhibits Ca^{2+} uptake faster than its inhibition of $(Ca^{2+} + Mg^{2+})$-ATPase activity, which is consistent with the explanation that Hg^{2+} inhibits the Ca^{2+}-ionophoric site.

These data suggest the separate identity of the two sites, namely, the ATP-hydrolysis and Ca^{2+}-ionophoric activity (Shamoo and MacLennan, 1975; Shamoo et al., 1976a,b).

To localize the Ca^{2+} ion gate within the $(Ca^{2+} + Mg^{2+})$-ATPase molecule, Shamoo and MacLennan's groups resorted to controlled tryptic digestion in order to separate fragments with various functions. Exposure of sarcoplasmic reticulum to trypsin in the presence of 1 M sucrose results in degradation of the 102,000-dalton $(Ca^{2+} + Mg^{2+})$–ATPase enzyme to two fragments of 55,000 and 45,000 daltons with subsequent appearance of fragments of 30,000 and 20,000 daltons (Thorley-Lawson and Green, 1975; Stewart *et al.*, 1976; Shamoo *et al.*, 1976a,b).

The various fragments were purified either by SDS column chromatography or by SDS–preparative gel electrophoresis (Shamoo *et al.*, 1976b). The use of antibodies against the various fragments indicated that the 55,000-dalton fragment was in large part exposed to the cytoplasm *in vivo*, whereas the 20,000- and the 45,000-dalton fragments were only partially exposed (Stewart *et al.*, 1976; Shamoo and Ryan, 1975). The 55,000-dalton fragment had ionophoric activity similar to the intact enzyme. The 45,000-dalton fragment had no or nonspecific ionophoric activity (Stewart *et al.*, 1976; Shamoo *et al.*, 1976b). The 50,000- and 30,000-dalton fragments but not the 45,000- and 20,000-dalton fragments have been shown to contain the site of phosphorylation and of $[^3H]$-N-ethylmaleimide binding, indicating the hydrolytic site in $(Ca^{2+} + Mg^{2+})$–ATPase. The 20,000-dalton fragment contains the site of Ca^{2+}-dependent and -selective ionophoric activity, as shown in Fig. 2 (Shamoo *et al.*, 1976b). However, the Ca^{2+} dependency or selectivity in the 20,000-dalton fragment was not so pronounced as that in the intact enzyme or in the 55,000-dalton fragment. Further digestion of the 20,000-dalton fragment with cyanogen bromide indicated that the Ca^{2+}-ionophoric activity is associated with a fragment of less than 2000 daltons (Shamoo *et al.*, 1976b).

Ruthenium red, a specific inhibitor of $(Ca^{2+} + Mg^{2+})$–ATPase activity (Vale and Carvalho, 1973), also inhibits the ionophoric activity of the intact enzyme and the 20,000-dalton fragment. Mercuric chloride inhibits the ionophoric activity of the intact enzyme (Shamoo *et al.*, 1975) and of the 20,000-dalton fragment. However, methylmercuric chloride, which inhibits $(Ca^{2+} + Mg^{2+})$–ATPase activity, does not inhibit the Ca^{2+}-ionophoric activity of either the intact ATPase or the 20,000-dalton fragment. The data which indicate the lack of in-hibition of Ca^{2+}-ionophoric activity in the intact enzyme or the 20,000-dalton fragment by methylmercury and the strong inhibition of the hydrolytic site of ATP hydrolysis which resides in the 30,000-dalton fragment are consistent with the amino acid analysis data. The amino acid analysis data show that the 20,000-dalton fragment, the site of the Ca^{2+} ionophore, contains no cysteine (Shamoo *et al.*, 1976b) or the least number of cysteine residues (Thorley-Lawson and Green, 1975), whereas the 30,000-dalton fragment is rich in cysteine. $ZnCl_2$ inhibits the ionophoric activity of the intact enzyme and of the 20,000-dalton fragment (Shamoo and MacLennan, 1974, 1975; Shamoo *et al.*, 1976b). The possibility that the Ca^{2+}-ionophoric activity associated with the 20,000-dalton fragment may be due to SDS or just due to 20,000-dalton protein associated with bound SDS was reduced by numerous control experiments (Shamoo *et al.*, 1976b).

These data suggest the separate identity of the two sites, namely the ATP-hydrolysis and Ca^{2+}-ionophoric activity (Shamoo and MacLennan, 1975; Shamoo *et al.*, 1976a,b).

From these data Shamoo *et al.* (1976b) concluded that BLM can serve as an assay tool for the Ca^{2+}-transporting site in the $(Ca^{2+} + Mg^{2+})$–ATPase

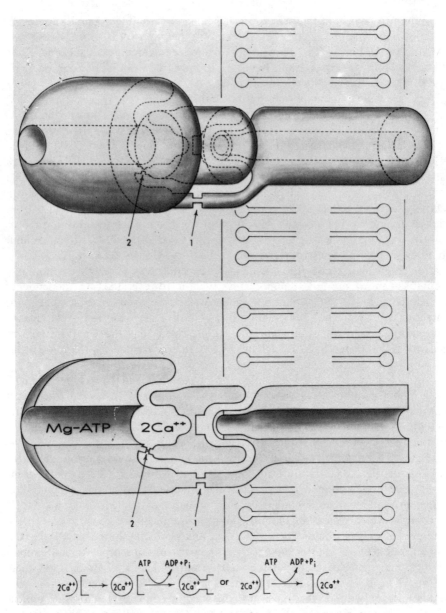

Fig. 2. Model for active transport of Ca^{2+} in sarcoplasmic reticulum.

molecule and that the activity resides in a discrete portion of the molecule of less than 20,000 daltons. They have also shown that the ATP-catalytic moiety resides in a separate molecular-weight fragment. The two sites, however, appear to be in proximity since both fragments originate from the 55,000-dalton fragment. The role of the 45,000-dalton fragment may be to form a large nonspecific channel across the membrane. The data are consistent with the following picture of the (Ca^{2+} + Mg^{2+})–ATPase action transport pump (Fig. 2): The 45,000-dalton fragment (the channel moiety) is embedded in the membrane. The 20,000-dalton fragment (the gate moiety) is intertwined through hydrophobic interaction with the 45,000- and 30,000-dalton (the catalytic moiety) fragments. The 20,000-dalton fragment is attached through a covalent bond susceptible to tryptic digestion to each of the 45,000- and 30,000-dalton fragments. This is consistent with the observation that Ca^{2+} transport is diminished when the 55,000-dalton fragment is cleaved into 30,000- and 20,000-dalton fragments (Shamoo *et al.*, 1976b). The bond between the 20,000- and 45,000-dalton fragments is the first bond susceptible to tryptic digestion followed by the bond between the 20,000- and 30,000-dalton fragments. The nonselective channel (i.e., 45,000 daltons) could serve in transport of counterions such as Mg^{2+} or K^+.

The active transport step occurs at the gate site (middle part of the 20,000-dalton fragment). The energy transduction step for active transport of Ca^{2+} can occur in two ways, as illustrated in Fig. 2. One model for energy transduction requires synchronous opening and closing of the gate with ATP hydrolysis, causing change of affinity to Ca^{2+} on one side only, and the other model requires a rotational movement of the gate (flip-flop).

As mentioned before, all the protein–BLM interactions reported in this section are probably due to interaction diagrammed in Fig. 1(b) or (c). However, those interactions have fortunately retained the selectivity and inhibition characteristics of the intact transport system. Therefore, our approach has led us to make some progress into the identification of the different components of the Ca^{2+}–ATPase transport system.

5.2. $(Na^+ + K^+)$–ATPase

Although this was the first system subjected to our enzymatic dissection methods, the results were not so clear-cut as in the case of Ca^{2+}–ATPase. Shamoo and Albers (1973) isolated a fraction which induced a Na^+-dependent conductance increase in the BLM by extraction of the trichloroacetic-acid-soluble 1-hr tryptic digest of the electroplax (*Electrophorus electricus*) microsome rich in $(Na^+ + K^+)$–ATPase. The microsomal fraction of the electric organ of the eel contained several other proteins besides $(Na^+ + K^+)$–ATPase. The two polypeptides known to constitute the $(Na^+ + K^+)$–ATPase molecule (about 100,000 and 50,000 daltons) were isolated by SDS–preparative gel electrophoresis (Shamoo, 1974; Shamoo and Myers, 1974; Shamoo *et al.*, 1974). The Na^+-dependent ionophoric activity was associated with the small polypeptide of $(Na^+ + K^+)$–ATPase; further digestion of the small polypeptide showed that the Na^+-dependent ionophore resides in a small fragment of the small polypeptide of $(Na^+ + K^+)$–ATPase. Cyanogen bromide digestion of the Lubrol-soluble purified $(Na^+ + K^+)$–ATPase resulted in a fraction having a K^+-dependent ionophoric activity (Shamoo and Ryan, 1975). The data on the isolation of Na^+-dependent ionophore from $(Na^+ + K^+)$–ATPase were difficult to reproduce from batch to batch; this problem is discussed in detail elsewhere (Shamoo and Ryan, 1975). What these data indicate is that the Na^+ transport site (Na^+ ionophore) is covalently bound to the transporting enzyme, and in order to release it a chemical cleavage of a covalent bond or bonds must occur.

5.3. The Acetylcholine Receptor

In 1971 Parisi *et al.* (1971) reported that the BLM conductance of artificial lipid membranes made from oxidized cholesterol in the presence of the proteolipid fraction from the electroplax of *Electrophorus electricus* and acetylcholine increases. The conductance increase was termed *cholinergic conductance.* The cholinergic conductance was inhibited by *d*-tubocurarine (Parisi *et al.,* 1972). Leuzinger and Schneider (1972) reported an increase in BLM conductance when "purified" acetylcholinesterase [which may contain acetylcholine receptor (AChR)] and acetylcholine were employed either on both sides or on opposite sides. Jain *et al.* (1973a,b) reported that a water-soluble impure acetylcholinesterase preparation (the assumption is that it may contain acetylcholine receptor) from eel electroplax, purchased from Sigma Chemical Company, when incorporated into oxidized cholesterol BLM, resulted in ion channel formation in response to acetylcholine and carbamylcholine. The channel conductance is apparent only when acetylcholine is added to the side of the membrane opposite to that to which the receptor protein is added. The results were amplified in a subsequent report (Jain, 1974). An impure detergent-solubilized AChR from neuromuscular junction of phrenic nerve of rat diaphragm increases the membrane conductance significantly even in the absence of acetylcholine (Kemp *et al.,* 1973). A fourfold increase in membrane conductance was observed when acetylcholine was introduced into the side opposite that of the AChR (Kemp *et al.,* 1973). An impure AChR from mouse brain was reported to cause discrete changes in conductance (Goodall *et al.,* 1974; Romine *et al.,* 1974). Exposure of the AChR to dithiothreitol, followed by *N*-ethylmaleimide, resulted in the blockage of an acetylcholine-induced conductance increase in BLM (Reader and DeRobertis, 1974). Levinson and Keynes (1972) were not able to purify the AChR as reported by DeRobertis' group (Latorre *et al.,* 1970; DeRobertis and de Plazas, 1970; DeRobertis *et al.,* 1971). They showed that the elution pattern obtained by DeRobertis' group using [^{3}H]acetylcholine was a property of the organic solvent elution pattern of Sephadex LH-20 regardless of the source of the protein.

Kometani *et al.* (1975) extracted an acetylcholine binding "proteolipid" from the electric organ of *Narka japonica* by using chloroform/ethanol (2:1, v/v), a method similar to that reported by DeRobertis and his co-workers (Parisi *et al.,* 1971, 1972, 1975; Reader and DeRobertis, 1974; Adragna *et al.,* 1975; Latorre, *et al.,* 1970; DeRobertis and de Plazas, 1970; DeRobertis *et al.,* 1971). The acetylcholine binding "proteolipid" isolated in this method is not a protein. Also, the acetylcholine binding "proteolipid" does not appear in the receptor-rich fraction of fractionated membrane (Kometani *et al.,* 1975).

Recently Parisi *et al.* (1975) have shown in a control experiment that their previous data of cholinergic conductance can be obtained with the use of acetylcholine on lipid bilayer containing negative phospholipids without the presence of the purified AChR or the "proteolipid." The increase in conductance was induced by choline as well as acetylcholine. The bilayer under such conditions was also cation selective, as has been previously claimed (Parisi *et al.,* 1975).

In a recent report by Shamoo and Eldefrawi (1975), the tryptic digest of carboxymethylcellulose-treated, purified AChR from the electroplax of *Torpedo californica* caused the BLM conductance to increase when acetylcholine was added. The conductance increase due to acetylcholine was cation selective and was prevented by curare. The tryptic digestion of the AChR did not change the ability of the receptor to bind acetylcholine SDS gel electrophoresis of the AChR, before and after tryptic digestion, indicating that this mild enzyme treatment hydrolyzed the receptor molecule subunits. Analytical gel electrophoresis without SDS indicated that the receptor remained intact after the treatment with trypsin.

The various reports of AChR interaction with lipid bilayer indicate that AChR with Ca^{2+} causes a nonselective increase in BLM conductance similar to that reported for other membrane proteins. But both the AChR and the mitochondrial Ca^{2+} binding protein show a divalent cationic dependency on Ca^{2+}, indicating some form of interaction between the membrane-associated proteins and the lipid bilayer. It is clear that the specific reconstitution of the AChR with the critical properties similar to that of the electrophysiological experiment has not yet been achieved.

6. The BLM as a Test System for Ionophorous Function of Isolated Membrane Proteins

There are two ways to incorporate a membrane protein into BLMs. The classical way is to disperse the protein into the aqueous solution and add it to a preformed BLM. Because intrinsic membrane proteins are generally not water soluble, they are originally dissolved in an organic solvent (this procedure is also used in the case of highly hydrophobic peptide antibiotics such as gramicidin), solubilized by detergent, succinilated or sonicated, and then added to the aqueous solution neighboring the BLM.

An alternative approach, developed by Montal (1976), involves either mixing detergent solutions of lipid and proteins and removing the detergent slowly by dialysis or sonicating a dispersion of lipid and protein in salt media, yielding lipoprotein vesicles. The aqueous dispersion lipid and protein are then partitioned into an organic phase with the aid of Ca^{2+}, and the resulting "proteolipid" is spread as a monolayer at the air–water interface and subsequently used to form planar bilayers by apposing two such monolayers. Alternatively, BLMs are formed by spreading the proteolipid in the organic solvent over the hole in the Teflon cup.

The two approaches reflect two alternative views on membrane biogenesis of protein. One view considers a sequential synthesis and assembly of lipid, and a special type of cytoplasmic protein, which is synthesized in the cytoplasm and then diffuses into the lipid membrane to become integrated into it (Bretcher, 1973). The other view favors a simultaneous synthesis and assembly of lipid and protein on the rough endoplasmic reticulum. The newly formed membrane portions are pinched off the endoplasmic reticulum and finally fuse with the plasma membrane (Palade, 1975).

In the following we shall review some attempts to incorporate putative transport molecules into BLMs. Most of these followed the sequential assembly method. Only the last one, rhodopsin, is incorporated according to the simultaneous assembly method favored by Montal.

6.1. Mitochondrial Membrane Proteins

Carafoli and his co-workers (Carafoli, 1975; Carafoli and Sottocasa, 1974; Prestipino *et al.*, 1974) reported that the mitochondrial Ca^{2+} binding glycoprotein caused an increase in conductance of a phosphatidylcholine black lipid membrane in the presence of calcium. The mitochondrial glycoprotein has been isolated and identified as water soluble and having a high affinity for Ca^{2+} (Lehninger, 1971; Gomez-Puyou *et al.*, 1972; Sottocasa *et al.*, 1972; Carafoli and Sottocasa, 1974; Carafoli, 1975). The increases in the electrical conductance of BLM were observed when 2–4 mM of Ca^{2+}, Mn^{2+}, or Sr^{2+} was used with the glycoprotein, but no such increase in BLM conductance was observed when monovalent cations were used. However, once the conductance increased it showed no selectivity between cations and anions. This indicates large nonselective conducting pathways when calcium induces glycoprotein incorporation into lipid bilayer (Prestipino *et al.*, 1974; Carafoli and Sottocasa, 1974; Carafoli, 1975). The glycoprotein properties of water solubility, Ca^{2+} binding, Ca^{2+}-dependent conductance, and lack of cation–anion selectivity indicate that the glycoprotein may function as a superficial receptor, analogous to the bacterial amino acid binding proteins, all of which are released by osmotic shock (Oxender and Quay, 1975). The glycoprotein probably contributes to the kinetic parameters of the process in intact mitochondria such as saturation kinetics, competitive inhibition by Sr^{2+}, and inhibition by La^{3+} and ruthenium red. The glycoprotein may be placed in series with a transmembrane (intrinsic) component responsible for the translocation of Ca^{2+}. This conclusion is further supported by the fact that the glycoprotein does not discharge Ca^{2+} from liposomes (Carafoli, personal communication).

6.2. Red Blood Cell Membrane Proteins

There are three major proteins associated with red blood cell membranes that have been isolated and well identified, namely spectrin (Marchesi *et al.*, 1969; Kirkpatrick, 1976), glycophorin (glycoprotein) (Marchesi *et al.*, 1972), and a protein known as band 3 from SDS gel electrophoresis (Rothstein *et al.*, 1976a,b). Band 3 has a molecular weight between 90,000 and 100,000 daltons based on SDS–acrylamide gel electrophoresis (Fairbanks *et al.*, 1971; Rothstein *et al.*, 1976a,b). All three proteins have been tested for ion transport mediation capability using various methods. Spectrin was shown to induce a nonselective increase in bilayer conductance in the presence of divalent cations (Kirkpatrick, 1976). Glycophorin is known to be a structural protein, and no known function has yet been attributed to it (Tosteson *et al.*, 1973). This glycoprotein from red blood cell membranes was solubilized in lithium diiodosalicylate (LIS). The purified glycoprotein was shown to increase the conductance of lipid bilayer membranes formed from red cell lipids dissolved in decane. The conductance increase due to glycophorin was the same whether NaCl or KCl solutions were employed. Moreover, the conductance increase in the presence of glycophorin was irreversible when the glycophorin-containing solution was replaced by glycophorin-free solution. It was also shown that concanavalin A produced an eightfold increase in conductance when added in the presence of Ca^{2+} to bilayers previously exposed to glycophorin (Tosteson *et al.*, 1973).

Lossen *et al.*, (1973) showed that the isolated "strongly bound" red blood cell membrane proteins (predominantly band 3 and glycophorin) caused a non-ion-dependent, nonselective increase in the conductance of oxidized cholesterol BLMs. Similarly, they showed (Brennecke *et al.*, 1975) that the "loosely bound" proteins of red blood cell membranes (predominantly spectrin) caused a similar increase in conductance with somewhat different kinetics. Neither the chemical basis of glycophorin interaction with bilayers nor its relationship to ion transport is yet clear. These results remain in the same category as the mitochondrial calcium (Prestipino *et al.*, 1974; Carafoli and Sottocasa, 1974; Carafoli, 1975), intact acetylcholine receptor (Shamoo and Eldefrawi, 1975), dopamine β-hydroxylase (Kafka *et al.*, 1978), and spectrin (Kirkpatrick, 1976), where all have been shown to induce nonselective increase in bilayer conductance with a wide ionic dependency. With such results we cannot associate any ion transport mediation capability with these materials. The results do indicate strong interactions between these membrane-bound proteins and the BLM.

The interaction of such proteins with BLMs can serve as an assay for determining the relationship of membrane-associated proteins and the lipid bilayer with certain ions such as Ca^{2+}. Evidence in support of this contention comes from Grant and McConnell (1974), who showed that glycoprotein penetrates the hydrophobic region of the bilayer in vesicles. Moreover, Lea *et al.* (1975) showed that a tryptic fragment of glycophorin—molecular weight of 3750 daltons— caused an increase in liposome permeability to K^+ and water and increased BLM conductance. This was explained in terms of the peptide causing local disordering of the bilayer membrane structure but not in terms of any relationship to the capability of this protein to mediate ion transport (Lea *et al.*, 1975).

The red blood cell is endowed with an extremely rapid, high-capacity, anion-passive permeation system that can exchange Cl^- and HCO_3^- (Passow, 1969; Gunn *et al.*, 1973). The anion transport has no intrinsic properties that are usually useful in identifying its components. Therefore, a large class of chemical agents were developed to inhibit anion transport (Zaki *et al.*, 1975; Cabantchik and Rothstein, 1974a,b; Cabantchik *et al.*, 1975a,b, 1976; Knauf and Rothstein, 1971; Rothstein and Cabantchik, 1974; Rothstein *et al.*, 1975, 1976a,b).

Cabantchik, Rothstein, and their co-workers (Knauf and Rothstein, 1971; Rothstein *et al.*, 1974, 1975, 1976a,b; Cabantchik and Rothstein, 1974a,b; Cabantchik *et al.*, 1976) have performed elegant work implicating that band 3 from red blood cell membranes is involved in anion transport. An equally impressive array of reports has been produced by Passow and co-workers (Zaki *et al.*, 1975).

No attempts have yet been reported to incorporate band 3 into a BLM, probably because band 3 consists of several proteins (Conrad and Penniston, 1976) which might produce conductance effects in BLMs which would override the Cl^--selective effect.

6.3. Gastric Mucosal Membrane Proteins

In an attempt to isolate anion-selective transport material, Sachs *et al.* (1974) extracted ionophoric material from mucosal membranes. The ionophoric material induced cation, anion, or nonselective conductances in phosphatidylcholine black lipid membranes.

In subsequent work, Goodall and Sachs (1977) aimed at the identification of the various fractions responsible for ionophoric activity using a vesicular membrane preparation rich in K^+–ATPase. These vesicles were shown to have an ATP-dependent H^+ uptake and $^{86}Rb^+$ extrusion. Further subfractionation of the gradient fraction by free-flow electrophoresis produced a vesicular preparation containing two major polypeptide regions on SDS–gel electrophoresis (100,000 and 80,000 daltons), further enriched in K^+–ATPase with $H^+:K^+$ exchange capability similar to the vesicular membrane preparation. The incorporation of this fraction into phosphatidylserine bilayers at 35°C resulted in a two to three order of magnitude increase in conductance and showed the following selectivity sequence:

$$H^+ > K^+ > Rb^+ > Cs^+ > Na^+ > Li^+ > Tl^+ > Cl^-$$

Except for Tl, the cation selectivity of the treated BLM was the same as that for the native vesicles. The addition of ATP, and not ITP or β-α-methylene ATP, caused a steady and regular increase in potential, indicative of pump activity. This work has two important new findings: first, the K^+–ATPase was nearly purified (two bands) without the use of detergents, and thus its effect on BLM is less vulnerable to skepticism, and, second, the BLM–K^+–ATPase properties closely resemble those of the native vesicle preparation.

6.4. Dopamine–β-Hydroxylase

Dopamine–β-hydroxylase (DBH), an enzyme which converts dopamine to norepinephrine, is localized within storage vesicles of adrenal medulla (chromaffin granules) and nerve cell body (Lovenberg *et al.*, 1975). While all the DBH appears to be localized within the norepinephrine-containing vesicles, some of the enzyme is in a soluble matrix which is found in the supernatant fraction after lysis of granules, and some of the enzyme is membrane bound. Of the DBH associated with bovine chromaffin granules, about 50% can be released by osmotic lysis (Winkler *et al.*, 1970). The remainder is associated with the vesicular membrane and can be solubilized by treatment with detergent. The soluble and membrane-bound fractions are purified by DEAE–cellulose and Sephadex G-200 column chromatography (Wallace *et al.*, 1973), yielding a tetrameric glycoprotein with a single 75,000-molecular-weight band on SDS–gel electrophoresis. As the substrate, dopamine, is present in the cytoplasm and the product, norepinephrine, is stored in the granule, it seemed possible that DBH was involved in the uptake mechanism of dopamine from the cytoplasm in addition to catalyzing its conversion to norepinephrine in the granule interior.

The purified membrane protein was therefore tested for ionophoric activity in a BLM (Kafka *et al.*, 1978). In nine out of ten experiments a conductance increase was observed when DBH was present on both sides of the BLM in the presence of dopamine and Ca^{2+}. Control experiments involving addition of the components to one compartment, absence of Ca^{2+} and/or dopamine, and addition of heat-denatured DBH, norepinephrine, or detergent were all negative. The time course for DBH-induced conductance followed an *autocatalytic* pattern; i.e., insertion of one conducting unit enhanced the rate of insertion of the next. The conductance increased in discrete steps, indicating channel formation. The current–voltage curve was linear, and the conductance showed no selectivity between potassium and chloride. A similar nonselectivity was shown by Shamoo and Eldefrawi with the acetylcholine receptor in the presence of calcium. The experiments of Kafka *et al.* (1978) do not provide a definite answer as to whether DBH has ionophoric activity in the intact system.

6.5. Rhodopsin

Rhodopsin is a chromophore-bearing protein of photoreceptor cells which are capable of converting the signal of a single photon into a neural impulse. Hagins (1972) has proposed that rhodopsin triggers the photoresponse by regulating the permeability of the disk membrane to a putative *messenger*—presumably Ca^{2+}. The released messenger diffuses to sites on the inner surface of the rod membrane, altering its permeability to Na^+ and blocking the dark current. Rhodopsin is therefore believed to be a receptor–ionophore system (similar to, for example, the acetylcholine receptor) that upon excitation activates an ion translocator for Ca^{2+}.

Various groups [for a review, see Montal (1976)] have attempted to prove this contention by incorporating rhodopsin into a BLM. A successful approach in rhodopsin incorporation seems to be the simultaneous assembly method developed by Montal and his co-workers. Rhodopsin is incorporated into planar bilayers by apposing two rhodopsin–lipid monolayers, which are formed from a hexane-soluble complex of rhodopsin and lipid, that exhibit in the organic phase spectral characteristics similar to those of native rhodopsin in disks. The rhodopsin-containing planar bilayer has a conductance of about $10^{-9}\ \Omega^{-1}$ in the dark, and upon illumination it increases to a variable extent (10^{-8}–$10^{-6}\ \Omega^{-1}$). Discrete events of about $0.4\ n\Omega^{-1}$ indicate channel formation. The conductance fluctuations are infrequent in the dark, but their frequency increases upon illumination. The reported data are not yet sufficient to assess whether the characteristics of ion transport such as ion specificity, kinetics, and light energy dependence measured in the bilayer system are consistent with the data obtained from the biological system (Hagins and Yoshikami, 1977). According to Hagins (private communication), the transduction of a light signal into a Ca^{2+} permeability change involves a biochemical step, probably energy mediated, which is lacking in the reconstitution experiments.

6.6. Immune Cytotoxic Factors

The lysis of foreign cells by the immune system can be accomplished by two classes of mechanisms: those that are humoral, involving substances found in the serum, and those that are cell mediated, involving the interaction of immuno-competent lymphocytes with specific sites on the target cell membrane.

Immunolysis is thought to be mediated by an initial breakdown of the permeability barrier to small ions, leading to subsequent cell swelling and colloid osmotic lysis (Blumenthal *et al.*, 1977). The mechanism of lysis follows the recep-tor–ionophore paradigm. The interaction of complement or killer lymphocytes with a "receptor" on the target cell surface (for example, a complex of an antibody with a cell surface antigen) triggers a cascade of events ending in the release of a putative cytotoxic factor which changes the target cell membrane permeability to small ions.

Kinsky (1972) and his associates have used liposomes in an extensive series of experiments on complement-mediated lysis. They have found that complement will release trapped glucose from antigenic liposomes if antibody is present. Their results suggest that nonmembrane target cell components are not necessarily involved in lysis. Other experiments also show that complement does not involve any detectable enzymatic reaction (i.e., one that breaks down target cell com-ponents). Although liposomes can be used to assay the lipid, antigen, and antibody specificity in complement lysis, planar bilayers have some advantages, especially in studies of the nature of the lesion incurred.

With BLMs it is more difficult, however, to establish a clear-cut example of complement action, because serum factors necessary for complement action will induce nonspecific detergent-like conductance changes in BLMs. The various attempts at BLM immunology have been reviewed by Blumenthal *et al.* (1977).

Only recently a *noise-free* complement system involving purified complement components C_{5b}–C_9 was applied in nanogram quantities to a BLM (Michaels *et al.*, 1976), giving rise to a stable and reliable conductance increase. In the case of lymphocyte-mediated cytotoxicity the cells can be directly applied to an *antibody-coated antigenic* BLM (Henkart and Blumenthal, 1975) in the absence of serum factors. As in the case of complement the results suggest that killer lymphocytes can act directly on the lipid portion of target cell membranes to cause a breakdown of the permeability barrier without involvement of an enzymatic reaction that breaks down target cell components. The experiments indicate that the lymphocyte surface might contain a component that upon activation induces permeability changes in a target cell membrane.

7. Coda

An important advantage of using the BLM as an assay for transport molecules is its exquisite sensitivity, but the blessing is not unmixed. In many cases the BLM will change its conductance when soluble proteins are added to the bathing solution at higher concentrations [due to some ill-defined detergent-like mechan-ism (Finkelstein, 1974b; Montal, 1976)]. Such nonspecific conductance changes presumably do not correspond to the physiological role of these proteins, which are not designed to be mediators of ion conductance. Because of this *detergency* effect, purified components are required for the characterization of induced conductance pathways.

The incorporation of transport molecules into BLMs has not approached the success story of the incorporation of transport molecules into liposomes (see Chapter 1 in this volume). The reason that most efforts of incorporation of transport molecules into BLMs have not been successful is that often the *brute force* method has been applied. Attempts are made to "ram" into a BLM a cell membrane preparation of dubious purity which is purported to be associated with a transport system. The "ramming through" is often successful in the sense that conductance changes are observed, but at the same time all the relevance is lost. The BLM experiments do not match the physiological situation in terms of specificity, kinetics, and action of inhibitors.

A number of approaches outlined in this chapter, however, show a different approach to the problem of incorporation of transport molecules into BLMs. Both the *dissection method* (Shamoo and Goldstein, 1977) and the *simultaneous assembly* method (Montal, 1976) are based on current notions about structure, function, and biogenesis of membrane components.

The methods for incorporating transport molecules into liposomes are more straightforward, but the information gained about their mode of action is limited. Whereas incorporation of transport molecules into BLMs is more troublesome, the degree of sophistication in the study of the mode of action of antibiotic ionophores in BLMs [see Shamoo (1975)] illustrates what can be learned about transport molecules and their receptors once they are successfully incorporated into BLMs.

ACKNOWLEDGMENTS. This chapter is based on work performed under contract with the U.S. Energy Research Development Administration at the University of Rochester Biomedical and Environmental Research Project and has been assigned Report No. UR-3490-1048. This chapter was also supported in part by NIH 1 ROL AM 17571, NIH 1 ROL AM 18892, NIH ES-01248, the Muscular Dystrophy Association (USA), and the Upjohn Company.

References

Adragna, N. C., Salas, P. J. I., Parisi, M., and DeRobertis, E., 1975, Curare blocks cationic conductance in artificial membranes containing hydrophobic proteins from cholinergic tissues, *Biochem. Biophys. Res. Commun.* **62**:110.

Alvarez, O., Diaz, E., and Latorre, R., 1975, Voltage dependent conductance induced by hemocyanin in black lipid films, *Biochim. Biophys. Acta* **389**:444.

Anderson, C. R., and Stevens, C. F., 1973, Voltage clamp analysis of acetylcholine produced end-plate current fluctuations at the frog neuromuscular junction, *J. Physiol. (London)* **235**:655.

Andreoli, T. E., 1974, Planar lipid bilayer membranes, in: *Methods in Enzymology*, Vol. 32 (S. Fleischer and L. Packer, eds.), Part B, pp. 513–538, Academic Press, New York.

Armstrong, C. M., 1975, Ionic pores, gates, and gating currents, *Q. Rev. Biophys.* **7**:179.

Bean, R. C., Shepherd, W. C., Chan, H., and Eichner, J., 1969, Discrete conductance fluctuations in lipid bilayer protein membranes, *J. Gen. Physiol.* **53**:741.

Blumenthal, R., 1975, The interaction of hemocyanin and lymphocytes with lipid bilayer membranes, *Ann. NY Acad. Sci.* **264**:476.

Blumenthal, R., and Shamoo, A. E., 1974, Ionophoric material derived from eel membrane preparations: II. Electrical characteristics, *J. Membr. Biol.* **19**:141.

Blumenthal, R., Weinstein, J. N., and Henkart, P., 1977, Lipid model membrane studies on immune cytotoxic mechanisms, in: *Proceedings of the 9th Rochester International Conference on Environmental Toxicity—Membrane Toxicity* (M. W. Miller, and A. E. Shamoo, eds.), pp. 495–508, Plenum Press, New York.

Brennecke, R., Lossen, O., and Schubert, D., 1975, Interactions between black lipid membranes and the loosely bound proteins of erythrocyte membranes, *Z. Naturforsch.* **30c**:129.

Bretcher, M. S., 1973, Membrane structure: Some general principles, *Science* **181**:622.

Cabantchik, Z. I., and Rothstein, A., 1974a, Membrane proteins related to anion permeability of human red blood cells. I. Localization of disulfonic stilbene binding sites in proteins involved in permeation, *J. Membr. Biol.* **15**:207.

Cabantchik, Z. I., and Rothstein, A., 1974b, Membrane proteins related to anion permeability of human red blood cells. II. Effects of proteolytic enzymes on disulfonic stilbene sites of surface proteins, *J. Membr. Biol.* **15**:227.

Cabantchik, Z. I., Balshin, M., Breuer, W., and Rothstein, A., 1975a, Pyridoxal phosphate. An anionic probe for protein amino groups exposed on the outer and inner surfaces of intact human red blood cells, *J. Biol. Chem.* **250**:5130.

Cabantchik, Z. I., Balshin, M., Breuer, W., Markus, H., and Rothstein, A., 1975b, A comparison of intact human red blood cells and resealed and leaky ghosts with respect to their interactions with surface labelling agents and proteolytic enzymes, *Biochim. Biophys. Acta* **382**:621.

Cabantchik, Z. I., Knauf, P. A., Oswald, T., Markus, H., Davidson, L., Breuer, W., and Rothstein, A., 1976, The interaction of an anionic photoreactive probe with the anion transport system of the human red blood cell, *Biochim. Biophys. Acta* **455**:526.

Carafoli, E., 1975, The interaction of Ca^{2+} with mitochondria, with special reference to the structural role of Ca^{2+} in mitochondrial and other membranes, *Mol. Cell. Biochem.* **8**:133.

Carafoli, E., and Crovetti, F., 1973, Interactions between prostaglandin E1 and calcium at level of mitochondrial membrane, *Arch. Biochem. Biophys.* **154**:40.

Carafoli, E., and Sottocasa, G., 1974, The Ca^{2+} transport system of the mitochondrial membrane and the problem of the Ca^{2+} carrier, in: *Dynamics of Energy-Transducing Membranes* (L. Ernster, R. W. Estabrook, and E. C. Slater, eds.), pp. 455–469, Elsevier, Amsterdam.

Celis, H., Estrada-O, S., and Montal, M., 1974, Model translocators for divalent and monovalent ion transport in phospholipid membranes. I. The ion permeability induced in lipid bilayers by the antibiotic X-537A, *J. Membr. Biol.* **18**:187.

Collander, R., and Barlund, M., 1933, Permeabilitats-studien an *Chara ceratophylla, Acta Bot. Fenn.* **11**:1.

Conrad, M. J., and Penniston, J. T., 1976, Resolution of erythrocyte membrane proteins by two-dimensional electrophoresis, *J. Biol. Chem.* **251**:253.

Del Castillo, J., Rodriquez, A., Romero, C. A., and Sanchez, V., 1966, Lipid films as transducers for detection of antigen–antibody enzyme–substrate reactions, *Science* **153**:185.

DeRobertis, E., and de Plazas, S. R., 1970, Acetylcholinesterase and acetylcholine proteolipid receptor: Two different components of electroplax membranes, *Biochim. Biophys. Acta* **219**:388.

DeRobertis, E., Lunt, G., and LaTorre, J. L., 1971, Multiple binding-sites for acetylcholine in a proteolipid from electric tissue, *Mol. Pharmacol.* **7**:97.

Ehrenstein, G., Lecar, H., and Nossal, R., 1970, The nature of the negative resistance in bimolecular lipid membranes containing excitability-inducing material, *J. Gen. Physiol.* **55**:119.

Ehrenstein, G., Blumenthal, R., Latorre, R., and Lecar, H., 1974, Kinetics of the opening and closing of individual excitability-inducing material channels in a lipid bilayer, *J. Gen. Physiol.* **63**:707.

Eisenberg, M., Hall, J. E., and Mead, C. A., 1973, The nature of the voltage-dependent conductance induced by alamethicin in black lipid membranes, *J. Membr. Biol.* **14**:143.

Fairbanks, G., Steck, T. L., and Wallach, D. F. H., 1971, Electrophoretic analysis of the major polypeptides of the human erythrocyte membrane, *Biochemistry* **10**:2606.

Feinstein, M. B., 1964, Reaction of local anesthetics with phospholipids; a possible chemical basis for anesthesia, *J. Gen. Physiol.* **48**:357.

Fettiplace, R., Andrews, D. M., and Haydon, D. A., 1971, The thickness, composition and structure of some lipid bilayers and natural membranes, *J. Membr. Biol.* **5**:277.

Fettiplace, R., Gordon, L. G. M., Hladky, S. B., Requena, J., Zingsheim, H. P., and Haydon, D. A., 1975, Techniques in the formation and examination of "black" lipid bilayer membranes, in: *Methods in Membrane Biology*, Vol. 3 (E. D. Korn, ed.), pp. 1–75, Plenum Press, New York.

Finkelstein, A., 1974a, Aqueous pores created in thin lipid membranes by the antibiotics nystatin, amphotericin B and gramicidin A: implicators for pores in plasma membranes, in: *Drugs and Transport Processes* (B. A. Callingham, ed.), pp. 241–245, Macmillan, New York.

Finkelstein, A., 1974b, Bilayers: Formation, measurements, and incorporation of components, in: *Methods in Enzymology*, Vol. 32 (S. Fleischer and L. Packer, eds.), Part B, pp. 489–500, Academic Press, New York.

Goldman, D. E., 1943, Potential, impedance and rectification in membranes, *J. Gen. Physiol.* **27**:37.

Goldstein, D. A., and Solomon, A. K., 1960, Determination of equivalent pore radius for human red cells by osmotic pressure measurement, *J. Gen. Physiol.* **44**:11.

Gomez-Puyou, A., Gomez-Puyou, M. T., Backer, G., and Lehninger, A. L., 1972, An insoluble Ca^{2+}-binding factor from rat liver mitochondria, *Biochem. Biophys. Res. Commun.* **47**:814.

Goodall, M. C., and Sachs, C., 1977, Reconstitution of a proton pump from gastric mucosa, *J. Membr. Biol.* **35**:285–301.

Goodall, M. C., Bradley, R. J., Saccomani, G., and Romine, W. O., 1974, Quantum conductance changes in lipid bilayer membranes associated with incorporation of acetylcholine receptors, *Nature (London)* **250**:68.

Gordon, L. G. M., and Haydon, D. A., 1972, The unit conductance of alamethicin, *Biochim. Biophys. Acta* **255**:1014.

Grant, C. W. M., and McConnell, H. M., 1974, Glycophorin in lipid bilayers, *Proc. Natl. Acad. Sci. USA* **71**:4653.

Gunn, R. B., Dalmark, M., Tosteson, D. C., and Wieth, J. O., 1973, Characteristics of chloride transport in human red blood cells, *J. Gen. Physiol.* **61**:185.

Hagins, W. A., 1972, The visual process, *Annu. Rev. Biophys. Bioeng.* **1**:131.

Hagins, W. A., and Yoshikami, S., 1977, Intracellular transmission of visual excitation in vertebrate retinal photoreceptors: Electrical effects of chelating agents introduced into rods by vesicle fusion, in: *International Symposium on "Vertebrate Photoreception"* (P. Fatt and H. B. Barlow, eds.), p. 97, Academic Press, New York.

Haydon, D. A., and Hladky, S. B., 1972, Ion transport across thin lipid membranes: A critical discussion of mechanisms in selected systems, *Q. Rev. Biophys.* **5**:187.

Henkart, P., and Blumenthal, R., 1975, Interaction of lymphocytes with lipid bilayer membranes: A model for lymphocyte-mediated lysis of target cells. *Proc. Natl. Acad. Sci. USA* **72**:2789.

Hille, B., 1970, Ionic channels in nerve membranes, *Prog. Biophys. Mol. Biol.* **21**:1.

Hladky, S. B., and Haydon, D. A., 1970, Discreteness of conductance changes in bimolecular lipid membranes in the presence of certain antibiotics, *Nature (London)* **225**:451.

Hladky, S. B., and Haydon, D. A., 1972, Ion transfer across lipid membranes in the presence of gramicidin A. I. Studies on the unit conductance channel, *Biochim. Biophys. Acta* **274**:294.

Hodgkin, A. L., and Katz, B., 1949, The effect of sodium ions on the electrical activity of the giant axon of the squid, *J. Physiol. (London)* **108**:37.

Hyman, E. S., 1971, Lecithin as a net neutral ionophore, *Biophys. Soc. Annu. Meet. Abstr.* **11**:A10.

Jain, M. K., 1972, *The Bimolecular Lipid Membrane: A System*, Van Nostrand Reinhold, New York.

Jain, M. K., 1974, Studies on a reconstituted acetylcholine receptor system: Effects of agonists, *Arch. Biochem. Biophys.* **164**:20.

Jain, M. K., White, F. P., Strickholm, A., Williams, A., and Cordes, E. H., 1973a, Cation pump versus Nernst potential (letter to editor), *J. Membr. Biol.* **11**:195.

Jain, M. K., Mehl, L. E., and Cordes, E. H., 1973b, Incorporation of eel electroplax acetylcholinerase into black lipid membranes. A possible model for the cholinergic receptor, *Biochem. Biophys. Res. Commun.* **51**:192.

Kafka, M. S., Blumenthal, R., Walker, G. A., and Pollard, M. B., 1978, The effect of dopamine-β-hydroxylase on the electrical conductance of bimolecular lipid membrane, *Membrane Biochemistry* **1**:279.

Kahn, C. R., 1976, Membrane receptors for hormones and neurotransmitters, *J. Cell Biol.* **70**:261.

Katz, B., and Miledi, R., 1971, Further observations on acetylcholine noise, *Nature (London) New Biol.* **252**:124.

Kemp, G., Dolly, J. A., Barnard, E. A., and Wenner, C. E., 1973, Reconstitution of a partially purified endplate acetylcholine receptor preparation in lipid bilayer membranes, *Biochem. Biophys. Res. Commun.* **54**:607.

Kimelberg, H. K., 1976, Protein–liposome interactions and their relevance to the structure and function of cell membranes, *Mol. Cell. Biochem.* **10**:171.

Kinsky, S. C., 1972, Antibody-complement interaction with lipid model membranes. *Biochim. Biophys. Acta* **265**:1.

Krikpatrick, F. H., 1976, Spectrin: Current understanding of its physical, biochemical and functional properties, *Life Sci.* **19**:1.

Kirschner, L. B., 1964, Phosphatidylserine as a possible participant in active sodium transport in erythrocytes, *Arch. Biochem. Biophys.* **68**:499.

Kirtland, S. J., and Baum, H., 1972, Prostaglandin-E1 may act as a calcium ionophore, *Nature (London)* **236**:47.

Knauf, P. A., and Rothstein, A., 1971, Chemical modification of membranes. II. Permeation paths for sulfhydryl agents, *J. Gen. Physiol.* **58**:211.

Kometani, T., Ikeda, Y., and Kasai, M., 1975, Acetylcholine-binding substance extracted by using organic solvent and acetylcholine receptor of electric organ of *Narke japonica, Biochim. Biophys. Acta* **412**:415.

Krasne, S., Eisenman, G., and Szabo, G., 1971, Freezing and melting of lipid bilayers and the mode of action of nonactin, valinomycin and gramicidin, *Science* **174**:412.

Ksenzhek, O. S., Omel'chenko, A. M., and Koganov, M. M., 1974, Discrete conductance induced in bilayer lipid membranes by sodium dodecyl sulfate, *Transl. Dokl. Biophys.* **218**:61.

Latorre, J. L., Lunt, G., and DeRobertis, E., 1970, Isolation of a cholinergic proteolipid receptor from electric tissue, *Proc. Natl. Acad. Sci. USA* **65**:716.

Latorre, R., Ehrenstein, G., and Lecar, H., 1972, Ion transport through excitability inducing material (EIM) channels in lipid bilayer membranes, *J. Gen. Physiol.* **60**:72.

Latorre, R., Alvarez, O., and Verdugo, P., 1974, Temperature characterization of the conductance of the excitability inducing material channel in oxidized cholesterol membranes, *Biochim. Biophys. Acta* **367**:361.

Lauf, P. K., 1975, Antigen–antibody reactions and cation transport in biomembranes: Immunophysiological aspects, *Biochim. Biophys. Acta* **415**:173.

Lea, E. J. A., Rich G. T., and Segrest, J. P., 1975, The effects of the membrane-penetrating polypeptide segment of the human erythrocyte MN-glycoprotein on the permeability of model lipid membranes, *Biochim. Biophys. Acta* **382**:41.

LeFevre, P. G., Habich, K. I., Hess, H. S., and Hudson, M. R., 1964, Phospholipid–sugar complexes in relation to cell membrane monosaccharide transport, *Science* **143**:955.

LeFevre, P. G., Jung, C. Y., and Chaney, J. E., 1968, Glucose transfer by red cell membrane phospholipids in $H_2O/CHCl_3/H_2O$ three-layer systems, *Arch. Biochem. Biophys.* **126**:677.

Lehninger, A. L., 1971, A soluble, heat-labile, high-affinity Ca^{++}-binding factor extracted from rat liver mitochondria, *Biochem. Biophys. Res. Commun.* **42**:312.

Leuzinger, W., and Schneider, M., 1972, Acetylcholine-induced excitation on bilayers, *Experientia* **28**:256.

Levinson, S. R., and Keynes, R. D., 1972, Isolation of acetycholine receptors by chloroform–methanol extraction: Artifacts arising in use of Sephadex LH-20 columns, *Biochim. Biophys. Acta* **288**:241.

Lossen, O., Brennecke, R., and Schubert, D., 1973, Electrical properties of black membranes from oxidized cholesterol and a strongly bound protein fraction of human erythrocyte membranes, *Biochim. Biophys. Acta* **330**:132.

Lovenberg, W., Goodwin, J. S., and Wallace, E. F., 1975, Molecular properties and regulation of dopamine–β-hydroxylase, in: *Neurobiological Mechanisms of Adaptation and Behavior* (A. J. Mandell, ed.), pp. 77–93, Raven Press, New York.

Marchesi, V. T., Steers, E., Jr., Tillack, T. W., and Marchesi, S. L., 1969, *Red Cell Membrane Structure and Function,* p. 117, Lippincott, Philadelphia.

Marchesi, V. T., Tillack, T. W., Jackson, R. L., Segrest, J. P., and Scott, R. E., 1972, Chemical characterization and surface orientation of major glycoproteins of human erythrocyte-membranes, *Proc. Natl. Acad. Sci. USA* **69**:1445.

Margoliash, E. G., Barlow, G. H., and Byers, V., 1970, Differential binding properties of cytochrome-c—possible relevance for mitochondrial ion transport, *Nature (London)* **228**:723.

Masuda, H., and deMeis, L., 1974, Calcium efflux from sarcoplasmic reticulum vesicles, *Biochim. Biophys. Acta* **332**:313.

Michaels, D. W., Abramovitz, A. S., Hammer, C. H., and Mayer, M. M., 1976, Increased ion permeability of planar lipid bilayer membranes after treatment with the C5b-9 cytolytic attack mechanism of complement, *Proc. Natl. Acad. Sci. USA* **73**:2852.

Minassian-Saraga, T. L., and Wietzerbin, J., 1970, The action of hexidecyltrimethyl ammonium bromide, *Biochem. Biophys. Res. Commun.* **41**:1231.

Montal, M., 1972, Lipid–polypeptide interactions in bilayer lipid membranes, *J. Membr. Biol.* **7**:245.

Montal, M., 1974, Formation of bimolecular membranes from lipid monolayers, in: *Methods in Enzymology*, Vol. 32 (S. Fleischer and L. Packer, eds.), Part B, pp. 545–554, Academic Press, New York.

Montal, M., 1976, Experimental membranes and mechanism of bioenergy-transductions, *Annu. Rev. Biophys. Bioeng.* **5**:119.

Montal, M., and Mueller, P., 1972, Formation of bimolecular membranes from lipid monolayers, and a study of their electrical properties, *Proc. Natl. Acad. Sci. USA* **69**:3561.

Moore, C., and Pressman, B. D., 1964, Mechanism of action of valinomycin on mitochondria, *Biochem. Biophys. Res. Commun.* **15**:562.

Moore, J. M., and Schechter, R. S., 1969, Transfer of ions against their chemical potential gradient through oil membranes, *Nature (London)* **222**:476.

Mueller, P., and Rudin, D. O., 1963, Induced excitability in reconstituted cell membrane structure, *J. Theor. Biol.* **4**:268.

Mueller, P., and Rudin, D. O., 1969a, Bimolecular lipid membranes: Techniques of formation, study of electrical properties, and induction of ionic gating phenomena, in: *Laboratory Techniques in Membrane Biophysics* (H. Passow and R. Stampfli, eds.), pp. 141–145, Springer, Berlin.

Mueller, P., and Rudin, D. O., 1969b, Translocators in bimolecular lipid membranes: Their role in dissipative and conservative bioenergy transductions, in: *Current Topics in Bioenergetics*, Vol. 3 (D. R. Sanadi, ed.), pp. 157–249, Academic Press, New York.

Mueller, P., Rudin, D. O., Tien, H. T., and Wescott, W. C., 1964, Formation and properties of bimolecular lipid membranes, *Recent Prog. Surf. Sci.* **1**:379.

Osterhout, W. J. V., and Stanley, W. M., 1932, The accumulation of electrolytes. V. Models showing accumulation and a steady state, *J. Gen. Physiol.* **15**:667.

Oxender, D. L., and Quay, S., 1975, Binding proteins and membrane-transport, *Ann. NY Acad. Sci.* **264**:358.

Pagano, R., and Thompson, T. E., 1967, Spherical lipid bilayer membranes, *Biochim. Biophys. Acta* **144**:866.

Pagano, R., Ruysschaert, J. M., and Miller, I. R., 1972, The molecular composition of some lipid bilayer membranes in aqueous solution, *J. Membr. Biol.* **10**:11.

Palade, G., 1975, Intracellular aspects of the process of protein synthesis, *Science* **189**:347.

Pant, H. C., and Conran, P., 1972, Keyhole limpet hemocyanin (KLM)–lipid bilayer membrane (BLM) interaction, *J. Membr. Biol.* **8**:357.

Papahadjopoulos, D., Jacobson, K., Nir, S., and Isac, T., 1973, Fluorescence polarization and permeability measurements concerning the effect of temperature and cholesterol, *Biochim. Biophys. Acta* **311**:330.

Parisi, M., Rivas, E., and DeRobertis, E., 1971, Conductance changes produced by acetylcholine in lipidic membranes containing a proteolipid from *Electrophorus, Science* **172**:56.

Parisi, M., Reader, T. A., and DeRobertis, E., 1972, Conductance properties of artificial lipidic membranes containing a proteolipid from *Electrophorus* response to cholinergic agents, *J. Gen. Physiol.* **60**:454.

Parisi, M., Adragna, C., and Sala, P. J. I., 1975, The influence of negative lipids on the interactions between artificial membranes and cholinergic drugs, *Nature (London)* **258**:245.

Passow, H., 1969, Passive ion permeability of the erythrocyte membrane, in: *Progress in Biophysics and Molecular Biology*, Vol. 19 (J. A. V. Butler and D. Noble, eds.), p. 424, Pergamon Press, Elmsford, N.Y.

Passow, H., 1971, Effects of pronase on passive ion permeability of human red blood cell, *J. Membr. Biol.* **6**:233.

Pressman, B. C., 1973, Properties of ionophores with broad range cation selectivity, *Fed. Proc.* **32**:1698.

Pressman, B. C., and deGuzman, N. T., 1974, New ionophores for old organelles, *Ann. NY Acad. Sci.* **227**:380.

Prestipino, G., Ceccarilli, D., Conti, F., and Carafoli, E., 1974, Interactions of a mitochondrial Ca^{2+}-binding glycoprotein with lipid bilayer membranes, *FEBS Lett.* **45**:99.

Rasmussen, M., 1975, Ions as "second messengers," in: *Cell Membranes* (G. Weismann and R. Claiborne, eds.), pp. 203–212, H.P. Publishing, New York.

Reader, T. A., and DeRobertis, E., 1974, The response of artificial lipid membranes containing a cholinergic hydrophobic protein from *Electrophorus* electroplax, *Biochim. Biophys. Acta* **352**:192.

Romine, W. O., Goodall, M. C., Peterson, J., and Bradley, R. J., 1974, The acetylcholine receptor. Isolation of a brain nicotinic receptor and its preliminary characterization in lipid bilayer membranes, *Biochim. Biophys. Acta* **367**:316.

Rosano, H. L., Duby, P., and Schulman, J. H., 1961, Mechanism of the selective flux of salts and water migration through non-aqueous liquid membranes, *J. Phys. Chem.* **65**:1704.

Rosenberg, B., and Pant, H. C., 1970, The semiconducting rectifier behaviour of a bimolecular lipid membrane, *Chem. Phys. Lipids* **4**:203. ,

Rothstein, A., and Cabantchik, Z. I., 1974, Protein structures involved in the anion permeability of the red blood cell membrane, in: *Comparative Biochemistry and Physiology of Transport* (L. Bolis, K. Bloch, S. E. Luria, and F. Lynen, eds.), pp. 354–362, North-Holland, Amsterdam.

Rothstein, A., Knauf, P. A., Cabantchik, Z. I., and Balshin, M., 1974, The location and chemical nature of drug "targets" within the human erythrocyte membrane, in: *Drugs and Transport Processes, a Symposium* (B. A. Callingham, ed.), pp. 53–72, Macmillan, New York.

Rothstein, A., Cabantchik, Z. I., Balshin, M., and Juliano, R., 1975, Enhancement of anion permeability in lecithin vesicles by hydrophobic proteins extracted from red blood cell membranes, *Biochem. Biophys. Res. Commun.* **64**:144.

Rothstein, A., Cabantchik, Z. I., and Knauf, P., 1976a, Mechanism of anion transport in red blood cells: Role of membrane proteins, *Fed. Proc.* **35**:3.

Rothstein, A., Takeshita, M., and Knauf, P. A., 1976b, Chemical modification of proteins involved in the permeability of the erythrocyte membrane to ions, in: *Biomembranes*, Vol. 3, *Passive Permeability of Cell Membranes* (F. Kreuzer and J. F. G. Slegers, eds.), pp. 393–413, Plenum Press, New York.

Sachs, F., and Lecar, M., 1973, Acetylcholine noise in tissue culture muscle cells, *Nature (London) New Biol.* **246**:214.

Sachs, G., Speeny, J. G., Saccomani, G., and Goodall, M. C., 1974, Characterization of gastric mucosal membranes. VI. The presence of channel-forming substances, *Biochim. Biophys. Acta* **332**:233.

Schneider, P. B., and Wolff, J., 1965, Thyroidal iodide transport. VI. On a possible role for iodide-binding phospholipids, *Biochim. Biophys. Acta* **94**:114.

Seufert, W. D., 1965, Induced permeability changes in reconstituted cell membrane structure, *Nature (London)* **207**:174.

Shamoo, A. E., 1974, Isolation of a sodium-dependent ionophore from $(Na^+ + K^+)$–ATPase preparations, *Ann. NY Acad. Sci.* **242**:389.

Shamoo, A. E., 1975, Carriers and channels in biological systems, *Ann. NY Acad. Sci.* **264**:1.

Shamoo, A. E., and Albers, R. W., 1973, Na^+-selective ionophoric material derived from electric organ and kidney membranes, *Proc. Natl. Acad. Sci. USA* **70**:1191.

Shamoo, A. E., and Eldefrawi, M. E., 1975, Carbamylcholine and acetylcholine-selective cation-selective ionophore as part of the purified acetylcholine receptor, *J. Membr. Biol.* **25**:47.

Shamoo, A. E., and Goldstein, D. A., 1977, Isolation of ionophores from ion transport systems and their role in energy transduction, *Biochim. Biophys. Acta,* **472**:13.

Shamoo, A. E., and MacLennan, D. H., 1974, A Ca^{++}-dependent and selective ionophore as part of the $Ca^{++} + Mg^{++}$–ATPase in sarcoplasmic reticulum, *Proc. Natl. Acad. Sci. USA* **71**:3522.

Shamoo, A. E., and MacLennan, D. H., 1975, Separate effects of mercurial compounds on the ionophoric and hydrolytic functions of the $Ca^{++} + Mg^{++}$–ATPase of sarcoplasmic reticulum. *J. Membr. Biol.* **25**:65.

Shamoo, A. E., and Myers, M., 1974, Na^+-dependent ionophore as part of the small polypeptide of the $(Na^+ + K^+)$–ATPase from eel electroplax membrane, *J. Membr. Biol.* **19**:163.

Shamoo, A. E., and Ryan, T. E., 1975, Isolation of ionophores from ion-transport systems, *Ann. NY Acad. Sci.* **264**:83.

Shamoo, A. E., Myers, M., Blumenthal, R., and Albers, R. W., 1974, Ionophoric material derived from eel membrane preparation. I. Chemical characteristics, *J. Membr. Biol.* **19**:129.

Shamoo, A. E., Thompson, T. E., Campbell, K. P., Scott, T. L., and Goldstein, D. A., 1975, Mechanism of action of "ruthenium red" compounds on Ca^{++} ionophore from sarcoplasmic reticulum $(Ca^{++} + Mg^{++})$–ATPase and lipid bilayer, *J. Biol. Chem.* **250**:8289.

Shamoo, A. E., MacLennan, D. H., and Eldefrawi, M. E., 1976a, Differential effects of mercurial compounds on excitable tissues, *Chem. Biol. Interact.* **12**:41.

Shamoo, A. E., Ryan, T. H., Stewart, P. S., and MacLennan, D. H., 1976b, Localization of ionophore activity in a 20,000 dalton fragment of the adenosine triphosphatase of sarcoplasmic reticulum, *J. Biol. Chem.* **251**:4147.

Solomon, A. K., Lionetti, F., and Curran, P. F., 1956, Possible cation-carrier substances in blood, *Nature (London)* **178**:582.

Sottocasa, G., Sandri, G., Panfili, E., deBernard, B., Gazzotti, P., Vasington, F. D., and Carafoli, E., 1972, Isolation of a soluble Ca^{2+} binding glycoprotein from ox liver mitochondria, *Biochem. Biophys. Res. Commun.* **47**:808.

Stewart, P. S., MacLennan, D. H., and Shamoo, A. E., 1976, Isolation and characterization of tryptic fragments of the ATPase of sarcoplasmic reticulum, *J. Biol. Chem.* **251**:712.

Szabo, J., Eisenman, G., and Ciani, S., 1969, The effects of the macrotetralide actin antibiotics on the electrical properties of phospholipid bilayer membranes, *J. Membr. Biol.* **1**:346.

Thorley-Lawson, D. A., and Green, N. M., 1975, Separation and characterization of tryptic digestion from the ATPase of sarcoplasmic reticulum, *Eur. J. Biochem.* **59**:193.

Tien, H. T., 1974, *Bilayer Lipid Membranes (BLM) Theory and Practice,* Dekker, New York.

Tosteson, M. T., Lau, F., and Tosteson, D. C., 1973, Incorporation of a functional membrane glycoprotein into lipid bilayer membranes, *Nature (London) New Biol.* **243**:112.

Tyson, C. A., Zande, H. V., and Green, D. E., 1976, Phospholipids as ionophores, *J. Biol. Chem.* **251**:1326.

Urry, D. W., Long, M. M., Jacobs, M., and Harris, R. D., 1975, Conformation and molecular mechanisms of carriers and channels, *Ann. NY Acad. Sci.* **264**:703.

Vale, M. G. P., and Carvalho, A. P., 1973, Effects of ruthenium red on Ca^{++} uptake and ATPase of sarcoplasmic reticulum of rabbit skeletal muscle, *Biochim. Biophys. Acta* **324**:29.

Van Zutphen, H., Merola, A. J., Brierly, G. P., and Cornwell, D. G., 1972, The interaction of nonionic detergents with lipid bilayer membranes, *Arch. Biochem. Biophys.* **152**:755.

Wallace, E. F., Krantz, M. J., and Lovenberg, W., 1975, Dopamine-β-hydroxylase: A tetrameric glycoprotein, *Proc. Natl. Acad. Sci. USA* **70**:2253.

Winkler, M., Mortnagl, M., and Smith, A. D., 1970, Membrane of the adrenal medulla, *Biochem. J.* **118**:303.

Wojtczak, L., 1974, Effect of fatty acids and acyl-CoA on the permeability of mitochondrial membranes to monovalent cations, *FEBS Lett.* **44**:25.

Wyssbrod, H. R., Scott, W. N., Brodsky, W. A., and Schwartz, I. L., 1971, Carrier-mediated transport processes, in: *Handbook of Neurochemistry,* Vol. 5 (A. Lajtha, ed.), Chap. 21, pp. 683–691, Plenum Press, New York.

Yafuso, M., Kennedy, S. J., and Freeman, A. R., 1974, Spontaneous conductance changes, multilevel conductance states and negative differential resistance in oxidized cholesterol black lipid membranes, *J. Membr. Biol.* **17**:201.

Yguerabide, J., and Stryer, L., 1971, Fluorescence spectroscopy of an oriented model membrane, *Proc. Natl. Acad. Sci. USA* **68**:1217.

Zaki, L., Fasold, H., Schuhmann, B., and Passow, H., 1975, Chemical modification of membrane proteins in relation to inhibition of anion exchange in human red blood cells, *J. Cell Physiol.* **86**:471.

Visualization and Counting of Receptors at the Light and Electron Microscope Levels

Eric A. Barnard

1. Receptors at the Cell Membrane

1.1. Introduction

Material elsewhere in this volume provides ample documentation of the existence of specific receptor molecules for a variety of signal-transmitting molecules acting on cells. The neurotransmitters and the circulating hormones are the most obvious categories of these messengers (although far from the only ones). We shall use the term *receptor* here to denote such structures in the cell plasma membrane, equipped specifically to recognize primary messenger molecules arriving at their cell and to trigger directly the cellular response to these or to initiate the required signal to other machinery that generates the response (e.g., a separate nucleotide cyclase). In contrast, messenger-binding components in the cell interior which are not concerned with the primary recognition of externally incident messengers, while they may be involved in some cases in mediating the cellular response (e.g., to thyroid or steroid hormones), present an entirely different aspect to the problem of receptor location and will not be considered here. Receptors involved in the immune response, which are a very special case, are also outside the scope of this chapter.

We shall consider, therefore, the usual case of a specific structure in the cell membrane to which the external signaling molecule in question binds in order to initiate the response. Usually this structure will be a particular membrane protein, although exceptions to this statement can occur if we stretch the concept to include among the messenger types the bacterial protein toxins that attack animal cells by binding to particular gangliosides (Cuatrecasas, 1973; Van Heyningen, 1974), since physiological equivalents of these may exist. The binding site on the receptor protein, for a physiological ligand which is itself a protein,

Eric A. Barnard ● Department of Biochemistry, Imperial College, London, England.

may in a few cases be a carbohydrate attachment, as in the binding of plant lectins to membrane glycoproteins, which appears to be a model for some physiological receptor actions (Stockert *et al.*, 1974). For all receptor molecules acting specifically at the cell membrane, important insights will be obtained if we can view them by some means *in situ*. In this chapter the ability to achieve this, and the techniques that can be applied, will be reviewed.

1.2. Information Required on the Distribution of Receptors

Basic spatial parameters needed for any analysis of receptor function in the cell membrane are the following:

1. The identity of the cells that bear the receptors. In some cases this is to be seen as a confirmation of the physiological evidence, as at the motor end plate. In other cases, as in the brain, the receptor locations will not be known until they are spatially mapped.

2. The actual location of the receptors—e.g., in one zone of the cell membrane, or dispersed, or on one cell surface or another (e.g., whether or not exclusively postjunctional, for a synaptic receptor). In the common case of a single neurone receiving multiple neuronal inputs, the delineation of each receptor zone on it would greatly aid our understanding of mechanisms of synaptic integration.

3. The total numbers of receptors present on one cell or at one synapse. Are these numbers fixed and of functional significance? Are they characteristically different in different cases (e.g., on diverse cell types that are targets for one hormone or transmitter)?

4. The *membrane density* of receptors, i.e., the number per square micrometer of membrane surface. This will be a very revealing quantity, for functional considerations. Thus, the nicotinic ACh receptor at the nerve-muscle synapse is present, as we shall see, at about 25,000 molecules/μm^2 in the sharply defined zones where it is situated, but this density falls to a minute fraction of that value just outside those zones. In contrast, the electrical relay of the impulse along the nerve axon is mediated by a sodium-channel component (strictly, not a receptor) which can be present, as in the case of a garfish nerve, at about only 35 molecules/μm^2 of membrane (Ritchie and Rogart, 1977).

5. Allied to item 4, the arrangement of receptors over the active zone. A value for the membrane density implies an averaging over a given area. Shorter-range densities, e.g., in island patches, or at a regular spacing, or nonuniformly, should be known for the membrane density value to be interpreted in detail. Whether the receptors have fixed locations in the membrane or have lateral mobility (Cuatrecasas and Hollenberg, 1976) is a further aspect of this question.

6. Distinction between types of receptor if more than one type may bind a given ligand. It must be recognized that the binding observed *in situ* with any ligand may not be specific for a single class of site. In all attempts at counting receptors, attention must be given to establishing whether one or more than one class of binding sites is present on a given cell and quantifying these. Where additional binding occurs, this may be extraneous, due to imperfect specificity of the probe chosen, and must therefore be known, so as to be avoided or subtracted, or it may be at a second type of receptor, as in the case of both synaptic and

nonfunctional, extrasynaptic, ACh receptors occurring on one *Aplysia* ganglion cell (Shain *et al.*, 1974).

For all of the above questions, localizations of a bound ligand that can be viewed *in situ* in the electron or light microscope will furnish information which goes beyond any that can be derived from biochemical binding studies (essential though the latter are, for determinations of affinity and specificity). In the usual case of receptors present in a heterogeneous cell population, both these types of study will be necessary if any of the above questions are to be answered.

It will be useful to distinguish between the *visualization* and the *counting* of receptors. By the former, we mean the use of any method whereby an image of the receptor, or the supramolecular assembly that contains it, is produced in the EM* or the light microscope. Such an image may be directly related to the geometry of the molecule or assembly, as in the highest resolution transmission EM images or the coated replicas produced after freeze-fracture. This feature is not in principle essential, however, since, for example, we can still be considered to have effectively visualized receptor molecules if we have obtained a pattern of developed silver grains in an EM autoradiograph due solely to radioactive disintegrations occurring on them, although the developed grains are scattered away from the receptor molecules themselves. The production of any image that can be traced back to the receptor sites themselves is sufficient. Receptors thus visualized may or may not be susceptible to the counting of the individual units. While some methods (e.g., isotopic labeling) permit measurement of the number of visualized sites, there are other images whose relationship to their origin is subject to so many variables that, at least with present technology, they cannot be quantified *in situ* in a meaningful manner. An example is the fluorescence label introduced by the binding of labeled antibodies (see Sec. 7). The minimum aim in the methodology to be recommended here is the visualization of recognized receptor molecules. The further aim, possible in some of the methods but not all of them, is to achieve this quantitatively, so that receptors can be counted in relation to perceived cell structure or ultrastructure to obtain parameters 1–6 listed above.

2. The Labeling of Receptors for Localization

2.1. Approaches

To see or count a receptor, one must label it in some manner. This usually means to attach a ligand to it. This is not, however, in principle, essential. In some cases, existing methods for ultrastructural study without *specific* labeling—e.g., freeze-fracture of membranes to reveal particles therein—can be directed at receptor-rich regions to obtain images of particles which can then in an indirect

---

*Nonstandard abbreviations: α-BuTX, α-bungarotoxin; DFP, diisopropyl fluorophosphate; EM, electron microscope; HCG, human chorionic gonadotropin; IgG, immunoglobulin G; LH, luteinizing hormone; MW, molecular weight; PAP, peroxidase–antiperoxidase complex; PrBCM, propylbenzilylcholine mustard; TMR, tetramethylrhodamine.

manner be related to the presence of receptors. The ligand-independent approach is, as detailed below, only applicable to certain favorable cases; in these, it can only be subsidiary to methods for the direct localization of those receptors. For the majority of receptor types, the introduction of a specific label is required to visualize them.

Such labeling must make use of the particular binding characteristics of the receptor to achieve specificity. The receptor–ligand complex used should also have a zero or extremely low dissociation rate, since it is obvious that freely reversible binding will be lost during the necessary removal of the excess unbound ligand and also during the processing necessary for the microscopic examination. Three alternative situations are encountered with respect to diffusion risks.

1. The binding is irreversible, i.e., covalent (as in the case of nitrogen mustard reagents used for the muscarinic ACh receptor; see Sec. 2.3), or pseudo-irreversible, i.e., of such great affinity that the preparation can be washed without significant diffusion of the ligand initially bound. In practice the latter means that the dissociation constant of the complex, in the conditions employed, must be below 10^{-10} M, as with the α-neurotoxins discussed below or some antibody cases (but see Sec. 2.3.2). For a covalent reaction, the possibility of subsequent hydrolytic loss of the labeled group must be excluded.

2. The binding is more reversible, i.e., the dissociation rate of the complex is significant, in the time periods involved in normal processing of tissue specimens. This will usually be true for dissociation constants greater than about 10^{-10} M. In reversible cases frozen specimen (dry-mount) methods are required. This greatly restricts the possibilities for the application of the ligand and for the subsequent treatment for visualization. Thus methods dependent on secondary treatments in aqueous media (e.g., by peroxidase substrate reaction) are ruled out when the primary attachment to the receptor is not irreversible or pseudo-irreversible.

3. Diffusion of the ligand from the receptor does not occur, but the converse problem of cytochemistry (especially in the EM), namely the possible redistribution of the receptor macromolecules themselves during the treatments applied, must be considered. This usually will not occur with the membrane-bound receptors at synapses, since they generally are tenaciously held on the membrane. However, in some cases there is a risk that the receptors may become free to diffuse laterally in the membrane (Singer and Nicolson, 1972). This would mean that a misleading localization over the surface is detected if such diffusion is affected by cell death or other steps in the processing for examination or if aggregation and segregation are induced by the ligand binding used for the detection. The latter process is exemplified by the *capping* of IgM molecules on the cell membrane of B lymphocytes, induced by the binding (De Petris and Raff, 1973). Thus while the complete segregation of receptors appears to require cross-linking (Singer, 1976), formation of patches may occur upon a simple binding to diffusible receptors. Further, if the ligand binding is performed *in vivo* or in metabolizing cells *in vitro*, endocytosis of the clustered region may ensue (De Petris and Raff, 1973), leading to a loss of receptors during the labeling. If lateral mobility of the receptor under study is a possibility, then these problems must be seriously considered. One experimental

test for mobility could be made where the ligand used can be fluorescently labeled: the technique of fluorescence photobleaching recovery (Peters *et al.*, 1974) permits lateral motion in the membrane of fluorescent macromolecules to be measured. One can thus check whether the ligand in use or the conditions applied induce redistribution of the receptors. Such use of fluorescent markers with receptors is described in Sec. 6.1.

Assuming a suitable ligand is bound to a receptor, we shall first consider (Sec. 2.2) the methods in use for revealing such a complex *in situ* and for counting the receptors. The types of ligand that have been labeled by these methods will then be listed (Sec. 2.3).

In discussing receptor detection, the term *primary attachment* will be used to designate the initial binding to the free receptor that occurs, this being, obviously, the receptor-specific step. In many cases this is the only binding needed, since this *primary ligand* can carry the detection device (e.g., isotope). In many other cases, however, further binding (e.g., of an antibody) to this complex is needed to introduce a particular type of detection device, and such steps (one or more in a given series) will be termed, collectively, the *secondary attachment* to the receptor.

2.2. *Methods of Labeling and Visualizing Receptors*

2.2.1. *Autoradiography*

If a suitable radioactive receptor-specific ligand is available at high specific radioactivity, autoradiography at the tissue or the subcellular level can give unique information on the receptor in question. While its resolution is not at the highest limits, the great advantage of autoradiography here is that it provides quantitative data, both relative and absolute. Silver grains can be related, by suitable methodology, to structural features, so that the numbers of receptors can be counted, for example, per unit area of synaptic membrane or in total number at a single synapse or cell. Quantitative tracing can be performed at both the light and the electron microscope levels. The operations involved in the autoradiography of receptors are discussed in Sec. 3.

2.2.2. *Electron-Dense Label Attachment*

Ferritin is the usual example of such a label. Attached covalently to a specific protein ligand for the receptor, it provides a marker for EM visualization at high resolution. Some relative quantitation is possible in certain cases. Its application will be reviewed in Sec. 5. Hemocyanine, an even larger appendage, is a second example of such a metal-containing EM marker.

Heavy atoms giving sufficient electron scattering in EM specimens might also be introduced onto protein ligands by conjugating metal groups to them (Kendall and Barnard, 1967). One can expect that suitable synthetic electron-opaque markers, smaller than the bulky ferritin attachment, will come into use for receptors, thus including the possibility of innocuous and sufficient labeling of smaller primary ligands (e.g., peptides) for particular receptors. Cytochrome c in mitochondrial membranes has been visualized by an iron chelate formed with it (Tsou et al., 1976). For receptors, closely adjacent sites of the ligand binding would be a requirement, however, to introduce sufficient local density for detection in the transmission EM. While several alternative theoretical treatments of scattering in the EM have been propounded, it is clear from any of them that multiple adjacent heavy atoms are always going to be needed for a center to stand out as labeled. The ferritin molecule contains about 2000 iron atoms (as a ferric hydroxyphosphate lattice) in a center of 5.5-nm diameter, so the contrast, even at a single binding site, is fully adequate in that case. A uranium chelate of the antibody molecule has been used as an alternative to ferritin labeling, but this required a superimposed bridging to osmium to form centers with sufficient opacity (Sternberger, 1972).

Polyethyleneimine has also been proposed as a smaller alternative to ferritin (Schurer et al., 1977), having a MW of about 40,000 instead of 600,000. This polymer can give rise to electron-dense particles by reaction at its many amino groups with, e.g., osmium tetroxide and uranyl acetate. In this scheme, an antibody, for example, is conjugated to the polymer by glutaraldehyde cross-linking of amino groups (as is used for ferritin), and after cellular binding the metal disposition is applied. Investigation of this and other devices for attaching a number of heavy atoms within a small radius to a specific ligand can be expected to develop new means of receptor localization.

2.2.3. Enzymatic Reaction Product Markers

This class comprises visualization, in either the light or electron microscopes, of reaction product deposited locally by an enzyme which is linked by a secondary attachment to the receptor molecule. This reaction product must be such that it is captured in insoluble form without the opportunity for diffusion even over the short distances distinguishable at the ultrastructural level: this requirement is stringent, and there is always doubt as to whether it has been met (Novikoff et al., 1972). The only commonly employed case is that of peroxidase, which can be linked to a receptor in one of several alternative secondary attachment series (Sec. 5). When peroxidase has been bound in some manner at the receptor site, the tissue is treated with 3,3′-diaminobenzidine and hydrogen peroxide (Graham and Karnovsky, 1966) and (usually) finally by osmium tetroxide. The insoluble reaction product (yellow-brown in the light microscope) is assumed to be deposited at the actual site of the enzyme and, when viewed in the EM, is seen in limiting conditions (very low substrate concentrations, short reaction time, and low levels of bound peroxidase) as discrete zones about 40 nm across, which increase in number as more peroxidase is bound, until eventually they become continuous; in those limiting conditions, the number of individual peroxidase-binding centers can be quantitated (Hinton et al., 1973) provided the receptor sites are themselves well enough separated. More usually, however, the method is nonquantitative (see Sec. 7).

A few other enzymes may be capable of a similar application, and new and superior cases may yet be found. Glucose oxidase has been proposed (Kühlmann and Avrameas, 1971). Catalase can be used in the same way as peroxidase and has been applied as a tracer in other types of study in the EM (Fahimi, 1975). Beef liver catalase has a MW of 240,000, whereas horseradish peroxidase has 40,000. The large size of the former will usually be a disadvantage, but in cases where confinement to external regions is important it may be of use in receptor tracing. The method is less sensitive than with peroxidase. Peroxidase has a number of advantages over other cases known so far and the benefit of a specific immunocytochemical methodology built upon it, as will be discussed in Sec. 5.

2.2.4. Fluorescent Markers

For light microscopy only, fluorescent labels (e.g., fluorescein or rhodamine dyes) are convenient for covalent attachment to ligands, especially proteins, and for visualization. Their application to receptors is covered in Sec. 6. This method is generally qualitative only.

2.2.5. X-Ray Microanalysis

Electron probe X-ray microanalysis permits the amount of an individual element to be measured over subcellular distances (Chandler, 1975). This could be applied to receptor site measurement whenever a foreign atom is introduced in some manner, as in a metal-labeled ligand (including ferritin). Bound iodinated ligands, if their binding sites are numerous, could also be counted. The method is most readily applied quantitatively in the scanning EM, at rather low resolution; it is also feasible, however, with a transmission EM image, using one of the several types of commercially available microanalyzers (Peters *et al.*, 1976), where an area 50–300 nm in diameter in a thin section can be analyzed for any significant element introduced. Autoradiographs can be quantitated over a three-dimensional structure by measurement of their deposited Ag in the scanning EM (Hodges and Muir, 1975). No application of X-ray emission microspectroscopy to receptor measurement has as yet been reported, but this development is presently under study for ACh receptors.

2.3. Ligands for Receptor Labeling

2.3.1. Selection of Primary Ligands

In general, it will be necessary before any ligand is chosen for microscopic visualization to have an adequate guide, from biochemical binding evidence, to its usefulness. Thus, there should be evidence of its receptor specificity: the receptor labeled should be identified by pharmacological or equivalent evidence on the response to that ligand, and its binding should have a major saturable component. Likewise, the binding should be prevented by, and only by, known specific agents for that receptor acting at low, pharmacologically effective concentrations and, where relevant, should be stereospecific.

When that evidence is available for a particular ligand, its affinity for the receptor on cell membranes will then be known, as a guide for its use for *in situ* mapping.

2.3.2. Ligands Available for Receptor Tracing

Obviously, the physiologically effective ligand for the receptor in question would be the first choice for localizing the latter, but in the majority of cases the binding thereof will be too reversible (see above). Thus, the binding of ACh or norepinephrine cannot itself be used for visualizing any of their types of receptor. A protein hormone as a ligand, on the other hand, is often likely to possess sufficient multiple points of binding for its receptor site that the affinity is great enough for the retention of the hormone on the receptor for microscopy. Some applications of this kind have already been made: iodinated or ferritin-labeled protein hormones have been thus employed in certain cases (Table I), and other protein hormones can be expected to be capable of localization by similar methods using applied labeled hormone. On the other hand, some small hormones do not apparently bind to their cell membrane sites so tightly, and only *in vivo* application followed

Table I
Receptors Localized Microscopically by Their Ligands[a]

Receptor type	Source	Ligand	Method[b]	References
Nicotinic ACh	Vertebrate skeletal muscles	α-Neurotoxins	LM	Barnard *et al.* (1971, 1975), Fambrough (1974), Bevan and Steinbach (1977)
	(various)		EM	Porter *et al.* (1973a,b), Albuquerque *et al.* (1974), Porter and Barnard (1975a,b,1976), Fertuck and Salpeter (1976)
			LM,EM(P)	Daniels and Vogel (1975), Bender *et al.* (1976), Engel *et al.* (1977), Lentz *et al.* (1977)
			LM (F)	Anderson and Cohen (1974), Ko *et al.* (1977)
	Muscle cells in culture		LM	Hartzell and Fambrough (1973), Prives *et al.* (1976)
			LM, EM	Land *et al.* (1977)
			LM (F)	Axelrod *et al.* (1976), Anderson *et al.* (1977)
			LM, EM (P)	Vogel and Daniels (1976)
			EM (Fer)	Hourani *et al.* (1974)
	Electroplaque		EM	Bourgeois *et al.* (1972)
			LM (F)	Bourgeois *et al.* (1971)
	Brain (rat, frog)		LM	Polz-Tejera *et al.* (1975)
	Tectum (chick)		EM (P)	Barnard *et al.* (1979)
	Retina (chick)		LM	Vogel and Nirenberg (1976)
			EM (P)	Vogel *et al.* (1977)

(cont.)

Table I (cont.)[a]

Receptor type	Source	Ligand	Method[b]	References
Muscarinic ACh	Brain (rat)	PrBCM	LM	Rotter et al. (1977a,b)
	Brain (rat)	QNB (rev)	LM	Kuhar and Yamamura (1975, 1976)
Opiate	Brain (rat)	Diprenorphine; etorphine (rev)	LM	Pert et al. (1976), Schubert et al. (1975), Atweh and Kuhar (1977)
Gonadotropins	Ovary (mammals)	Chorionic gonadotropin; prolactin	LM LM,	Petrusz (1979), Ashitaka et al. (1973)
			EM	Abel (1979)
			LM (P)	Petrusz (1974)
	Brain (rat)		LM (P)	Petrusz (1975)
	Ovary (mammals)	Luteinizing hormone	LM, EM	Abel (1979), Rajaniemi and Vanha-Perttula (1973), Barofsky et al. (1971)
			EM (P)	Abel (1979)
Oxytocin	Oviduct (rat)	Oxytocin (rev)	LM	Soloff et al. (1975)
Vasopressin	Pituitary	Vasopressin (rev)	EM (P)	Castel (1978)
FSH	Testes (rat)	FSH	LM, EM	Orth and Christensen (1977)
β-Melanotropin (MSH)	Melanoma cells (mouse)	MSH (rev)	LM	Varga et al. (1976a)
			LM (F)	Varga et al. (1976b)
			EM (Fer)	DiPasquale et al. (1978)
Insulin	Lymphocytes (human)	Insulin (rev)	LM	Goldfine et al. (1977)
	Liver, fat cells (rat)		EM (Fer)	Jarett and Smith (1975)
			EM	Bergeron et al. (1977)
Releasing hormones for LH and thyrotropin	Pituitary and brain (rat)	Hormone (rev)	LM EM (P) EM (Fer)	Stumpf and Sar (1977) Sternberger (1978) Hopkins and Gregory (1977)
Tetanus toxin	Muscle, spinal cord	Toxin	EM	Price et al. (1977)
	Nerve (rat)		LM (P)	Fedinec et al. (1970)
	Muscles (mouse)		LM	Wernig et al. (1977)
Cholera toxin	Lymphocytes (rat)	Toxin	LM (F)	Craig and Cuatrecasas (1975)
	Brain cell cultures		LM, EM (P)	Manuelidis and Manuelidis (1976)
	Lymphocytes, intestine (mammals)		EM (P)	Hansson et al. (1977)
Serotonin	Brain (rat)	LSD (rev)	LM	Diab et al. (1971)
β-Adrenergic	Brain (rat)	Antagonists (rev)	LM (F)	Atlas and Segal (1977)

[a] Abbreviations: EM, electron microscope; F, fluorescence; Fer, ferritin; LM, light microscope; PrBCM, propyl-benzilylcholine mustard; P, peroxidase; QNB, quinuclidinylbenzilate; rev, binding is readily reversible; LSD, lysergic acid diethylamide.

[b] All studies are autoradiographic, except where noted. For α-neurotoxins, only autoradiographic studies that were quantitative are listed (except for brain and retina).

by solid-phase processing methods (Sec. 3.2.2) would give, by current techniques, any chance of correctly localizing them; reports published until quite recently of the autoradiography of labeled small hormones by wet autoradiographic techniques are now known to be unreliable and are not listed here. Sometimes it is assumed, e.g., for [^{125}I]insulin (Table I), that fixation, e.g., by glutaraldehyde, will leave a protein ligand attached to its binding site in insoluble form and without translocation occurring during the fixation process, but this needs to be proved in each instance. The counting of active, labeled membrane preparations before and after fixation would give some evidence on this question.

In practice, for neurotransmitter receptors, the best possibility usually is to find an extraneous blocking agent of the receptor that is held firmly enough and with sufficient specificity. This has been best realized in the case of natural polypeptides from snake venoms that are blocking agents for neuromuscular junction receptors, i.e., the α-neurotoxins, such as α-bungarotoxin (BuTX) (Table I). These toxins are, of course, equally effective blockers at nicotinic ACh receptors in electroplaques, and they may be of use in other nicotinic synapses such as those in the CNS (Polz-Tejera *et al.*, 1975). However, the specificity in the CNS or ganglia of those receptors for α-neurotoxins has been reported to be very different from the muscle case (Bursztajn and Gershon, 1977; Miledi and Szce-paniak, 1975; Patrick and Stallcup, 1977), so that each such case must be investi-gated physiologically first before a meaningful visualization is substantiated. The α-neurotoxins, perfected by evolution to bind to the nicotinic ACh receptor at muscle end plates, are exceptional in their potency and selectivity. For studies with [3H]-α-BuTX binding to muscle, followed at the light and electron micro-scope levels, it has indeed been shown, by the localizations seen, by correlation with biochemical evidence on the binding and by preblockade of the labeling by specific antagonists, that an α-neurotoxin can be employed for microscopic localization of these receptors with unambiguous identification of the sites counted (Porter *et al.*, 1973a,b; Porter and Barnard, 1975a).

The next best option would be the use of an affinity-labeling reagent that is based chemically upon a specific ligand for the receptor in question. Since such a reagent must react covalently with a suitable acceptor group adjacent to it on the receptor when it is bound yet minimize interfering reaction elsewhere, these cases are difficult to construct. One such reagent that has been applied for *in situ* tracing is a nitrogen mustard (PrBCM) based on choline that alkylates the muscarinic ACh receptor (Burgen *et al.*, 1974). The reaction is, therefore, irrevers-ible; nonspecific bimolecular reaction elsewhere was shown to be sufficiently low and was corrected for by a parallel labeling after preblocking the receptor sites with 10^{-6} M atropine. Application of labeled PrBCM in light microscopic on the receptor when it is bound yet minimize interfering reaction elsewhere, these cases are difficult to construct. One such reagent that has been applied for this purpose.

Somewhat less satisfactory, but still of value in a number of cases, is the use of a low-molecular-weight antagonist or agonist, binding reversibly but with sufficient affinity, as discussed in Sec. 2.1. Examples are given in Table I, with this reversible case denoted *rev.* When seeking to apply this type of ligand for a neurotransmitter receptor, special attention must be paid to the requirement, discussed above, for correlating its binding with a physiological effect at the receptor level; some ligands that block or activate the receptor bind, in addition, to a high affinity site of uptake of the neurotransmitter. In other cases, there may be binding of a drug both to a receptor for it and to an enzyme metabolizing it. As an example, [^{3}H]-GABA binds to GABA receptors *and* to a GABA uptake system, the latter in part distinguishable by its Na$^+$ dependence (Olsen *et al.*, 1975; Enna and Snyder, 1977). Such ligands of dual action should be avoided in receptor localization.

The specimens for microscopy with a reversible ligand must be handled frozen at all stages to prevent loss of bound label; even with such precautions, it is doubtful whether any ligand with a dissociation constant (K_D) larger than 10^{-9}–10^{-10} M can be safely employed, since at the concentration that will be required for saturation, nonspecific binding at larger numbers of weaker sites is then likely to contribute proportionately too much. For a given value of K_D, the dissociation rate constant of the receptor–ligand complex can vary with different ligands, of course, and that constant, as well as K_D, must be very small. In some cases that dissociation rate will be determinable from studies on the binding of the ligand to membranes of the tissue in question and is then the proper guide to the risk of diffusion.

As an example of an applicable reversible ligand, one can cite the case of QNB (see Table I), which has $K_D = 10^{-11}$ M for the muscarinic ACh receptor of rat brain and 2×10^{-11} M for that of guinea pig intestine (Yamamura and Snyder, 1974). Various high-affinity and specific drugs for other receptors should become available for autoradiographic, etc., tracing from the biochemical studies of drug binding to membranes now being widely conducted, since the requirements for good performance in the two systems are basically the same. For β-adrenergic receptors, [^{3}H]-(−)alprenolol and [^{125}I]iodohydroxybenzylpindolol have been proposed as specific ligands and have been used biochemically (Alexander *et al.*, 1975; Atlas *et al.*, 1977; Bylund *et al.*, 1977); they have $K_D = 3 \times 10^{-8}$ M and 2×10^{-10} M, respectively. The latter should be applicable to the autoradiographic tracing of β-adrenergic receptors. An affinity-labeling drug has also been described for this receptor (Atlas *et al.*, 1976), and a fluorescent reversible ligand for it has been applied microscopically, revealing patches on cell surfaces, by Atlas and Levitzki (1978). Drugs of potential applicability employed in biochemical studies of receptors on membranes include strychnine ($K_D = 1.5 \times 10^{-9}$ M) for glycine receptors (Young and Snyder, 1974) and spiroperidol ($K_D = 2.5 \times 10^{-10}$ M) for dopamine receptors (Burt *et al.*, 1976), while for α-adrenergic receptors [^{3}H]WB-4101 (U'Prichard *et al.*, 1977) has $K_D = 4.8 \times 10^{-10}$ M and [^{3}H]dihydroergocryptine has $K_D = 2.4 \times 10^{-9}$ M on rat brain membranes (Williams and Lefkowitz, 1976). The latter drug, however, can also bind to brain receptors for serotonin and for dopamine (Tittler *et al.*, 1977); with the latter (using excess phentolamine to block the α-adrenergic receptors), $K_D = 5 \times 10^{-10}$ M and wash-out was slow. This case exemplifies another commonly observed problem with ligands for neurotransmitter receptors, namely that they often bind well to more than one type of receptor, and specificity studies at the biochemical level are, again, needed to determine if selective binding can be obtained.

3. Cell and Tissue Autoradiography

3.1. Problems of Application of a Labeled Ligand

When a radioactively labeled ligand for a receptor, meeting the requirements in Sec. 2, is to be used for tracing *in situ*, i.e., by autoradiography, the following additional variables must be considered.

3.1.1. Mode of Application

The ligand must be able to gain access to all of the receptor sites. For cells in culture or suspension, simple addition of the ligand to them, if the receptors are on the cell surface, poses no problem of access. For tissues to be taken from an animal, the most usual procedure for autoradiographic studies is to inject the ligand *in vivo*. The alternatives, direct application to excised tissue fragments or to tissue sections, may engender several different problems. Thus, the higher concentration of the labeled ligand necessarily encountered can increase non-specific labeling. For reversible ligands, application of the label in solution and the subsequent washing are contraindicated, and the only satisfactory methods for such specimens are *in vivo* administration and rapid freezing of the tissue at death for autoradiography, which must be by dry-mount methods (Stumpf, 1970). Even for ligands of very low reversibility, the receptor affinity may be affected after cell death or during the cytological processing, whereas the complex, once formed *in vivo*, is less likely to be so affected subsequently. Regional distributions, e.g., across the brain, are easier to follow after administration of the ligand by perfusion in the intact animal, either by the blood or artificially. Even for very localized tracing in the EM, tissue blocks *in vitro* thick enough to maintain un-injured zones will often not permit full penetration of the ligand in a reasonable time. For example, with [³H]estradiol applied to uterus slices *in vitro*, Stumpf (1970) noted by autoradiography that in normal conditions the ligand did not penetrate at all more than 200–300 μm from the surface. In contrast, the intact circulation can perform an efficient distribution, e.g., throughout a thick muscle. The circulation can bring a pseudo-irreversible ligand, at extremely low concentrations, to all of the receptors until they are saturated and then wash away the excess.

When *in vivo* labeling is to be so used, it must be verified that the receptors at different locations are, indeed, all saturated with the ligand. They may not all be equally well served by the circulation; alternatively, some dissociation of the complex may occur if the equilibration period needed is lengthy. Parallel data on biochemical uptake can usually demonstrate any large deviations. Thus, for the opiate receptors in mammalian brain, Pert *et al.* (1975) showed that after intravenous injection of [³H]diprenorphine, 98% of the brain radioactivity, when subsequently analyzed biochemically, was at sites having the stereospecific binding characteristics of the opiate receptor, but with [³H]naloxone used similarly only 30–60% was (varying with the brain region), the rest being non-specific. Hence, autoradiography with the latter type of drug applied could give a misleading picture. With many ligands, the blood–brain barrier may prevent circulatory delivery to the brain.

When labeled α-neurotoxin, for nicotinic ACh receptors, was applied for quantitative autoradiography (Barnard *et al.*, 1971; Porter *et al.* 1973a,b; Fig. 1), such a check was made by showing that the uptake at muscle end plates so measured was equal to that which could be measured biochemically. The [^{3}H]-BuTX used was applied *in vivo* to lethality, and diaphragm muscles so labeled were tested by subsequently further exposing them to [^{3}H]-BuTX *in vitro* in saturating conditions, without producing any increase in the uptake of the label at end plates, proving that the receptors were saturated when counted autoradiographically in the light or electron microscope specimens.

It cannot be assumed, however, for all muscles in all species that an α-neurotoxin will have complete access to the receptor *in vivo*. Such a test must be made in each case. In the case of these pseudo-irreversible toxins, topical application may also be feasible, since adequate washing can be given. Bourgeois *et al.* (1972) cut frozen sections of *Electrophorus* electric organ, thawed them, and labeled them with a solution of [^{3}H]-α-cobratoxin for light microscope autoradiography and labeled cells similarly for EM sectioning. Fertuck and Salpeter (1976) applied [^{125}I]-BuTX solution topically to the exposed sternomastoid muscle of a mouse until tetanic muscle contraction was blocked. For autoradiography of end plates in general it cannot be assumed that conditions that saturate receptors with α-toxin at one end plate, e.g., by *in vivo* injection, will do so in other muscles and other species. Preferred conditions for labeling muscles of any type with a pseudo-irreversible toxin are summarized in Sec. 3.1.4.

3.1.2. Nonspecific Labeling

In many cases, labeling of tissues or cells *in vitro*, or by a topical application after exposure in an animal preparation, is often necessary, for the reasons given above, to ensure saturation of the receptors. Tissues immersed in a solution of almost any radioactive substance, at concentrations adequate for ready penetration throughout a tissue specimen, will show some nonspecific labeling in autoradiographs unless truly adequate washing is given. With irreversible or pseudo-irreversible ligands, the best such washing procedure is a brief postlabeling exchange with a large excess of the same ligand unlabeled, which will normally displace the adsorbed excess isotope responsible, followed by washing with the medium alone.

With [^{3}H]-BuTX applied at concentrations of about $1-2 \times 10^{-7}$ M *in vitro* to thin intact muscles such as mouse diaphragm, that type of washing eliminated nonspecific binding, leaving only end-plate binding; this was shown by the concurrence of autoradiographic and biochemical measurements of uptake into the end plates and non-end-plate zones (Porter *et al.*, 1973a). In these studies, however, it was shown in thicker muscles (such as the rat diaphragm) or at higher concentrations of labeled BuTX that nonreceptor binding can persist through normal washing procedures. This persistence in washed whole muscle fibers appears to be due to the entry of the toxin into undamaged muscle cells by endocytosis; it thus becomes resistant to external washing, and to exchange with external toxin, and appears irreversibly bound. At least some of it does, in fact, become bound to a sarcoplasmic component which is not an ACh receptor (Chiu et al., 1973). This additional, endocytotic uptake of BuTX has been demonstrated in biochemical studies (Libelius, 1975; Libelius et al., 1975); it is prominent in incubations at 37° and is blocked at 4°. The nonspecific character of this uptake is shown by its slowness, its nonsaturability, its stimulation by polycations, and the failure of the cholinergic ligands (e.g., 10^{-2} M decamethonium) added to the incubation medium to inhibit it (Libelius, 1974). There is always a risk that this type of phenomenon may be present with other radioactive ligands used for binding to cell surface receptors, and similar criteria should be used in eliminating it in autoradiographic studies. With a hormone in vivo, for example, label due to its catabolism, e.g., in lysosomes, must be distinguished from its initial sites of receptor activation.

A further requisite for a specific binding of an irreversible or pseudo-irreversible isotopic ligand is that a pretreatment with the same ligand solution in nonradioactive form should abolish the labeling seen in autoradiographs. This is the case after pretreatment with α-BuTX, subsequent application in vivo or in vitro of labeled BuTX, and light microscope or EM autoradiography (Porter et al., 1973a,b). This is also another test for saturation in the conditions of specific labeling.

For in vitro labeling, it must be borne in mind that the diffusion out of the tissue of an applied ligand—radioactive or, for preblocking, unlabeled—during washing is often slower than the diffusion in, so that (especially in the case of polypeptide ligands) a pool of slowly diffusing ligand may remain after washing, capable of reacting further later at free receptor sites. This problem, and its solution, is exemplified by the case of [^{3}H]-BuTX in muscles, where, to wash out the excess without further reaction occurring (in any experiment where blockade of the receptors is not intended to be complete, as in competition experiments) it is necessary to add, for example, 10^{-4} M d-tubocurarine to all of the washes, due to such a residual pool of [^{3}H]-α-BuTX (Barnard et al., 1977). This factor, if such a wash-out system is not used, can complicate interpretation of experiments (Fertuck and Salpeter, 1976) in which the labeling exposure is terminated at a physiological end point, such as the cessation of tetanic response. While α-BuTX has a MW of 8000, even a charged molecule as small as decamethonium can give delayed wash-out from muscle (Creese and MacLagan, 1970), so that a more general caution regarding this possibility for ligands in tissues is indicated.

For a reversible binding, it is necessary to assess the extent of nonspecific labeling by competition with unlabeled ligands that bind to the receptor in question. Further, if the ligand is stereoisomeric, a biologically inactive isomer should be used as a control for specificity. As an example, Pert et al. (1976), preparing autoradiographs after in vivo administration of [^{3}H]diprenorphine to trace opiate receptors, showed that prior injection of nonradioactive levallorphan (5 μg/g of body weight in rats) blocked the labeling, and its inactive dextro isomer had no effect. These results, and a selective distribution of grain densities in brain regions, argued for a specific binding of this opiate antagonist. For in vitro application of a reversible radioactive ligand, competition by a physiologically saturating level of another, unlabeled ligand for the receptor should abolish labeling, as a control.

In general, for autoradiographic studies of any type of ligand, it is desirable where feasible to perform a parallel labeling of similar cells, or membranes derived from them, on a biochemical scale. Measurements should thus be made to confirm that only receptor-specific labeling occurs in the conditions of application actually to be used, by means of the various criteria of specificity discussed above. Absence of specific uptake, by these criteria, in other tissues or regions or cell types known not to possess these receptors, where applicable, is an additional useful check to be made at the biochemical level. Examples of these checks on autoradiographic data will be given below.

3.1.3. Tissue Processing for Autoradiography of Receptors

Preparation of an autoradiograph for either the light or the electron microscope will require in most cases fixation for preservation of structure and section cutting. The treatments involved must be evaluated with respect to the possible loss of binding of the labeled ligand or its redistribution.

As noted above, the use of a reversible, labeled ligand will require dry-mount methods of autoradiography (discussed in Sec. 3.2.2). For a binding assumed to be effectively irreversible and occurring at a receptor which is firmly membrane bound and therefore indiffusible, fixation may often be postponed until after the reaction with ligand, so that the native affinity is maintained. It is assumed that the receptor–ligand complex is stable to fixation and to the embedding procedures. For electron microscopy it is desirable to give as light as possible a treatment in glutaraldehyde, and a secondary fixation in OsO_4 at a later stage. Tissue or membrane specimens containing the receptor should be labeled and treated, with and without the same procedures, to verify that loss of binding is not introduced. Such checks are vital when quantitative results are claimed.

3.1.4. Application to a Pseudo-Irreversible Reaction at a Receptor

Ligand application for the quantitative study of ACh receptors of vertebrate muscle by EM autoradiography using radioactive α-neurotoxin can demonstrate the requirements described above. A recommended procedure for labeling the receptors to saturation is as follows.

1. The muscle in question is exposed in an anesthetized or spinal whole-animal preparation and rapidly arranged in a perfusion chamber for tension recording.

2. The labeled toxin (e.g., $[^3H]$-BuTX) is applied (at about $2-4 \times 10^{-7}$ M) in physiological medium at room temperature, recording the response to well-spaced short bursts of high-frequency nerve stimulation. When all response is abolished, the toxin exposure is maintained for a further period twice that already elapsed. It is checked then that the muscle contractile response to a direct stimulus is unimpaired.

3. The muscle is superfused with medium and then with unlabeled toxin

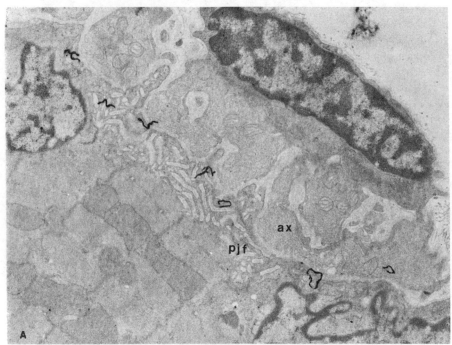

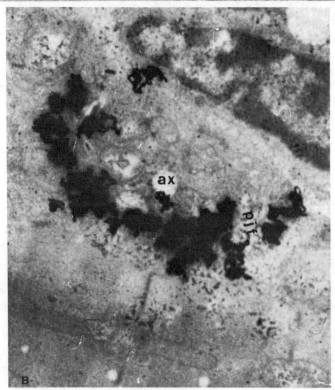

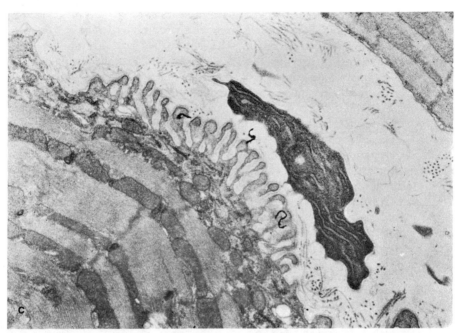

Fig. 1. (A) Electron microscope autoradiograph of a section through a bat diaphragm end plate, labeled with [³H]-α-BuTX. Note that the silver grains lie over the postjunctional folds or within the limits of scattered radiation from the folds. Axon, ax; postjunctional folds, pjf. (B) An overexposed EM autoradiograph over a labeled end plate to show that the labeling is confined essentially to the postjunctional fold region and in fact is concentrated (apart from radiation scatter) over the upper (crest-containing) sector of that region. (C) EM autoradiograph of an end-plate section from a mouse diaphragm muscle labeled with [³H]-BuTX 5 days after denervation. The nerve terminal has been totally removed (a part of a Schwann cell [dense] has come to occupy one part of its position). The postjunctional folds are seen to retain their normal structure; this picture shows clearly the denser membrane (after OsO₄ treatment) at the crest and upper walls of the folds, which is due to the surface packed with receptors there. The three grains in this field are associated with that zone. When the density of the labeled sites was calculated from a large number of such fields, it was found to be the same at the crest zone (about 25,000 per μm² of membrane) in both the normal and denervated muscle end plates, so that no receptors (within the experimental error of the method) are associated with the presynaptic membrane, and no loss of receptors occurs from the end plate when receptors are appearing in the extrajunctional muscle membrane. [From Porter and Barnard (1975b).]

for 10 min at 20-fold higher concentration than for the labeling. Washing with medium is continued for 10 min.

 4. The muscle is lightly fixed by application of 0.2% glutaraldehyde/0.2 M sucrose/0.1 M Na phosphate (pH 7.4), by intravascular perfusion where the preparation permits, and otherwise by superfusion and then immersion for 30 min at 4°.

 5. Small innervated strips are isolated by needle dissection under the dissecting or Nomarski microscope, observing the nerves and obtaining a known orientation. Embedding is, e.g., in Spurr's low-viscosity resin; postfixation in OsO₄ can be given briefly prior to embedding or, preferably, by vapor on the cut sections. Autoradiography follows.

 6. The same procedure is applied to muscle that is first treated with unlabeled

toxin for the same period of time, followed by a wash-out for 30 min and then labeling. Sections of end plates from this muscle are examined similarly.

7. Parallel (e.g., contralateral) muscles are labeled similarly and removed either (a) unfixed or (b) fixed by the same procedures. Further samples (c) are preblocked by unlabeled toxin as in 6, and then labeled. All washes are as above. Yet others (d) are labeled and washed in the muscle as in 2 and 3 above and then exposed *in vitro*, after dissection, to a fivefold excess of the labeled toxin, followed by thorough washing. Each of these specimens is exhaustively extracted with 1.5% Triton X-100 medium and fractionated on a Sepharose 6B gel column (Chiu *et al.*, 1973). This separates the peak of toxin–receptor complex, the total radioactivity of which is determined in each case. It is thus checked [(a) vs. (c) and (a) vs. (d)] that saturation of the receptors occurred during the labeling and that the labeling was specific. Likewise [(a) vs. (b)], the effect of the fixatives is evaluated and should be negligible. With thick muscles, it may be necessary to use after labeling only the superficial zones of the muscle. An example of an EM autoradiograph from a muscle after treatment by these procedures to label the ACh receptors at end plates is given in Fig. 1.

3.2. Autoradiographic Methods

3.2.1. Treatments in Aqueous Media

For essentially irreversible ligands, the tissue specimens, handled in aqueous media throughout, are best sectioned frozen in the case of light microscope autoradiography and after suitable low-temperature embedding for the EM. Autoradiography for the light microscope is by conventional methods. For quantitative EM work, the photographic emulsion (see Table II for the properties of the types in common use) is best applied to the sections by an automatic dip-

Table II
Resolution and Sensitivity in EM Autoradiography[a]

Isotope	Emulsion	Developer	HD (Å)	Sensitivity, s
^{3}H	L4	Gold–EA	1450[b]	0.27[c]
		Microdol X	1600[b]	0.12[c]
		D 19	1450[b]	0.22–0.24[c,d]
	NTE-2	Dektol	1000[e]	0.14[c]
^{125}I	L4	Gold–EA	800[e]	0.55[f]
	NTE-2	Dektol	550[e]	~0.35[f]

[a] All data are for sections of 1000-Å thickness. There is little difference, if any, in these values, for 500-Å thickness, but there may be for greater thicknesses, especially for ^{125}I. Due to the significant and nonlinear dependence in some cases of s upon the grain density, the s-values have not been averaged over a wide range of grain density but have been given for a fixed grain density (observed) of 0.5 grains/μm^2 of emulsion, i.e., in the commonly encountered range. Values are for a monolayer of emulsion in each case. EA = Elon–ascorbic acid.
[b] Salpeter *et al.* (1969).
[c] Salpeter and Szabo (1972).
[d] Barnard (unpublished data, for recent batches of L4).
[e] Salpeter *et al.* (1977).
[f] Salpeter and Szabo (1976).

coating instrument* forming a monolayer. Procedures for other steps and for observation of the labeled areas and of background are detailed in the references cited in the rest of this section.

3.2.2. Dry-Mount Methods

These are essential for reversibly bound ligands. The labeled tissue is quenched in, e.g., a fluorinated hydrocarbon medium cooled in liquid nitrogen and ideally is not allowed to thaw or come into contact with a solvent or with water vapor until emulsion exposure is completed. In one version, described by Appleton (1966) and Horowitz (1974), dry photographic emulsion is pressed onto the frozen section or smear and exposed at about $-80°$. It is difficult to accomplish this without transient thawing, and it is technically easier to thaw-mount the frozen sections on the emulsion-coated slide and expose at $2°$ (Gerlach and McEwen, 1972; Stumpf, 1976); for a diffusible ligand, however, some risk of translocation may be incurred. In another approach (Stumpf, 1976), frozen sections are cut and then freeze-dried and dry-mounted on emulsion-coated slides. It must be investigated whether the receptor binding and the resolution are affected by freeze-drying.

Dry-mount methods have not been described so far for autoradiography of receptors in the EM. They have been used (Table I) in light microscopy for receptors for ACh (muscarinic) (Kuhar and Yamamura, 1975), opiates (Pert *et al.*, 1976), and releasing hormones (Stumpf and Sar, 1977) as well as for nuclear binding sites of estradiol (Keefer *et al.*, 1976).

3.3. Interpretation of EM Autoradiographic Data on Receptors

The great advantage of the autoradiographic method is that it is quantitative. Quantitation in the light microscope is dealt with in Sec. 4. The assignment of identified receptors to particular cell types or regions in the brain, etc., can be valuable at that level and requires reliable counting procedures for this to be valid. In most cases, however, the most useful autoradiographic data on receptors will come from studies at the ultrastructural level, where these are feasible. For all such studies, quantitative analysis is necessary to use the information in the autoradiograph effectively. It is essential to use such analysis in order to identify accurately the labeled structures as well as to determine their relative contents of receptors.

3.3.1. Assignment of Silver Grains in Autoradiographs to Most Probable Locations of the Labeled Receptors

The grains must be counted and related to either areas over particular cellular structures or to lengths along the membrane which is taken as containing

---

*Directions for constructing a semiautomatic dip-coating device are given by Kopriwa (1967). An improved and convenient instrument has recently been designed by Dr. David Border in the Department of Biochemistry, Imperial College, and is available commercially from Codoc Ltd., 16 Stacey Avenue, London, N18, England.

the receptors. Multiplication by the section thickness will then yield the density per unit volume of the element concerned, or per unit area of the membrane. Three approaches are possible for assigning developed grains in the autoradiographs to their originating radioactive sources.

3.3.1.1. Single-Source Zonal Method. In this, the simplest approach, it is taken as known from other evidence that the receptors are entirely on a particular structure, usually the cell membrane. Grains are counted within the limits of radiation scatter from that membrane as a radioactive source. These limits are reasonably well known for each isotope, emulsion, and processing routine, being conveniently expressed in half-distance (HD) values (Salpeter *et al.*, 1969). Some of these are quoted in Table II. Fifty percent of the scattered grains will be within a zone 1 HD wide around the source (i.e., on both sides of a membrane containing it) and about 90% up to 3 HD. The total of the grains within a chosen limit (*or*, preferably, double the total within 1 HD away) is used to give the grain density per square micrometer of the membrane. This is the basis of the density derived by Bourgeois *et al.* (1972) for the ACh receptors along the innervated membrane of *Electrophorus* electroplaques. A parallel simple method is often used for receptor density per unit volume of cytoplasm or per organelle* and can provide relative analyses in simple situations. It provided absolute measurements in the simple case of the extrasynaptic muscle cell membrane (Porter *et al.*, 1973b; Fertuck and Salpeter, 1976). It is important to note that this method can only give the mean density over the entire membrane (or cytoplasmic volume) counted; i.e., uniformity has to be assumed. It also can only be used for very simple geometries, e.g., a nonfolded, uninterrupted membrane source.

3.3.1.2. Density Histogram Analysis. The zonal method cannot be used when nearby structures may be labeled also, e.g., the membrane of an adjacent cell, or when it is desired to test without assumption models for the distribution of receptors. For absolute receptor densities in situations of any complexity, a more detailed analysis is necessary. In this second approach, one plots a histogram of the density of grains in zones of increasing HD for each structure to be tested as the source. This is illustrated in Fig. 2 for the ACh receptors at the mouse neuromuscular junction. In Fig. 2(a), the distribution on either side of the postsynaptic membrane is shown for silver grains due to [^{3}H]-BuTX-labeled receptors, taken from a large number of sections through different labeled end plates (Porter *et al.*, 1973b). The distributions must be expressed as a normalized grain density, this having been corrected for area changes, due to the dilution on moving outward radially, by the point-grid method (Weibel *et al.*, 1966). A square lattice of points is overlaid on the print, and the area in each HD zone from the postulated source is found by the number of points falling within it (Porter *et al.*, 1973b) to give the grains per unit area, expressed relative to that at the origin. [A refinement is to make a least-squares fit of the histogram of those points (Fertuck and Salpeter, 1976)]. Salpeter *et al.* (1969) showed how sources along a line or band

---

*The calculation of relative grain densities over organelles or other identified structures requires the relative areas of those structures, which can be determined by their intersections with a random grid of circles of a diameter dependent on the resolution (Williams, 1973). The number of grains per structure is divided by the number (normalized) of its overlying circles to give the grain density. This, again, assumes uniformity of label over each labeled structure. An example of such analysis is given by Gould and Dawson (1976).

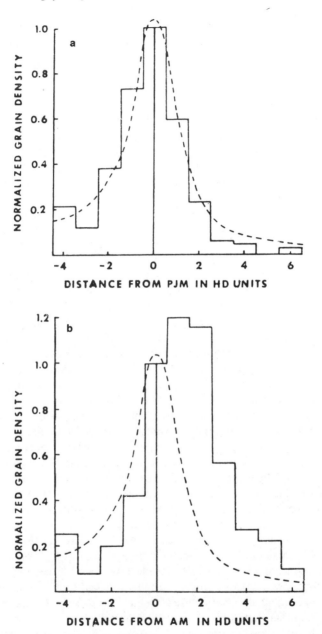

Fig. 2. (a) Histogram showing the distribution of silver grains around the postjunctional membrane (PJM) for mouse diaphragm end plates labeled with $[^3H]$-α-BuTX. Note that the experimental distribution agrees closely with the theoretical distribution curve (broken line) for a source located on the PJM. HD (half-distance) units are a measurement of autoradiographic resolution. For further details, see the text. Negative values of HD designate distances on the axon side of the PJM (HD = 0) and positive values on the muscle side. (b) Histogram of grain densities around the axonal membrane (AM) for mouse diaphragm end plates labeled with $[^3H]$-α-BuTX. The grain scatter in this case does not fit the theoretical curve for a source centered on the axonal membrane. Instead the grain densities are concentrated over the muscle side of the axonal membrane (positive HD units) corresponding to the crests of the postjunctional folds. [Taken from Porter *et al.* (1973b).]

with various geometries would generate theoretical distributions of grains. One such distribution for label in a membrane source is superimposed on the density histograms of Fig. 2, being derived for a line source running over a surface having the slight curvature observed for the nerve terminal inserted into the surface of mouse diaphragm muscle fibers. This explains the shallower curve on the axonal (negative HD) side.

It is seen that a density histogram plotted with the postsynaptic membrane [Fig. 2(a)] as the source fits the theoretical curve for that source very well. In contrast, with the presynaptic membrane used as the source [Fig. 2(b)], the fit is very poor, and it is clear that, while allowing for the experimental error of the method, the labeling must be at least predominantly on the postsynaptic membrane, and the distribution is consistent with it *all* being on that membrane. Although some authors have considered that in such methods the resolution (1 HD = 1600 Å in this case) is too low to distinguish sources separated only by the synaptic cleft, in practice this is not the case: since the postsynaptic membrane is convoluted, its contours differ widely enough, even in the region where the receptors are concentrated, from those of the presynaptic membrane for a distinction to be made clearly in such paired histograms. Comparison of the distributions for two lines that do not, for much of the zone in question, run parallel adds an extra analytical element that reduces the resolution limit error in practice. Such a device can be applicable to other cases of receptor autoradiography. Further, particular zones of a folded membrane can be compared to determine a nonuniform receptor distribution by a variant of this method, as is exemplified in Sec. 3.4.

3.3.1.3. Hypothetical Grain Analysis Method. The third analytical approach uses the *hypothetical grain* method of Blackett and Parry (1973, 1977). In this, there is placed over the print a transparent screen containing a grid of circles (radius 1.7 HD) denoting hypothetical silver grains, each associated with a random set of points (say 10, at varying distances and directions) that represent theoretically permissible sites of the originating nuclear disintegration (Fig. 3, inset). Counting a large number of these grains and sites over structures of interest in the micrographs gives their distribution of grains obtained hypothetically if the radioactivity content had been uniform over the section. An appropriate computer program then varies the relative contents in the structures until their hypothetical grain distribution best matches (by χ^2 test) the observed distribution of real grains. This method is particularly useful where different structural elements or a number of separate, replicate structures (e.g., vesicles) dispersed in each section are truly labeled. In that case the density histogram method is not applicable, and statistical models for theoretical distribution are difficult to set up *ab initio*. Labeled receptors present in different locations over, say, complex neuronal structures in one section, such that cross-scatter occurs between them, are best analyzed by this method. This method can also be used for receptors confined to a membrane even if it is sterically complex. An example is shown in Fig. 3, where 9 of the grains arising from sources on a highly curved membrane are related to it by this method. For the geometry of Fig. 3, which could, with trivial changes, be taken to represent a labeled boundary enclosing an element of a neurone 1 that penetrates a larger element of cell 2, Blackett and Parry (1977) found that of 487 grains (each taken in a circle of 1.7-HD radius from its center), 54 would actually be over the labeled

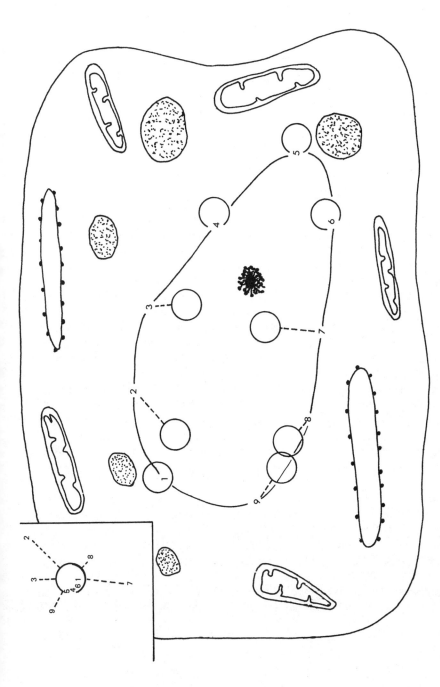

Fig. 3. Diagram showing the location of hypothetical grains produced by nuclear disintegrations occurring in labeled molecules on a membrane—in this example, the nuclear membrane of the cell. The inset shows the diagram used to generate random locations of disintegrations (numbered 1–9) for each grain: the circle represents the position of an observed grain, diameter 3.4 HD units, so that there is a 50% chance that the originating disintegration occurred within the circle, others being generated at appropriate random distances and angles. This diagram is placed with each point in turn on the membrane, at equispaced points on it, recording the grains that then occur in the circles as membrane-originated grains in the statistical analysis. [From Blackett and Parry (1977).]

membranes, 59 in the enclosed cell 1, and the rest over various portions of the outer cell 2 element. This illustrates the type of distribution that would be found and that can be analyzed statistically to show that it is attributable ($P \sim 0.6$) to a source entirely on the separating boundary. The method, in this form, assumes uniformity of the sites over the element in question, but a membrane, for example, could instead be taken in zones in the analysis and computer programs for various patterns or gradients of labeling used in fitting the distribution.

3.3.2. Calculation of Receptor Density

In the three methods listed, the grain density is obtained either per unit length of a membrane or per unit area over a cellular structure or zone. In the density histogram method for a membrane source, all of the grains contributing within the area of the theoretical curve for that source [Fig. 2(a)] are summed in computing the density over the membrane. Off-tissue background levels are subtracted throughout. Lengths of membranes of complex shape are measured using a map measurer, by the number of intersections of all membrane profiles with lines in a parallel array (Weibel *et al.*, 1966) or a square network in standard stereological procedures, or (best of all) by an electronic digitizer.

The following variables must be considered in converting the grain density thus observed to a density of labeled receptors.

1. The stoichiometry of the reaction must be assumed. What is measured is the number of binding sites, not the number of receptor macromolecules, and the results should always be expressed in terms of these active centers. It is assumed, obviously, that the binding has proceeded to saturation.

2. The thicknesses of the actual sections used should be determined. Reflectance colors are not a sufficient guide to this. The most convenient method for measuring section thickness is to use an interference microscope [e.g., the Zeiss (Jena) Inter-phaco, with which we find the error in measurement of sections at 1000Å to be $< \pm 50 \text{Å}$]. By mounting in a liquid of known refractive index, e.g., water, the refractive index of the section need not be assumed. A benchtype interferometer (e.g., the Varian Angstrom-scope) can, of course, be used if available. Whether significant variation in thickness occurs across a single section can also be checked in the interference microscope. For a membrane location, it is assumed for the calculation that the section is cut perpendicularly across the membrane; sections are selected where the sharpness of its image, or the geometry seen, shows this to be so. A section thickness of 1000Å is optimal for most cases using ^3H or ^{125}I.

3. The specific activity of the isotope used, S, is expressed in curies per millimole. For ^{125}I, this must be corrected for the period t, the autoradiographic exposure time in days.

4. The sensitivity s of the emulsion must be determined. This is the fraction of the disintegrations that produce one observed grain. This varies between roughly 0.1 and 0.6 with the isotope (^3H or ^{125}I), the emulsion type, and the developing process used (Table II). s can vary somewhat between batches of the same emulsion. Salpeter and Szabo (1972) found that s for ^3H varies with the dose of radiation experienced by the emulsion, this effect being particularly marked

when Microdol X is used as the developer. The same is true for ^{125}I (Fertuck and Salpeter, 1974a). A convenient method for determining s is to use a $[^3H]$methacrylate block (Amersham), cutting 500–1000-Å-thick sections on the ultramicrotome and determining their ^3H content by scintillation counting and their area to yield the dose per square micrometer. Autoradiography with these sections under the experimental conditions, for varying exposure periods, yields s as a function of grain density. For ^{125}I, iodinated albumin has been used (Fertuck and Salpeter, 1974a). This, however, requires layers of the protein to be formed on slides and measured throughout for thickness. The grain density over a suitable zone in the tissue autoradiograph, processed similarly, is employed with the calibration curve in determining the relevant value of s.

5. Latent image fading must be negligible. Checks on this are obtained in the exposures of standard sources described above.

The number of labeled sites per unit area of membrane, or per unit volume, is then obtained, by an obvious calculation, from

$$\frac{GN_0}{stS \times 3.20 \times 10^{15}}$$

where G is the observed grain density and N_0 is Avogadro's number expressed per millimole (6.02×10^{20}).

3.3.3. Isotopes and Resolution

For receptor cases in general, only ^3H and ^{125}I can be introduced into receptor ligands easily with the required high specific activity and resolution for EM autoradiography. Labeling by ^3H is more generally accessible. ^{125}I has the advantage of higher resolution (Table II) and higher sensitivity (Fertuck and Salpeter, 1974a). It has the disadvantage of its half-life (60 days) and the frequent chemical instability of iodinated derivatives.* With the new NTE-2 (Kodak) emulsion, the resolution with ^{125}I can be decreased to 500 ± 70 Å (Salpeter *et al.*, 1977), making receptor localizations more precise than hitherto. With the more usual L4 (Ilford) emulsion, sensitivity is more than twice as high as with NTE-2. It is considerably greater (with L4) for gold–Elon-ascorbate or D19 developer than for Microdol X (Fertuck and Salpeter, 1974a) with either isotope. The dose dependence is also less with the former developers.

3.4. Applications to Synaptic Receptors

Studies of nicotinic ACh receptors along the highly folded membranes of vertebrate fast-twitch muscle fibers can be taken to illustrate the application of quantitative EM autoradiography. This method for receptor counting was shown to be feasible in that case by Porter *et al.* (1973a,b) using $[^3H]$-BuTX on mouse

---

*High specific activities can be introduced by ^{125}I, but recently those that can be safely introduced by $[^3H]$ into a small protein have been raised to 50–200 mC/mmol by the use of $[^3H]$proprionyl labeling groups, which can modify, for example, one to three lysines in α-BuTX without change of biological activity (Dolly, Nockles, Lo, and Barnard, unpublished data).

diaphragm end plates (Fig. 1). As shown in Fig. 2, the grain density was shown by analysis to be preferentially associated with the postsynaptic membrane. It was noted, however, in those reports that the possibility of an additional, lower level of labeling on the axonal membrane could not be excluded with the resolution of the method. Other methods of defining the localization further could, however, be used. Proteolytic digestion removed the cleft substance and separated the nerve terminal from the muscle end plate (Barnard et al., 1975) but did not decrease the postsynaptic density observed. Removal of the nerve terminal by denervation (Porter and Barnard, 1975b) also left the postsynaptic density unchanged [Fig. 1(c)]. It was concluded that the ACh receptors are postsynaptic, to the error limits of the measurement, estimated to be about 10%, it being noted that the latter fraction could still possibly be presynaptic.

The analysis illustrated in Fig. 2 yields a mean density of receptor binding sites over the entire postsynaptic membrane, 8500 μm^2 (Porter et al., 1973b). An alternative mode of analysis is possible, however, for labeled sites along a folded membrane in which the normalized grain density is measured in zones progressively down the folds from crest to base (Fig. 4). When this was first attempted, it was believed that an apparent asymmetry seen was not significant within the limits of the technique used (Porter et al., 1973b), but when more data were collected the effect became more pronounced and statistically highly significant (Albuquerque et al., 1974), as represented by the solid line in Fig. 4.

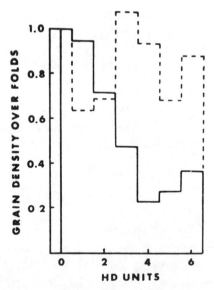

Fig. 4. Grain density distributions exclusively over the region of the postjunctional folds for mouse diaphragm end plates labeled with [^{3}H]-α-BuTX (—) or [^{3}H]-DFP (---). Grain densities are plotted in relation to the distance down the folds from the axonal membrane (HD = 0). The region of fold crests is included largely in the first three bins of the histogram. Note that the grain densities representing ACh receptor sites ([^{3}H]-α-BuTX labeling) are concentrated in the fold crest region, while those representing acetylcholinesterase (see the text) are distributed more or less evenly over the folds.

The grain density is plotted in Fig. 4 in distances down the surface of every fold encountered: if the receptor sites were uniformly placed over the entire postsynaptic membrane, that grain density (solid line) would be, by calculation, nearly constant from 0 to 6 HD units from the axonal membrane. It is seen that, in contrast, the crests of the folds are heavily labeled, while the depths of the folds are very poorly labeled. The observed pattern was found, in both mouse and bat, to be best fitted by a model in which about the upper 30% of the membrane in the crest region (see Fig. 5)—which includes an upper zone on the walls of the folds—is packed with receptor ACh binding sites at about $25,000 \, \mu m^2$, while all of the remaining lower regions of the folds have only about one-tenth of this level (Porter and Barnard, 1975a,b; see Table III). The value of about 30% for the crest surface fraction corresponds to the fraction of the postsynaptic membrane in such muscles that appears "thickened" in osmicated or metal-stained tissue electron micrographs (see Fig. 10), supporting the view that this 30% crest fraction is specialized for receptor protein packing. This statement was specifically confined by the authors cited to the mammalian muscle types so far studied (Table III). It is obvious that in end plates of other muscle types where the folding is less frequent or the folds are shallower, and assuming that the lower portions of the folds are, again, devoid of receptors, the specialized zone would there occupy a much greater fraction of the total postsynaptic surface. In fact, in frog slow-twitch fibers we have observed that this is true of the dense zone, as has Rosenbluth (1974) in frog tadpole tail muscles: in these studies the specialized crest zone was more readily visualized morphologically by permanganate–acetate staining. A fuller range of autoradiographic studies is needed to confirm the

Table III

Densities of Receptors (as α-BuTX Binding Sites) on the Postsynaptic Membrane (PSM) of Mammalian Neuromuscular Junctions

Muscle and fiber type	%Dense region[a] on PSM	Density of sites/μm^2		References
		Mean over PSM	Crests[b]	
Mouse diaphragm (red)	28	8400	25,000[c]	Porter and Barnard (1975a)
Mouse diaphragm (red), denervated	28	9200	25,000[c]	Porter and Barnard (1975b)
Bat diaphragm (red)	31	8800	25,000[c]	Porter and Barnard (1975a)
Mouse sternomastoid (white)	31	8800[d]	30,500	Porter and Barnard (1975a) Fertuck and Salpeter (1976)

[a] The fraction of this membrane that is seen as electron-dense or "thickened" in suitably processed electron micrographs (see the text).

[b] The crests of the postjunctional folds, defined as in Fig. 3.

[c] In the references cited, this value was estimated as $20,000–25,000/\mu m^2$; the more exact figure is a mean value of $25,000/\mu m^2$, by further analysis based on (1) fit of models of various zones of labeling and (2) fit of a model in which the dense region is labeled at 100%, and the rest of the fold surface at 10%, of the level of labeling. Both methods give a best fit of the data with this level at a mean of $25,000/\mu m^2$.

[d] Estimated by an indirect method, also based on EM autoradiography and stereology.

presumption that the extent of the morphologically distinct "thickened" zone, both at the crests of the folds and over longer interfold sectors of the postsynaptic membrane (where these occur), is always a visual demarcation of the receptor-laden regions.

That the ACh receptors are concentrated at the crests of the folds was deduced, concurrently, by Fertuck and Salpeter (1974b), using $[^{125}]$-BuTX on mouse sternomastoid muscle, on the basis of qualitative EM autoradiographs (using NTE-2 emulsion) showing heavy labeling around the crests. In a quantitative study (Fertuck and Salpeter, 1976) of the same system, they used the density histogram method and comparison with a theoretical curve for a band along the tops and upper walls of the folds. The grain density was assigned to the observed electron-dense membrane, giving a labeled site density of 30,500 μm^2 there. This is in agreement with the other studies noted above, within the error of the latter (Table III).

The distribution of AChE active centers, also determined by EM auto-radiography, is, in contrast to the ACh receptors, approximately uniform over the entire fold surface (Fig. 4, broken line). In freeze-fracture studies, several authors [see Porter and Barnard (1975a) for references] have found characteristic intramembranous particles concentrated in the same crest zone, which may be presumed to contain the receptors. A summary of the distribution found of receptors and of AChE at the end plate, in relation to transmitter release sites, is diagrammed in Fig. 5.

4. Counting Receptors per Cell or per Synapse

In dispersed cells, as in homogeneous cultures, receptor counting can be done simply by measuring in bulk the extent of binding of a ligand. However, for cells which are in tissues or which, if separable, are not homogeneous, in situ methods are required. Light microscope autoradiography is a suitable method for relative counts of receptors in such situations, where whole cells or large zones are to be compared. In certain conditions it can be refined to yield absolute numbers of labeled centers.

4.1. Light Microscope Autoradiography of Receptors

The requirements for a radioactive ligand are the same for autoradiography in the light microscope as in the EM, as discussed in Sec. 3. Where the ligand is reversible, dry-mount autoradiography in some form is needed.

Methods for grain counting in the light microscope in general have been described, e.g., by Rogers (1967), Horowitz (1974), Stumpf (1976), and others. Calibration curves to correct for deviations from nonlinearity in the relationship to disintegrations must be made in each system studied.

A number of applications of relative grain counting to receptor density comparisons are listed in Table I. Regional distributions of receptors are usefully obtained by this method. Thus, the muscarinic ACh receptor in the brain was mapped by dry-mount autoradiography using injected $[^3H]$-QNB, a reversible

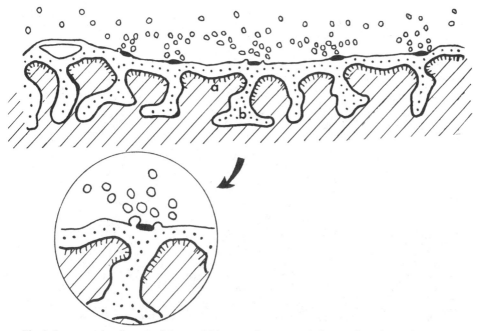

Fig. 5. Interpretative diagram of the amphibian muscle motor end plate to show the spatial relationships between sites of transmitter release and concentration of ACh receptors and acetylcholinesterase.
The basic morphology was traced from micrographs of frog muscle end plates. The vesicles characteristically stack opposite the mouths of the postjunctional folds. Dense bars on the presynaptic
membrane opposite the fold mouths represent ridges (seen in freeze-etch preparations). The striations
along the postjunctional membrane at the fold crest correspond to the receptor-rich region (a) of
the subneural apparatus. Dotted lines show the extent of visible cleft substance and, it is presumed,
of the acetylcholinesterase (see Fig. 4). The inset shows, in detail, the fusion of vesicles with the presynaptic membrane. The dense presynaptic ridge prevents the vesicles from releasing transmitter
directly into the receptor-poor, acetylcholinesterase-rich, region of the fold depths (b). Instead, the
ACh is released from vesicles at specific release sites on either side of the presynaptic ridge, so that it is
expelled onto the receptor-rich crests of the folds. It is proposed that upon release from receptors on
the fold crests the ACh diffuses into the fold depths where it is hydrolyzed by the acetylcholinesterase
located there. [Taken from Porter and Barnard (1975a).]

antagonist (Kuhar and Yamamura, 1975; 1976) *or* by autoradiography on thaw-
mounted sections lightly fixed in glutaraldehyde and reacted with [^{3}H]-PrBCM,
an irreversible antagonist (Rotter *et al.*, 1977a,b), as illustrated in Fig. 6. The
regional variation in the forebrain was about the same in the two different methods;
Rotter *et al.* (1977a,b) also revealed a heavy concentration of these receptors in
the hypoglossal nuclei of the lower medulla which fell abruptly at its boundaries.
In the hippocampus (Rotter *et al.*, 1977b), high grain densities were observed in
layers rich in synaptic connections, while virtually no labeling was seen in cellular
or fiber layers, showing that synaptic receptors are being visualized. An interesting
finding was that after unilateral axotomy [Fig. 6(c)] the labeling per square
micrometer of the sections due to atropine-sensitive PrBCM binding was (after
6 days) reduced by 47%. The study showed that the synaptic contacts and the
receptors are lost simultaneously, and both changes reverse together on regeneration.

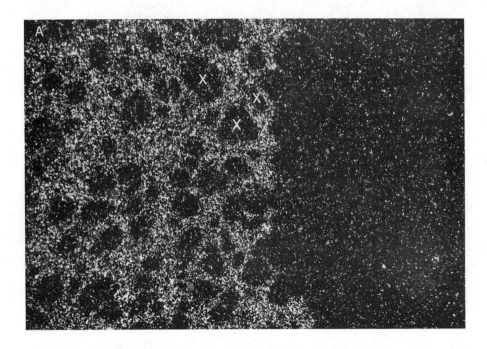

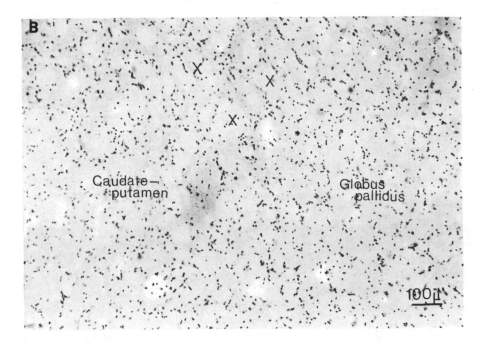

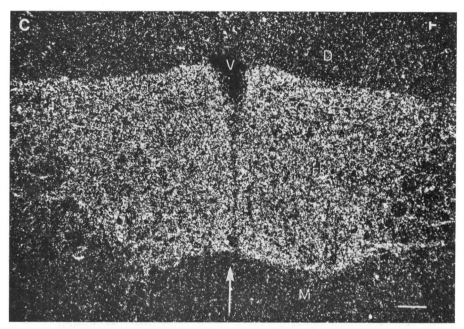

Fig. 6. Labeling of the muscarinic ACh receptor in rat brain by [³H]propylbenzilylcholine mustard. (A) Distribution of silver grains in autoradiographs (dark field illumination) of sections across the junction of caudate–putamen (intensely labeled) and globus pallidus (background level only). (B) The same area under bright field illumination. (C) Coronal section through the caudal medulla, where the hypoglossal nucleus has its maximum cross-sectional area; autoradiograph, dark field illumination. There is an intense deposition of grains over the hypoglossal nuclei. The labeling falls abruptly at all the boundaries of the hypoglossal neuropil and over the caudal part of the fourth ventricle. Six days after unilateral sectioning of the hypoglossal nerve, the labeling on the operated side in an equivalent section fell to 47% of the intensity on the unoperated side. H, hypoglossal nucleus; D, dorsal vagus nucleus; M, medial lemniscus; V, caudal part of the fourth ventricle; arrow indicates midline; scale bar, 100 μm. [From Rotter *et al.* (1977a,b).]

For the nicotinic ACh receptor, densities in [³H]-α-BuTX-labeled mammalian muscles were measured autoradiographically by Barnard *et al.* (1971), showing the confinement of label to the end plates in innervated fibers, blockade by reversible ligands, and the 1:1 relationship to the total cholinesterase-like active centers reacting with DFP, and giving absolute measurements of the total receptors per end plate (see Sec. 4.3). Conditions for specific labeling in various muscle types were further established by Porter *et al.* (1973a). An illustration of [¹²⁵I]-α-BuTX-labeled end plates is given in Fig. 7. [¹²⁵I]-α-BuTX was used by Hartzell and Fambrough (1972) in obtaining densities of labeled sites along 2–14 day denervated, whole-mounted fibers of rat diaphragm to show a nonlinear relationship to ACh sensitivities. The absolute receptor densities could not be obtained reliably, however, by the light microscope autoradiography under these conditions (Fambrough, 1974) but were obtained by scintillation counting of labeled muscle specimens. That gave a mean density of $630/\mu m^2$ (14 days); the same method was

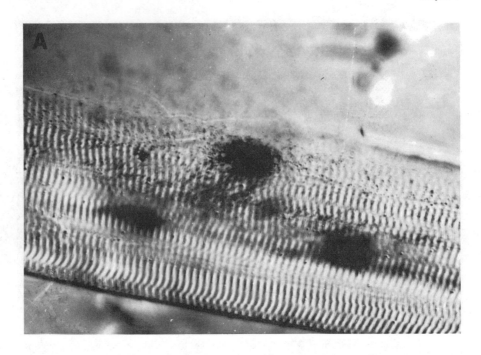

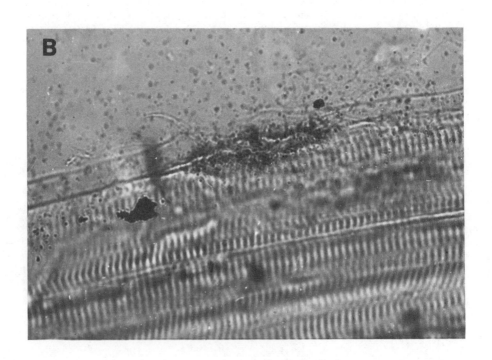

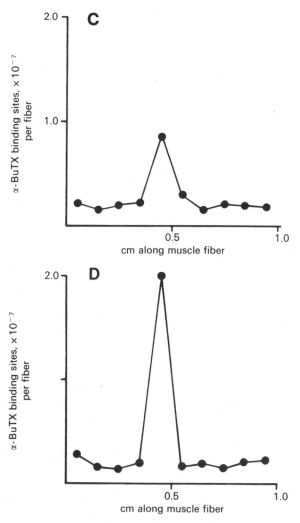

Fig. 7. Autoradiographs of fibers from [^{125}I]-α-BuTX-treated human intercostal muscle: (A) normal; (B) from a myasthenia gravis patient, processed identically; note the reduction in the grain density at the end plate and the enlargement in its area (× 600). (C) and (D) Radioactivity content of 1-mm segments of bundles of 20–50 fibers from myasthenic (C) and (D) normal muscle. Note the peaking of binding, which corresponds to the end-plate zone, and the reduced binding there in the myasthenic end plate. Both analyses indicate a smaller number of active receptors at the myasthenic end plate. [Adapted from Ito *et al.* (1978).]

found by Libelius (1974) to yield 400 receptors/μm^2 for 6-day denervated mouse extensor digitorum longus muscle. In contrast, Howe *et al.* (1976) used auto-radiography to obtain a maximal value for denervated mouse gastrocnemius muscles of only 80 extrajunctional receptors/μm^2. However, 6-μm sections were used there, and it was assumed that the grain yield was independent of section thickness and that grain densities over those sections represented the receptors on the surface of the muscle fibers (which had been 50 μm in diameter); both these assumptions must lead to a considerable underestimate of the receptor density actually on the membrane.

An application of the combination of light microscope autoradiography with scintillation counting for quantitation is shown in Fig. 7. Ito *et al.* (1978) have found that at end plates in $[^{125}I]$-α-BuTX-treated muscle fibers from myasthenia gravis patients (B) the grain density is lower, while the end-plate area labeled is larger than in control fibers (A). In such fibers the radioactivity was also estimated directly, showing the peak of binding due to the normal end plates (D) and its decrease in the myasthenic case (C). This decrease was also found by Drachman *et al.* (1976) by $[^{125}I]$-α-BuTX binding, measured both in bulk on muscle samples and by relative grain counts in end-plate auto-radiographs. The decrease is in accord with qualitative evidence from peroxidase-labeled BuTX binding (see Sec. 5.1.2) for loss of receptors in this condition.

Nicotinic ACh receptors have also been sought in the CNS by $[^{125}I]$-α-BuTX binding (Polz-Tejera *et al.*, 1975). α-BuTX binding in the brain, while demonstrated in a number of biochemical studies, has not as yet been shown to correspond to inhibition of observed nicotinic responses there (Miledi and Szcepaniak, 1975), and further evaluation is required. In chick retina, Vogel and Nirenberg (1976) found specific $[^{125}I]$-α-BuTX binding, and in the retinal tectum of amphibians or birds a synaptic uptake of α-BuTX seen in autoradiographs or with peroxidase does correlate with its trasmission blockade (Freeman, 1977; Barnard *et al.*, 1979).

Opiate receptors in the brain were traced by Pert *et al.* (1976), Bird *et al.* (1976), and Schubert *et al.* (1975), using injected ^{3}H-labeled opiates (see Sec. 3.1.1) and diffusible-ligand (dry-mount) autoradiographic techniques. A medullary region showed both high grain density and electrophysiological sensitivity to opiate agonists and antagonists (Atweh and Kuhar, 1977). A comparison can be made with the sites of endogenous enkephalins (believed to be among the natural transmitters at these synapses), tracing these by immunofluorescent staining (Simantov *et al.*, 1977; Elde *et al.*, 1976). In many brain regions the two localizations correspond, but in a number of opiate binding zones the enkephalins were not seen. There are clearly risks in attempting the localizations of the stores of a small peptide; possible loss of some enkephalins in the formalin fixation and other aqueous treatments used or a lack of penetration of the antibody to all sites must be considered, as well as the multiple classes of sites for the ligands (Lord *et al.*, 1977) which may denote that different peptide transmitters are stored at various locations.

4.2. Absolute Enumeration of Total Receptors

To count the total number of receptor sites (as labeled ligand molecules bound) per cell or per synapse in a heterogeneous cell population, absolute autoradiographic methods are required. Two different approaches are possible for this purpose. In the first, the density of the receptor sites is estimated per unit area of membrane, and the total surface area labeled at each density observed is estimated to give the total of sites by simple multiplication. In EM autoradiography, the mean density of label over a surface can be estimated reasonably reliably if the calibration of the grain yield is performed carefully, as described above. Bourgeois *et al.* (1972) used this method to calculate the number of toxin-binding receptor sites per *Electrophorus* electroplaque. They determined the absolute density of ^{3}H-labeled toxin-binding sites, with certain assumptions relating to the surface area of membrane involved due to the complex stereology there. A revised grain yield determination (Changeux *et al.*, 1975) has given 50,000 $\pm$ 16,000 sites/μm^2 of postsynaptic membrane (higher than at muscle end plates). This, together with estimates of total areas, gave a value of about 2×10^{11} sites/cell (Changeux, 1975). This would be of the order of 10^6 sites/synapse (lower than most end plates). Surface receptor densities for denervated electroplaques were similarly examined by Bourgeois *et al.* (1973). For the insulin receptor on the plasma membrane of hepatocytes, visualized by light microscope and EM autoradiography, Bergeron *et al.* (1977) used the observed grain density and approximate estimations of ^{125}I-source efficiency and cell surface area to obtain a value of the order of 1×10^5 receptors/cell.

When EM autoradiographic densities are used to make an enumeration per cell, in any system, summations have to be made over the surface of a relatively large and complex object from counts on sections through it, so that assumptions on the stereology and the zones of distribution of label are implicit. In light microscope autoradiography, where much larger areas can be covered, estimates of total area can still be seriously in error insofar as true membrane surface is concerned, and, further, the correction for the high self-absorption of ^{3}H or ^{125}I radiation in the thick specimens with nonuniform labeling is very difficult to make and varies greatly with the conditions. Fambrough (1974) found that absolute densities of ACh receptor sites in denervated muscle fibers determined autoradiographically were in error due to these factors.

The second approach overcomes these problems by using an absolute internal autoradiographic standard (Barnard *et al.*, 1971). It is limited to specimens that can be separated as single units, e.g., dispersed or cultured cells, separated muscle fibers, microdissected large neurones, etc. The reaction with DFP, which reacts irreversibly with AChE and other bound serine esterases present in the cell, is employed to saturation of those sites. By using ^{32}P in this reagent, β-track autoradiography can be performed in the light microscope. The number of tracks per cell gives the absolute number of DFP molecules reacted without assumptions as to geometry, since there is no self-absorption for these tracks (Barnard, 1970). After this calibration of the cell type to be studied, the latter is labeled for receptors, e.g., with [^{3}H]-α-BuTX, and again, separately, with [^{3}H]-DFP. In sections examined by light microscope autoradiography, the ratio of the densities of these two ^{3}H labelings is readily determined. Since the absolute number of groups introduced by DFP is already known, this gives at once the absolute number of toxin molecules bound per cell or per synapse. This method does not require the sites of DFP reaction to be identified as long as they react to saturation. It is also independent of their situation in the cell and local density as long as they are sampled in the same sections as the receptors. Representative values obtained by this method for vertebrate muscle ACh receptors are listed in Table IV. It has been found that for a given species, muscle type, and fiber size the end plate has a fixed number of ACh receptors. This

Table IV

Numbers of ACh Receptors per End Plate in a Selection of
Muscle Fiber Types[a]

Muscle[b]	Receptors per end plate (millions)
Mouse	
Sternomastoid (white fibers)	87 ($\pm$3)
Diaphragm	30 ($\pm$4)
Extraocular (slow-twitch fibers)	11[c]
Rat	
Diaphragm	53 ($\pm$3)
Sternomastoid (white fibers)	90 ($\pm$4)
Frog (*Rana pipiens*)	
Sartorius	30[c]
Chicken	
Post. latissimus dorsi	
12 weeks	36 ($\pm$2)
8 days	2 ($\pm$0.1)
Biventer (slow tonic fibers), 12 weeks	27 ($\pm$2)

[a] Determined by the ratio of grain counts from [^{3}H]-BuTX and [^{3}H]-DFP labelings and β-track counts on isolated fibers after [^{32}P]-DFP labeling. Data from Porter *et al.* (1973a), Porter and Barnard (1976), and Jedrzejczyk *et al.* (1973). Counts were made on 100–900 fields for each type of ^{3}H-labeled end plate and on at least 50 fibers for each β-track measurement, with the standard error of the mean of the determination given in parentheses. For the mouse diaphragm and the chicken posterior latissimus dorsi cases, variation in a population (10 or 16 animals) was also studied and shown, for a given mean muscle fiber diameter, to be insignificant. The other values are for 2 animals of each type.
[b] All are from adult animals, except for the chickens.
[c] Values approximate only.

number varies from 10^6 to 10^8 active centers in different types. It is greater in larger fibers of the same muscle, as during growth (Fig. 8). It is greater when the postsynaptic folding is increased. This suggests that a constant fold crest density of receptors characterizes all of these types, so that as the area of the folds increases, a larger total number is accommodated.

The ACh receptor number is unchanged in end plates of muscles affected by muscular dystrophy in the chicken (Porter and Barnard, 1976). While it changes considerably with fiber type (in normal muscle end plates), the maximum sensitivity to focally applied ACh remains constant [Albuquerque *et al.* (1974); see also the values quoted in Table 4 of Barnard *et al.* (1975)]; this is interpreted as a constant responsiveness of receptors present at a constant packing on the synaptic crests.

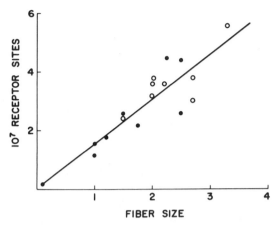

Fig. 8. Numbers of ACh receptors at individual end plates in relation to the diameter of their muscle fibers. The receptors per end plate (ordinate) were measured as α-BuTX binding sites by autoradiography and the β-track internal standard method for absolute counting. The diameter of the fibers bearing these end plates was measured (expressed in relative units). Using the fast-twitch fibers of the post. lat. dorsi *or* biceps muscles of chickens of ages ranging from 1 week to 14 months, a wide range of muscle fiber sizes was obtained. The closed circles represent normal muscles: the correlation coefficient for the fit of the least-squares line drawn is 0.85. The open circles represent the corresponding muscles from related chickens homozygous for muscular dystrophy; these points fit the same line, showing that when allowance is made for fiber size, there is no difference in the number of ACh receptors in dystrophic muscle. Each point represents the mean for 2–4 birds. The total number of fibers used from normal, or dystrophic, birds was 254. [Taken from Porter and Barnard (1976).]

4.3. Direct Determination of Receptor Occupancy Relations

The light microscope autoradiographic method is well adapted for determining the relationship between physiological response and number of occupied receptors. Changes in receptor site density, due to progressive blockade by an antagonist or to biological modulation of receptor numbers, can be monitored; the ability to do this by microscopy for the same cell and the same region as recorded electrophysiologically greatly reduces the range of error. The studies of this kind reported to date have all been with the ACh receptors at muscle end plates. Hartzell and Fambrough (1972) and Land *et al.* (1977) related the sensitivity to iontophoretically applied ACh with the observed membrane density of extrajunctional receptors. Unless the ACh applied is always saturating, this does not yield the conductance change produced per ACh-occupied receptor. It did, however, show that the Hill coefficient for the receptor–ACh binding is 1.7 (Land *et al.*, 1977).

For innervated muscles the margin of safety for transmission was determined directly by relating decreasing numbers of free receptors at an end plate (after toxin blockades) to the evoked contractile response (Barnard *et al.*, 1971). In a similar manner the degree of neurally evoked depolarization (in rat extensor muscles) was related to the fraction of the receptors available by measuring the latter autoradiographically after increasing degrees of occlusion by [^{3}H]-BuTX and recording intracellularly from the same end plates (Albuquerque *et al.*, 1973; Barnard *et al.*, 1975). Inhibition of the response to ACh was found to begin at 20–25% blockade of the receptor population and to be complete at about 75% blockade. It was thus seen that the number of spare receptors at the neuromuscular junction is very small.

5. Electron Microscope Methods for Visualization of Receptors

As noted in Sec. 2.2, the main detection devices for the EM, available for a ligand bound to its receptor, comprise radioactivity (dealt with in Sec. 3), electron density enhancement (e.g., with ferritin), and enzymatic product deposition (i.e., with peroxidase). We discuss here the last two methods as well as other, less common devices. With these two methods, most of the applications employ immunocytochemistry in some form, the use of which with receptors *in situ* will be exemplified in reviewing these techniques.

5.1. Peroxidase Cytochemistry of Receptors

5.1.1. Peroxidase Methods

The methods in use can be classified (Table V) as:

1. Direct linkage of peroxidase to a primary ligand for the receptor. For this to be feasible, such a ligand must be a protein. These cases involve, therefore, protein toxins and protein hormones.

2. Linkage of peroxidase to an antibody directed against the receptor itself. This is hardly practicable as yet, since it requires soluble, pure receptor to start the procedure, but this is an immunocytochemical technique which is becoming available in the case of the nicotinic ACh receptors.

Table V
Reaction Sequences for Use with Peroxidase and Equivalent Detection Devices [a]

Method	Step 1	Step 2	Step 3	Step 4	Step 5
1. Direct	(Ligand)-D				
2. IC, direct	(Antireceptor-Ab)-D				
3. IC, indirect	Ligand	(Antiligand-Ab)-D			
4. IC, sandwich	Ligand	Antiligand-Ab$_1$	(Anti-IgG$_1$-Ab$_2$)-D		
5. Unlabeled					
antibody a	Ligand	Antiligand-Ab$_1$	Anti-IgG$_1$-Ab$_2$	AP$_{(AD)}$[b]	P$_{(D)}$[b]
antibody b	Ligand	Antiligand-Ab$_1$	Anti-IgG$_1$-Ab$_2$	PAP	—

[a] The steps shown for each method are applied in turn. D denotes the detection device (peroxidase, ferritin) linked covalently to the entity in parentheses preceding it. IC = immunocytochemical; Ab$_1$ = antibody, raised in species 1 (rabbit), etc.; IgG$_1$ = immunoglobulin of species 1, etc.; P = peroxidase; AP = antibody to peroxidase.

[b] D denotes that while this method is shown, and usually used for peroxidase, it can be extended to other detector molecules.

3. Linkage of peroxidase to an antibody (Nakane and Pierce, 1966) against the primary ligand. This is the original principle introduced by Coons (1956) with fluorescent labeling. It has the disadvantages that the attachment of the large peroxidase appendage usually introduces some nonspecific binding and loss of binding of the antibody and that unless the derivative used is homogeneous, competition by unlabeled antibody species can seriously interfere.

4. Binding to a first antibody–ligand complex of a second labeled antibody directed against the first (e.g., goat against rabbit IgG). This is the sandwich principle of Coons (1956). It has the same disadvantages as noted for method 3, but the advantages are that it gives some amplification (through multiple binding) and that only one type of antibody (goat anti-rabbit-IgG) has to be purified and labeled, for use with all tissue antigens, and this type is available in ample quantities. Sheep or swine sera are often used as the second species in place of the goat.

5. A triple-sandwich method (see Table V) in which the third layer is an antibody directed against horseradish peroxidase, with the final addition of peroxidase (Mason *et al.*, 1969). The essential feature of this method is that no antibody need be chemically labeled. In another commonly used version of this technique, due to Sternberger [reviewed by Sternberger (1972) and Moriarty *et al.* (1973)], the last two steps are combined, in the addition of PAP (peroxidase–antiperoxidase complex) for that stage (5b, Table V; Fig. 9). PAP is used as the immunospecifically purified complex (Petrali *et al.*, 1974).

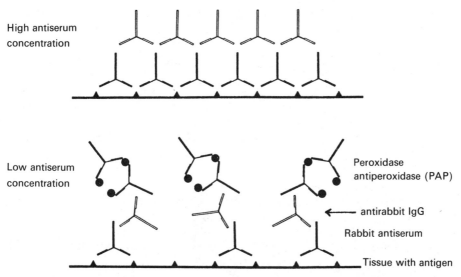

Fig. 9. Schematic representation of the PAP method (5b of Table V) and the possibility of saturation of antigen in it. In the lower diagram, the receptor sites are covered by a firmly bound primary ligand (triangles), and the sandwich is shown of rabbit antibody directed against that ligand, then heterologous antibody directed against rabbit IgG, and finally the PAP complex of peroxidase (circles) and rabbit antiperoxidase to give the detection. In the upper diagram, when the receptor sites are more abundant (e.g., when not largely destroyed or extracted by tissue processing), the antibody to the primary ligand may find sufficient ligand sites for its two valences to be satisfied, so that no free binding sites on that antibody remain available for the secondary attachment of anti-IgG. Hence, staining can be obstructed in this method if the primary ligand is too abundant. [Adapted from Bigbee *et al.* (1977).]

The general advantage of the use of peroxidase for localizing receptors is the extreme sensitivity obtained, due to the enormous multiplication by enzyme turnover. Petrali *et al.* (1974) have estimated that a peptide hormonal antigen is revealed by its antibody at 16,000–100,000 times lower concentration of the latter with a peroxidase method than with radioimmunoassay. In the EM, single molecules of an antigen on an erythrocyte membrane have been reported to be visualized by the unlabeled antibody peroxidase method as the separate points of deposit observed at very high antibody dilution (Sternberger, 1972; Petrali *et al.*, 1974). Further, the PAP complex was seen at the highest resolution to be cyclic, with a diameter of about 200 Å, recognizable for counting even in the absence of product deposition. The latter method is likely to be practicable only in favorable cases permitting very high resolution and with sufficient numbers of receptor sites present for recognition of the PAP complexes. It does, however, introduce the possibility of quantitation with the peroxidase method in those cases, if a parallel section developed for the enzymic stain, at a limiting low concentration of the initial antibody, gives numbers of sites that correspond. The observation that the number of sites of staining increased as the first antibody concentration was increased (Sternberger, 1972), rather than an increase in the degree of staining of a final number, confirms that such mutual checks are necessary, since when more stain is formed, artifactual localizations are obviously encountered. With that limiting first antibody titration method, on the other hand, there is a risk of not occupying all of the sites, especially if they vary in accessibility.

The peroxidase methods offer the advantage of sharp localization pictures at high resolution. The caution noted above (Sec. 2.2.3) on the apparently precise resolution seen applies, however, since this varies with the amount of initial antibody applied and with the extent of enzymatic reaction that is conducted. The location of PAP cyclic structures would overcome this problem (Petrali *et al.*, 1974) as long as they are not overlaid by deposited stain. The easiest use of this PAP recognition device, in practice, is to employ a dilution of the initial antibody and an extent of the staining reaction which produce sufficient stain for the regions concerned to be discerned but not too great an amount to prevent associated PAP circles being identified; stain deposits without the latter would then be ignored in defining the sites of the binding.

The special advantages of the unlabeled antibody versions of the method are the following: there is no competition between a labeled and an unlabeled antibody; the peroxidase is unhindered by a covalent attachment; and, where used, the PAP complex is prepared in immunospecifically pure form, removing a source of nonspecific reaction. Specificity in the reaction in part depends on the factors in the method that have been noted and in part upon the selectivity of the antibodies used (Petrusz *et al.*, 1976). Checks on these two aspects must be made by (1) controls with nonimmune sera or IgG, (2) trial omission of the anti-IgG (to give complete prevention of staining), (3) trial preabsorption with the antigen (likewise), (4) tests with increasing dilutions of the first antibody solution, as noted above, and (5) trial preabsorption with a tissue devoid of the antigen or omission of its application to the tissue under study.

Nonspecific adsorption of any of the antibodies used is minimized by using normal (sheep or goat) serum as their diluent (Sternberger, 1972) or 0.1 % bovine serum albumin (Petrusz, 1974). If there is some staining in the absence of the ligand for receptor but not in the absence of the unlabeled antibody, such staining is probably attributable to traces of host IgG in the tissue and may be overcome by a method using blockade by the monovalent fragment of anti-IgG (Sternberger, 1972). After the enzymatic staining, exposure to OsO_4 vapor is usually advantageous, both for further fixation and for contrast enhancement of the stain for both light and EM specimens, since the polymer deposited is osmiophilic. Otherwise, contrast of the stain in light microscopy is enhanced by use of a Kodak Wratten 46 filter. Alternatively, instead of the brown stain with diaminobenzidine, a blue color can be obtained by using benzidine or tetramethylbenzidine (Hardy and Heimer, 1977), and the latter compound has been claimed to lack the carcinogenicity that is a hazard with the other chromogens in use.

Intracellular penetration of the IgG–peroxidase conjugate (MW 190,000, or greater if higher than 1 : 1) has been reported to be poor into fixed cells (Gonatas *et al.*, 1974). After strongly cross-linking fixatives such as glutaraldehyde, inward diffusion of proteins is severely hampered, and misleading surface localizations can be produced. Gonatas *et al.* have attempted to overcome this by using the monovalent proteolytic fragment of antibody (FAB, MW 40,000). When labeled with ^{125}I this gave good penetration of plasma cells for EM autoradiography. When FAB labeled with peroxidase (which may be used with multiple substitution) was used, this was not found satisfactory by these authors but was found so by Nakane (1975) on frozen sections of fixed tissues. This situation can be improved upon by the attachment to FAB of a *microperoxidase*, i.e., a heme-containing octapeptide (for example) that possesses weak but sufficient peroxidase activity (Kraehenbuhl *et al.*, 1975). A conjugate of this peptide (ratio 3:1) with FAB, MW 50,000, employed in a secondary attachment in fixed tissue prior to embedding, was found satisfactory for light and EM peroxidase staining. These studies illustrate the importance of investigating the degree of conjugation and the penetrability of the conjugate and the likely future usefulness of antibody fragments and of microperoxidase in tracing intracellular receptors.

5.1.2. Applications of Peroxidase Cytochemistry to Receptors

So far few applications to receptors have been made of the peroxidase methods, despite their great power for displaying pictorially, and with very high sensitivity, the sites of bound components. An irreversible or pseudo-irreversible binding of the primary ligand is required, due to the aqueous processing involved, and so far only α-BuTX, as a ligand for ACh receptors, has been exploited in depth for this purpose (see Table I).

Direct conjugation of peroxidase to a protein such as α-BuTX can be made by one of two approaches. The first of these employs a cross-linking reaction between protein side chains (—NH_2 or —COOH) in each partner (Nakane and Pierce, 1966; Avrameas and Ternynck, 1971). A bifunctional reagent such as difluorodiphenyl sulfone or a carbodiimide or a diisocyanate or glutaraldehyde, etc., is used, preferably in a two-step (Avrameas and Ternynck, 1971) reaction sequence. The yield of the required conjugate is always very low, however, and some decrease in the activity of each protein may be introduced; a check must be made for this. The second, newer version links the protein to the carbohydrate portion (about 18% of the molecule, by weight) of horseradish peroxidase via periodate oxidation of hydroxyls therein, followed by Schiff base formation with NH$_2$ groups of the protein and borohydride reduction to stabilize the product (Nakane and Kawaoi, 1974). The carbohydrate-linking version minimizes self-coupling of each protein and is free of any effect on the enzymatic activity of the peroxidase. The latter method has been applied to form a 1:1 peroxidase conjugate of α-BuTX by Lentz et al. (1977) and Engel et al. (1977) and has been studied further in detail by Barnard et al. (1979). It is important to check that the binding of the toxin (or other protein ligand used) is not impeded by the attachment of the peroxidase (MW 40,000) molecule. We have observed that this does occur (resulting in a great reduction in affinity) when the linkage is protein–protein, using glutaraldehyde or toluene-diisocyanate for bridging. The carbohydrate bridge conjugate, on the other hand, has been shown to produce receptor blockade with a rate and an affinity sufficiently similar to that of native α-BuTX for cyto-chemical application (Barnard et al., 1979). It is necessary to isolate chromato-graphically the 1:1 α-BuTX–peroxidase species from the mixture formed, since the conjugates containing several toxin molecules (or 1:1 conjugates on which the borohydride reduction had been extended)* show an increased rate of dissociation for the complex. Testing for the latter phenomenon can be made by wash-out of treated nerve-muscle preparations; if present, it can seriously impair the localization, due to loss of toxin in the processing. Another problem encountered is the multiplicity of horseradish peroxidase species present in the preparations of this enzyme unless special separation methods are used (Shannon et al., 1966). The conjugates of BuTX with the three isoenzymes commonly present can be separated chromatographically; they could give rise to different affinities of the bound toxin, which could be significant if quantitation is attempted. A 1:1 α-BuTX–peroxidase complex with native toxin properties can be prepared and shown to be free of all these complications (Barnard et al., 1979).

Using α-BuTX–peroxidase conjugate on rat and human muscles, Engel et al. (1977) found that an intense stain deposition occurred only on the membrane at the crests of the end-plate postsynaptic folds, but in some folds this was also in parts of their lower regions. The presynaptic membranes were also stained, but more lightly, a phenomenon also noted as significant by Lentz et al. (1977) in mouse and amphibian end plates treated similarly. Engel et al. (1977) note that this presynaptic localization may be artifactual; this point is discussed below.

The sandwich immunocytochemical method (4 in Table V) has also been applied with α-BuTX by linking to peroxidase a goat anti-rabbit-IgG antiserum and using this after applying to toxin-treated muscles a rabbit antibody against α-BuTX. Daniels and Vogel (1975) and Bender et al. (1976) so stained mammalian end plates specifically, as seen by light microscopy, and in the EM found, as above, that staining was confined to the crests of the folds, with little in their depths.

---

*The concentration of sodium borohydride used is important in this regard. Nakane and Kawaoi (1974) specify 5 mg thereof with 5 mg of peroxidase in 3 ml. We have found that this level damages the α-BuTX–peroxidase complex, increasing greatly its dissociation rate from the receptor. The use of 0.5 mg or less of borohydride removes this problem, while still allowing full reduction of the Schiff base bonds.

Significant stain was, again, observed on the presynaptic plasma membranes, but this was attributed by Daniels and Vogel (1975) to diffusion from the nearby postsynaptic membrane. It was, in particular, noted that similar staining occurred on any other structure close to the latter, namely nerve membrane, Schwann cell processes, and synaptic cleft substance. Nuclear membranes near rat diaphragm end plates were, likewise, consistently stained (Ringel *et al.*, 1975). Frog muscle end plates, with wider and shallower folds, had stain in the lower regions thereof also. The staining at all sites was blocked by cholinergic ligands (at 10^{-3} M concentration). These pictures, and those of Engel *et al.* (1977), serve to confirm the location of ACh receptors at the crests of the folds deduced from the auto-radiographic studies with α-BuTX, discussed in Sec. 3.3. They also underline the risk of local diffusion of some of the peroxidase reaction product, e.g., to the presynaptic or nuclear membrane. The EM autoradiographic studies discussed in Sec. 3.3, especially after denervation (Porter and Barnard, 1976), have shown that presynaptic sites of α-BuTX binding do not exist above the level of the sensi-tivity of the autoradiographic technique. Bender *et al.* (1976), on the other hand, consider the presynaptic localization seen with the peroxidase–BuTX methods to be probably real. It must be borne in mind, in this regard, first that blockade by specific ligands does not reveal diffusion artifact in staining, since when the true sites are blocked there will be no source for the diffused product, which will likewise disappear. Second, the factors determining the local deposition of the enzymatic reaction product will include the rate of reaction of the substrates and hence their concentrations in the end plate, the local concentration of the bound enzyme, the rate of formation of the polymeric, the insoluble form of the final oxidized product of diaminobenzidine from its monomer, and the affinities for that product of the various surfaces in the vicinity. In ideal conditions the polymer is formed rapidly enough to be captured at the peroxidase molecule itself (Seligman *et al.*, 1973). Since, however, the reaction can continue to form additional deposits after individual enzyme sites are covered by a layer of final product (Sternberger, 1972), it is clear that the ideal condition is not necessarily met, probably due to a finite rate of polymerization of the product relative to its diffusion time to nearby membranes. Limiting extents of enzyme reaction will minimize such errors, and recognition of PAP complexes, as noted above, where feasible will avoid them.

Applications made of these methods include the demonstration of ACh receptors all along the sarcolemma of denervated (but not innervated) rat dia-phragm fibers (Ringel *et al.*, 1975), correlating with the known supersensitivity to ACh there. Fibers in a denervated state in muscles of humans with lower motor neurone disease were recognized by this method. This study employed the sandwich immunocytochemical method and a modification that avoided the frozen sections originally used by Daniels and Vogel (1975). Engel *et al.* (1977), to avoid frozen sections, used instead the direct peroxidase conjugate of α-BuTX; they used muscles from myasthenia gravis patients and found that some or most end plates had lost much of the staining, as had those from rats with receptor-induced autoimmune chronic myasthenia gravis. A similar finding, by the α-BuTX sandwich method, was made by Bender *et al.* (1976), who also used this staining method to show that some (44%) myasthenic sera could block junctional ACh receptors, while 68% could block extrajunctional ACh receptors. Since, in such antibody methods, a question must always arise as to the penetrability of the antibody molecule to the interior of a structure such as a muscle end plate, a useful study was made by Zurn and Fulpius (1976) to test this case. They showed that an IgG-α-BuTX conjugate could cover all of the end-plate ACh receptors, as evidenced by the blockade by this conjugate (but not by nonspecific IgG) of autoradiographically seen reaction of $[^{125}I]$-α-BuTX on mouse diaphragm. Likewise, $[^{125}I]$-IgG, directed against α-cobratoxin, labeled end plates that had been pretreated with that toxin.

A third peroxidase method applicable to the ACh receptor is the PAP unlabeled antibody method (5b in Table V), using antibody to α-BuTX. This has recently been shown to be satisfactory on vertebrate muscles (Lang et al., 1979). Its advantage is the absence of any chemical manipulation of the toxin or an antibody. Further, the same PAP sequence can be applied to an antibody to any toxin, so that the subcellular localization of other types of neurotoxin— e.g. β-bungarotoxin, botulinum toxin, tetanus toxin, etc.—could be feasible by making their specific antibodies. A low-molecular-weight toxin that binds pseudo-irreversibly can be attached to a macromolecular immunogen to form an antibody for its tracing after binding. Another useful application of this method is to combine the peroxidase label with an isotopic one. ^3H or ^{125}I can be introduced into the toxin, with the result that the receptors binding, for example, α-BuTX can be counted by means of autoradiography and then visualized on the same or parallel sections for more precise determination of the locations corresponding to those counts (Lang et al., 1979), combining the benefits of both these approaches.*

For protein hormone binding sites, peroxidase has been applied via unlabeled antibody methods in the case of gonadotropins (see Table I). In each case, pseudo-irreversible binding of the hormone at target cells was found, permitting formalin or glutaraldehyde fixation and aqueous treatments. Nontarget tissues such as muscle and liver gave no staining. The unlabeled antibody methods were found superior in that chemical conjugation of peroxidase to sheep anti-rabbit-IgG antiserum gave variable staining and a component of nonspecific stain (Petrusz, 1974, 1979). While binding of HCG, LH, or prolactin was demonstrated all over the cytoplasm of various ovarian cells, the initial phase of binding can be demonstrated to be at the cell membrane (Abel, 1979; Petrusz, 1979). Support for this comes from similar autoradiographic studies (Ashitaka et al., 1973; Rajaniemi and Vanha-Perttula, 1973). Subsequently the occupied sites there become internalized, and the hormone follows cytoplasmic pathways. Corroborative evidence that the membrane sites are receptors has been obtained by the binding of radioactive LH to membrane fragments, measured biochemically (when $K_D = 10^{-11}$ M), and by the blockade of the cellular uptake by the physiological antagonist prostaglandin F2α (Abel, 1979).

---

*The technique for combining dry-mount autoradiography with peroxidase cytochemistry in the light microscope has been described by Keefer et al. (1976) and for EM combined localizations by Gonatas et al. (1974); the latter authors showed that a peroxidase-labeled FAB fragment, bound specifically, could be visualized in the same EM section as the $[^{125}I]$-FAB.

5.2. Tissue Preservation for Immunocytochemistry of Receptors

In the peroxidase methods reviewed above, it is usual to apply a fixative to the tissues and to embed them in conventional media for sectioning for light or for electron microscopy in order to preserve and view cellular structure. These procedures can decrease or selectively destroy receptor visualization, either by loss of the primary binding of the ligand or by loss of antigenicity of a component in the immunocytochemical cases. Glutaraldehyde fixation, which gives superior preservation, destroys these affinities of some hormones and antibodies, probably due to its cross-linking action, but buffered formalin is gentler; a paraformaldehyde–lysine–periodate has been recommended (McLean and Nakane, 1974). Even that fixation destroys some antigenicity. Ligand binding to a receptor is best accomplished in the native state, e.g., *in vivo* or fresh topical application, and a subsequent mild fixation is preferable to prior fixation, since the complex is more resistant to denaturation. As noted in Sec. 3.3, the α-BuTX–receptor complex, once formed, is resistant to glutaraldehyde fixation. Thus, the safest method for EM study is to apply, if feasible, both the primary ligand and the peroxidase-labeling step to native tissue, followed by mild fixation, section preparation, and then enzymatic staining and postosmication. Where long exposures to aqueous media are involved in the peroxidase treatment, an initial fixation step may be needed after the primary ligand binding stage. Use of a glutaraldehyde concentration as low as 0.1% was found necessary by Vogel and Daniels (1976) to preserve α-BuTX and antibody binding, and still gave sufficient fixation of muscle cells. The same considerations apply to EM immunocytochemistry using ferritin labeling.

Peroxidase cytochemistry is often performed for light microscopy on paraffin-embedded, sectioned tissue and for the EM on sections from methacrylate, epoxy, or equivalent blocks. Positive staining at predicted sites is achieved after such treatments in most cases, but it is likely that the embedding and the fixation processes in fact remove most antigenic sites, so that one is dealing with a small residual fraction of sites. Since the method is generally nonquantitative and extremely sensitive, this situation will usually not be recognized, but it incurs severe risks of selective losses. When frozen or vibratome sections are used instead, despite poorer preservation these losses are avoided and much greater reaction can be discovered. Thus, Bigbee *et al.* (1977) found that an antigen that had been positive in the PAP reaction in embedded sections ceased to stain, paradoxically, at higher concentrations of the initial antibody applied to frozen sections, but was highly positive at very low concentrations. This is attributable to much greater numbers of antigenic sites in a frozen section, binding more primary antibody and therefore giving bivalent, rather than monovalent, attachment of the antiperoxidase antibody (Fig. 9). Both the bridge and the PAP forms of the unlabeled antibody method are susceptible to this problem if the receptor–ligand antigenic sites occur in clusters, so that the immunocytochemical series cannot proceed to completion. At very low antibody concentration this effect will be overcome, but some sites may be missed. High dilutions of antibody and a broad dilution range should therefore be tested when localizing a receptor by this method. The labeled antibody and direct peroxidase methods are free of this problem.

Where the peroxidase is introduced in the first stages, as in the direct application of a peroxidase-linked ligand *in vivo* or to intact cells, subsequent fixation may depress its enzymatic activity. This occurs with the usually used paraformaldehyde fixative (Malmgren and Olsson, 1977); 1.5% glutaraldehyde (4 hr, 4°) gives minimal losses of peroxidase activity. However, it should be checked, wherever possible by biochemical parallel reactions, that the fixation treatment used does not reduce any of the bindings involved in the cytochemical demonstration. Where fixation must be used, for preservation of structure, before peroxidase–antibody is introduced, the problems of penetrability noted above (Sec. 5.1.1) must also be considered.

5.3. Ferritin Labeling

Ferritin has frequently been attached to antibodies (Singer and Schick, 1961) for EM visualization of various antigens, and the same techniques have been applied in a few cases to receptor localization. The methods developed for antibodies employ glutaraldehyde or some similar cross-linking reagent (Siess *et al.*, 1971). For receptors one could use, again, either the direct attachment of ferritin to a protein ligand, or its attachment to an antiligand antibody for use in one of the methods 2–4 of Table V, or antiferritin antibody (D = ferritin in method 5). The preparation of active IgG–ferritin conjugates is an established procedure, so it can usually be adapted where a ligand is bound pseudo-irreversibly. In contrast, the preparation of a new conjugate of ferritin with a protein requires a full study of the separation of the products and of their affinities. A difficulty often encountered is that free ferritin is not readily removed from its conjugate, since the size difference is too small for resolution on a gel filtration column, the usually used procedure. Ferritin–ferritin conjugates are also formed and not separated. Risks of confusion thus arise due to nonspecific sites of adsorption of ferritin in the tissue. Additional checks are necessary by, e.g., agar block precipitin titration, which should also demonstrate that ferritin is confined to the precipitin bands, and by immunoferritin assay in the EM where feasible. It is necessary to apply to the tissue, as a control, free ferritin in the same manner as the conjugate to exclude simple binding of the former, which is known to occur in some special cases. Some reported localizations by ferritin conjugates have shown results, in the absence of this test, that are highly ambiguous. Orth and Christensen (1977) have pointed out that a reported distribution of injected ferritin-labeled FSH in the vesicles of rat Sertoli cells is called into question by the similar distribution seen after the injection of ferritin alone. The linkage of ferritin to a protein ligand by glutaraldehyde is generally via a Schiff base, and this can be unstable; reduction to stabilize it can cause loss of the biological potency of the ligand. Cuatrecasas and Hollenberg (1976) noted both of these problems with the conjugate with insulin as a receptor marker. The use of a spacer group and of a two-step conjugation method, instead, for ferritin is less prone to cause inactivation, especially of larger proteins, but each case must be carefully investigated.

An unlabeled antibody method, equivalent to that used for peroxidase but using antiferritin and heterologous anti-IgG antibodies, would be of value in some receptor cases but has thus far not been applied to them. A general feature of the ferritin methods is that the sites revealed must be readily accessible to this very large molecule. For receptor sites on membranes, this will usually not be a problem. Protein A (which binds to IgG specifically), conjugated to ferritin, should be more satisfactory than the IgG–ferritin conjugate (Templeton *et al.*, 1978) for use with receptors in methods 2 or 4.

For EM localization of ACh receptors in cultured muscle cells, α-BuTX was conjugated to ferritin with glutaraldehyde by Hourani *et al.* (1974). A gel-filtered peak of conjugate blocked ACh depolarization, but at 6–10 times higher concentration than with α-BuTX. High salt concentration was needed to separate some native α-BuTX adsorbed to ferritin, which otherwise strongly interfered, a cautionary note for this technique. Nonspecific binding was shown to be absent by the prevention of labeling in cultures pretreated with free α-BuTX. However, the specific labeling seen was very variable, with 80% of cell surfaces examined being unlabeled. In the author's laboratory, purified ferritin–α-BuTX conjugates formed by any of several coupling procedures gave only highly reversible binding to ACh receptors in muscles or in solution. It may be difficult to attach this enormous appendage to the small toxin molecule without reducing greatly the stability of its receptor complex. Such reversibility in a ferritin complex would rule it out for reliable microscopic studies. Methods based on the unlabeled antibody principle, *or* using a ferritin-labeled heterologous antibody in method 4, seem more promising for α-BuTX. When conjugated to ferritin for receptor visualizations, α-MSH is partly active, with receptor complex half-life of 1 hr (DiPasquale *et al.*, 1978), but a 9-residue releasing hormone retains full activity (Hopkins and Gregory, 1977).

Ferritin-labeled anti-IgG for use in the sandwich method (4 in Table V) has been successfully applied for location of other cell surface components, e.g., those binding lectins. Thus, oligosaccharide receptors of slime mold cells that are specific for certain cell-recognition lectins of those cells have been studied by application of the antilectin rabbit antibodies (Chang *et al.*, 1975); lectin on the cell surface was then located by the ferritin-labelled goat anti-rabbit-IgG. Similar lectins, perhaps mediating cell interactions, can be demonstrated on the surface of developing muscle cells (Teichberg *et al.*, 1975; Nowak *et al.*, 1976). A direct ferritin conjugate of concanavalin A retains its lectin activity and has frequently been used for EM demonstration of surface hexose-containing structures. For example, this method can be used to show in the EM the surface diffusion of such lectin "receptors" into interfaces where cell agglutination is occurring due to cross-linking by the applied lectin (Singer, 1976). In all of these studies the lectin affinities and the antibody bindings persisted after initial glutaraldehyde fixation and final EM embedding procedures, as is usually found in the application of ferritin (although this resistance has only been judged qualitatively).

5.4. *Other Labels Applicable for Transmission and Scanning EM Studies of Receptors*

It has already been noted that, besides ferritin and the enzymatic type of marker, it is possible to link to suitable proteins, including antibodies, several other types of EM markers containing heavy metals (Sec. 2.2.2). These have not, as yet, been applied to receptor ligand labeling, by direct or antibody methods. Another possible approach would be the recognition of receptor-rich regions in the scanning EM. While the resolution is lower than in the transmission EM, the ability of the scanning EM to survey intact surfaces in three dimensions can be of special value for some receptor studies, which are largely of cell surface phenomena. Large clusters or arrays of receptors could be visualized and measured if recognized by some bound specific marker. Ferritin and hemocyanine, attached through ligands, can be used for this purpose, as has been shown with labeled lectins on cell membranes, e.g., by Braten (1976). Tobacco mosaic virus is a large distinctive marker in the scanning EM and can be attached to an antibody for recognizing therein a cell surface component (Nemanic et al., 1974). PAP complexes, if clustered, may also be so recognized. Even fluorescent antibody can be recognized, by a technique as yet little developed, in the scanning EM (Soni et al., 1975; Hough et al., 1976).

Synaptic surfaces, in particular, can yield useful information at the level of the scanning EM. The topography of receptor-rich zones on them could be mapped in three dimensions if such labels for this technique can be applied. This has not as yet been reported, but methods for the direct viewing of those surfaces in the scanning EM have been developed, as illustrated in Fig. 10. The synaptic surfaces pose obvious problems of accessibility for viewing; Shotton et al. (1979) have shown that in the case of the vertebrate motor end plate, direct viewing of the entire postsynaptic membrane surface is feasible in the scanning EM after collagenase digestion of fresh muscle to remove the cohesive connective tissue overlay. A specimen such as that seen in Fig. 10 is well adapted to complete receptor mapping by labeling techniques, either in the scanning EM or in transmission EM viewing of surface replicas suitably prelabeled. In neural tissue, presynaptic surfaces might be viewed similarly by the use of synaptosomes.

Where receptors are particularly dense on a membrane and are already identified, ligand-independent applications of electron microscopy can be made for determining their precise arrangement and number. A known case is the image reconstruction technique for ordered, unstained EM specimens, as developed by Unwin and Henderson (1975) for the case of the bacterial rhodopsin arrays, which, in conjunction with membrane X-ray diffraction, revealed the detailed geometry of the receptors in that membrane. X-ray diffraction and negative staining can be applied to studies of particle distributions in cases such as the ACh receptors highly concentrated on the synaptic membranes of *Torpedo* electroplaques [reviewed by Cohen and Changeux (1975)]. Membranes containing arrays of receptors less extensive than those of the bacterial purple membrane can still be analyzed for detailed structure by negative staining, optical diffraction of the EM micrographs, and image reconstruction based thereon. This has been shown to be possible in the case of *Torpedo* and *Narcine* electroplaque membrane fragments by Conti-Tronconi et al. (1978) and Ross et al. (1977). The latter authors, by this method and by X-ray diffraction of the membranes, deduced that the ACh receptors are transmembrane structures of length 110 Å, projecting 55 Å above the bilayer, and are packed in a hexagonal (oblique) lattice. These are, however, special cases of previously identified receptor arrays favorable for these techniques. Transmission EM studies at very high resolution have likewise

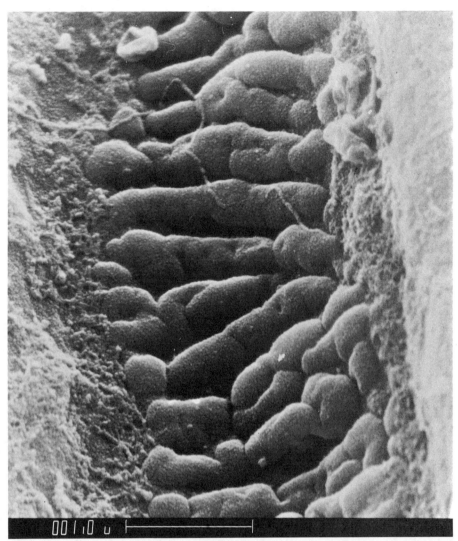

Fig. 10. Direct view of the crests of the postsynaptic folds. Frog cutaneous pectoris muscle was treated with collagenase (24 hr) to remove connective tissue and allow the nerve terminals to be retracted, followed by specimen preparation (with osmication but without metal coating) for scanning electron microscopy of the exposed postsynaptic surface. In frog fast-twitch muscle the folds are often well aligned, as seen here (cf. Fig. 5 for a cross section of such a zone), although often, too, more irregular, as in the mammalian case. The regions facing upwards here are the receptor-packed crests. [Photograph supplied by Dr. D. Shotton (Imperial College).]

revealed an asymmetric morphology and, very approximately, the numbers of arrayed particles believed to be ACh receptors at neuromuscular junctions of the earthworm and the frog (Rosenbluth, 1972, 1974), but a direct identification has not been made by a marker method.

On the other hand, the freeze-fracture and freeze-etch techniques (Da Silva and Branton, 1970), revealing intramembraneous particles in the transmission

EM, could be combined with specific chemical or immunochemical markers to identify receptors in those particles. Ferritin is frequently applied as a marker in freeze-etched specimens in other studies (Da Silva *et al.*, 1973), often being used conjugated to an antibody. Insulin receptors have been visualized on the plasma membranes of liver and fat cells by means of ferritin (Jarett and Smith, 1975); ferritin linked to insulin via a dextran spacer became bound to the cell membranes and was recognized in freeze-etched membrane replicas. Ferritin molecules were seen therein at $90/\mu m^2$ (either dispersed or in patches), comparable to the 25–60 insulin sites/μm^2 calculated (assuming a plasma membrane location) from biochemical uptake to liver cells. Autoradiography, also, can be combined with freeze-fracture, as shown by Fisher and Branton (1976); the freeze-fractured, shadowed replicas are coated while frozen with a monolayer of nuclear emulsion, exposed at $-80°$, thawed and photographically processed, and mounted on grids for EM examination. Applications of this highly promising approach to receptor identification have not yet been reported. Its advantages would be that a known chemical label is directly related to observable particles in the membrane and that the continuously frozen state avoids diffusion in fixation and embedding procedures for conventional transmission EM autoradiography.

In summary, the various methods described earlier for labeling receptor ligands, or antibodies to them, can, in fact, be expected to be capable of important extensions, by combination with freeze-fracture, to reveal the particles containing the receptor sites. Such a development will necessarily provide more precise locations and numbers of these receptors.

6. Fluorescence Marker Methods

For many topographical types of study of receptor distribution, immediate visualization in the light microscope by means of introduced fluorescence is more convenient than other methods. The disadvantages are the low resolution and the lack of quantitation (discussed in Sec. 7).

6.1. Fluorescence Labeling

Reagents containing fluorescein or tetramethyl-rhodamine (TMR) have been commonly in use. The rhodamine dyes have the advantage of less spectral overlap of their emission with the autofluorescence of cells, which is frequently considerable, especially with unfixed or lightly fixed specimens. The various alternative sequences for labeling described for peroxidase (Sec. 5.1.1) are applicable, also, to a fluorochrome, except for method 5. The same remarks on their respective advantages apply.

One useful principle that remains to be fully explored is the combination of autoradiography with either fluorescence or peroxidase labeling. If both an isotope and the second label are introduced on the same ligand or the two methods are independently directed to the same target, quantitation can be combined with optical or (for peroxidase) EM visualization. Thus, Keefer *et al.* (1975, 1976) localized different cell types in the pituitary that bound both [^{3}H]estradiol *and* an antibody to one of four pituitary hormones (using the peroxidase bridge method). In another approach, α-BuTX labeled both with ^{3}H and via a peroxidase bridge has been found to be useful for quantitation of receptors viewed at high resolution.

A unique feature of the fluorescent label for receptors is the ability to bleach it in a selected zone on a cell by use of a laser microbeam. This is the basis of the fluorescence photobleaching recovery technique (Peters *et al.*, 1974; Axelrod *et al.*, 1976). The lateral diffusion of the labeled complex in the membrane can be measured by the rate of reappearance of the fluorescence. This was shown in the case of fluorescent α-BuTX on rat myotubes in culture by Axelrod *et al.* (1976) [Figs. 11(B) and (C)]. The recovery technique permitted the distinction to be made between the diffuse ACh receptors throughout the membrane area, which are mobile (diffusion coefficient 1.6×10^{-10} cm^2 sec^{-1} at 35°, the same or greater than that for lectin receptors in this and other membranes), and the ACh receptors in patches, which are largely immobile (see Sec. 8). This technique would also permit the diffusibility of other receptors tagged by fluorescence to be followed in the membrane.

An attempt was made in that study (Axelrod *et al.*, 1976) to count the sites by fluorescence microspectrophotometry using a solution standard. This gave a density in 5–9 day rat myotubes of 1500 ± 500 toxin sites/μm^2 of membrane (excluding clusters), which is considerably greater than the 200–600/μm^2 found (Land *et al.*, 1977) for most of these cells by light microscope autoradiography (and 110–260 by EM autoradiography) and the revised (Fambrough, 1974) autoradiographic values for rat and chick myotubes of Hartzell and Fambrough (1973). Likewise, L6 myotubes have mostly 15–200 sites/μm^2 (Land *et al.*, 1977), but no fluorescence was detectable by the same rhodamine–BuTX method (Axelrod *et al.*, 1976). These comparisons suggest that the quantitation of fluorescence is still open to large errors.

6.2. Application of Fluorescent Labeling to Receptors

Nicotinic ACh receptors in muscles, cell cultures, or electroplaques have been labeled by α-neurotoxins conjugated with a fluorochrome [Table I; Fig. 11(A)]. Such a conjugation with BuTX was described in detail by Suszkiw and Ichiki (1976), but a mixture of products was obtained, and only one of the fluorescein-toxin derivatives separated was considered suitable for application. The latter species, however, had $K_D = 4.4 \times 10^{-7}$M with electroplaque membranes and showed measurable reversibility. This illustrates the difficulties of such conjugation, this affinity probably being too low for reliable marking of all receptor sites. Anderson and Cohen (1974) used similar conjugates with either fluorescein or TMR, unfractionated, and these have potencies of 1–3% or 5–10%, respectively, of the native α-BuTX (Anderson *et al.*, 1977). Recently, a different preparation of TMR–α-BuTX conjugates has been described (Ravdin and Axelrod, 1977); one of these, monolabeled, gave better labeling of electroplaque membranes than the others. This was the derivative applied to cultured muscle cells in the photobleaching recovery experiments (Axelrod *et al.*, 1976) discussed above. It has also recently [see Fig. 11(A)] been found highly suitable for ACh receptor tracing and mobility studies in vertebrates: the pure mono-TMR–α-BuTX has a reaction rate with ACh receptor ~20% that of native α-BuTX but has sufficient affinity to be

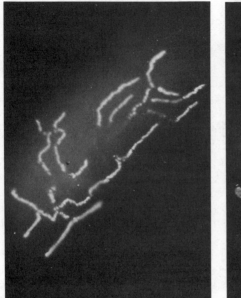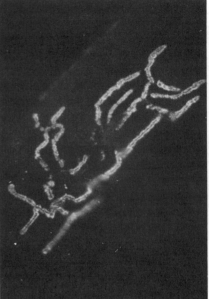

Fig. 11A. Illustration of fluorescent labeling of ACh receptors at the end plate. On the *left*, a frog sartorius muscle is seen after labeling by TMR–α-BuTX using a homogeneous 2:1 conjugate which has about 20% of the association rate of native α-BuTX with the receptor but is as irreversible as the native toxin (data of A. V. Le and E. A. Barnard). A single end plate on a muscle fiber is seen. The bright elongated zones represent the labeling along the fold crests; the periodicity just discernible corresponds to the spacing of the secondary folds, which are at about 1-μm intervals in these end plates. On the *right*, the same end plate was stained cytochemically for acetylcholinesterase, using an acetylthiocholine–copper method; the reaction product is viewed in dark-field illumination. Note that the reaction product occupies exactly the same zones as the receptors but over wider limits. This may be due to the location of the enzyme in the basement lamina of the cleft or to product diffusion before capture.

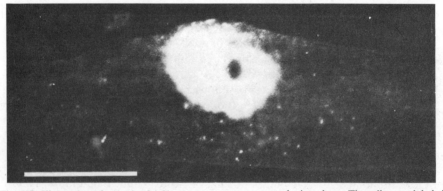

Fig. 11B. Illustration of a patch of ACh receptors on a rat myotube in culture. The cells were labeled by TMR–α-BuTX, showing patches of receptors (about 1 patch per 400 μm of tube length, on average); one of these is shown on the faintly labeled diffuse background on the whole membrane. The dark central spot was induced by a focused laser bleach (30 min prior to photography). Such a bleached area remains distinct, due to the immobility of the receptors in the clustered state. Bar 25 μm. [From Axelrod *et al.* (1976).]

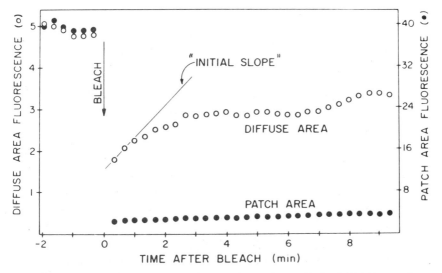

Fig. 11C. Measurements of fluorescence intensities before and after this local bleaching on the same myotube are shown in the lower diagram (note the different scales for the two zones). After the bleaching flash, the dark spot produced disappeared if in a diffuse area, but if in a patch, it persisted (as here). The average effective diffusion coefficient was so measured for each area; for diffuse areas on unbranched myotubes (at $35°$) it was 1.6×10^{-10} cm^2 sec^{-1}, whereas for their patches it was $<10^{-12}$. [From Axelrod *et al.* (1976).]

pseudo-irreversibly bound. Fluorescent toxin labeling has been used to demonstrate the receptor-rich patches that occur on embryonic cells in culture [Fig. 11(B)], as discussed in Sec. 8. In such an application (Anderson *et al.*, 1977), the specificity of the labeling seen was shown by qualitative observation of its prevention by pretreatment with active α-BuTX (1 μg/ml, 20 min) or cotreatment with *d*-tubocurarine or carbachol (100 μg/ml).

The sandwich immunocytochemical method (Table V) has been used with a fluorochrome-conjugated heterologous antibody in several types of qualitative, low-resolution location of receptors. Bourgeois *et al.* (1971) used this technique originally with an α-cobratoxin as antigen to show fluorescence on the innervated side of the electroplaque only. Other applications of fluorescent antibodies to receptor demonstration have been made using hormones as the primary ligand for various hormone receptors (Table I). Direct fluorescent conjugates of cholera toxin have also been used to visualize its receptors on lymphocytes (Craig and Cuatrecasas, 1975).

7. Possibilities of Quantitation of Receptors in Immunocytochemical and Other Nonradioisotopic Techniques

Quantitation is of great importance in the tracing of receptors microscopically. Subjective assessments of labeling as "high" or "low" with any method of visualization in the light or electron microscopes can be treacherous, especially for receptors, which are usually confined to the narrow zone of a membrane.

Two basic types of difficulty are involved:

1. Technical difficulties in performing actual measurements. Thus, photometry or densitometry of images produced by, e.g., peroxidase-deposited stain or fluorescence emission, will usually not be based on a known or regular relationship to the amount of product responsible in the case of a tissue specimen. This is due, first, to serious systematic errors, such as *distributional error* [reviewed by Pearse (1972)], inherent in the cytological case. Second, there are errors due to saturation effects produced by the overlap of dense centers, or (for fluorescence) to quenching due to the microenvironment, and a number of other optical problems.

2. Departures from a stoichiometric relationship between the amount of product or label bound and the number of receptors. These errors can become very serious in direct labeling methods, and they can vary between receptors at different sites, so that relative measurements are then meaningless. The causes of such departures are considered in Sec. 7.1.

For autoradiographic tracing, the first class of difficulty does not arise, since nuclear disintegration is truly independent of the environment, and, with the precautions discussed for this technique in Sec. 3, the second difficulty can also be avoided in most cases. Quantitation of receptors can usually be meaningfully achieved with autoradiography.

For the other methods, the first class of difficulty can be minimized in certain cases by an optical scanning device such that a very small probe area systematically scans the optical field, feeding an integrating microspectrophotometer or densitometer (Pearse, 1972; Ploem, 1975). This removes distributional error (but not any errors due to opaque deposits, overlapping granules, etc.). Computerized systems for correcting for noise of various origins can make the instrumentation available sophisticated. No case where these instruments can be applied to the measurement of a marker for receptors on cells has yet been reported.

For fluorescent light emission, distributional error does not arise, and improved instrumentation [reviewed in Hijmans and Schaeffer (1975)], with scanning, can solve some of the other difficult technical problems of microspectrofluorimetry. A number of attempts have been made recently to quantitate cytochemical fluorescence using as optical standards fluorescently labeled microbeads or chicken red cells [see the papers by various authors in Hijmans and Schaeffer (1975)]. However, the systems used are only applicable to dispersed, flattened cells. Other factors will greatly complicate the amount of light emission from a receptor-bound fluorochrome, such as the degree of fluorescent quenching and the concentration dependence of the quantum efficiency. Quenching of fluorescence is likely to occur when the fluorescent probe becomes bound and immobilized, and the quantum efficiency can vary with the local density of the fluorochrome groups; these effects will be of unknown magnitude and will vary between different receptor locations in the same specimen, being highly dependent on the environment. In tissue sections, quantitations developed on dispersed systems cannot be assumed to hold. For all of these reasons, quantitation of fluorescently labeled receptors on cells, with the possible exception of a few especially favorable cases, is not at present attainable.

7.1. Quantitation in Electron-Dense Marker Techniques

With the peroxidase or ferritin labels, some attempts at quantitation have been made. In some cases, comparison can be made very approximately by the highest dilutions of antiserum giving significant counts of particles. In a few favorable cases, counting by the more direct methods described above, such as by counting PAP complexes or ferritin particles, may be feasible. There are reasons to think that, in general, quantitation of receptors by these methods will be misleading or very difficult to achieve:

1. Where, as is more usual, an antibody is used, one does not know the stoichiometry of its binding and whether this is uniform over all the receptors. Antibody is bivalent, and this in itself can give rise to different levels of the marker becoming bound at different site densities (see Fig. 9). Anti-IgG antibody is not directed against a single determinant in IgG, and more than one molecule of antibody can be bound to an IgG molecule as antigen, giving a further variable.

2. For antibody-based methods, it must be assumed for quantitation that all of the types of antigenic site involved resist fully the cytological processing treatments required for examination in the microscope and that no loss of any of the bound antibodies (which in many cases have a dissociability that is not negligible) by wash-out occurs. It will very rarely be possible to guarantee that these conditions hold. Thus, when fixed, embedded, sectional material is used, and if a comparison with nonembedded material is feasible, it is found that in the latter usually a much greater dilution of the antibody solution is needed to obtain detectable staining (see Sec. 5.2). Generally, only a very small percentage of the binding sites for the antibody remain, but the method is so sensitive that these are visualized. Since at different locations the inactivation is likely to vary, serious errors are incurred when such processing is employed.

3. For peroxidase, limiting low antibody concentrations are needed to permit counting (Sec. 5.1), and it is then uncertain that all of the sites are occupied.

4. If the receptor binding sites are close together, it may not be sterically possible to build up the marker assembly on each site. The sandwich and bridge methods with peroxidase or the binding of ferritin by any method will each give a complex larger than 15 nm in diameter for each marker molecule bound. This is a minimum value, but it gives a maximum density—if the labeling molecules *can* be introduced sufficiently to become close-packed—of 4800 receptors/μm^2 that can be so labeled. Some receptor binding sites will be much denser than this. All receptors at vertebrate end plates have 25,000 sites/μm^2 and at electroplaques up to 50,000 (Sec. 2.3). Counting these by means of either the peroxidase or the ferritin method, even if one could perform this counting when the groups are close-packed, would give a grossly misleading value.

5. For peroxidase, even if the sites of deposit of reaction product are counted when these are at their smallest, each still occupies a sphere of about 40 nm in diameter. The receptors must be much farther apart than this for the sites to be counted separately, so that the upper limit for counting will be set at below 600 binding sites/μm^2. With the PAP method of counting, the upper limit will not be more than about twice this value. Both the direct binding of peroxidase and the unlabeled antibody methods would be subject to these limitations.

We see, therefore, that receptors which are at all dense in their surface concentration cannot be counted by these methods. For those cases where the receptors are sparse, the steric obstructions will be absent, but factors 1–3 will still need very careful investigation before counting can be meaningful.

8. Conclusions

The tracing and counting of receptors by microscopy of various forms is an approach that is complementary to, and not competitive with, the study of receptor binding to cell suspensions or membrane fractions. While the latter type of study is needed for determining specificity and quantitative affinities, the *in situ* approach can yield a wealth of additional information, as outlined above, that will be needed for an understanding of receptor function. For tissues, resolution at this level is essential. In the central nervous system, the tracing of receptors, which has only just begun, offers a path through its complexity and can be expected to become a primary tool. It will be important to relate there the localization of transmitter stores (by specific immunocytochemistry) to the labeled receptor sites; a start has been made in the case of opiate-like peptide receptors (Sec. 4.1).

The detailed differentiation of a cell membrane with respect to its zones of concentration of a receptor, exemplified by the data obtained on the crests of the muscle synaptic folds (Sec. 3.4), can only be appreciated by means of microscopic methods, and this is an application that will undoubtedly repay much further extension. One example of the knowledge of receptors now being so gained in other systems is the study of *receptor clusters* in membranes not making synapses. It has been found by visualizations in both the light and electron microscope that nonsynaptic receptors of diverse types are usually not uniformly distributed over the membrane containing them, but form clusters, sometimes over one region of the cell and in others interspersed over all of the surface. This was shown for adrenergic receptors by a fluorescent antagonist (Atlas and Levitzki, 1978), among other types. But the most detailed description of this phenomenon has been provided by the use of labeled α-neurotoxins for the extrajunctional ACh receptors of cultured or embryonic or denervated muscle cells. These receptors have been shown by fluorescent [Fig. 11(B)] or autoradiographic microscopy to occur in irregularly spaced and discrete clusters (Sytkowski *et al.*, 1973; Prives *et al.*, 1976; Vogel and Daniels, 1976; Anderson *et al.*, 1977; Ko *et al.*, 1977). As noted in Sec. 6.1, fluorescent toxin labeling and photobleaching recovery have shown that there are two populations of receptors on such cells, namely those in such clusters, which are largely immobile, and the rest, which are ubiquitous, diffuse, and mobile (Axelrod *et al.*, 1976). The density of the latter population of receptors is, by autoradiography, about 4% that in the clusters (Anderson *et al.*, 1977; Ko *et al.*, 1977). Correlation between the clusters and sensitivity to ACh can be shown approximately (Hartzell and Fambrough, 1973). However, while primary cultures of rat myotubes exhibit the clusters, cloned L6 myotubes show an apparently uniform receptor density by light microscope autoradiography (Land *et al.*, 1977), although they, too, can be innervated in culture. Innervation causes a high accumulation of receptors under the nerve-muscle contact (Anderson *et al.*, 1977) by redistribution (Anderson and Cohen, 1977).

The observed receptor movement is similar to that seen in modulation of binding sites on cells for lectins and immunoglobulins.

Insight into the function of a receptor will assuredly be aided by the mapping of its distribution in as detailed a manner as possible and as quantitatively as possible and by the development of methods for determining its movements in the membrane.

ACKNOWLEDGMENTS. This chapter was prepared while the author's work was supported by a Programme Grant from the Medical Research Council, U.K.

References

Abel, J. H., 1979, Sites of binding and metabolism of gonadotropic hormones in the mammalian ovary, in: *Drugs, Hormones and Membranes* (R. W. Straub, ed.), Raven Press, New York, in press.

Albuquerque, E. X., Barnard, E. A., Jansson, S.-E., and Wieckowski, J., 1973, Occupancy of the cholinergic receptors in relation to changes in the endplate potential, *Life Sci.* **12**:545.

Albuquerque, E. X., Barnard, E. A., Porter, C. W., and Warnick, J. E., 1974, The density of acetylcholine receptors and their sensitivity in the postsynaptic membrane of muscle endplates, *Proc. Natl. Acad. Sci. USA* **71**:2818

Alexander, R. W., Davis, J. N., and Lefkowitz, R. J., 1975, Direct identification and characterization of β-adrenergic receptors in rat brain, *Nature (London)* **358**:437.

Anderson, M. J., and Cohen, M. W., 1974, Fluorescent staining of acetylcholine receptors in vertebrate skeletal muscle, *J. Physiol. (London)* **237**:385.

Anderson, M. J., and Cohen, M. W., 1977, Nerve-induced and spontaneous redistribution of acetylcholine receptors in cultured muscle cells, *J. Physiol. (London)* **268**:757.

Anderson, M. J., Cohen, M. W., and Zorychta, E., 1977, Effects of innervation on the distribution of acetylcholine receptors of cultured muscle cells, *J. Physiol. (London)* **268**:731.

Appleton, T. C., 1966, Resolving power, sensitivity and latent image fading of soluble-compound autoradiographs, *J. Histochem. Cytochem.* **14**:414.

Ashitaka, Y., Tsong, Y. Y., and Koide, S. S., 1973, Distribution of tritiated human chorionic gonadotropin in superovulated rat ovary, *Proc. Soc. Exp. Biol. Med.* **142**:395.

Atlas, D., and Levitzki, A., 1978, Fluorescent visualization of β-adrenergic receptors on cell surfaces, *FEBS Lett.* **85**:158.

Atlas, D., and Segal, M., 1977, Simultaneous visualization of noradrenergic fibres and β-adrenoreceptors in pre- and postsynaptic regions in rat brain, *Brain Res.* **135**:347.

Atlas, D., Steer, M. L. and Levitzki, A., 1976, Affinity label for β-adrenergic receptor in turkey erythrocytes, *Proc. Natl. Acad. Sci. USA* **73**:1921.

Atlas, D., Hanski, E., and Levitzki, A., 1977, Eighty thousand β-adrenoreceptors in a single cell, *Nature (London)* **268**:144.

Atweh, S. F., and Kuhar, M. J., 1977, Autoradiographic localization of opiate receptors in rat brain. III. The telencephalon, *Brain Res.* **134**:393.

Avrameas, S., and Ternynck, T., 1971, Peroxidase-labelled antibody and Fab conjugates with enhanced intracellular penetration, *Immunochemistry* **8**:1175.

Axelrod, D., Ravdin, P., Koppel, D. E., Schlessinger, J., Webb, W. W., Elson, E. L., and Podleski, T. R., 1976, Lateral motion of fluorescently labeled acetylcholine receptor in membranes of developing muscle fibers, *Proc. Natl. Acad. Sci. USA* **73**:4594.

Barnard, E. A., 1970, Location and measurement of enzymes in single cells by isotopic methods. I, *Int. Rev. Cytol.* **29**:213.

Barnard, E. A., Wieckowski, T., and Chiu, T. H., 1971, Cholinergic receptor molecules and cholinesterase molecules at mouse skeletal muscle junctions, *Nature (London)* **234**:207.

Barnard, E. A., Dolly, J. O., Porter, C. W., and Albuquerque, E. X., 1975, The acetylcholine receptor and the ionic conductance modulation system of skeletal muscle, *Exp. Neurol.* **48**:1.

Barnard, E. A., Coates, V., Dolly, J. O., and Mallick, B., 1977, Binding of α-bungarotoxin and cholinergic ligands to acetylcholine receptors in the membrane of skeletal muscle, *Cell Biol. Int. Rep.* **1**:99.

Barnard, E. A., Dolly, J. O., Lang, B., Lo, M., and Shorr, R. G., 1979, The application of specifically-acting toxins to the detection of functional components common to peripheral and central synapses, in: *Advances in Cytopharmacology*, Vol. 3 (B. Ceccarelli and F. Clementi, eds.), Raven Press, New York, in press.

Barofsky, A. L., Lipner, H. J., and Armstrong, D. T., 1971, Localization of radioactivity in immature rat ovaries following physiological doses of ^{125}I-labelled bovine luteinizing hormone, *Proc. Soc. Exp. Biol. Med.* **138**:1062.

Bender, A. N., Ringel, S. P., Engel, W. K., Vogel, Z., and Daniels, M. P., 1976, Immunoperoxidase localization of α-bungarotoxin, *Ann. NY Acad. Sci.* **274**:20.

Bergeron, J. J. M., Levine, G., Sikstrom, R., O'Shaughnessy, D., Kopriwa, B., Nadler, N. J., and Posner, B. I., 1977, Polypeptide hormone binding sites *in vivo*: Initial localization of ^{125}I-labeled insulin to hepatocyte plasmalemma as visualized by electron microscope radioautography, *Proc. Natl. Acad. Sci. USA* **74**:5051.

Bevan, S., and Steinbach, J. H., 1977, The distribution of α-bungarotoxin binding sites in mammalian skeletal muscle developing *in vivo*, *J. Physiol. (London)* **267**:195.

Bigbee, J. W., Kosek, J. C., and Eng, L. E., 1977, Effects of primary antiserum dilution on staining "antigen-rich" tissues with the peroxidase–antiperoxidase technique, *J. Histochem. Cytochem.* **25**:443.

Blackett, N. M., and Parry, D. M., 1973, A new method for analysing electron microscope auto-radiographs using hypothetical grain distributions, *J. Cell. Biol.* **57**:9.

Blackett, N. M., and Parry, D. M., 1977, A simplified method of "hypothetical grain" analysis of electron microscope autoradiographs, *J. Histochem. Cytochem.* **25**:206.

Bourgeois, J. P., Tsuji, S., Boquet, P., Pillot, J., Ryter, A., and Changeux, J. P., 1971, Localization of the cholinergic receptor protein by immunofluorescence in eel electroplax, *FEBS Lett.* **16**:92.

Bourgeois, J. P., Ryter, A., Menez, A., Fromageot, P. Boquet, P., and Changeux, J. P., 1972, Localization of the cholinergic receptor protein in *Electrophorus* electroplax by high resolution auto-radiography, *FEBS Lett.* **25**:127.

Bourgeois, J. P., Popot. J. L., Ryter. A., and Changeux, J. P., 1973, Consequences of denervation on the distribution of the cholinergic (nicotinic) receptor sites from *Electrophorus electricus* revealed by high resolution autoradiography, *Brain Res.* **62**:557.

Braten, T., 1976, Detection of concanavalin A sites on the surface of algal gametes and zygotes with the scanning electron microscope, *J. Ultrastruct. Res.* **57**:226.

Burgen, A. S. V., Hiley, C. R., and Young, J. M., 1974, The properties of muscarinic receptors in mammalian cerebral cortex, *Br. J. Pharmacol.* **51**:279.

Bursztajn, S., and Gershon, M. D., 1977, Discrimination between nicotinic receptors in vertebrate ganglia and skeletal muscle by alpha-bungarotoxin and cobra venoms, *J. Physiol. (London)* **269**:17.

Burt, D. R., Creese, I., and Snyder, S. H., 1976, Properties of [3H]haloperidol and [3H]dopamine binding associated with dopamine receptors in calf brain membranes, *Mol. Pharmacol.* **12**:800.

Bylund, D. B., Charness, M. E., and Snyder, S. H., 1977, Beta adrenergic receptor labelling in intact animals with ^{125}I-hydroxybenzylpindolol, *J. Pharmacol. Exp. Ther.* **201**:644.

Castel, M., 1978, Immunocytochemical evidence for vasopressin receptors, *J. Histochem. Cytochem.* **26**:581.

Chandler, J. A., 1975, Electron probe x-ray microanalysis in cytochemistry, in: *Techniques of Biochemical and Biophysical Morphology*, Vol. 2 (D. Glick and R. M. Rosenbaum, eds.), pp. 307–438, Wiley, New York.

Chang, C. M., Reitherman, R. W., Rosen, S. D., and Barondes, S. H., 1975, Cell surface location of discoidin, a developmentally regulated carbohydrate binding protein from *Dictyostelium discoideum, Exp. Cell Res.* **95**:136.

Changeux, J. P., 1975, The cholinergic receptor protein from fish electric organ, in: *Handbook of Psychopharmacology*, Vol. 6 (L. L. Trepon, S. D. Trepon, and S. H. Snyder, eds.), pp. 235–301, Plenum Press, New York.

Changeux, J. P., Benedetti, L., Bourgeois, J. P., Brisson, A., Cartaud, J., Devaux, P., Grünhagen, H., Moreau, M., Popot, J. L., Sobel, A., and Weber, M., 1975, Some structural properties of the cholinergic receptor protein in its membrane environment relevant to its function as a pharmacological receptor, *Cold Spring Harbor Symp. Quant. Biol.* **25**:211.

Chiu, T. H., Dolly, J. O., and Barnard, E. A., 1973, Solubilization from skeletal muscle of two components that specifically bind α-bungarotoxin, *Biochem. Biophys. Res. Commun.* **51**:205.

Cohen, J. B., and Changeux, J. P., 1975, The cholinergic receptor protein in its membrane environment, *Annu. Rev. Pharmacol.* **15**:83.

Conti-Tronconi, B. M., Gotti, C., and Clementi, F., 1978, Molecular organisation of cholinergic receptor in post-synaptic membrane of *Torpedo marmurata*, in: *The Biochemistry of Myasthenia Gravis and Muscular Distrophy* (G. G. Lunt and R. M. Marchbanks, eds.), pp. 111–119, Academic Press, New York.

Coons, A. H., 1956, Histochemistry with labelled antibody, *Int. Rev. Cytol.* **5**:1.

Craig, S. W., and Cuatrecasas, P. S., 1975, Mobility of cholera toxin receptors on rat lymphocyte membranes, *Proc. Natl. Acad. Sci. USA* **72**:3844.

Creese, R., and MacLagan, J., 1970, Entry of decamethonium in rat muscle studied by autoradiography, *J. Physiol. (London)* **210**:363.

Cuatrecasas, P., 1973, Gangliosides and membrane receptors for cholera toxin, *Biochemistry* **12**:3558.

Cuatrecasas, P., and Hollenberg, M. D., 1976, Membrane receptors and hormone action, *Adv. Protein Chem.* **30**:251.

Daniels, M. P., and Vogel, Z., 1975, Immunoperoxidase staining of α-bungarotoxin binding sites in muscle endplates shows distribution of acetylcholine receptors, *Nature (London)* **254**:339.

Da Silva, P. and Branton, D., 1970, Membrane splitting in freeze-etching. Covalently bound ferritin as a membrane marker, *J. Cell Biol.* **45**:598.

Da Silva, P., Moss, P. S., and Fudenberg, H. H., 1973, Anionic sites on the membrane-intercalated particles of human erythrocyte ghost membranes: Freeze-etch localization, *Exp. Cell Res.* **81**:127.

De Petris, S., and Raff, M. C., 1973, Normal distribution, patching and capping of lymphocyte surface immunoglobulin studied by electron microscopy, *Nature (London) New Biol.* **241**:257.

Diab, I. M., Freedman, D. X., and Roth, L. J., 1971, ^{3}H-lysergic acid diethylamide: Cellular autoradiographic localization in rat brain, *Science* **173**:1022.

DiPasquale, A., Varga, J. M., Moellman, G., and McGuire, J. S., 1978, Synthesis of a hormonally active conjugate of α-MSH, ferritin and fluorescein, *Anal. Biochem.* **84**:37.

Drachman, D. B., Kao, I., Pestronk, A., and Toyka, K. V., 1976, Myasthenia gravis as a receptor disorder, *Ann. NY Acad. Sci.* **274**:226.

Elde, R., Hökfelt, T., Johansson, O., and Terenius, L., 1976, Immunohistochemical studies using antibodies to leucine-enkephalin: Initial observations on the nervous system of the rat, *Neuroscience* **1**:349.

Engel, A. G., Lindstrom, J. M., Lambert, E. H., and Lennon, V. A., 1977, Ultrastructural localization of the acetylcholine receptor in myasthenia gravis and in its experimental autoimmune model, *Neurology* **27**:307.

Enna, S. J., and Snyder, S. H., 1977, Influences of ions, enzymes and detergents on α-aminobutyric acid binding in synaptic membranes of rat brain, *Mol. Pharmacol.* **13**:442.

Fahimi, H. D., 1975, Fine-structural cytochemical localization of peroxidatic activity of catalase, in: *Techniques of Biochemical and Biophysical Morphology*, Vol. 2 (D. Glick and R. M. Rosenbaum, eds.), pp. 197–246, Wiley, New York.

Fambrough, D. M., 1974, Acetylcholine receptors. Revised estimates on extrajunctional receptor density in denervated rat diaphragm, *J. Gen. Physiol.* **64**:468.

Fedinec, A. F., Debrow, M. M., Gardner, D. P., and Sternberger, L. A., 1970, Immunohistochemical localization of tetanus toxin by the unlabeled antibody enzyme method, *J. Histochem. Cytochem.* **18**:684.

Fertuck, H. C., and Salpeter, M. M., 1974a, Sensitivity in electron microscope autoradiography for ^{125}I, *J. Histochem. Cytochem.* **22**:80.

Fertuck, H. C., and Salpeter, M. M., 1974b, Localization of acetylcholine receptor by ^{125}I-labelled α-bungarotoxin binding at mouse motor endplates, *Proc. Natl. Acad. Sci USA* **71**:1376.

Fertuck, H. C., and Salpeter, M. M., 1976, Quantitation of junctional and extra-junctional acetylcholine receptors by electron microscope autoradiography after ^{125}I-α-bungarotoxin binding at mouse neuromuscular junctions, *J. Cell Biol.* **69**:144.

Fisher, K. A., and Branton, D., 1976, Freeze-fracture autoradiography: Feasibility, *J. Cell Biol.* **70**:453.

Freeman, J. A., 1977, Possible regulatory function of acetylcholine receptor in maintenance of retino-tectal synapses, *Nature (London)* **269**:218.

Gerlach, J. L., and McEwen, B. S., 1972, Rat brain binds adrenal steroid hormone: Radioautography of hippocampus with corticosterone, *Science* 175:1133.

Goldfine, I. D., Smith, G. J., Wong, K. Y., and Jones, A. L., 1977, Cellular uptake and nuclear binding of insulin in human cultured lymphocytes: Evidence for potential intracellular sites of insulin action, *Proc. Natl. Acad. Sci. USA* 74:1368.

Gonatas, N. K., Stieber, A., Gonatas, J., Gambetti, P., Antoine, J. C., and Avarameas, S., 1974, Ultrastructural autoradiographic detection of intracellular immunoglobulins with iodinated Fab fragments of antibody: The combined use of ultrastructural autoradiography and peroxidase cytochemistry for the detection of two antigens (double labelling), *J. Histochem. Cytochem.* 22:999.

Gould, R. M., and Dawson, R. M. C., 1976, Incorporation of newly formed lecithin into purified nerve myelin, *J. Cell Biol.* 68:480.

Graham, R. C., Jr., and Karnovsky, M. J., 1966, The early stages of absorption of injected horse-radish peroxidase in the proximal tubules of mouse kidney: Ultrastructural cytochemistry by a new technique, *J. Histochem. Cytochem.* 14:291.

Hansson, H. A., Holmgren, J., and Svenneholm, L., 1977, Ultrastructural localization of cell membrane GMl ganglioside by cholera toxin, *Proc. Natl. Acad. Sci. USA* 74:3782.

Hardy, H., and Heimer, L., 1977, A safer and more sensitive substitute for diaminobenzidine in the light microscopic demonstration of retrograde and anterograde axonal transport of horse-radish peroxidase, *Neurosci. Lett.* 5:23.

Hartzell, H. C., and Fambrough, D. M., 1972, Acetylcholine receptors: Distribution and extra-junctional density in rat diaphragm after denervation with acetylcholine sensitivity, *J. Gen. Physiol.* 60:248.

Hartzell, H. C., and Fambrough, D. M., 1973, Acetylcholine receptor production and incorporation into membranes of developing muscle fibers, *Dev. Biol.* 30:153.

Hijmans, W., and Schaeffer, M., 1975, Fifth international conference on immunofluorescence and related staining techniques, *Ann. NY Acad. Sci.* 254:1.

Hinton, D. M., Petrali, J. P., Meyer, H. G., and Sternberger, L. A., 1973, The unlabeled antibody enzyme method of immunohistochemistry: Molecular immunocytochemistry of antibodies in the erythrocyte surface, *J. Histochem. Cytochem.* 21:978.

Hodges, G. M., and Muir, M., 1975, Quantitative evaluation of autoradiographs by x-ray spectro-scopy, *J. Microsc. (Oxford)* 104:173.

Hopkins, C. R., and Gregory, H., 1977, Topographical localization of the receptors for luteinizing hormone-releasing hormone on the surface of dissociated pituitary cells, *J. Cell Biol.* 75:528.

Horowitz, S. B., 1974, The ultra-low temperature autoradiography of water and its solutes, *Methods Cell Biol.* 8:249.

Hough, P. V. C., McKinney, W. R., Ledbetter, M. C., Pollack, R. E., and Moos, H. W., 1976, Identi-fication of biological macromolecules *in situ* at high resolution via the fluorescence excited by a scanning electron beam, *Proc. Natl. Acad. Sci. USA* 73:317.

Hourani, B. T., Torain, B. F., Henkart, M. P., Carter, R. L., Marchesi, V. T., and Fischbach, G. D., 1974, Acetylcholine receptors of cultured muscle cells demonstrated with ferritin-α-bungarotoxin conjugates, *J. Cell. Sci.* 16:473.

Howe, P. R. C., Telfer, J. A., and Austin, L., 1976, Binding sites for ^{125}I-labeled-α-bungarotoxin in normal and denervated mouse muscle, *Exp. Neurol.* 52:272.

Ito, Y., Miledi, R., Vincent, A., and Newsom-Davis, J., 1978, Acetylcholine receptors and end-plate electrophysiology in myasthenia gravis, *Brain* 101:345.

Jarett, L., and Smith, R. D., 1975, Ultrastructural localization of insulin receptors on adipocytes, *Proc. Natl. Acad. Sci. USA* 72:3526.

Jedrzejczyk, J., Rymaszewska, T., Wieckowski, J., and Barnard, E. A., 1973, Dystrophic chicken muscle: Altered synaptic cholinesterase, *Science* 180:406.

Keefer, D. A., Stumpf, W. E., Petrusz, P., and Sar, M., 1975, Simultaneous autoradiographic and immunohistochemical localization of estrogen and gonadotropin in the rat pituitary, *Am. J. Anat.* 142:129.

Keefer, D. A., Stumpf, W. E., and Petrusz, P., 1976, Quantitative autoradiographic assessment of ^{3}H-estradiol uptake in immunocytochemically characterized pituitary cells, *Cell Tissue Res.* 166:25.

Kendall, P., and Barnard, E. A., 1967, Thiolation for labelling of proteins, using catalysis by silver and imidazole combined, *Biochim. Biophys. Acta* 188:10.

Ko, P. K., Anderson, M. J., and Cohen, M. W., 1977, Denervated skeletal muscle fibers develop discrete patches of high acetylcholine sensitivity, *Science* **196**:540.

Kopriwa, B. M., 1967, A semi-automatic instrument for the radioautographic coating technique, *J. Histochem. Cytochem.* **14**:923.

Kraehenbuhl, J. P., Racine, L., and Galardy, R. E., 1975, Localization of secretory IgA, secretory component, and α-chain in the mammary gland of lactating rabbits by immunoelectron microscopy, *Ann. NY Acad. Sci.* **254**:190.

Kuhar, M. J., and Yamamura, H. I., 1975, Light autoradiographic localization of cholinergic muscarinic receptors in rat brain by specific binding of a potent antagonist, *Nature (London)* **253**:560.

Kuhar, M. J., and Yamamura, H. I., 1976, Localization of cholinergic muscarinic receptors in rat brain by light microscopic radioautography, *Brain Res.* **110**:229.

Kühlmann, W., and Avrameas, S., 1971, Glucose oxidase as an antigen marker for light and electron microscopic studies, *J. Histochem. Cytochem.* **19**:361.

Land, B. R., Podleski, T. R., Salpeter, E. E., and Salpeter, M. M., 1977, Acetylcholine receptor distribution on myotubes in culture correlated to acetylcholine sensitivity, *J. Physiol. (London)* **269**:155.

Lang, B., Barnard, E. A., Cavanagh, J., and Dolly, J. O., 1979, in preparation.

Lentz, T. L., Mazurkiewicz, J. F., and Rosenthal, J., 1977, Cytochemical localization of acetylcholine receptors at the neuromuscular junction by means of horseradish peroxidase-labelled α-bungarotoxin, *Brain Res.* **132**:423.

Libelius, R., 1974, Binding of ^{3}H-labelled cobra neurotoxin to cholinergic receptors in fast and slow mammalian muscles, *J. Neural Transm.* **35**:137.

Libelius, R., 1975, Evidence for endocytotic uptake of cobra neurotoxin in mouse skeletal muscle, *J. Neural Transm.* **37**:161.

Libelius, R., Eaker, D., and Karlsson, E., 1975, Further studies on the binding properties of cobra neurotoxin to cholinergic receptors in mouse skeletal muscle, *J. Neural Transm.* **37**:165.

Lord, J. A. H., Waterfield, A. A., Hughes, J., and Kosterlitz, H. W., 1977, Endogeneous opioid peptides: Multiple agonists and receptors, *Nature (London)* **267**:495.

Malmgren, L. T., and Olsson, Y., 1977, A more sensitive method for histochemical demonstration of horseradish peroxidase, *Proc. Int. Soc. Neurochem.* **6**:282.

Manuelidis, L., and Manuelidis, E. E., 1976, Cholera toxin-peroxidase: Changes in surface labelling of glioblastoma cells with increased time in tissue culture, *Science* **193**:588.

Mason, T. C., Phifer, R. F., Spicer, S. S., Swallow, R. A., and Dreshin, R. B., 1969, An immunoglobulin-enzyme bridge method for localizing tissue antigens, *J. Histochem. Cytochem.* **17**:563.

McLean, I. W., and Nakane, P. K., 1974, Periodate–lysine–paraformaldehyde fixative: A new fixative for immunoelectron microscopy, *J. Histochem. Cytochem.* **22**:1077.

Miledi, R., and Szcepaniak, A., 1975, Effect of *Dendroaspis* neurotoxin on synaptic transmission in the spinal cord of the frog, *Proc. Roy. Soc. London Ser. B* **190**:267.

Moriarty, G. C., Moriarty, C. M., and Sternberger, L. A., 1973, Ultrastructural immunocytochemistry with unlabelled antibodies and the peroxidase–antiperoxidase complex: A technique more sensitive than radioimmunoassay, *J. Histochem. Cytochem.* **21**:825.

Nakane, P. K., 1975, Recent progress in the peroxidase-labeled antibody method, *Ann. NY Acad. Sci.* **254**:203.

Nakane, P. K., and Kawaoi, A., 1974, Peroxidase-labelled antibody. A new method of conjugation, *J. Histochem. Cytochem.* **22**:1084.

Nakane, P. K., and Pierce, G. B., 1966, Enzyme-labeled antibodies: preparation and application for localization of antigens, *J. Histochem. Cytochem.* **14**:929.

Nemanic, M. K., Shannon, J. M., Carter, D. P., and Wofsy, L., 1974, Immunospecific labeling of a cell surface antigen with a marker visible in the scanning electron microscope, *J. Histochem. Cytochem.* **22**:295.

Novikoff, A. B., Novikoff, P. M., Quintana, N., and Davis, C., 1972, Diffusion artefact in 3,3′-diaminobenzidine cytochemistry, *J. Histochem. Cytochem.* **20**:745.

Nowak, T. P., Haywood, P. L., and Barondes, S. H., 1976, Developmentally regulated lectin in embryonic chick muscle and a myogenic cell line, *Biochem. Biophys. Res. Commun.* **68**:650.

Olsen, R. W., Lee, J. M., and Sar, M., 1975, Binding of α-aminobutyric acid to crayfish muscle and its relationship to receptor sites, *Mol. Pharmacol.* **11**:566.

Orth, J., and Christensen, A. K., 1977, Localization of ^{125}I-labelled FSH in the testes of hypophysectomized rats by autoradiography at the light and electron microscope levels, *Endocrinology* **101**:262.

Patrick, J., and Stallcup, W. B., 1977, Immunological distinction between acetylcholine receptor and the α-bungarotoxin-binding component on sympathetic neurons, *Proc. Natl. Acad. Sci. USA* **74**:4689.

Pearse, A. G. E., 1972, *Histochemistry, Theoretical and Applied*, Vol. 2, 3rd ed., Churchill, London.

Pert, C. B., Kuhar, M. J., and Snyder, S. H., 1975, Autoradiographic localization of the opiate receptor in rat brain, *Life Sci.* **16**:1849.

Pert, C. B., Kuhar, M. J., and Snyder, S. H., 1976, Opiate receptor: Autoradiographic localization in rat brain, *Proc. Natl. Acad. Sci. USA* **73**:3729.

Peters, P. D., Yarom, R., Dorman, A., and Hall, T., 1976, X-ray microanalysis of intracellular zinc: EMMA-4 examinations of normal and injured muscle and myocardium, *J. Ultrastruct. Res.* **57**:121.

Peters, R., Peters, J., Tews, K. H., and Bähr, W., 1974, A microfluorimetric study of translational diffusion in erythrocyte membranes. *Biochim. Biophys. Acta* **367**:282.

Petrali, J. P., Hinton, D. M., Moriarty, G. C., and Sternberger, L. A., 1974, The unlabeled antibody enzyme method of immunocytochemistry: Quantitative comparison of sensitivities with and without peroxidase–antiperoxidase complex, *J. Histochem. Cytochem.* **22**:782.

Petrusz, P., 1974, Demonstration of gonadotropin-binding sites in the rat ovary, by an immunoglobulin-enzyme bridge method, *Eur. J. Obstet. Gynecol. Reprod. Biol. Suppl.* **4**:53.

Petrusz, P., 1975, Localization and sites of action of gonadotropins in brain, in: *Anatomical Neuroendocrinology* (W. E. Stumpf and L. D. Grant, eds.), pp. 176–184, Karger, Basel.

Petrusz, P., 1979, Gonadotropin receptors, in: *Drugs, Hormones and Receptors* (R. W. Straub, ed.), Raven Press, New York, in press.

Petrusz, P., Sar, M., Ordronneau, P., and Di Meo, P., 1976, Specificity in immunocytochemical staining, *J. Histochem. Cytochem.* **24**:1110.

Ploem, J. S., 1975, in Hijmans and Schaeffer, 1975, pp. 4–20.

Polz-Tejera, G. Schmidt, J., and Karten, H. J., 1975, Autoradiographic localization of α-bungarotoxin binding sites in the CNS, *Nature (London)* **258**:349.

Porter, C. W., and Barnard, E. A., 1975a, The density of cholinergic receptors at the endplate postsynaptic membrane: Ultrastructural studies in two mammalian species, *J. Membr. Biol.* **20**:31.

Porter, C. W., and Barnard, E. A., 1975b, Distribution and density of cholinergic receptors at the motor endplates of a denervated mouse muscle, *Exp. Neurol.* **48**:542.

Porter, C. W., and Barnard, E. A., 1976, Ultrastructural studies on the acetylcholine receptor at motor endplates of normal and pathologic muscles, *Ann. NY Acad. Sci.* **274**:85.

Porter, C. W., Chiu, T. H., Wieckowski, J., and Barnard, E. A., 1973a, Types and locations of cholinergic receptor-like molecules in muscle fibres, *Nature (London) New Biol.* **241**:3.

Porter, C. W., Barnard, E. A., and Chiu, T. H., 1973b, The ultrastructural localization and quantitation of cholinergic receptors at the mouse motor endplate, *J. Membr. Biol.* **14**:383.

Price, D. L., Griffin, J. W., and Peck, J., 1977, Tetanus toxin: Evidence for binding at presynaptic nerve endings, *Brain Res.* **121**:379.

Prives, J., Silman, I., and Amsterdam, A., 1976, Appearance and disappearance of ACh receptors during differentiation of chick skeletal muscle *in vitro*, *Cell* **7**:543.

Rajaniemi, H., and Vanha-Perttula, T., 1973, Attachment to the luteal plasma membranes: An early event in the action of luteinizing hormone, *J. Endocrinol.* **57**:199.

Ravdin, P., and Axelrod, D., 1977, Fluorescent tetramethyl-rhodamine derivatives of α-bungarotoxin: Preparation, separation and characterization, *Anal. Biochem.* **80**:585.

Ringel, S. P., Bender, A. N., Festoff, B. W., Engel, W. K., Vogel, Z., and Daniels, M. P., 1975, Ultrastructural demonstration and analytical application of extrajunctional receptors of denervated human and rat skeletal muscle fibres, *Nature (London)* **255**:730.

Ritchie, J. M., and Rogart, R. B., 1977, The binding of labelled saxitoxin to the sodium channels in normal and denervated mammalian muscle, and in amphibian muscle, *J. Physiol. (London)* **269**:341.

Rogers, A. W., 1967, *Techniques of Autoradiography*, Elsevier, Amsterdam.

Rosenbluth, J., 1972, Myoneural junctions of two ultrastructurally distinct types in earthworm body wall, *J. Cell Biol.* **54**:566.

Rosenbluth, J., 1974, Substructure of amphibian motor endplate, *J. Cell Biol.* **62**:755.

Ross, M. J., Klymkowsky, M. W., Agard, D. A., and Stroud, R. M., 1977, Structural studies of a membrane-bound acetylcholine receptor from *Torpedo californica, J. Mol. Biol.* **116**:635.

Rotter, A., Birdsall, N. J. M., Burgen, A. S. V., Field, P. M., and Raisman, G., 1977a, Axotomy causes loss of muscarinic receptors and loss of synaptic endplates in the hypoglossal nucleus, *Nature (London)* **266**:734.

Rotter, A., Birdsall, N. J. M., Field, P. M., and Raisman, G., 1977b, personal communication.

Salpeter, M. M., and Szabo, M., 1972, Sensitivity in electron microscope autoradiography. I. The effect of radiation dose, *J. Histochem. Cytochem.* **20**:425.

Salpeter, M. M., and Szabo, M., 1976, An improved Kodak emulsion for use in high resolution electron microscope autoradiography, *J. Histochem. Cytochem.* **24**:1204.

Salpeter, M. M., Bachmann, L., and Salpeter, E. E., 1969, Resolution in electron microscope autoradiography, *J. Cell Biol.* **41**:1.

Salpeter, M. M., Fertuck, H. C., and Salpeter, E. E., 1977, Resolution in electron microscope autoradiography, III. Iodine-125, the effect of heavy metal staining and reassessment of critical parameters, *J. Cell Biol.* **72**:161.

Schubert, P., Höllt, V., and Herz, A., 1975, Autoradiographic evaluation of the intracerebral distribution of ^{3}H-etorphine in the mouse brain, *Life Sci.* **16**:1855.

Schurer, J. W., Hoedemaeker, P. J., and Molenaar, I., 1977, Polyethyleneimine as tracer particle for (immuno)electron microscopy, *J. Histochem. Cytochem.* **25**:384.

Seligman, A. M., Shannon, W. A., Jr., Hoshino, Y., and Plopinger, R. E., 1973, Some important principles in 3,3'-diaminobenzidine ultrastructural cytochemistry, *J. Histochem. Cytochem.* **21**:756.

Shain, W., Green, L. A., Carpenter, D. O., Sytkowski, A. J., and Vogel, Z., 1974, *Aplysia* acetylcholine receptors: Blockade by and binding of α-bungarotoxin, *Brain Res.* **72**:225.

Shannon, L. M., Kay, E., and Lew, J. Y., 1966, Peroxidase isoenzymes from horseradish roots. I. Isolation and physical properties, *J. Biol. Chem.* **241**:2166.

Shotton, D. M., Heuser, J. E., Reese, B. F., and Reese, T. S., 1979, Postsynaptic membrane folds of the frog neuromuscular junction visualized by scanning electron microscopy, *Neuroscience*, in press.

Siess, E., Wieland, O., and Miller, F., 1971, A simple method for the preparation of pure and active α-globulin-ferritin conjugates using glutaraldehyde, *Immunology* **20**:659.

Simantov, R., Kuhar, M. J., and Snyder, S. H., 1977, Opioid peptide enkephalins: Immunohistochemical mapping in the rat central nervous system, *Proc. Natl. Acad. Sci. USA* **74**:2167.

Singer, S. J., 1976, The fluid mosaic model of membrane structure: some applications to ligand-receptor and cell–cell interactions, in: *Surface Membrane Receptors* (R. A. Bradshaw, W. A. Frazier, M. C. Merrell, D. I. Gottlieb, and R. A. Hogue-Angeletti, eds.), pp. 1–24, Plenum Press, New York.

Singer, S. J., and Nicolson, G. L., 1972, The fluid mosaic model of the structure of cell membranes, *Science* **175**:720.

Singer, S. J., and Schick, A. F., 1961, The properties of specific stains for electron microscopy prepared by the conjugation of antibody molecules with ferritin, *J. Biophys. Biochem. Cytol.* **9**:519.

Soloff, M. S., Rees, H. D., Sar, M., and Stumpf, W. E., 1975, Autoradiographic localization of radioactivity from [^{3}H]oxytocin in the rat mammary gland and oviduct, *Endocrinology* **96**:1475.

Soni, S. C., Kalnins, V. T., and Haggis, G. H., 1975, Localisation of caps on mouse lymphocytes by scanning electron microscopy, *Nature (London)* **255**:717.

Sternberger, L. A., 1972, The unlabelled-antibody-peroxidase and the quantitative-immunouranium methods in light and electron immunohistochemistry, in: *Techniques of Biochemical and Biophysical Morphology*, Vol. 1 (D. Glick and R. M. Rosenbaum, eds.), pp. 57–88, Wiley, New York.

Sternberger, L. A., 1978, Receptors for luteinizing hormone-releasing hormone, *J. Histochem. Cytochem.* **26**:542.

Stockert, R. J., Morell, A. G., and Scheinberg, I. H., 1974, Mammalian hepatic lectin, *Science* **186**:365.

Stumpf, W. E., 1970, Tissue preparation for the autoradiographic localization of hormones, in: *Introduction to Quantitative Cytochemistry*, Vol. 2 (G. L. Wied, and G. F. Bahr, eds.), pp. 507–526, Academic Press, New York.

Stumpf, W. E., 1976, Techniques for the autoradiography of diffusible compounds, *Methods Cell Biol.* **13**:171.

Stumpf, W. E., and Sar, M., 1977, Steroid hormone target cells in the periventricular brain: relationship to peptide hormone producing cells, *Fed. Proc.* **36**:1973.

Suszkiw, J. B., and Ichiki, M., 1976, Fluorescein conjugated α-bungarotoxin: Its properties and interaction with acetylcholine receptors, *Anal. Biochem.* **73**:109.

Sytkowski, A. J., Vogel, Z., and Nirenberg, M. W., 1973, Development of acetylcholine receptor clusters on cultured muscle cells, *Proc. Natl. Acad. Sci. USA* **70**:270.

Teichberg, V. I., Silman, I., Beutsch, D. D., and Resheff, G., 1975, A β-D-galactoside binding protein from electric organ tissue of *Electrophorus electricus*, *Proc. Natl. Acad. Sci. USA* **72**:1383.

Templeton, C. L., Douglas, R. J., and Vail, W. J., 1978, Ferritin conjugated protein A, *FEBS Lett.* **85**:95.

Tittler, M., Weinreich, P., and Seeman, P., 1977, New detection of brain dopamine receptors with [^{3}H]dihydroergocryptine, *Proc. Natl. Acad. Sci. USA* **74**:3750.

Tsou, K. C., Mela, L., Gupta, P. D., and Lynn, D., 1976, Mitochondrial ultrastructure study with the new cytochrome c binding agent 4-4'-diamino-2-2'-bipyridyl ferrous chelate. *J. Ultrastruct. Res.* **54**:235.

Unwin, P. N. T., and Henderson, R., 1975, Molecular structure determination by electron microscopy of unstained crystalline specimens, *J. Mol. Biol.* **94**:425.

U'Prichard, D. C., Greenberg, D. A., and Snyder, S. H., 1977, Binding characteristics of a radio-labeled agonist and antagonist at central nervous system alpha-noradrenergic receptors, *Mol. Pharmacol.* **13**:454.

Van Heyningen, W. E., 1974, Gangliosides as membrane receptors for tetanus toxin, cholera toxin and serotonin, *Nature (London)* **249**:415.

Varga, J. M., Saper, M. A., Lerner, A. B., and Fritsch, P., 1976a, Non-random distribution of receptors for melanocyte-stimulating hormones on the surface of mouse melanoma cells, *J. Supramol. Struct.* **4**:45.

Varga, J. M., Moellman, G., Fritsch, P., Godavska, E., and Lerner, A. B., 1976b, Association of cell surface receptors for melanotropin with the Golgi region in mouse melanoma cells, *Proc. Natl. Acad. Sci. USA* **73**:559.

Vogel, Z., and Daniels, M. P., 1976, Ultrastructure of acetylcholine receptor clusters on cultured muscle fibers, *J. Cell Biol.* **69**:501.

Vogel, Z., and Nirenberg, M., 1976, Localization of acetylcholine receptors during synaptogenesis in retina, *Proc. Natl. Acad. Sci. USA* **73**:1806.

Vogel, Z., Malone, G. T., Ling, A., and Daniels, M. P., 1977, Identification of synaptic acetylcholine receptor sites in retina with peroxidase-labeled α-bungarotoxin, *Proc. Natl. Acad. Sci. USA* **74**:3268.

Weibel, E. R., Kistler, G. S., and Scherle, W. F., 1966, Practical stereological methods for morphometric cytology, *J. Cell Biol.* **30**:23.

Wernig, A. Stöver, H., and Tonge, D., 1977, The labelling of motor endplates in skeletal muscle of mice with ^{125}I-tetanus toxin, *Arch. Pharmacol.* **298**:37.

Williams, L. T., and Lefkowitz, R. J., 1976, Alpha-adrenergic receptor identification by [^{3}H]dihydroergocryptine binding, *Science* **192**:791.

Williams, M. A., 1973, Electron microscope autoradiography: its application to protein biosynthesis, in: *Techniques in Protein Biosynthesis*, Vol. 3 (P. N. Campbell and J. R. Sargent, eds.), pp. 126–190, Academic Press, New York.

Yamamura, H. I., and Snyder, S. H., 1974, Muscarinic cholinergic receptor binding in the longitudinal muscle of the guinea pig ileum with [^{3}H]quinuclidinyl benzilate, *Mol. Pharmacol.* **10**:861.

Young, A. B., and Snyder, S. H., 1974, Strychnine binding in rat spinal cord membranes associated with the synaptic glycine receptor: Co-operativity of glycine interactions, *Mol. Pharmacol.* **10**:790.

Zurn, A. D., and Fulpius, B. W., 1976, Accessibility to antibodies of acetylcholine receptors in the neuromuscular junction, *Clin. Exp. Immunol.* **24**:9.

Problems and Approaches in Noncatalytic Biochemistry

R. D. O'Brien

1. Introduction

Most of the story of biochemistry in the last half-century is the story of enzymes. The demonstration by Sumner in 1926 that enzymes are proteins, coupled with the vigorous explorations of metabolic pathways that grew out of the nutritional beginnings of biochemistry, led to the extraordinary depth of knowledge that we have now, both of the routes of metabolism and the enzymes which catalyze the individual steps in these routes, constituting a network of systems of fearful complexity. One of the great advantages for those who study enzymes is the availability of a catalytic step to monitor. The result is that every molecule of enzyme present in a tissue or in a purified preparation can be made to yield thousands or millions of molecules of reaction product in moderate times, and this enormous amplification factor made it possible by relatively crude techniques to demonstrate the presence of low concentrations of enzyme and to achieve purifications when relatively arbitrary fractionations were pursued, coupled with assays of enzymatic activity in the relatively dilute fractions derived therefrom. It is not difficult to make even dilute enzyme preparations yield product in the millimolar to micromolar range, which is comfortably within the competence of colorimetric and other spectrophotometric techniques to detect.

In the relatively new era of nucleic acid biochemistry, the catalytic phenomenon was not so evident. Happily, the great length of DNA permitted electron microscopy to be of assistance in selected cases, but in general the existence of an active template is still measured operationally, i.e., by demonstrating the catalytic production of a highly specific product when appropriate enzymes (such as DNA-dependent RNA–polymerase) are supplied by the investigator.

It is within the last 10 years particularly that biochemists have developed a detailed interest in the question of molecules which neither have innate catalytic activity nor are able to be coupled with enzymes in order to achieve catalytic activity at second hand. These molecules are those which, in the intact cell, promote a variety of activities, such as the pumping of ions, the stereospecific

R. D. O'Brien ● Section of Neurobiology and Behavior, Cornell University, Ithaca, New York. Present affiliation: Provost, University of Rochester, Rochester, New York.

control of the entry of organic substances (as in the "permeases"), and the extra-ordinary array (much of it still vague in detail) of recognition processes which lead on to extremely important noncatalytic events, such as the change in permeability to ions of a membrane (as in nicotinic acetylcholine receptor, and probably many chemoreceptors and mechanoreceptors) or the transportation of a macromolecule from one cell location to another (as in the case of the steroid hormone receptors) or the synthesis of second messengers (as in many systems mediated by cyclic AMP). The study of such noncatalytic molecules has two underlying difficulties. One is that the process to be explored is not catalytic, and consequently, if one is to "see" the phenomenon of physiological interest, one has to work with techniques which are able to visualize either the macromolecule itself or stoichiometric quantities of things with which the macromolecule reacts. Second, the essential phenomenon, whether it be a change in permeability or migration from one part of a cell to another, is something which is only observable in the intact cell. Consequently, when the biochemist seeks to characterize the molecular nature of the events and begins by his customary technique of disrupting the cell as a prelude to fractionation and purification, the phenomenon in which he is interested disappears.

The result is that the biochemist is thrown back upon relatively indirect approaches to assaying these macromolecules as fractionation proceeds. The indirect techniques almost invariably rely upon utilizing known facts about the physiology or pharmacology of the macromolecule of interest in order to find a large or small *probe*, for instance, a binding molecule (ligand) which can be made radioactive or fluorescent and whose binding to the macromolecule of interest is followed through fractionation. This indirect approach is fraught with uncertainties, which we shall explore below, and its ultimate validity is only satisfactorily proved if the purified material which is derived from its use can be incorporated into a relatively simple system, upon which it can bestow the property originally sought. Thus if one were looking for a receptor which, in the intact cell, made the cell permeable to Na^+ when transmitter substance X was applied to the cell, then one would hope that the purified receptor could, when incorporated into an appropriate membrane, bestow Na^+ permeability upon the membrane when transmitter X was applied. This reconstitution phenomenon is itself a serious problem, excellently covered in Chapters 1 and 6 in this volume.

Let us consider some of the uncertainties about the indirect approach to monitoring activity. The most grievous one is that if one is using the ability to "recognize" a given ligand as an index of activity, then what one purifies is the component which recognizes the ligand. If one is fortunate, the recognition site may be an intrinsic part of the total macromolecule or complex, which is thereby purified as a whole. But there is reason to expect that there will be cases of receptors and related macromolecules whose recognition sites constitute but one portion of the whole, and then the purification necessarily leads to a pure recognition system and that is all; the baby will have been thrown out with the bath water.

How does one tackle this dilemma? One procedure is to take advantage of the fact that in very many regulatory systems (and these receptors and gates are almost invariably regulatory) there are multiple sites in the whole system which are the targets of regulatory ions and compounds. The multiplicity of such sites is the consequence of the fact that allosteric mechanisms are commonly to be found, and this in turn is a corollary of the fact that extremely diverse agents, such as different drugs and metal ions, are often found to have a regulatory role. One can then use a multiplicity of probes to follow the purification. For instance, in our current attempts to isolate and purify the sodium gate, we are utilizing four different labeled ligands. Two of them are drastically different chemically, but both appear to stabilize the gate in the open configuration; these are the proteinaceous *Condylactis* toxin and the steroid veratridine. In addition we are labeling two unrelated nonproteinaceous toxins, saxitoxin and tetrodotoxin, both of which stabilize the gate in the closed configuration. In addition, labeled sodium may be used. By increasing the number and nature of ligands of known physiological interest, one can minimize the chance that one is purifying a unique recognition site, and if all the recognition sites turn out to be within a single protein, one has additional (though far from complete) confidence that one is on the right track.

If the above approach is unsuccessful, one is left with the fearful need to use as an assay the ability to reconstitute the phenomenon itself, and the difficulties then become indeed severe.

The use of this "probe" technique to detect the presence of interesting macromolecules in particular fractions needs to be accompanied by a series of precautionary considerations, to all of which there has to be a satisfactory response if one is to be assured that one is dealing with the anticipated macromolecule. In simple cases, one may readily answer such questions as to whether the item reacting with the probe is present in the right tissue, absent in the wrong tissue, present in the appropriate amount, has the affinity which is anticipated, and suffers interference by the expected agents. Every one of these criteria has a potential problem in it, and these are discussed in Sec. 3.

2. Measurement

Except in those few cases where total purification has been achieved, there are probably no good examples where direct measurement of the macromolecule of interest can be made, either by quantitatively measuring its effectiveness in a reconstituted system or by measuring simultaneously some series of properties which belong uniquely to it, and not to contaminants. One is almost invariably driven to relatively indirect measurements which depend on the ability of the macromolecule to bind some sort of ligand and then to explore the extent of such binding by changes in the physical state of the ligand (i.e., bound or unbound) or changes in the properties of the ligand (such as changes in its fluorescent characteristics) or, less commonly, of the target.

2.1. Histological Techniques

This problem is dealt with in detail in Chapter 7, so a few comments only will be made at this point.

An apparently straightforward technique is to expose various tissues to the radioactive ligand of interest and subsequently expose sections of slices to X-ray plates in order to detect the location of radioactive areas of concentration. This technique was extremely successful in the case of steroidal hormones, in which it was clearly established that the labeled hormone concentrated within cells of the target tissue (e.g., the uterus) and not in other locations (O'Malley and Means, 1974). Nevertheless, there is a serious risk of artifacts when one uses a diffusible ligand which is reversibly bound; the problem has been reviewed by Stumpf and Roth (1968).

In spite of the success of this particular application, there is a major problem which may prove insuperable in other cases. It is that in some cases the localization of the ligand may be determined by purely physical properties of the various tissues exposed, and the target itself may either not have an unusually high affinity for the ligand, or it may constitute such a minor component of the target organ that the physical properties of other nontarget organs may be the major factors in determining such binding. In a sense, this is simply an extension of the problem of nonspecificity, which is dealt with in detail in Chapter 5 and which is particularly hard to tackle in a radioautography, because one is not usually able to characterize the nature of the bound ligand. Two cases where the technique would be anticipated to be unsuccessful are as follows.

A study was carried out with the objective of finding whether a derivative of grayanotoxin, called α-dihydrograyanotoxin II, which we shall abbreviate as G II, was a suitable probe for the sodium gate. G II had been shown to promote sodium permeability in axons (Seyama and Narahashi, 1973). Biochemical studies (Soeda *et al.*, 1975) showed indeed that radioactive G II bound very satisfactorily to membranes derived from lobster axons. Unfortunately, it bound equally satisfactorily to membranes derived from both neural and nonneural sources, and a more detailed analysis showed that the "binding" was nonsaturable and presumably represented simply a partitioning from an aqueous phase into any available lipid phase. Clearly in a case like this, the histological localization of radioactive G II in any particular tissue would have absolutely no relevance to the localization of the sodium gate.

The second example is of the psychogenic agent, lysergic acid diethylamide, which we shall abbreviate as LSD. There is absolutely no question that LSD exerts its effect upon humans via the central nervous system, and extraordinarily small doses (such as 100 μg) produce profound central effects without other disturbances. But if one measures the concentration of radioactive LSD, after intravenous administration, it is observed (Idänpään-Heikkila and Schoolar, 1969) by autoradiography that the brain has far lower concentrations than liver, kidney, lung, adrenals, thymus, or salivary glands. Obviously the gross distribution of LSD is being determined by purely physical characteristics, and the fact that only an extraordinarily small percentage of the drug obtains access to the brain has absolutely no connection with the fact that the only target of interest is in the brain. Once again, radioautographic techniques would have been completely useless, used in a straightforward way, in determining the target of the drug.

These same precautions apply to the attempts to use radioautography to determine the target of toxicants. Elegant techniques have been developed by which sections of whole mice and insects can be prepared and studied radioautographically, and it is common to find graphic and important differences in the distribution of labeled toxicants in different tissues (Kurihara *et al.*, 1970). Unfortunately, the results need have no correlation with the location of the target site and provide information only about the physical characteristics of the toxicant and of the various body tissues.

2.2. Physical Separation: General Considerations

The commonest kinds of techniques involve the use of a radioactive ligand, coupled with a method which distinguishes between the free ligand and that which is bound to macromolecules. With such techniques, if one can circumvent the problem of nonspecificity discussed in Chapter 5, one may be able to determine not only the total number of binding sites available at saturation but their affinity for the ligand. The analysis of such observations to reveal the amounts and affinities are dealt with at length in Chapter 4.

The problem of measuring ligand bound to receptor is primarily one of maximizing a signal-to-noise ratio; the signal is the radioactivity (or other index, such as fluorescent emission) of the ligand bound to receptor, and the noise is the background of unbound ligand.

Should the investigator be fortunate enough to have an irreversible ligand, or one which can be rendered irreversible by (for instance) photoactivation [see the review by Glazer (1976)], then there is no problem. Another common approach to making irreversible agents is to make haloacyl analogues of acyl agents which are irreversible. Thus the β-adrenoceptor antagonist practolol $CH_3C(O)R$ is the basis for making the irreversible chloropractolol, $ClCH_2C(O)R$, where R is $HN(phenyl)OCH_2CH(OH)CH_2NH-isoPr$ (Erez *et al.*, 1975). In such cases one can incubate with ligand and then wash out all unbound ligand, and then the ability to measure the receptor–ligand complex can be computed simply from the number of complexes which can be squeezed into the counting vial and the specific activity of the ligand. Thus if the ligand has a fairly good specific activity of 1 Ci/mmol, and the investigator will settle for 100 cpm above background, in a counting system of 50% efficiency, then he can detect 200 dpm (disintegrations per minute); and since each curie gives 2.2×10^{12} dpm, he can count $200/2.2 \times 10^{12} = 91 \times 10^{-12}$ Ci, which corresponds to 91×10^{-12} mmol or 91 fmol of receptor. In a tissue with the electroplax-like level of 1 nmol of receptor per gram of original tissue, the 91 fmol correspond to only 91 μg of tissue, so that obviously one has superb sensitivity.

But the number of situations in which an irreversible label is available will be few. If one uses ligands such as the hormones or pheromones or transmitters themselves, these are by definition reversible. The following discussion will deal exclusively with reversible ligands. One's ability to detect such reversible binding is limited; one can peer through a "window" of binding activities. The lower edge of the window is set, as we shall see, by the quantity of receptor present and its binding constant K_d, and the upper edge of the window by the quantity of receptor and the specific activity of the ligand. Let us work through the case for equilibrium dialysis and then treat other physical separations as variants of that technique.

If one considers binding of a ligand whose free concentration is (L) to a receptor whose total concentration is (R_0), then the relation between the concentration of free R and the complexed form LR is given by the dissociation constant K_d, defined as

$$K_d = \frac{(L)(R)}{(LR)} \qquad (1)$$

and since $(R) = (R_0) - (LR)$, one can show that the fraction of receptor in the bound form is given by

$$\frac{(LR)}{(R_0)} = \frac{1}{\{[K_d/(L)] + 1\}} \qquad (2)$$

This equation describes a situation in which there is only one kind of binding, without cooperativity or nonspecific binding.

There are basically four kinds of techniques to separate bound and free ligands. The most popular and simple procedure is equilibrium dialysis, in which one takes advantage of the fact that only the free form of appropriate ligands can diffuse out of a dialysis bag. The second is the speeded-up version of this, in which one achieves separation by pressure, using such things as filtration or ultrafiltration to squeeze out solvent and free ligand and retain primarily bound ligand. The third involves some form of chromatographic separation of free from bound, of which the prototype is the technique of Hummel and Dryer (1962) in which the macromolecule is passed through a Sephadex column which has been equilibrated with free ligand. One observes the emergence of a peak of radioactivity where macromolecule-plus-ligand comes through, followed by a trough which results from the depletion of ligand. Finally, there is the technique of centrifugation, in which one centrifuges a mixture of bound and free ligand and measures either the pelleted bound form or the residual material in the supernatant, whose depletion is an index of the amount that was bound in the pellet. All these techniques depend on the same fundamental principle, i.e., the physical separation of the free and the bound forms, and therefore the same considerations of signal-to-noise ratio, sensitivity, and accuracy all apply, but to varying extents.

To explore the suitability of different techniques, let us consider two fairly extreme situations. One of these is perhaps optimal, representing as close a packing of receptors as one can reasonably ask. It is the nicotinic receptor of *Torpedo* receptor, in which the receptor makes up more than 50% of the postsynaptic membrane, with a corresponding density approximating 30,000 receptors/μm^2 (Miledi *et al.*, 1971). The other is the sodium gate of lobster axon, with a density which has been variously reported, but for which values per square micrometer of 36 for lobster nerve and 49 for crab nerve have been reported (Keynes *et al.*, 1971). The values of 30,000 and 36 come close to bracketing the range of those receptors for which simple techniques with reversible ligands are possible.

With the above densities, one finds that 1 g of fresh *Torpedo* electroplax contains about 1 nmol of nicotinic receptor, and 1 g of lobster walking-leg nerve contains about 7000 cm^2 of axon membrane (Moore *et al.*, 1967) and therefore 29 pmol of receptor. It is noteworthy that in spite of the fact that the site densities differ by a factor of $50{,}000/25 = 2000$ the moles per gram differ only by $1000/29 = 34$. The reason is, of course, that in spite of the close packing of the active (post-synaptic) membrane of *Torpedo* electroplax, a fairly low percentage of the tissue is made up of these membranes.

2.3. Equilibrium Dialysis

In equilibrium dialysis, one measures (at equilibrium) the excess of counts (Δ cpm) per milliliter inside the dialysis bag over the counts per milliliter in the external bath. The excess represent counts due to ligand which is bound to receptor. A practical·problem is that even with large numbers of receptors in the bag, the Δ cpm becomes "lost" in the counts in the bath when the K_d is much worse than 10^{-6} M. The concern is therefore to find an acceptable value of Δ/bath, which in essence is the signal-to-noise ratio. For instance, if one can squeeze into a 1-ml dialysis bag 1 nmol of acetylcholine receptor (which would have to be obtained from 1 g of electroplax) with a K_d of 10^{-6} M, then from Eq. (2) one finds that half of the receptor (i.e., 0.5 nmol) is in the bound form with 10^{-6} M ligand. Unfortunately, 1 ml of the bath contains 1 nmol of ligand, so that the value of Δ/bath is $0.5/1 = 0.5$; thus the background is twice as high as the binding. For a practical example, if the ligand were 1 Ci/mmol, counted at 50% efficiency, the bath per milliliter would have 1.1 million cpm, and the bag would have 1.6 million cpm, for a Δ of 0.5 million cpm. Thus, if one's estimate of the bath were 5% low and of the bag were 5% high, one's error in estimating the Δ of 0.5 million cpm would be 27%.

Figure 1 shows plots of the way in which the Δ/bath value varies. (In the figure I have used fractional binding in place of Δ in order to provide a generalized case.) Looking first at the curve for a given K_d, for instance, 10^{-6} M, it is apparent that the best value of signal-to-noise ratio, which is the same as Δ/bath, is obtained at 10^{-8} M ligand, and at lower L concentrations no improvement occurs. In comparison with the example in the previous paragraph, the bath (background) value at 10^{-8} M has fallen 100-fold to 11,000 cpm, but the Δ value has fallen 50-fold to 11,000 cpm, so that the new Δ/bath is 1. With further reduction in L the Δ/bath values remain constant at 1. Going in the other direction, if one *increases* L to from 10^{-6} to 10^{-5} M, the value of Δ/bath drops from 0.5 to 0.09, so that one would be engaged in the hopeless task of measuring a Δ value of 0.99 million cpm against a background of 11 million cpm.

The general conclusion: the signal-to-noise ratio is almost minimal when the bath value is 1/100th of the K_d and is little improved by reducing this ratio.

There are some obvious corollaries. If one has a K_d of 10^{-5} M, which is a ten times worse binding constant than that just considered, then conditions become far worse. The ratio Δ/bath reaches a maximum of 0.1. With the quantities of receptor specified above, one can never observe a Δ value which is better than 0.99 million against a background of 11 million; thus one has reached the lower limit of the window unless, to compensate for the poor fraction that is bound, one multiplies greatly (e.g., tenfold) the *amount* of receptor. This leads to the second generality: unless one has extraordinarily large amounts of receptor, dissociation

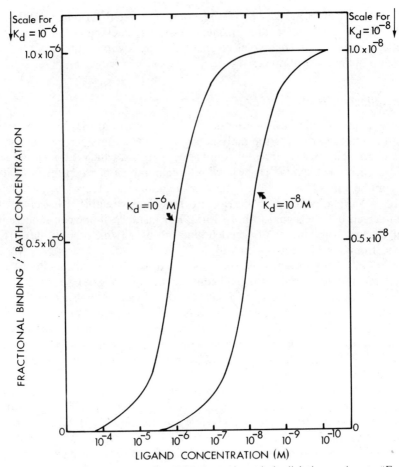

Fig. 1. Effect of ligand concentration upon signal-to-noise ratio in dialysis experiments. "Fractional binding" is the value $(LR)/(R_0)$, i.e., the fraction of the binding at saturation, calculated from Eq. (2). The optimum bath concentration is $K_d/100$; thereafter, further reductions in bath concentration affect equally the fraction bound (signal) and the bath value (noise).

constants above 10^{-6} M are very hard to measure, and indeed it may be very hard to determine whether binding occurs at all.

When we turn to receptors with very low binding constants (high affinities) and large receptor concentrations, the problems of receptor quantity and high background disappear. Instead one reaches at some point the problem of accurately measuring bath concentrations. Thus for the above ligand of specific activity 1 Ci/mmol, if the bath concentration is reduced to 10^{-12} M (which would be appropriate for studying K_ds in that range), the bath count per milliliter would be 1.7 cpm, which is well below measurability in customary systems. Of course it is easy to show for such a situation that binding exists (unlike the situation when the K_d is too low); this could be done at, for instance, a bath concentration of 10^{-9} M, giving a cpm/ml (counts per minute per milliliter) of 1700 and (if the receptor had a $K_d = 10^{-12}$ M, for instance) virtually all of the receptor would be in the bound form. The problem would be only in titrating the receptor binding for K_d measurement.

A common situation is that quantity of receptor is a severe limitation. Then direct study (by equilibrium dialysis or its equivalent) by reversible ligands may frequently be impossible, even when extremely favorable dissociation constants are indicated by physiological findings, as is often the case for receptors for hormones, pheromones, and gustatory or olfactory receptors. For example (Caprio, 1975), the 8 barbels (whisker-like projections) of the catfish have 20,000 taste buds and can detect such low concentrations of alanine (such as 0.1 nM) that a $K_d = 1$ nM is plausible. An attractive candidate for isolation? No!

The 20,000 taste buds can be computed as having 110 receptive cells each and a total receptor membrane area of 6×10^5 μm^2. If the packing were as favorable as for nicotinic receptor, in which the 30,000 receptors/μm^2 constitute over 50% of the sensitive membrane, then the receptors from 8 barbels would number 1.8×10^{10}. If the 8 barbels (weighing 5 g) could be stripped of receptors and these could be squeezed into 5 ml, one would have (bearing in mind Avogadro's number, i.e., 1 mol has 6×10^{23} molecules) 5-pM receptor. Then in a 1 nM bath as above, with a cpm/ml of 1700, half of the receptors (which we can envisage as put into a 4-ml bag) would be bound, and the extra counts in the bag would be 0.5 pM against a background of 1000 pM, a hopeless case. Somehow one must increase the receptors in the bag by a factor of about 2000-fold if the extra counts are to be as large as the background.

The conclusion from these considerations is that even superbly high-affinity systems can often give inadequate bindings. I fear this situation will be common in many fascinating systems, including other gustatory receptors, pheromone receptors, and other highly localized sensors.

2.4. Other Physical Separations

I have gone into a lot of detail on equilibrium dialysis because the same principles extend to other quite different assay systems. For example, there are two variants on the standard centrifugal assay. In both, one studies a precipitable binding species with a soluble radioactive ligand. They are mixed and then centrifuged to bring all bound ligand into the pellet. One approach then is clean analytically but of low sensitivity; it involves counting a sample of the supernatant, which can be done with great precision and without contamination. Unfortunately, this method seeks to measure a small difference between large numbers, i.e., between the ligand concentration originally added and that left in the supernatant. An example: If one has a handsomely large 1 nmol of receptor with a K_d of 10^{-6} M in a 15-ml volume and uses the above-recommended ligand concentration of $K_d/100$ or 10^{-8} M, then the amount bound will be about $\frac{1}{100}$ of the receptor amount, or 10 pmol. The amount of ligand originally in the tube will be 15×10^{-8} mmol or 150 pmol. Thus the total reduction in the supernatant (and in any sample of it) will decline by 10 in 150, which is 7%, a decline which is only marginally measurable if one has typical errors of sampling and counting of 5% or more.

The centrifugal technique can alternatively be utilized with counting of the appearance of precipitable radioactivity (bound ligand) rather than by disappearance of free ligand. In principle, the sensitivity of such a procedure can be extra-ordinarily high. If one could harvest the pellet with zero contamination, then the sensitivity would be limited only by the specific activity of the ligand. For example, in the above example in which 10 pmol was bound, if one could convey the whole precipitate into a counting vial without any entrapped supernatant, then with a ligand of only 100 mCi/mmol, one could obtain (at 50% efficiency) 1100 cpm.

But the practical problem is that one can never reduce the supernatant contamination to zero. And unfortunately it is very hard to hold the contamination constant. Recall that we are talking of reversible binding, so washing the pellet is impossible. Attempting to pour off the supernatant in a fully reproducible way is fairly close to hopeless, because any tissue preparation will give one a relatively greasy centrifuge tube inside, which will give quite extensive holdup of supernatant. If in the above example there were 0.2 ml of entrapped or otherwise unremoved supernatant, it would contribute (neglecting the depletion by the pellet) about 2 pmol of ligand, or 20% of the truly bound reactivity. Thus, if in various tubes the unremoved supernatant varied from 0.1 to 0.3 ml, the scatter of readings would be from 110% to 130% of the true value.

The situation does not improve if one reduces the ligand concentration. The reduction in binding and in supernatant is linear once one has passed the optimal condition (ligand concentration $= K_d/100$) so the percent contamination problem is unaffected.

Turning now to filtration techniques, the principles are precisely the same as for the centrifugation technique. But with suitable systems, such as Whatman glass fiber paper filters in a Millipore filtering system, it is fairly easy to achieve filtration in 5 sec and minimize the problems of desensitization described below. In this case also, one physically separates particulate (bound) and soluble (unbound) ligand and can choose to measure the depletion of free ligand (by assaying the filtrate) or the appearance of bound ligand (by assaying the filter).

Measurement of the filtrate is more precise and can be performed readily by filtering directly into scintillation vials and then adding cocktail and counting. It is not self-evident that one can filter the sample almost to dryness and that the filtrate will then have precisely the free concentration of ligand that was initially in equilibrium with receptor. What about the fact that as filtration proceeds the concentration of total receptor progressively increases? Is the equilibrium thereby shifted, so that the concentration of ligand in the early filtrate is different from that in the late filtrate? One can show that this is not the case by considering what happens if one does the reverse, i.e., adds to an equilibrium mixture additional volumes of ligand solution with the same concentration of the *free* ligand. Using Eq. (1), and solving the quadratic for computing (LR) initially and after adding any volume of extra ligand at concentration (L), shows that (LR) decreases proportionately to the dilution, e.g., decreases tenfold for a tenfold dilution. Therefore there is no dissociation of LR when this special kind of dilution is carried out, and therefore the free ligand concentration (L) is not changed when one adds (or removes) ligand solution.

But this technique of filtrate counting suffers from the defect of insensitivity described above for centrifugal techniques—one is measuring a loss of a little radioactivity from a large amount of radioactivity. As before, measuring the filter is far better in principle, and in practice the problems of contamination by filtrate are not only reduced in amount, but the replicability of the contamination is far better. Thus, although one can achieve a lesser percent variance in the measurement of filtrates from experimental and control preparations (whose difference gives the binding), it is frequently better to tolerate a larger percent variance in precipitate-plus-filter for experimentals and controls, because in the latter case the controls (i.e., the filters plus some inactivated form of receptor) are typically about half the experimentals (i.e., the filters with active receptor), rather than about ten times the experimentals in filtrate-counting techniques. Furthermore, if one counts precipitate-plus-filter, one can harvest receptor from large initial volumes and yet count the receptor with a single vial of scintillation solution.

The merits of simple precipitate-plus-filter techniques were reactivated by the report of Suter and Rosenbusch (1976) for the case of succinate and carbamyl phosphate binding to aspartic transcarbamylase. They used nitrocellulose filters (we find that Whatman glass fiber filters are good for membrane-bound receptors) and did not wash the precipitates at all. But when necessary they did prepare for every ligand concentration a blank involving inactivated enzyme. They used the oxidized form of the enzyme; in our work with nicotinic receptor we use boiled receptor for the blank. Of course, the blank measures nonspecific binding to the protein and also holdup in the filter itself, which is usually the major component. Consequently, careful replication is needed for such factors as the time of retention of filters on the apparatus and the transmembrane pressure during filtration. Spivak and Taylor (1977) have gone to great pains to minimize variance from this and related mechanical sources. By avoiding washing, the technique is fully applicable to both reversible and irreversible systems, for it measures *amount bound* when in equilibrium with the ligand level of the filtrate. If the binding is either truly or operationally irreversible within the time span of the experiment, one can of course wash the filters and so vastly reduce the blanks by perhaps 100-fold and may even avoid the tiresome need for boiled or otherwise inactivated receptor. In this way one can improve the sensitivity enormously.

2.5. Negative Binding

In equilibrium dialysis, one may see apparent "negative binding"; that is, the cpm/ml of bag contents is less than the cpm/ml of the bath. This may be caused by the presence of inert solid matter (such as heavy membrane fragments) which excludes liquid, so that the 1 ml of the bag contents contains less than 1 ml of radioactive solution. The effect will be seen only when there is little or no true binding present, and is worst in fairly concentrated preparations. When true binding is observed in such preparations (e.g., of 20% tissue homogenates), the binding is probably underestimated because of the exclusion effect.

Similar exclusion can occur with less concentrated preparations in those special cases where vesicles are formed which can exclude the radioactive ligand. We believe that has occurred when the quaternary compound muscarone was studied with brain homogenates, which we therefore presume to contain vesicles impermeable to muscarone.

There are two ways of establishing whether events such as the two cited above are occurring. One is to check with some readily excludable material for which binding is not anticipated, e.g., [^{3}H]inulin. This should show the same value of "% negatively bound" as was observed for the ligand originally under study. The other approach is to study the % negatively bound at various ligand concentrations. The % negatively bound is of course constant for all concentrations, quite unlike any true binding phenomena.

If exclusion is extensive, then binding to receptor may be masked. Such a masking can be revealed if one repeats the experiment with a potent inhibitor or a large excess of nonradioactive ligand. This would then increase the apparent negative binding only if the originally observed negative binding was due to a combination of some binding to receptor plus some exclusion.

2.6. The Magnitude of the Off-Time

I have emphasized the sharp difference between the possibilities of detection of binding in reversible and irreversible systems. But what does *reversible* mean? Clearly, it is not an absolute term: It means only "reversible within the context of the experiment." Let us consider a reaction of binding of ligand to receptor:

$$L + R \underset{K_{-1}}{\overset{K_1}{\rightleftharpoons}} RL, \qquad K_d = \frac{k_{-1}}{k_1} \qquad (3)$$

Suppose reversal is rather slow and is accomplished in a matter of several minutes; for instance, suppose the half-time ($t_{1/2}$) for the reaction $RL \rightarrow R + L$ is 4 min. Obviously if the experiment which separates free and bound ligand is a slow one, for instance, a centrifugal technique lasting an hour, or an equilibrium dialysis technique lasting one-half day, the reaction is operationally reversible. But if one uses a fast filtration technique which is completed in 5 sec, the reaction is operationally irreversible as long as one uses a no-washing approach.

What sorts of off-times can one reasonably expect to encounter in ligand binding? From the analogous case of the binding of substrates to enzymes (Eigen and Hammes, 1963), the on-step is invariably very fast, quite often reaching the diffusion-limited value of 10^9 M^{-1} sec^{-1} (for a large, slowly diffusing species). This is scarcely surprising, because as the ligand approaches the receptor, the fit of one to another has not been expressed. Let us consider three cases: ultratight binding, with $K_d = 10^{-10}$ M; tight binding, with $K_d = 10^{-6}$ M; and loose binding, $K_d = 10^{-3}$ M. Bearing in mind that $K_d = k_{-1}/k_1$ and that $t_{1/2} = 0.7/k_{-1}$, then if the on-rate is diffusion limited, $k_1 = 10^9$ M^{-1} sec^{-1}, and it follows that for these three cases (1) ultratight binding $k_{-1} = 0.1$ sec^{-1} so $t_{1/2} = 7$ sec; (2) tight binding $k_{-1} = 10^3$ sec^{-1} so $t_{1/2} = 0.7$ msec; and (3) loose binding $k_{-1} = 10^6$ sec^{-1} so $t_{1/2} = 0.7$ μsec. Thus, in these circumstances the reaction will only be treatable by irreversible techniques (i.e., will appear to be effectively irreversible) if one has ultratight binding and a swift procedure completed in 1 sec or less. Even with stopped-flow techniques, the tight and loose bindings will be seen as reversible.

Of course, if one has a slow on-step, the above considerations will become more favorable. Some substrate–enzyme bindings have on-rates which are up to

1000 times slower than 10^9 M^{-1} sec^{-1}; but such slowness may relate to the fact that the *effective* binding is what is measured by enzymatic techniques, and this may involve orientational changes which, being a prerequisite for enzymatic action, are rate limiting and yet follow the initial binding. By contrast, all the procedures described in this chapter measure the initial binding of ligands to receptors.

Sometimes knowledge about the physiological response times can be relevant to these considerations. For instance, the duration of the excitatory postsynaptic potential in nicotinic cholinergic systems is in the order of 1 msec. This implies that the $t_{1/2}$ for the k_{-1} step is in this same order of magnitude. It has been pointed out that [assuming R in Eq. (3) is the closed and RL the open form of receptor] this implies a k_{-1} value of about 0.7×10^3 sec^{-1} and thus a K_d of 0.7 μM if the on step is diffusion limited or an even larger value if the on step is not so limited. This argument would appear to negate the significance of reported K_d values in the 10-nM range. But (as usual) there is a counterargument. If the Magleby and Stevens (1972) conception holds, so that the formation of the complex RL is a precursor to activation of RL (= closed receptor) to RL* (= open receptor), then the physiologically imposed upper limit on K_d disappears:

$$L + R \rightleftharpoons \underset{\text{closed}}{LR} \longrightarrow \underset{\text{open}}{LR^*} \tag{4}$$

3. Relation of in Vivo to in Vitro Properties

All receptors act by regulating a cellular process. Because this regulating character is lost on homogenization, one must use indirect evidence to show that the broken cell or solubilized preparation still contains viable receptor. If one can show that one has retained in the *in vitro* preparation precisely the properties observed *in vivo*—such as reversibility or otherwise of reactive agents, response only to appropriate drugs and at appropriate concentrations, location only in appropriate tissues and subcellular fractions, binding and unbinding in the anticipated time, and the correct number of receptors per unit weight or area—then one has reason to believe that the receptor has been preserved. However, the following discussion points to situations in which some of these desired criteria may not be met, and yet preservation may in fact have been achieved.

3.1. Reversibility

A minimal requirement is that the degree of reversibility observed *in vitro* should approximate that observed *in vivo*. Thus if the response caused by a ligand is readily reversible *in vivo*, one would certainly anticipate the same to be true *in vitro*. Such parallelisms have been observed in the cases, for instance, of α-bungarotoxin (irreversible in both cases) and *naja* toxin (slowly reversible in both cases), both these agents being specific to the nicotinic acetylcholine receptor. The reverse situation is sometimes not so clearly evident. Thus some agents which are only poorly reversible in intact tissues may owe this to difficulties in the wash-off procedure or to relatively slow off-rates. Thus curare is readily demonstrable to be a reversible agent for a nicotinic acetylcholine receptor *in vitro*, but *in vivo* it is not readily washed out.

3.2. Location

It would appear to be a minimal requirement that the macromolecule of interest should be present in the appropriate tissues only. For example, the fact that acetylcholine binding in the appropriate range was *not* observed in rat spleen, liver, kidney, and lung was additional evidence that the binding which was indeed observed in electroplax was associated with the neural activity of that tissue (O'Brien and Gibson, 1974). And the fact that α-dihydrograyanotoxin II was bound to rat liver and kidney as well as to lobster nerve was partial evidence against its having a role in binding to the sodium gate (Soeda *et al.*, 1975). However, caution must be exercised in interpreting the finding of ligand binding in unexpected places. For instance, Eisenfeld *et al.* (1976) found that [^{3}H]estradiol bound to liver cytosol but then went on to show that the binding material was a true receptor, in the sense that the receptor–estradiol complex could migrate to the nucleus and attach to the chromatin, just as in gonadal tissues.

3.3. Specificity

Other things being equal, one would expect a simple relationship between the rank order of potency of agents *in vivo* and *in vitro*. Obviously some cases of this should be observed in geometrical and optical isomers of compounds. Thus only amino acids which interfere with glutamate receptors *in vivo* showed blockade of L-glutamate to synaptic membranes (Michaelis *et al.*, 1974). Perhaps the most extreme case of this is in that of the dopamine receptor (Creese *et al.*, 1976a) in which the prime case for the evidence that the receptor played a role in schizophrenia was a rather close correlation of the sequence of haloperidol affinities found *in vitro* (demonstrated to be the best indicator of dopamine receptors) and of clinical potencies measured *in vivo*, employing a fairly long series of compounds (Fig. 2). Following the same logic, the observation by DeRobertis *et al.* (1969) that, after moderate purification, the material from brain which they believed to be acetylcholine receptor reacted (as judged by light scattering) with amphetamine, eserine, and strychnine as well as with atropine was a serious detriment to their claim that they were indeed observing the "real" receptor.

Appropriate caution needs to be taken in such interpretation. If an intact receptor, for instance, consists of subunits which interact, and if purification should lead to a dissociation and a loss of such subunits, then an interaction could be lost, and a drastic change in drug sensitivity would be noticed as purification proceeds. The conclusion should not be that the original binding was not to the macromolecule of interest but simply that purification had led to a separation of interacting subunits. However, once such a loss of specificity is found, the onus is upon the investigator to demonstrate that he is still on the trail of the "correct" macromolecule.

It is still relatively infrequent in the literature for the investigator to survey a reasonably large series of drugs which are inactive *in vivo* against the proposed

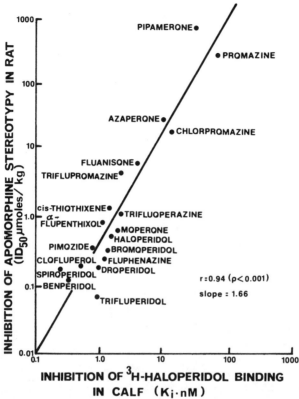

Fig. 2. Ability to antagonize schizophrenia and haloperidol bindings. Relation of ability of anti-schizophrenic drugs to block haloperidol binding to membranes from the striatum of calf brain and to inhibit the stereotyped response of rats to apomorphine. [Redrawn from Creese *et al.* (1976a).]

macromolecule. Such relatively extensive observations can often pose awkward problems. For instance, in the study of an axonic macromolecule which we have called the axonic cholinergic binding macromolecule, or ACBM, the ability to bind labeled α-bungarotoxin seemed to imply that we were dealing with an acetylcholine receptor. The view was supported by the fact that labeled nicotine was bound with an affinity quite comparable to that observed in *Torpedo* electroplax. However, when an extensive series of agents was explored as blockers of nicotine binding to axonic membranes (Table I) the pattern of drug interference (or *drug profile* as I have called it elsewhere) was profoundly unlike that of the classic acetylcholine receptors, either nicotinic or muscarinic. This observation, coupled with the finding that the binding of α-bungarotoxin was reversible (Jones *et al.*, 1977) rather than irreversible as in the case of the nicotinic receptor, clearly indicated that the ACBM was a macromolecule only distantly related to the vertebrate nicotinic acetylcholine receptor. The view was strongly reinforced by the observation that acetylcholine bound with rather poor affinity to the ACBM, so that in fact its binding could only be demonstrated by its interference with the binding of labeled nicotine.

Table I

Drug Interference with Binding of Nicotine to an Axonic Macromolecule[a]

	% Inhibition		
	Lobster nerve homogenate	Horseshoe crab nerve homogenate	Rat kidney homogenate
Acetylcholine + 10^{-4} M Tetram	51	38	28
Tetram (10^{-4} M)	44	43	42
Curare	78	70	−1
Carbamylcholine	11	14	−2
Succinylcholine	14	2	−6
Atropine	80	76	3
Pilocarpine	48	10	2
Norepinephrine	31	−3	4
Serotonin	28	4	3
γ-Aminobutyric acid	24	−5	21
α-Bungarotoxin	74	82	60
α-Bungarotoxin (50 μM)	85	96	68

[a] From Jones *et al.* (1977) with statistical data omitted. Nicotine: 10^{-7} M; drugs: 10^{-5} M except as indicated.

A rudimentary kind of specificity might seem to be that the binding of radioactive X should be blocked by unlabeled X and that only the appropriate enantiomorph should block effectively. For instance, unlabeled thyroid-stimulating hormone blocked the binding of the labeled form (Melidi and Nussey, 1975). But it should be pointed out that if the concentration of labeled X is much below the dissociation constant, then addition of (for instance) a 100-fold excess of unlabeled X will increase the total binding (labeled and unlabeled) by 100-fold, but because now only 1% of the total ligand will be labeled, the amount of labeled-and-bound ligand will be precisely the same as that without addition of the unlabeled X. Conclusion: For this, and for other kinds of competitive displacement where the K_d of the displacer is not far better than that of the displaced ligand, one needs concentrations of unlabeled ligand well in excess of the K_d.

3.4. Dissociation Constants

As a first approximation, one might expect that the affinity which a ligand displays for an isolated macromolecule should be comparable to that which is observed *in vivo*. Any serious discrepancy leads to doubts as to the relation between the binding and the macromolecule being the purported one. This is particularly important because of the prevalence of nonspecific bindings (see Chapter 5), and indeed at sufficiently high concentration of ligands, nonspecific binding is the rule rather than the exception. This is enshrined in the jocular comment in our laboratory, which we claim to have inscribed above every door, that "everything binds to everything." In general, observations of dissociation constants larger than 10^{-4} M lead to profound suspicions that one is not dealing with a physiologically interesting observation. This rule of thumb may well have to be overturned in the relatively rare cases in which one is dealing with macromolecules which interact physiologically with high concentrations of ligands. For instance, transport molecules for common cations presumably operate in concentration ranges which are related to those observed for such cations in normal tissues, and this will be in the millimolar range for sodium or potassium or calcium ions. But if one is dealing with the binding of hormones or transmitters or pheromones or drugs or toxicants, it is extremely unlikely that dissociation constants in this realm should be encountered, because in most cases such agents are not regarded as effective unless they are potent at rather low concentrations. An agent which is effective at 10 mg/kg of body weight, and which has a molecular weight of 300, has an average concentration in the body of 30 μM. One would anticipate that the binding constant for such an agent would be well below the millimolar range.

In general, the anticipation seems to be in accordance with observations. Examples of K_d values in the nanomolar range include binding of haloperidol (Table II), of hydroxybenzylpindolol to the β-adrenergic receptor of turkey erythrocytes (Aurbach *et al.*, 1974), of LSD to the serotonin receptor of rat brain (Bennett and Aghajanian, 1975), of human chorionic gonadotropin to bovine corpora lutea (Haour and Saxena, 1974), of estradiol to the uterine receptor (Sanborn *et al.*, 1971), and of prostaglandin-E to bovine corpora lutea (Rao, 1974).

But there is an extremely important category of observation which flies in the face of this principle. For some agents, the observed binding constant is extraordinarily small in comparison with the concentration which is physiologically effective. This case is well illustrated in the case of the nicotinic acetylcholine receptor, in which the observations of binding constants in the region of 10 nM were originally dismissed as being of no physiological interest, because the physiologically active concentrations were in the region of 1 μM.

In such a situation, one possibility is that the binding which is observed *in vitro* is truly of no physiological interest. But an alternative and more interesting situation may hold if positive cooperativity exists in the system. The existence of such positive cooperativity is not simply a guess but can be (in favorable situations) positively identified by the existence of Scatchard plots of binding which show a maximum. Let us consider such a situation, using for an example the two-state (MWC) model of Monod *et al.* (1965); essentially the same quantitative behavior can be demonstrated if one uses instead the induced-fit model of Koshland *et al.* (1966).

The MWC model proposes that the system is made up of two configurations of the macromolecule, the R-state, which has a high affinity for a ligand but represents a small fraction of the total population in the absence of ligand, and the T-state, which has a substantially lower affinity but represents a majority of the population in the absence of ligand. Other prerequisites are that the system can exist in only one of these two states, that it contains at least two units (*protomers*), each of which bears a binding site, and that the two protomers are symmetrically disposed toward each other. In these circumstances it is easy to demonstrate that

Table II

Antischizophrenic Drugs : Comparison of Affinities for [³H]Haloperidol and [³H]Dopamine Binding Sites with in Vivo Pharmacological Potencies[a]

	Inhibition of [³H]haloperidol binding, K_d (nM)	Average clinical daily dose (μmol/kg)	Inhibition of [³H]dopamine binding, K_d (nM)
Spiroperidol	0.25 ± 0.02 (4)	0.058	1400 ± 190 (3)
Benperidol	0.33 ± 0.02 (4)	0.060	4100 ± 540 (4)
Clofluperol	0.50 ± 0.03 (4)	0.077	360 ± 20 (3)
(+)Butaclamol	0.55 ± 0.09 (8)	2.14	70 ± 10 (10)
Fluspirilene	0.60 ± 0.13 (4)	0.066	1400 ± 220 (4)
Pimozide	0.80 ± 0.07 (4)	0.108	4100 ± 1140 (4)
Trifluperidol	0.95 ± 0.19 (3)	0.096	740 ± 20 (3)
Droperidol	1.0 ± 0.10 (4)		880 ± 80 (3)
α-Flupenthixol	1.1 ± 0.22 (4)	0.099	180 ± 30 (8)
Fluphenazine	1.2 ± 0.12 (6)	0.168	180 ± 30 (5)
Bromoperidol	1.4 ± 0.15 (4)	0.153	600 ± 90 (3)
cis-Thiothixene	1.4 ± 0.11 (4)	0.393	540 ± 140 (6)
Haloperidol	1.5 ± 0.14 (9)	0.152	650 ± 90 (4)
Moperone	1.9 ± 0.26 (4)	0.802	1200 ± 160 (4)
Triflupromazine	2.1 ± 0.12 (4)	4.59	530 ± 80 (5)
Trifluoperazine	2.1 ± 0.34 (4)	0.297	740 ± 80 (5)
Fluanisone	3.8 ± 0.80 (4)	3.44	800 ± 180 (4)
Penfluridol	5.6 ± 1.40 (7)	0.466	1600 ± 310 (4)
Azaperone	10.0 ± 0.6 (4)		1700 ± 290 (4)
Chlorpromazine	10.3 ± 0.2 (5)	12.0	900 ± 200 (7)
Thioridazine	14.0 ± 0.2 (5)	12.6	1780 ± 332 (4)
Pipamperone	31.3 ± 5.2 (4)	11.1	4900 ± 500 (4)
Promazine	71.5 ± 3.2 (4)	33.3	7100 ± 1640 (8)
Clozapine	100 ± 6 (6)	24.6	1890 ± 340 (5)
Promethazine	238 ± 32 (4)		12000 ± 3600 (7)
Correlation with [³H]haloperidol binding		$r = 0.87$ $P < 0.001$	$r = 0.58$ $P < 0.01$
Correlation with [³H]dopamine binding		$r = 0.27$ $P > 0.05$	

[a]From Creese *et al.* (1976b).

if one progressively increases the ligand concentration from zero, the fraction of the molecule in the R-state progressively increases until at concentrations near saturating it becomes the major species. Taking such a simple two-protomer system, the principal parameters are the c value, which describes the ratio of the dissociation constants of the R- and the T-state, and the L value, which describes the ratio of the concentrations of the two states which exists in the absence of ligand.

Such a model is readily applied to a variety of gating or receptor systems. The implication is that the R-state represents an open configuration, insofar as the flux of ions or other materials of interest is concerned, and the T-state represents the inactive or closed state. Consequently, application of a suitable ligand (such as a hormone or neurotransmitter) converts the macromolecule from primarily T to primarily R, and this is the event which the physiologist sees as the opening of a gate by the ligand. A variation of this theme can be derived from a somewhat different model proposed by Magleby and Stevens (1972). The variation suggests that conversion to the R-state is a precondition for opening, but some fluctuating process occurs by which R, a still-closed state, flips into an R*-state which is open:

$$\underset{\text{closed}}{T} \rightleftharpoons \underset{\text{closed}}{R} \rightleftharpoons \underset{\text{open}}{R^*} \tag{5}$$

In those cases where the MWC model is accurate, there is a complex connection between the microscopic binding constants (that is, the actual physically measured binding constants to the R-state and T-state separately) and the fraction of molecules in the R-state, which is given by

$$\bar{R} = \left(1 + \frac{L(1 + c\alpha)^n}{(1 + \alpha)^n}\right)^{-1} \tag{6}$$

where L is the equilibrium constant in the absence of ligand $[L = (T)/(R)]$, c is the ratio of dissociation constants ($c = K_R/K_T$), n is the number of protomers, and $\alpha = x/K_R$, where x is the ligand concentration.

The situation is illustrated for a simple case in Fig. 3, in which the relation between ligand concentration and the fraction of the molecules in the R-state is plotted. For this particular model, fairly plausible values for c and L were selected, as shown in the legend, and it can be seen that there is a 100-fold discrepancy between the binding constant for the R-state, K_R (which is that normally measured in binding studies; the binding constant for the T-state is much harder to find experimentally), and the concentration which gives half of the maximum possible number of macromolecules in the R-state and which is presumably what the physiologist would report as the apparent binding constant of K_{app}.

It should be noted that in such systems the maximum number in the R-state will not normally reach 100% of all the molecules; thus there is a difference between the concentration of ligand which will give 50% of the molecules in the R-state and the concentration of ligand which will produce half of the maximal possible number of molecules in the R-state. The maximum amount in the R-state, expressed as a fraction of the total, is $\bar{R}_{max} = (1 + Lc^n)^{-1}$.

In these relatively complicated cases, it is clear that one can only be persuasive about the accuracy of one's microscopic binding constants if one can separately measure the appropriate constants (K_R, L, and c) and show that they can account for the observed K_{app}. This is not an easy proposition.

Quite another reason which may account for discrepancies between microscopic binding constants and K_{app} has been pointed to in the case of nicotinic acetylcholine receptor and doubtless may apply to other systems. This is the phenomenon of desensitization. The physiological situation is commonly observed that application of unusually high concentrations of an activating substance or exposure to moderate concentrations for unusually long times may lead to an

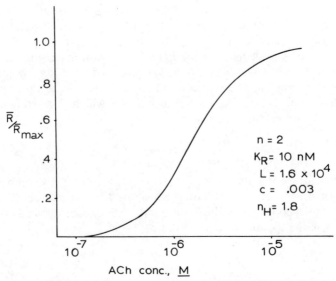

Fig. 3. Relation of concentration for half-maximal response to dissociation constants in a cooperative system. Calculations are based on the Monod–Wyman–Changeux model; n = number of protomers, K_R is the dissociation constant for the R-state, L is the equilibrium constant $(T)/(R)$ in the absence of ligand, $c = K_R/K_T$, and n_H is the Hill coefficient. The primary point is that although the microscopic binding constant K_R is 10 nM, the concentration for a half-maximal response (i.e., $R/R_{max} = 0.5$) is greater than 1 μM. [Based on Gibson (1976).]

altered state of the receptor, so that it no longer becomes sensitive to a standard dose of ligand. Such an effect is clearly exhibited in the experiment illustrated in Fig. 4, in which a testing dose of acetylcholine, suitable to produce a fixed response in a normal situation, becomes much less effective for a short period of time following the application of a prolonged pulse of acetylcholine. This phenomenon of desensitization may represent a protective mechanism against an overresponse to an overexposure and consequently can be expected to be of a rather general nature. One rationalization of such a phenomenon involves the conversion of the receptor to a proposed desensitized state. Katz and Thesleff (1957) concluded from a kinetic exploration of this phenomenon (in physiological preparations) that it implied that the desensitized form had a substantially greater affinity for the acetylcholine than did the native form of the receptor. Thus the progress of desensitization should be marked by an increase in affinity for the transmitter.

If the desensitization phenomenon existed *in vitro*, one would anticipate that prolonged exposure to the ligand (as occurs in equilibrium dialysis, for instance) might well mean that the binding constants which were observed were those described in the desensitized state and would not have simple physiological significance for the native receptor. Supporting evidence that prolonged exposure does lead to such a change has been reported by Weber et al. (1975), who showed that the rate of reaction of a preparation with α-bungarotoxin was increased as the period of preincubation with agents such as carbamylcholine or acetylcholine

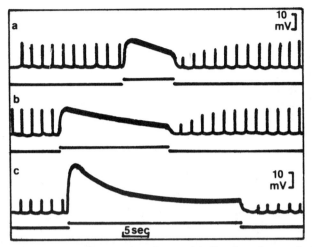

Fig. 4. Desensitization of frog muscle end plates by acetylcholine. In each case, the left-hand spikes show the constant depolarization of the end plate caused by a constant test pulse of acetylcholine, applied by an electrode. After delivering (middle part) variously lengthy applications of a *conditioning dose* of acetylcholine to induce desensitization, the response (right part) to test pulses is much reduced but soon recovers to normal. [Redrawn from Katz and Thesleff (1957).]

was increased. Calculations, based on a simple presupposition, derived from these data indicated that the increase of affinity was not extremely large, being on the order of about fivefold. But the direction was very clear. Since then we have performed direct studies upon acetylcholine binding and demonstrated that indeed there are substantial changes in the amount of binding of acetylcholine if one uses quick assays (with binding times of a few seconds) as compared to relatively slow ones (with binding times of a few minutes or more).

Another reason for favoring swift binding assays is the possible deterioration of a receptor after isolation. One example is uterine estrogen receptor, which loses 85% of its specific (that is, capable of being competed for by diethylstilbestrol) receptor for estradiol after 2 hr at 30–37°C (Peck *et al.*, 1973).

3.5. Detergents

Receptors occur in one of two situations: they are membrane bound or they are cytoplasmic. If they are membrane bound, then the biochemist will want to remove them from the membrane (i.e., solubilize them) in order to establish their size and composition and to attempt subsequent reconstitution. There are a few cases in which receptors which are known to be (or else ought to be) membrane bound can be obtained in a soluble form without detergent; the acetylcholine receptors of squid optic ganglion (Kato and Tattrie, 1974) and house fly head (Mansour *et al.*, 1977) are cases in point. The very fact of this easy removal raises suspicion as to their true receptor nature. As far as we know, such receptors face outwards in the cell membrane, and occupancy by their agonists leads to either an internal metabolic effect (such as the production of cyclic AMP) or a direct change in the ability of ions to penetrate the membrane. It is hard to see how this can occur unless the receptor passes through the membrane. A possible explanation might be that the recognition site of the receptor (which is what is commonly isolated as "the receptor") half-penetrates the membrane from the outside and conjoins some effector site, such as an ionophore or adenyl cyclase, which half-penetrates the membrane from the inside.

But most receptors under current study, including all neurohormonal, gustatory, olfactory, and pheromonal receptors, are membrane bound. It is well known that the use of detergents (especially ionic ones) for solubilization is fraught with danger; Ephraim Racker has been quoted as saying that a whole human being can easily be solubilized with the ionic detergent sodium dodecyl sulfate and that the resultant proteins migrate together as a single peak on electrophoresis. Virtually all receptor solubilizations (unlike enzyme solubilizations) have utilized nonionic detergents, because only then are their binding properties retained. Consequently, unless otherwise stated, the term *detergent* will from now on refer only to nonionic detergents. The most common such detergent is Triton X-100.

When a receptor is detergent-solubilized, the question arises as to its configurational relation to its native state. There can be little doubt that such solubilized forms bind substantial amounts of detergent, even after painstaking attempts to remove it. One way to assess the amount of binding is to compute the density of the receptor (often expressed as its reciprocal, the partial specific volume $\bar{v}$) and then to measure $\bar{v}$ by comparing sedimentation coefficients in sucrose gradients with H_2O or D_2O. Using this approach, it has been computed that crude nicotinic acetylcholine receptor solubilized by 1% Triton X-100, and in the presence of this Triton concentration, binds 0.23 g of detergent per gram of receptor. The concentration of applied detergent can change the amount of detergent bound to a receptor and hence its sedimentation characteristics. For example, the relative sedimentation velocities of this same receptor in Triton X-100 of 0.01, 0.03, 0.1, and 1.0% were, respectively, 0.59, 0.57, 0.52, and 0.46 (Gibson *et al.*, 1976).

Many receptors change their binding characteristics relatively modestly upon solubilization. When large changes occur, one wonders if the solubilized form has not been seriously damaged. A case of a drastic change on solubilization, either by deoxycholate or Lubrol-PX, is in the case of canine cardiac β-adrenergic receptor. The particulate form shows two binding sites of K_d, 1 μM and 0.1 μM. But solubilization, with or without purification, gave a single species with a K_d of 5 μM (Lefkowitz *et al.*, 1972). A less severe case was observed for the prolactin receptor of the rabbit mammary gland; solubilization increased the affinity for human growth hormone by fivefold (Shiu and Friesen, 1974).

The primary question, when one deals with solubilized receptor, is whether the native state is most accurately approximated when the detergent binding is minimized. On first thought, it might seem that because the native receptor is detergent-free, then maximal removal of detergent will give the "best" preparation. This view finds reinforcement in those cases in which it can be shown that detergents inhibit important properties; thus Triton X-100 inhibits acetylcholine binding to purified nicotinic receptor in a concentration-dependent way (Edel-

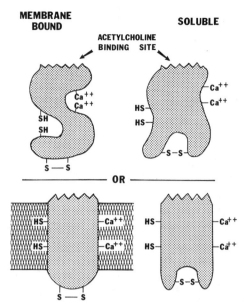

Fig. 5. Proposed configurational changes caused by detergent solubilization of nicotinic receptor. The changes are proposed to account ·for the new sensitivity to SH-oxidizing reagents and insensitivity to S-S reducing reagents that occurs on solubilization by detergents. [From O'Brien and Gibson (1977).]

stein *et al.*, 1975). On the other hand, centrifugation of this same receptor in sucrose gradients with various Triton X-100 concentrations leads to progressive aggregation at low Triton concentrations as judged by sedimentation velocities, suggesting an artifactual aggregation. And some properties of this receptor undergo drastic changes upon solubilization by detergent; thus the soluble form becomes newly sensitive to oxidizing agents and newly insensitive to reducing agents (O'Brien and Gibson, 1977). At the very least, one has to conclude that substantial configurational changes occur to account for such new sensitivities, as proposed in Fig. 5. But at worst the receptor might turn itself entirely inside out! In the membrane, most of the receptor's periphery is buried in the lipid bilayer and must have its hydrophobic residues directed outwardly. In detergent, the hydrophobic portion of the detergent provides a substitute for this cozy lipophilic environment. But if one subsequently strips off the detergent, the receptor must necessarily be forced to direct its polar residues outwards and bury its hydrophobic residues, or else it must clump with neighboring receptor molecules to minimize the water to which its surface must otherwise be exposed. In cases of this kind, the configuration may be "more native" in high than in low concentrations of detergent.

References

Aurbach, G. D., Fedak, S. A., Woodard, C. J., Palmer, J. S., Hauser, D., and Troxler, F., 1974, β-Blocking agent with high affinity site, *Science* **186**: 1223.

Bennett, J. L., and Aghajanian, G. K., 1975, D-LSD binding to brain homogenates: possible relationship to serotonin receptors, *Life Sci.* **15**:1935.

Caprio, J., 1975, High sensitivity of catfish taste receptors to amino acids, *Comp. Biochem. Physiol. A* **52**:247.

Creese, I., Burt, D. R., and Snyder, S. H., 1976a, Dopamine receptors and average clinical doses, *Science* **194**:546.

Creese, I., Burt, D. R., and Snyder, S. H., 1976b, Dopamine receptor binding predicts clinical and pharmacological potencies of antischizophrenic drugs, *Science* **192**:481.

DeRobertis, E., Gonzalez-Rodriguez, J., and Teller, D. N., 1969, The interaction between atropine sulphate and a proteolipid from cerebral cortex studied by light scattering, *FEBS Lett.* **4**:4.

Edelstein, S. J., Beyer, W. B., Eldefrawi, A. T., and Eldefrawi, M. E., 1975, Molecular weight of the acetylcholine receptors of electric organs and the effect of Triton X-100, *J. Biol. Chem.* **250**:6101.

Eigen, M., and Hammes, G. G., 1963, Elementary steps in enzyme reactions—as studied by relaxation spectrometry, *Adv. Enzymol.* **25**:1.

Eisenfeld, A. J., Aten, R., Weinberger, M., Haselbacher, G., Halpern, K., and Krakoff, L., 1976, Estrogen receptor in the mammalian liver, *Science* **191**:862.

Erez, M., Weinstock, M., Cohen, S., and Shtacher, G., 1975, Potential probe for isolation of the β-adrenoceptor, chloropractolol, *Nature (London)* **255**:635.

Gibson, R. E., 1976, Ligand interactions with the acetylcholine receptor from *Torpedo californica.* Extensions of the allosteric model for cooperativity to half-of-site activity, *Biochemistry* **15**:3890.

Gibson, R. E., O'Brien, R. D., Edelstein, S. J., and Thompson, W. R., 1976, Acetylcholine receptor oligomers from electroplax of *Torpedo* species, *Biochemistry* **15**:2377.

Glazer, A. N., 1976, Photochemical labeling, in: *The Proteins*, Vol. 2 (E. Neurath and R. Hill, eds.), pp. 76–103, Academic Press, New York.

Haour, F., and Saxena, B. B., 1974, Characterization and solubilization of gonadotropin receptor of bovine corpus luteum, *J. Biol. Chem.* **249**:2195.

Hummel, J. P., and Dryer, W. J., 1962, Measurement of protein-binding phenomena by gel filtration, *Biochim. Biophys. Acta* **63**:530.

Idänpään-Heikkila, J. E., and Schoolar, J. C., 1969, LSD: Autoradiographic study on the placental transfer and tissue distribution in mice, *Science* **164**:1295.

Jones, S. W., Galasso, R. T., and O'Brien, R. D., 1977, Nicotine and α-bungarotoxin binding to axonal and non-neural tissues, *J. Neurochem.* **29**:803.

Kato, G., and Tattrie, B., 1974, Solubilization of the acetylcholine receptor protein from *Loligo opalescens* without detergents, *FEBS Lett.* **48**:26.

Katz, B., and Thesleff, S., 1957, A study of the "desensitization" produced by acetylcholine at the motor end-plate, *J. Physiol. (London)* **138**:63.

Keynes, R. D., Ritchie, J. M., and Rojas, E., 1971, The binding of tetrodotoxin to nerve membranes, *J. Physiol. (London)* **213**:235.

Koshland, D. E., Jr., Némethy, G., and Filmer, D., 1966, Comparison of experimental binding data and theoretical models in proteins containing subunits, *Biochemistry* **5**:365.

Kurihara, N., Nakajima, E., and Shindo, H., 1970, Whole body autoradiographic studies on the distribution of BHC and nicotine in the American cockroach, in: *Biochemical Toxicology of Insecticides* (R. D. O'Brien and I. Yamamoto, eds.), pp. 41–50, Academic Press, New York.

Lefkowitz, R. J., Haber, E., and O'Hara, D., 1972, Identification of the cardiac beta-adrenergic receptor protein: Solubilization and purification by affinity chromatography, *Proc. Natl. Acad. Sci. USA* **69**:2828.

Magleby, K. L., and Stevens, C. F., 1972, A quantitative description of end-plate currents, *J. Physiol. (London)* **223**:173.

Mansour, N. A., Eldefrawi, M. E., and Eldefrawi, A. T., 1977, Isolation of putative acetylcholine receptor proteins from house fly brain, *Biochemistry* **16**:4126.

Melidi, S. Q., and Nussey, S. S., 1975, A radioligand receptor assay for the long-acting thyroid stimulator, *Biochem. J.* **145**:105.

Michaelis, E. K., Michaelis, M. L., and Boyarsky, L. L., 1974, High-affinity glutamic acid binding to brain synaptic membranes, *Biochim. Biophys. Acta* **367**:338.

Miledi, R., Molinoff, P., and Potter, L. T., 1971, Isolation of the cholinergic receptor protein of *Torpedo* electric tissue, *Nature (London)* **229**:554.

Monod, J., Wyman, J., and Changeux, J.-P., 1965, On the nature of allosteric transitions: A plausible model, *J. Mol. Biol.* **12**:88.

Moore, J. W., Narahashi, T., and Shaw, T. I., 1967, An upper limit to the number of sodium channels in nerve membrane?, *J. Physiol. (London)* **188**:99.

O'Brien, R. D., and Gibson, R. E., 1974, Two binding sites in acetylcholine receptor from *Torpedo marmorata* electroplax, *Arch. Biochem. Biophys.* **165**:681.

O'Brien, R. D., and Gibson, R. E., 1977, The binding of acetylcholine to its nicotinic receptor, in: *Cholinergic Mechanisms and Psychopharmacology* (D. J. Jenden, ed.), pp. 1–23, Plenum Press, New York.

O'Malley, B. W., and Means, A. R., 1974, Female steroid hormones and target cell nuclei, *Science* **183**:610.

Peck, E. J., Jr., DeLibero, J., Richards, R., and Clark, J. H., 1973, Instability of the uterine estrogen receptor under *in vitro* conditions, *Biochemistry* **12**:4603.

Rao, C. V., 1974, Characterization of prostaglandin receptors in the bovine corpus luteum cell membranes, *J. Biol. Chem.* **249**:7203.

Sanborn, B. M., Rao, B. R., and Korenman, S. G., 1971, Interaction of 17 β-estradiol and its specific uterine receptor, *Biochemistry* **10**:4955.

Seyama, I., and Narahashi, T., 1973, Increase in sodium permeability of squid axon membranes by α-dihydrograyanotoxin II, *J. Pharmacol. Exp. Ther.* **184**:299.

Shiu, R. P., and Friesen, H. G., 1974, Solubilization and purification of a prolactin receptor from the rabbit mammary gland, *J. Biol. Chem.* **249**:7902.

Soeda, Y., O'Brien, R. D., Yeh, J. Z., and Narahashi, T., 1975, Evidence that α-dihydrograyanotoxin II does not bind to the sodium gate, *J. Membr. Biol.* **23**:91.

Spivak, C. E., and Taylor, D. B., 1977, A filter assay for binding of labeled ligands to membrane-bound receptors, *Anal. Biochem.* **77**:274.

Stumpf, W. E., and Roth, L. J., 1968, High resolution autoradiography of H^3-estradiol with unfixed, unembedded 1.0 μ freeze-dried frozen sections, *N. Engl. Nucl. Atomlight* **65**:1.

Suter, P., and Rosenbusch, J. P., 1976, Determination of ligand binding: Partial and full saturation of aspartate transcarbamylase, *J. Biol. Chem.* **251**:5986.

Weber, M., David-Pfeuty, T., and Changeux, J.-P., 1975, Regulation of binding properties of the nicotinic receptor protein by cholinergic ligands in membrane fragments from *Torpedo marmorata*, *Proc. Natl. Acad. Sci. USA* **72**:3443.

Index

Abbreviations, 249
ACBM, *see* Binding, of drug, axonic, cholin-
    ergic
Acenocoumarin, 200
Acetylcholine(ACh), 47, 50, 51, 59, 64, 75,
    97, 102, 106, 235, 265, 323, 324, 326,
    330, 331
  action, 194-195
  antagonist, 101
  binding, 111, 115-117
    kinetics of, 117-118
  in electroplaque of eel, 114
  in end-plate of muscle, 93
  -like drugs, 93-110
    response by, 93-110
  and liposome, 14
  muscarinic, 254-255, 297
  nicotinic, 254-255
  noise, 94
  receptor of, 13-17, 20, 64, 68, 93, 232-233,
    274, 277, 280-283, 289
    and drug-binding, 110-118
    nicotinic, 297
  response to, 95
  and transport, 14
Acetylcholinesterase, 115, 232, 281
  inhibitor of, 36
Acetyl-β-methylcholine, 45, 51
ACh, *see* Acetylcholine
AChE, *see* Acetylcholinesterase
Action, biological, 34-35
  of drug, 34-35
  structure relationship, 48
Activation—occupation model, 37
Adair model, 179
Adenosine monophosphate, cyclic (cAMP), 42,
    43, 59
Adenosine triphosphatase, 8-10, 12, 13, 17,
    18, 21, 22
  calcium ion transport, 17, 21, 227-231
  fragment, 219
  mitochondrial, 18, 22

Adenosine triphosphate (*cont'd*)
  sodium and potassium ion transport, 17,
    21, 231
  transport of ions, 17, 21, 227-231
Adenosine triphosphate, synthesis, 11
Adenylate cyclase, 40, 41, 67, 68, 208, 332
  representation, schematic, 41
Adrenaline, 62, 63
β-Adrenergic receptor, 62, 168, 208
Agonist, 52, 66-67, 72
  affinity, 61, 197
  binding, 67
  displacement, 80
  efficacy defined, 126
  structure—action relationship, 48
  weak, 127
Alamethicin, 14, 19, 222
Alanine transporter, 10
Alprenolol, 169
  dissociation rate from frog erythrocyte,
    169
γ-Aminobutyric acid (GABA), 257
Aminopterin, 60
Ammonium, 127
  quaternary, 39
AMP, *see* Adenosine monophosphate
Amphetamine, 324
Amphotericin B, 222
Amplifier system, 40
Anesthetic agent
  and axon, 108
  and channel blocking, 107-110
  gaseous, 35
  local, 107-110
  quaternary, 109
Antagonist, 48, 52, 57, 59, 66-67, 73, 197
  binding, 67
  competitive, 65
Antibiotic, 221-222
  ionophorous, 216, 222
  see also separate antibiotics
Antibody, 301